T0201142

ADVANCES IN NETWORK CLUSTERING AND BLOCKMODELING

Wiley Series in Computational and Quantitative Social Science

Computational Social Science is an interdisciplinary field undergoing rapid growth due to the availability of ever increasing computational power leading to new areas of research.

Embracing a spectrum from theoretical foundations to real world applications, the Wiley Series in Computational and Quantitative Social Science is a series of titles ranging from high level student texts, explanation and dissemination of technology and good practice, through to interesting and important research that is immediately relevant to social / scientific development or practice. Books within the series will be of interest to senior undergraduate and graduate students, researchers and practitioners within statistics and social science.

Behavioral Computational Social Science

Riccardo Boero

Tipping Points: Modelling Social Problems and Health

John Bissell (Editor), Camila Caiado (Editor), Sarah Curtis (Editor), Michael Goldstein (Editor), Brian Straughan (Editor)

Understanding Large Temporal Networks and Spatial Networks: Exploration, Pattern Searching, Visualization and Network Evolution

Vladimir Batagelj, Patrick Doreian, Anuška Ferligoj, Nataša Kejžar

Analytical Sociology: Actions and Networks

Gianluca Manzo (Editor)

Computational Approaches to Studying the Co-evolution of Networks and Behavior in Social Dilemmas

Rense Corten

The Visualisation of Spatial Social Structure

Danny Dorling

ADVANCES IN NETWORK CLUSTERING AND BLOCKMODELING

Edited by

Patrick Doreian

University of Pittsburgh, USA
University of Ljubljana, Slovenia

Vladimir Batagelj

University of Ljubljana and Institute of Mathematics, Physics and Mechanics, Ljubljana, Slovenia
University of Primorska, Andrej Marušič Institute, Koper, Slovenia
NRU High School of Economics, Moscow, Russia

Anuška Ferligoj

University of Ljubljana, Slovenia
NRU High School of Economics, Moscow, Russia

Registered Offices
John Wiley & Sons, Inc., 111 River Street, Hoboken, NJ 07030, USA
John Wiley & Sons Ltd, The Atrium, Southern Gate, Chichester, West Sussex, PO19 8SQ, UK

Editorial Office
9600 Garsington Road, Oxford, OX4 2DQ, UK

For details of our global editorial offices, customer services, and more information about Wiley products visit us at www.wiley.com.

Wiley also publishes its books in a variety of electronic formats and by print-on-demand. Some content that appears in standard print versions of this book may not be available in other formats.

Library of Congress Cataloging-in-Publication Data

Names: Doreian, Patrick, editor. | Batagelj, Vladimir, 1948- editor. |
    Ferligoj, Anuška, editor.
Title: Advances in network clustering and blockmodeling / Patrick Doreian,
    Vladimir Batagelj, Anuška Ferligoj.
Description: Hoboken, NJ : Wiley, 2020. | Series: Wiley series in
    computational and quantitative social science | Includes bibliographical
    references and index.
Identifiers: LCCN 2019024475 (print) | LCCN 2019024476 (ebook) | ISBN
    9781119224709 (cloth) | ISBN 9781119224686 (adobe pdf) | ISBN
    9781119224679 (epub)
Subjects: LCSH: Social networks–Mathematical models. | Sociometry.
Classification: LCC HM741 .A35 2020 (print) | LCC HM741 (ebook) | DDC
    302.301/13–dc23
LC record available at https://lccn.loc.gov/2019024475
LC ebook record available at https://lccn.loc.gov/2019024476

Cover Design: Wiley
Cover Image: Courtesy of Vladimir Batagelj

Set in 10/12pt TimesLTStd by SPi Global, Chennai, India
Printed and bound in Singapore by Markono Print Media Pte Ltd

10 9 8 7 6 5 4 3 2 1

We dedicate this book to the memories of Sue Freeman and Lin Freeman because of their impacts on our lives in both professional and personal realms. Together, they formed a fine couple living a rich and full life. Together, they provided a kind of role model worthy of emulation.

Lin founded and was a longtime editor of Social networks. For decades it was the premier journal of the field. While joined by other network venues, it remains the primary home for scholars publishing papers about social networks. He had a clear vision for the role this journal could play as a venue for publishing important work and how it could help to define our field. Especially important was the breath of his commitment to encouraging interdisciplinary work across multiple fields. He was very catholic in encouraging multiple approaches and different ways of doing research. But he was clear that this work had to be rigorous and done within sound research designs.

Both Lin and Sue were social active at the annual Sunbelt Social Network conferences both in sessions and, more importantly, outside sessions when a lot of ideas are generated in free-flowing discussions about our field and potential future directions.

Editors

Brief Contents

Contents

List of Contributors

Mauricio Barahona
Department of Mathematics, Imperial College London, London, UK; EPSRC Centre for Mathematics of Precision Healthcare, Imperial College London, London, UK

Vladimir Batagelj
Professor Emeritus, Faculty of Mathematics and Physics, University of Ljubljana, Ljubljana, Slovenia; Institute of Mathematics, Physics and Mechanics, Ljubljana, Slovenia; University of Primorska, Andrej Marušič Institute, Koper, Slovenia; NRU High School of Economics, Moscow, Russia

Steve Borgatti
LINKS Center for Social Network Analysis, Gatton College of Business and Economics, University of Kentucky, Lexington, KY, USA

Marjan Cugmas
Faculty of Social Sciences, University of Ljubljana, Ljubljana, Slovenia

Jean-Charles Delvenne
ICTEAM and CORE, Université Catholique de Louvain, Belgium

Patrick Doreian
Professor Emeritus, Department of Sociology, University of Pittsburgh, Pittsburgh, USA; Faculty of Social Sciences, University of Ljubljana, Ljubljana, Slovenia

Martin Everett
School of Social Sciences, University of Manchester, Manchester, United Kingdom

Anuška Ferligoj
Faculty of Social Sciences, University of Ljubljana, Ljubljana, Slovenia; NRU High School of Economics, Moscow, Russia

Luka Kronegger
Faculty of Social Sciences, University of Ljubljana, Ljubljana, Slovenia

Renaud Lambiotte
Mathematical Institute, University of Oxford; Entités de recherche: Institut Namurois des systèmes complexes (naXys), University of Namur, Namur, Belgium

Andrej Mrvar
Faculty of Social Sciences, University of Ljubljana, Ljubljana, Slovenia

Carl Nordlund
The Institute for Analytical Sociology,
Linköping University, Sweden; Center for
Network Science, Central European
University, Hungary

Tiago P. Peixoto
Department of Mathematical Sciences and
Centre for Networks and Collective
Behaviour, University of Bath, United
Kingdom; ISI Foundation, Turin, Italy

Martin Rosvall
Department of Physics, Umeå University,
Umeå, Sweden

Michael Schaub
Institute for Data, Systems and Society,
Massachusetts Institute of Technology;
Department of Engineering Sciences,
University of Oxford

Lovro Šubelj
Faculty of Computer and Information
Science, University of Ljubljana, Ljubljana,
Slovenia

Vincent Traag
Centre for Science and Technology Studies,
Leiden University, Leiden, The Netherlands

Aleš Žiberna
Faculty of Social Sciences, University of
Ljubljana, Ljubljana, Slovenia

Anja Žnidaršič
Faculty of Organizational Sciences,
University of Maribor, Maribor, Slovenia

1

Introduction

Patrick Doreian[3,4], Vladimir Batagelj[1,2,5], and Anuška Ferligoj[3,5]

[1]IMFM Ljubljana
[2]IAM, University of Primorska, Koper
[3]FDV, University of Ljubljana
[4]University of Pittsburgh
[5]NRU HSE Moscow

This book focuses on network clustering regardless of the disciplines within which a network was established. In the initial conception for the book, our attention was driven primarily by concerns regarding blockmodeling and community detection as they applied to social networks. But as we looked further into this general topic to invite potential contributors, we realized that the domain was much broader. The wide variety of approaches contained in this volume exemplifies this diversity. For us, as we assembled this volume, this was an exciting learning experience, one we hope will be experienced by readers of this book.

There is no single best approach to network clustering. Put differently, there is no cookie-cutter approach fitting all such data sets. Yes, there are adherents of one (their) approach who think this is the case. As shown in the chapters that follow, none of the authors of the contributed chapters share this very narrow view. This is a wide-open realm with multiple exciting approaches. We reflect further on this in the concluding chapter. Here, we describe briefly the contents of the following chapters merely as an introduction to them. In our view, each chapter merits close attention.

1.1 On the Chapters

As the book is concerned with network clustering, Chapter 2 offers an extended examination of the network clustering literature. Identifying the citation network for this literature turned out to be a complex task. Methods for doing this are described in detail. The initial search used the Web of Science (WoS) to identify documents using search terms. There were multiple searches, with the first being for 2015 and the final network was obtained for February 2017. This literature expanded dramatically and in a complex fashion.

While a citation network is composed of links between works treated as nodes, there is more to consider when other types of units are included. These include authors, journals, and keywords. As part of a more general strategy, one starting from the citation network but using additional information, multiple two-mode networks were constructed. This included an authorship network with works × authors, a journal network featuring works × journals, and a keyword network with works × keywords. They help give more insight into the one-mode citation network having only scientific productions.

Chapter 2 lists the most cited works, the most citing works, the most used keywords, and a discussion of the "boundary problem" as it relates to citation networks. One message from this chapter is that the way the boundary of a citation network is established affects greatly the data analytic results that follow. This implies using *great care* in constructing citation networks – not all citation networks will suffice for meaningful analyses.

Chapter 2 presents a wide range of analyses of the citation network for the network clustering literature. Components of the network were established. Both critical project main paths and key-route paths were identified. In the analysis of this network, a clear transition between the blockmodeling and the community detection literatures was revealed. Another technique used in Chapter 2 is the identification of link islands which have higher levels of internal cohesion as a way identifying some important subnetworks of this literature. The largest of these link islands featured works from the blockmodeling and community detection literatures. Since the transition from the former to the latter, it seems the two have developed independently. This seems problematic given our belief in the utility of useful ideas flowing between fields and sub-fields.

The other islands discussed in Chapter 2 come from the fields of engineering geology, geophysics, as well as electromagnetic fields and their impacts on humans. One of the searches used in the searches of the WoS database included the terms "block model" and "block". The latter crops up in other literatures, a surprise for us. In considering these other link islands, another surprise awaited the authors of this chapter. We are used to debates in the social network literature regarding the difference between static and temporal approaches to studying networks. This divide is present also in the natural sciences.

Chapter 2 also contains an examination of authors and measures of productivity within research groups, collaboration, and an examination of citations among authors contributing to the network clustering literature. Again, the stark division between the community detection and blockmodeling literatures was clear. Examined also are citations between journals publishing articles in the broad area of the network clustering literature. The methodological details of doing this, as outlined in this chapter, merit further attention.

Bibliographic coupling, which occurs when two works both cite a third work in their bibliographies, was also examined. This included a sustained assessment as to how this coupling is measured. These tools were applied to the network clustering literature, especially for the largest identified island. This included an examination of the most frequently used keywords in the social networks literature and the physicist-driven approach to studying social networks. While some keywords were the same, there were considerable differences, again illustrating the different concerns in these two literatures.

Chapter 3 provides an overview of "classical" clustering including both the clustering of networks and clustering in networks. The clustering problem is considered as a discrete optimization problem which turns to be, in most cases, NP hard. Therefore, local optimization or greedy methods are usually used for solving it. These methods can be adapted also for clustering in networks (or clustering with relational constraints). The hierarchical agglomerative clustering method can be extended for efficiently clustering large sparse networks. This is illustrated with an analysis of normalized author citation networks from the network clustering literature from Chapter 2.

Chapter 4 describes different approaches to community detection. The authors ask very useful questions which led them to provide helpful guidelines for researchers contemplating network clustering within a community detection framework. It starts with a bold claim that there is no precise definition of a community. We concur. Their focused review of community detection methods makes it clear that researchers need to have a clear idea as to why any method is selected prior to its use. The authors point to the problem that the same term can have different meanings in different subfields, reflecting what was found in Chapter 2. They argue "community detection should not be viewed as a well-defined problem but rather an umbrella term with many facets." This delightful image is equally applicable to blockmodeling within social network analysis!

Four different approaches to community detection are outlined. The first uses the cut-based perspective. The second is a clustering approach maximizing the internal density of clusters and the third is the stochastic equivalence perspective. Finally, there is a dynamical perspective focusing on the impact of communities and dynamic processes to establish a dynamically relevant coarse-grained partitions of network structure. Four sections follow which provide precise descriptions of the fundamental properties of these approaches and the results stemming from adopting them.

Their discussion makes it clear that there is no single "best" community detection algorithm and that there can be multiple equally valid partitions of a network depending on which of the four considered approaches is used. Again, this sentence holds fully when "blockmodeling" is used instead of "community detection". Of course, this applies to all the network clustering approaches presented in this volume.

Chapter 5 provides an extensive discussion of label propagation as a heuristic method initially proposed for community detection. There is natural segue between Chapters 4 and 5. Label propagation is a partially supervised machine learning algorithm assigning labels to previously unlabeled data points. At the start of the algorithm, subsets of nodes have labels, which amounts to a clustering of them. These labels are propagated to the unlabeled points throughout the course of the algorithm. Nodes carry a label denoting the community to which they belong. Membership in a community changes based on the labels that the neighboring nodes possess as the labels diffuse through the network.

The author is clear that while it is not the most accurate or the most robust clustering method, a label propagation algorithm is simple to implement and is exceptionally fast. Networks with hundreds of millions of nodes can be analyzed readily. The early work on this approach is described with the basic ideas presented for simple undirected networks. However, the author points out that it can be used for many more types of networks, including those with multiple edges between nodes, two-mode networks, and signed networks. It can be used also to identify and delineate overlapping clusters: it is not restricted to establishing only partitions of nodes, a useful property. Nested hierarchies of groups of nodes can be identified also.

Issues regarding the number of clusters are discussed along with updating labels, which can be done with either synchronous propagation or asynchronous propagation. Depending on the structure of the network, each can produce undesirable outcomes, which are discussed in detail. Reaching an equilibrium with nodes having stable labels is critical and ways for achieving this are discussed.

Advances in label propagation methods are described. They include adding constraints to the objective function to prevent trivial solutions, using preferences to adjust the propagation strength of nodes, and improving algorithmic performance by promoting its stability and reducing its complexity. Simple examples with planted communities are provided and used to show the subtlety of the method and the choices made when using it. A connection is made with the blockmodeling literature when using structural equivalence, consistent with the general conception motivating the book to apply and connect different methods for establishing clusters of network nodes.

Chapter 6 marks a transition from community detection issues to blockmodeling concerns. There is now an abundant amount of valued network data being collected and made available. As the initial work on blockmodeling dealt with binary networks, extending and adapting this approach to handle valued networks is important. The authors show that creating this extension is far from straightforward because subtle issues arise as to how valued network data can be treated prior to blockmodeling them. The authors discuss the difference between the traditional indirect approach and the direct approach to blockmodeling. In the former, networks are transformed into arrays expressing the similarity, or dissimilarity, of the nodes. These measures are then used to cluster the nodes. Their Figure 6.1 lays out the relevant decision points. In contrast, the direct approach eschews such transformations. Within the rubric of generalized blockmodeling, network data are analyzed directly. For valued networks, the authors cleave to the latter approach by making strategic adaptations to handle valued data in useful ways. This includes homogeneity blockmodeling championed by one of the authors and deviation generalized blockmodeling promoted by the other contributing author.

Two well-known empirical data sets were selected for a detailed examination of the issues involved in blockmodeling valued data. The simplest of the two is a friendship network. The second involves trade flows between nations, one raising the issue of relational capacity. This has major implications for the nature of discerning the relevant useful transformations. As with the previous chapters in this volume, close attention is paid to the choices that must be made regarding the appropriateness of methods given the data being analyzed and the criteria for making these choices. Their Figure 6.2 is particularly important as a way of guiding researchers to make appropriate decisions. It could be generalized more broadly and adapted for the other network clustering approaches presented in this volume.

By presenting detailed analyses of the selected empirical networks using different approaches and a variety of transformations, the authors show, and examine, the different outcomes resulting from making different strategic choices. The results have interest value in their own right, and lead to a set of useful recommendations about these choices. Some open problems are discussed briefly.

Chapter 7 continues the consideration of blockmodeling but tackles a very different issue, namely, measurement error. The premise for blockmodeling is that using these methods reveals the structural features of networks at both the macro and micro levels. The presence of measurement error complicates these analyses. In the worst case scenario it can render blockmodeling results useless. Three types of measurement errors are discussed. One takes the form of having errors in the recorded ties. The second is item non-response and the third is actor non-response.

Actor non-response is the primary focus of this chapter. Typically, and regrettably, within social network analysis, the standard response to this problem is to discard all information about the non-respondents, including information about the ties directed to them by respondents. This discarded information can be used to recover (most of) the network. The authors contend that this must be done and provide strong evidence supporting this claim. The question that follows is simple to state: how is this done?

The authors present seven ways for using such data as imputation methods for recovering the network from the ravages of actor non-response: reconstruction, imputation using the mean of the incoming ties, imputation with modal values of the incoming ties, reconstruction combined with using the incoming modal values, imputation of the total mean, imputations using the median of the three nearest neighbors based on incoming ties, and null tie imputation.

The authors examine the relative merits of these ways of recovering network data using four known empirical networks. Five steps are involved. The first establishes a partition of the known network within the indirect blockmodeling approach. The second step creates "observed" networks by randomly removing some actors (at various levels ranging from 1% to over 45%). Step three involves using each of the imputation methods to generate recovered networks. The fourth step is the clustering of each recovered network using the exact same clustering method as in the first step. The final step compares the partitions for all pairs of known and recovered networks. Two criteria were used in the comparisons. One is the Adjusted Rand Index for comparing two partitions. The other is the proportion of correctly identified blocks by position in the blockmodel. These criteria are stringent and there are clear differences regarding the adequacy of imputation methods. As with the foregoing chapters, recommendations are made regarding the best ways for recovering network data given the presence of actor non-response.

Chapter 8 addresses the clustering of signed networks, a topic that has garnered considerable attention within both the blockmodeling and community detection literatures. The authors adopt a formal approach and start within the structural balance perspective, which is a substantively driven approach to studying signed networks. The basics of this approach are reviewed. Some of the early theorems are restated with some new proofs provided. One critical feature of structural balance theory states that a signed network is balanced if all its cycles are balanced. But cycles can vary in length, a feature that complicates algorithms used for determining the extent to which graphs are imbalanced. The authors use the concept of chords, which allows cycles to have two subcycles which simplifies computing the sign of a cycle.

One prominent feature of the balance theoretic approach centers on what has come to be called the "structure theorems". Initially, if a signed graph was balanced, its nodes could be partitioned into two clusters such that all the positive ties were in one cluster and all the negative ties went between the two clusters. Later, this was extended to any number of clusters having this property. Here, the authors call the former strong structural balance and the latter weak structural balance. One interesting theorem in this chapter states that signed graphs are weakly structurally balanced if and only if all chordless cycles are weakly structurally balanced.

The authors then turn to consider the clustering of signed networks in a more general fashion. For strong structural balance, they couple this approach to spectral theory. They rework an early concept of switching (from a 1958 paper) to prove theorems relating to balance in signed networks using this concept. For weak structural balance, one allowing for more than two clusters, additional ideas have emerged. The structure theorems point to a blockmodel where diagonal blocks are positive (with primarily positive ties) and non-diagonal blocks are negative (with primarily negative ties). Yet empirical networks can come in the form of having off-diagonal positive blocks and, far more rarely, on-diagonal negative blocks. As the authors note, more research is required regarding the distribution of signed blocks in a blockmodel. They consider also community detection issues for signed networks and note that one of the main concepts of this approach, modularity, has problems when there are negative links in the network. They provide ways of addressing this problem.

The authors address the critical problem of studying temporal networks in a dynamic context and, as an empirical example, study the international system with signed relation between nations. They display timeline graphs showing variations in the levels of imbalance in the international system using different methods. They confirm that signed networks move both towards balance and away from balance depending on contexts. Their results add to the solid argument *against* the early presumption of structural balance theorists that signed networks always move towards balance. Also displayed are partitions of nations for various time points, which are interpreted in interesting ways.

Chapter 9 presents a summary of work on multimode network clustering and illustrates the results of using different methods on a single well-known early two-mode network. One of the conceptions behind this volume is bringing together ideas from multiple disciplines. Actually, the authors of this chapter engage in this process explicitly. While they focus on two-mode networks in the form of actors × events as a bipartite network, the implications of the materials in this chapter extend much further. Although such two-mode networks have been considered in earlier chapters, especially Chapter 2, having an integrated discussion of the ways in which they can be analyzed is particularly useful. They note that both binary and valued two-mode networks can be analyzed within a common rubric. Multiple such methods are discussed in the chapter.

The authors establish a conceptual link to community detection, make some definitions to help link the two literatures, and note that community detection is a special case of blockmodeling. The same point was made in Chapter 4. Some authors focused on community detection methods might disagree! Here, the core community detection notion of modularity is provided and, more importantly, the authors extend this to two-mode networks. Their first presented partition (of actors) concerns group assignment maximizing modularity.

Given a two-mode network, denoted A, it is straightforward to create two projections for actors, using AA', and events, using $A'A$. They challenge the presumption that evidence is lost in this dual projection. Without doubt, as the authors note, this holds when projections are dichotomized or if only one projection is used. However, they challenge the claim that dual projection loses information even when both projections are used in their undichotomized forms. Elsewhere, they have provided strong evidence that this is not the case and have promoted what they call the dual-projection approach. They present further partitions of the actors of the considered empirical network using dual-projection, using core-periphery notions and for dual-projection community detection. The latter led to two more partitions, one with two clusters and one with four clusters of actors.

In their spirit of integration, the authors include a consideration of signed two-mode networks and a consideration of spectral methods. Two more partitions of the actors are presented. As with previous chapters, it seems reasonable to have multiple valid partitions of a network. The authors finish with suggestions regarding the analyses of more complex data structure involving more modes and extend this to temporal evolution of two-mode networks.

Chapter 10 is devoted to blockmodeling linked networks and provides another segue in this volume. This time it is from Chapter 9. The term "linked networks" features a set of one-mode networks where the nodes from the one-mode networks are linked through two-mode networks.

This can be done in a variety of ways, including the coupling of networks linked over multiple time points. In dealing with these configurations, the author distinguishes analyses of separate networks, using a conversion approach under which all the one-mode networks are converted to a single level by joining them through the two-mode networks, and using a genuine multilevel approach. The results of examining both the conversion to a single level and using the multilevel approach are presented. Comparisons of them demonstrate clearly how the linked blockmodeling approach has greater potential value.

Two empirical examples are used. One concerns a coauthorship network at two time points while the other features participants in a fair-trade exchange for TV programs. In both examples, results of different partitions are presented with insightful comparisons of the results. As with all methods, parameter settings require consideration. For these analyses, this concerns the weighting of null and complete blocks for the scientific citation network. The final reported results provide a very coherent result with strikingly clear differences in the partitions for two distinct time points, which provide useful interpretations of the dynamics of scientific collaboration. The empirical results resulting from using the genuine multilevel method for the trade fair are equally compelling. As with other chapters, the author provides a provisional agenda for future work.

Chapter 11 provides a self-contained introduction to using Bayesian inference to extract the large-scale modular structure from network data. In terms of the foregoing content of this chapter, the modules are clusters (or groups) identified in the network. Rather than focus on deterministic blockmodeling, Chapter 11 deals with Bayesian stochastic blockmodeling. A major focus is on estimating probabilistic models to shed light on the network mechanisms generating the observed network(s). An overarching feature is to distinguish genuine structure from randomness.

In this context, Figure 11.1 is especially provocative, with three displays of a randomly generated network having three separate orderings of the nodes. Two of them appear to show clear – but different – blockmodel structures, exactly the sort that those using deterministic blockmodeling would take as evidence of structure. While these blockmodels could be accepted as "real" and could be "interpreted", this would reveal nothing about the generative process creating the network. This might rattle the cages of some social network blockmodelers. The author invites readers to think probabilistically and couple two ideas. One is to think about mechanisms that could generate networks. The other is to use the network data to discern which mechanism was the most likely to have generated the network. This leads directly to notion of stochastic blockmodels within which known (prescribed) modular structures are generated according to probabilistic rules. Then, given network data, Bayesian inference is used to infer the modular structure of observed networks.

The author provides formal discussions of a wide variety of prior distributions and how data affect them to create posterior distributions. Many empirical applications are used to illustrate the outcomes stemming from using the methods described in formal detail. This includes the subtleties of model selection and the establishment of efficient estimation procedures. The ultimate outcome is the establishment of modular structures that are supported by statistical evidence.

Chapter 12 also focuses on a dynamical perspective. Both this chapter and Chapter 11 are concerned with modular structures having a coarse grain, along with the rich interplay between network structures and network dynamics. However, the authors of Chapter 12 take a very different approach compared with the one contained in Chapter 11. Their concern is centered on the dynamical processes occurring on a fixed network structure. They do not consider, at least initially, the question of how and why networks occur. Throughout their presentation, they focus on consensus dynamics and diffusion processes both substantively and as guiding examples. Later, they consider diffusion and consensus as dual processes, an important extension.

For modeling dynamics, they use ordinary differential equations in which the actors have attributes that can be changed by the operation of social processes operating over a fixed network. While discrete-time versions could be used, they use continuous-time models throughout their chapter. Also, of great interest, they consider processes having different network time scales under which some variables change slowly while others change more quickly. They illustrate this with a modular network having k modules with strong within coupling and weak between coupling. More complicated structures are examined also.

The authors extend this approach to consider signed networks and use the early work on structural balance theory while restricting their attention to strong balance as described in Chapter 8. The early structural balance literature was fixated on the notion that signed networks always move towards balance. This claim is repeated here. A far more important issue is the examination of how (and why) signed networks move towards balance at some points in time and move way from balance at other times. It would seem that the author's use of using differential equations, as is done in Chapter 12, for signed networks holds immense promise for examining these dynamics, especially with the inclusion of different time scales.

Later in their chapter, the authors turn to using dynamical processes to reveal network structure within the community detection framework. They employ a genetic algorithm framework to do this with a variety of extensions, all of which hold considerable promise. As with previous chapters, some open problems are stated with interesting methodological and substantive implications if they are pursued in a dynamic framework using differential equations.

Chapter 13 is the final contributed chapter, one that examines scientific coauthorship networks. A blockmodeling approach is adopted to understand the structure of these networks with a view to understanding the dynamics of scientific knowledge production. The data used feature collaboration among Slovene scientists using a rich temporal database. While blockmodeling can be used to discern the structure of these networks at multiple points in time, one particularly interesting question is whether these blockmodels, especially the composition of positions (clusters) and the relationships between positions (blocks) are stable over time. The authors of this chapter present a methodology for measuring the stability of such blockmodels over time. Of particular interest is the stability (or not) of cores. For this important task, a variety of indices are proposed for assessing this stability.

Science is dynamic in many ways. Of particular interest in Chapter 13 is the changing relationships between researchers through time. Over the course of their careers, the collaborative behavior of researchers changes as new problems engage their interests and collaborative partners change. Also, some researchers depart while new researchers enter the scientific system. The authors of this chapter consider first one discipline that has a core-periphery structure (with cores, a semi-periphery, and a periphery) at the two time periods they consider. They developed a visualization for transitions between these positions. This is then extended to consider 43 disciplines. For each identified discipline, they identify changes in the number of cores, the average size of them, and the relative sizes of the semi-periphery and the periphery.

The authors use a variety of different indices for measuring the stability of cores for all the studied scientific disciplines. They establish a partition of disciplines into three clusters. The smallest cluster has eight disciplines for which the cores are stable. The next smallest is a cluster of 13 disciplines whose cores are unstable. The largest cluster has 22 disciplines that are located between these extremes. One implication is that science in not monolithic. While is it obvious that disciplines have different concerns in terms of content, these results reveal clearly how their collaboration structures vary greatly. The authors present results showing why this is the case.

1.2 Looking Forward

Even though the single focus of network clustering defines the impulse behind this volume, the topic has many facets within which many approaches have been adopted. In looking at the contributed chapters, there is great diversity in the topics considered and the approaches taken by the contributing authors. This was expected and was core to constructing this volume. There are many points of consistency across the chapters along with apparent disagreements. The former is great. The latter is not a problem for there will always be diverse views in this literature. We compliment all the contributing authors for their willingness to contribute in an open-minded and engaged fashion. While many academic disciplines have been riven by deep divides across which no compromise is possible, our hope is that the consideration of the network clustering literature presented in this volume will allow us to rise above such foolish divides. The contributed chapters suggest this is very possible.

So, to our readers of this volume, we hope you will enjoy the contributions in each of the contributed chapters. Each has great merit. We will return to some of the general issues raised within each of the chapters, as well joining issues from these chapters, in the concluding chapter.

2

Bibliometric Analyses of the Network Clustering Literature

Vladimir Batagelj[1,2,5], Anuška Ferligoj[3,5], and Patrick Doreian[3,4]

[1] IMFM Ljubljana
[2] IAM, University of Primorska, Koper
[3] FDV, University of Ljubljana
[4] University of Pittsburgh
[5] NRU HSE Moscow

2.1 Introduction

Partitioning networks is performed in many disciplines, as is evidenced by the chapters of this book. The data we consider here are from the *network clustering literature*. Our focus here is the *large* set of publications identified in the area of graph/network clustering and blockmodeling, and included in the Web of Science[1] (WoS) through February 2017. The two dominant approaches for clustering networks are found in the "social" social network literature and the literature featuring physicists and other scientists examining networks. Blockmodeling is an approach that partitions the nodes of a network into *positions* (clusters of nodes) with the *blocks* being the sets of relationships within and between positions. The result is a simplified image of the *whole* network. Community detection, associated with the work of physicists studying networks, aims to identify *communities* composed of nodes *having a higher probability of being connected to each other than to members of other communities*. In identifying the literature featuring the clustering of networks we ensured the inclusion of both of these approaches.

The rest of the chapter is structured as follows: Section 2.2 outlines steps in the collection of data and cleaning them together with constructing measures and identifying specific

---

[1] The origins of, and the rationale for, collecting such data are found in the work of Garfield 16, [1].

productions. Section 2.3 presents several approaches to identifying network features including components, critical main paths, and key-route paths for analyzing citation networks. Section 2.4 examines line islands as clusters in the network clustering literature. Section 2.5 focuses on authors, productivity, collaboration, and bibliometric coupling. The chapter concludes with suggestions for future work.

2.2 Data Collection and Cleaning

We view scientific productions as *works* and sought the citation links connecting them. Citations from later works to prior works can be viewed as "votes" from researchers in their scientific fields regarding the value of earlier scientific works. Given our focus on network clustering literature, we obtained data from the WoS (now owned by Clarivate Analytics) by using the following terms in a general query:

```
"block model*" or "network cluster*" or "graph cluster*" or
"community detect*" or "blockmodel*" or "block-model*" or
"structural equival*" or "regular equival*"
```

We limited the search to the WoS Core Collection because other data bases from WoS do not permit exporting CR fields (which contain citation information). Some works appear only in the WoS CR field as a reference and lack a description in the collected data set. We call such works *cited-only* works. Additionally, we collected, using WoS and Google, some information about cited-only nodes with large indegrees (highly cited works) to add such descriptions to the collected data set. When a description of a node was unavailable in these sources, we manually constructed a description for them.[2]

Our first WoS search was completed on May 16, 2015. It was updated on January 6, 2017 for 2014–2017. A further updating for 2015–2017 was completed on February 22, 2017. We applied the new WoS2Pajek 1.5 [3] to convert WoS data into **Pajek** networks.[3] Preliminary results regarding the size of the data set are shown in Table 2.1. In slightly less than two years, the number of works increased by 56%, the number of authors by 38%, the number of journals by 40%, and the number of records by 136%. Clearly, partitioning networks is a *rapidly expanding* area of research in multiple areas given the increases in the number of works, authors, and journals. Of some interest is that the increase in authors was less than the increase in the number of works. The decrease in the final number of keywords is due to the replacement of keyword phrases with the constituting words.

While a citation network is simply composed of links between works treated as nodes, there is more to consider when other units are included. These include authors, journals, and keywords.

---

[2] There are two approaches to dealing with the resulting data: (i) manually filtering the hits and preserving only those matching the criteria or (ii) using all obtained hits while considering non-topic hits as noise. Given the enormous amount of work required for the first option, we used the second one.

[3] Most of the analyses featured in the chapter were done in **Pajek** (see [5]) and R [27]. For a highly accessible introduction to **Pajek**, see [26].

Table 2.1 Sizes of networks on clustering literature

	2015/05/16	2017/01/06	2017/02/23
Number of works	75249	112114	117082
Number of authors	44787	60419	62143
Number of journals	8993	12271	12652
Number of keywords	10095	12715	10269
Number of records	2944	5472	6953

As part of a more general strategy, the following two-mode networks were constructed: (i) an author network, **WA** as works × authors, (ii) a journal network **WJ** featuring works × journals, (iii) a keyword network **WK** with works × keywords, and (iv) a one-mode citation network **Ci** featuring only scientific productions. Additional information was obtained considering some useful partitions: (i) *year* of works by publication year, (ii) a *DC* partition distinguishing works having a complete description ($DC = 1$) and cited-only works ($DC = 0$), and (iii) a vector of the number of pages, *NP*. The dimensions of the studied networks (shown in the right-most column of Table 2.1) are the number of works, $|W| = 117,082$, the number of contributing authors, $|A| = 62,143$, the number of journals where these works appear, $|J| = 12,652$, and the number of keywords employed to characterize works, $|K| = 10,269$. All these networks share the set of works (papers, reports, books, etc.), W.

Another problem complicating data collection is that different data sources use different conventions for their data items. The usual *ISI name* of a work (field CR) has the form:

```
LEFKOVITCH LP, 1985, THEOR APPL GENET, V70, P585
```

All its elements are upper case. AU denotes author, PY is for publication year, SO denotes journals (with an allowance for at most 20 characters), VL is for Volume, and BP denotes the beginning page. The format is:

AU + ', ' + PY + ', ' + SO[:20] + ', v' + VL + ', p' + BP

In WoS, the same work can have different ISI names! To improve the precision of identification of works (entity resolution, disambiguation), the program **WoS2Pajek** supports also *short names* with the format:

LastNm[:8] + '_' + FirstNm[0] + '(' + PY + ')' + VL + ':' + BP

For example: LEFKOVIT_L(1985)70:585

For last names with prefixes, e.g. VAN, DE, … the space is deleted. Unusual names start with ∗ or $. A citation network, **Ci**, is based on the citing relation where w **Ci** z means work, w, cites work, z.

For correcting equivalent data items, there are two options: (i) make corrections in the local copy of original data (WoS file) or (ii) make the equivalence partition of nodes and shrink the set of works accordingly in all networks. We used the second option. For works with large counts (≥ 30), we prepared lists of possible equivalent items and manually determined equivalence classes. Using a simple program in Python, we produced a **Pajek** partition file, worksEQ.clu, and shrank sets of works using **Pajek**. Using the partition $p = worksEQ, p : V \rightarrow C$, we used

`Pajek` to shrink the citation network *cite* to *citeR*. As a byproduct, we obtained a partition $q : V_C \to V$, such that $q(v) = u \Rightarrow p(u) = v$. It was necessary to shrink also the partitions *year*, *DC* and the vector *NP*. This can be done in `Pajek` as follows. Given a general mapping $s : V \to B$, we seek a mapping $r : V_C \to B$ such that if $q(v) = u$, then $s(u) = r(v)$. Therefore, $r(v) = s(u) = s(q(v)) = q * s(v)$ or equivalently $r = q * s$.

In `Pajek`, given a mapping $q : V_C \to V$, the mapping r is determined for a partition s by:

```
select partition q as First partition
select partition s as Second partition
Partitions/Functional Composition First*Second
```

or for a vector s by

```
select partition q as First partition
select vector s as First vector
Operations/Vector+Partition/
Functional Composition Partition*Vector
```

For the partition *worksEQ*, we computed the "reduced" networks **CiR**, **WAr**, **WKr**, **WJr** and the partitions *YearR* and *DCr* as well as the vector *NPr*. Their sizes are shown in Table 2.2. For example, the network **WAr** has 179,049 nodes: 116,906 works and 62,143 authors.

For cited only works we have only information about their first author and no information about keywords. So, we have to limit our analysis about authors or keywords to works with complete descriptions ($DC > 0$). The sizes of corresponding networks are shown in Table 2.3.

In principle, citation networks are acyclic: earlier works cannot cite later works. Yet works appearing at the same time can cite each other. As the methods we use require a citation network to be acyclic, such ties must be located. More generally, strong components need to be identified. There were five in the network we studied, all in the form of reciprocal dyads. These are shown in Figure 2.1. Methods for identifying strong components and ways of treating them prior to analyzing citation networks are described in [7].

Table 2.2   Sizes of "reduced" networks

Network	Nodes	Arcs
WAr	179049 = 116906 + 62143	132776
WKr	127175 = 116906 + 10269	88965
WJr	129558 = 116906 + 12652	117044
CiR	116906	195784

Table 2.3   Sizes of networks with complete descriptions

Network	Nodes	Arcs
WAc	19071 = 5695 + 13376	21562
WKc	15964 = 5695 + 10269	88953
WJc	7451 = 5695 + 1756	5815
CiC	5695	38400

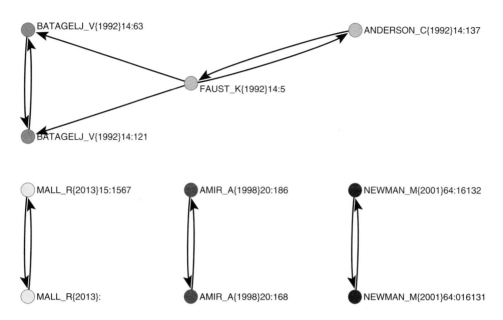

BATAGELJ_V{1992}14:63

ANDERSON_C{1992}14:137

FAUST_K{1992}14:5

BATAGELJ_V{1992}14:121

MALL_R{2013}15:1567

AMIR_A{1998}20:186

NEWMAN_M{2001}64:16132

MALL_R{2013}:

AMIR_A{1998}20:168

NEWMAN_M{2001}64:016131

Figure 2.1 Dyadic strong components.

2.2.1 Most Cited/Citing Works

It is straightforward to identify the works in a citation network receiving the most citations.[4] Similarly, identifying works with the greatest outdegrees is straightforward.

Table 2.4 lists the 60 most cited works (indegree in **CiR**). Heading the list are seven works produced in the physicist approach to networks featuring community detection. The top ranked document is by Girvan and Newman[5] as a research paper in the *Proceedings of the National Academy of Sciences* (US) (*PNAS*) in 2002. The second ranked paper written by Fortunato is a long survey paper on community detection in graphs, appearing in *Physics Reports* in 2010. In third place is a 2004 paper on community detection for very large networks by Clauset, Newman, and Moore in *Physical Review E*. The most cited paper from the social sciences, at rank 7, is by an anthropologist whose data attracted the attention of the aforementioned physicists. The next highest document from the social sciences is the Wasserman and Faust book of 1994. The other "social" social network productions in this list primarily feature works devoted to blockmodeling, albeit the earlier productions in this area. This is suggestive of the domination, in recent years, by the approach adopted by physicists when studying social networks to identify communities as clusters.

In Table 2.5 the top ten citing works (outdegree in **CiR**) are listed. They consist of books, theses, and survey papers. Only two of the items in this table come from the social sciences. The role of survey papers was studied in [7] with an emphasis on their secondary role in the *production* of scientific knowledge.

---

[4] The results reported here follow in the tradition outlined in [11].
[5] We have adopted the convention of citing only methodologically relevant items for the methods we consider in this chapter. Such frequently cited papers can be identified easily in the relevant literature.

Table 2.4 The most cited works in the network clustering literature

Rank	Citations	Work	Rank	Citations	Work	Rank	Citations	Work
1	1096	GIRVAN_M(2002)99:7821	21	292	NEWMAN_M(2003)45:167	41	145	BURRIDGE_R(1967)57:341
2	969	FORTUNAT_S(2010)486:75	22	292	LANCICHI_A(2009)80:056117	42	145	LANCICHI_A(2011)6:0018961
3	712	CLAUSET_A(2004)70:066111	23	286	NEWMAN_M(2004)69:1	43	139	GREGORY_S(2010)12:103018
4	638	BLONDEL_V(2008):P10008	24	259	GUIMERA_R(2005)433:895	44	139	LESKOVEC_J(2010):
5	621	NEWMAN_M(2004)69:026113	25	251	ALBERT_R(2002)74:47	45	138	BOCCALET_S(2006)424:175
6	578	NEWMAN_M(2006)103:8577	26	244	DUCH_J(2005)72:027104	46	137	GUIMERA_R(2004)70:025101
7	553	ZACHARY_W(1977)33:452	27	236	LUSSEAU_D(2003)54:396	47	129	NEWMAN_M(2004)70:056131
8	544	PALLA_G(2005)435:814	28	216	SHI_J(2000)22:888	48	127	BRANDES_U(2008)20:172
9	489	FORTUNAT_S(2007)104:36	29	216	LORRAIN_F(1971)1:49	49	126	BREIGER_R(1975)12:328
10	416	WATTS_D(1998)393:440	30	215	REICHARD_J(2006)74:016110	50	126	NOWICKI_K(2001)96:1077
11	412	DANON_L(2005):	31	211	HOLLAND_P(1983)5:109	51	125	ROSVALL_M(2007)104:7327
12	380	NEWMAN_M(2004)38:321	32	206	WHITE_H(1976)81:730	52	124	VONLUXBU_U(2007)17:395
13	369	LANCICHI_A(2008)78:046110	33	199	AHN_Y(2010)466:761	53	122	NEWMAN_M(2001)64:026118
14	351	WASSERMA_S(1994):	34	168	KERNIGHA_B(1970)49:291	54	119	REICHARD_J(2004)93:218701
15	329	NEWMAN_M(2006)74:036104	35	163	AIROLDI_E(2008)9:1981	55	118	ARENAS_A(2008)10:053039
16	326	ROSVALL_M(2008)105:1118	36	161	NEWMAN_M(2010):	56	118	ERDOS_P(1959)6:290
17	319	RAGHAVAN_U(2007)76:036106	37	157	SCHAEFFE_S(2007)1:27	57	116	FREEMAN_L(1979)1:215
18	307	LANCICHI_A(2009)11:033015	38	155	GOOD_B(2010)81:046106	58	116	FREEMAN_L(1977)40:35
19	306	RADICCHI_F(2004)101:2658	39	150	KARRER_B(2011)83:016107	59	113	NEWMAN_M(2001)98:404
20	304	BARABASI_A(1999)286:509	40	150	LANCICHI_A(2009)80:016118	60	112	SHEN_H(2009)388:1706

Table 2.5   The most citing works of the network clustering literature

Rank	Citations	Document	Rank	Citations	Document
1	1095	PRUESSNE_G(2012):1	6	417	NEWMAN_M(2003)45:167
2	863	BOCCALET_S(2006)424:175	7	398	FORTUNAT_S(2010)486:75
3	839	FOUSS_F(2016):1	8	327	HOLME_P(2015)88:e2015-60657-4
4	476	ARABIE_P(1992)43:169	9	321	SIBLEY_C(2012)12:505
5	456	TURCOTTE_D(1999)62:1377	10	310	FRANK_K(1998)23:171

2.2.2   The Boundary Problem for Citation Networks

For any network study, the boundary of the network must be determined with great care. In some studies, the context determines the boundary in a straightforward fashion. However, for citation networks the problem is far more ambiguous in that judgments must be made. It is reasonable to exclude cited-only works with indegree 1 for this indicates minimal notice. More generally, to get rid of the influence of sporadic citations, some threshold in terms of citations received for inclusion is necessary. To examine this the following counts were established.

The network **CiR** has 116,906 nodes and 195,784 arcs. The counts for the lowest number of received citations are 0 (4070), 1 (93,248), 2 (10,694), 3 (3352), and 4 (1610). Most nodes are cited only once (indegree = 1). We "solved" the boundary problem by including in our networks those nodes with $DCr > 0$ or $indeg > 2$. These criteria determined a subnetwork, denoted as **CiB**, with 13540 nodes and 82238 arcs.

With the network boundary determined, obtaining a general description is straightforward prior to completing any analyses. Table 2.6 lists journals whose articles were cited the most. The left panel comes from the **WJr** network while the right panel comes from **WJc** (defined for *only* those documents having complete descriptions). Unsurprisingly, the counts for the journals differ substantially as the two networks differ greatly in size. More consequentially, the orders of the journals differ. Journals from the social sciences are marked in boldface.

For the much larger network, **WJr**, the dominant journals are *PNAS* and *Nature* with over 1000 citations. Both Lecture Notes in Computing Sciences and *Science* contained more than 900 citations. Three physics journals follow. The top-ranked social science journal *Social Networks* is in tenth place. The remaining journals cover many disciplines.

For the network with only complete descriptions for the works, **WJc** (works are the hits dealing with the research topic), there are dramatic changes. *PNAS* drops to ninth place and Nature drops to 19th place. Many other journals drop out of the list. In contrast, both *Physica A* and *Physics Review E* retain their high rankings. *Social Networks* moves up to fourth place. Other journals in the right panel replace those dropping out of the left panel.

These differences reinforce the importance of solving the boundary problem appropriately. While strong cases can be made for using either **WJr** or **WJc**, it is clear that setting different boundaries can lead to dramatically different outcomes. One obvious question is whether having more information about productions is worth it. In terms of *interpreting* citation patterns and, more generally, understanding science dynamics, we contend that having more information is preferred. As a general point, when results are reported, the ways in which boundaries for networks are established *must* be made clear.

Most journals demand the use of keywords which become part of the information about works. When keywords are not parts of works, they can be constructed from titles. Composite

Table 2.6 The most used journals in two works × journals networks

Rank	Frequency in **WJr**	Journal	Frequency in **WJc**	Journal
1	1058	P NATL ACAD SCI USA	223	LECT NOTES COMPUT SC
2	1014	NATURE	175	PHYS REV E
3	941	LECT NOTES COMPUT SC	151	PHYSICA A
4	908	SCIENCE	122	**SOC NETWORKS**
5	667	PHYSICA A	88	PLOS ONE
6	639	PHYS REV E	56	LECT NOTES ARTIF INT
7	616	PHYS REV LETT	56	J GEOPHYS RES-SOL EA
8	549	BIOINFORMATICS	45	P NATL ACAD SCI USA
9	548	NUCLEIC ACIDS RES	40	SCI REP-UK
10	522	**SOC NETWORKS**	39	J STAT MECH-THEORY E
11	519	J GEOPHYS RES-SOL EA	33	NEUROCOMPUTING
12	428	B SEISMOL SOC AM	30	PHYS REV LETT
13	400	TECTONOPHYSICS	28	COMM COM INF SC
14	398	GEOPHYS J INT	27	APPL MECH MATER
15	348	NEUROIMAGE	27	BMC BIOINFORMATICS
16	342	J GEOPHYS RES	27	EUR PHYS J B
17	342	J BIOL CHEM	27	GEOPHYS J INT
18	336	J MOL BIOL	25	PROCEDIA COMPUT SCI
19	330	PHYS REV B	25	BIOINFORMATICS
20	321	IEEE T PATTERN ANAL	24	INFORM SCIENCES
21	285	**AM J SOCIOL**	23	IEEE DATA MINING
22	274	PATTERN RECOGN	23	KNOWL-BASED SYST
23	272	**AM SOCIOL REV**	23	**J MATH SOCIOL**
24	260	GEOPHYS RES LETT	21	**SOC NETW ANAL MIN**
25	249	GEOLOGY	21	ADV INTELL SYST
26	239	**SCIENTOMETRICS**	20	MATH PROBL ENG
27	229	LECT NOTES ARTIF INT	20	EXPERT SYST APPL
28	224	EARTH PLANET SC LETT	19	EPL-EUROPHYS LETT
29	220	BIOCHEMISTRY-US	19	INT J MOD PHYS B
30	214	APPL ENVIRON MICROB	19	TECTONOPHYSICS
31	212	J CHEM PHYS	19	ANN STAT
32	207	J NEUROSCI	19	NATURE
33	207	J AM STAT ASSOC	18	IEEE T KNOWL DATA EN
34	205	J GEOPHYS RES-SOLID	18	PATTERN RECOGN LETT
35	201	J AM CHEM SOC	18	**AM J SOCIOL**
36	187	J PHYS A-MATH GEN	17	ADV MATER RES-SWITZ
37	185	**ADMIN SCI QUART**	17	PURE APPL GEOPHYS
38	184	CELL	16	DATA MIN KNOWL DISC
39	184	PURE APPL GEOPHYS	16	GEOPHYS RES LETT
40	181	INFORM SCIENCES	16	IEEE T PATTERN ANAL
41	171	BIOPHYS J	16	**SCIENTOMETRICS**
42	170	**PSYCHOMETRIKA**	15	INT CONF ACOUST SPEE
43	167	IEEE T KNOWL DATA EN	14	NEW J PHYS
44	165	EUR PHYS J B	14	J CLASSIF
45	159	EXPERT SYST APPL	14	IEEE T MICROW THEORY
46	159	GEOL SOC AM BULL	14	**PSYCHOMETRIKA**
47	158	EUR J OPER RES	13	SCI WORLD J
48	154	IEEE T INFORM THEORY	13	J COMPUT SCI TECH-CH
49	144	PATTERN RECOGN LETT	13	PLOS COMPUT BIOL
50	142	**J PERS SOC PSYCHOL**	13	ADV COMPLEX SYST

The bolded journals come from the social sciences.

Table 2.7   The most used keywords

Rank	Freq.	Keyword	Rank	Freq.	Keyword	Rank	Freq.	Keyword
1	1204	network	25	291	earthquake	48	186	similarity
2	1064	community	26	281	protein	49	184	multi
3	1533	detection	27	276	stochastic	50	181	evolution
4	1499	model	28	270	overlap	51	176	mining
5	1177	graph	29	268	fault	52	166	functional
6	1135	cluster	30	265	equivalence	53	165	behavior
7	1104	algorithm	31	241	prediction	54	164	simulation
8	1082	complex	32	240	organization	55	163	state
9	1080	social	33	237	interaction	56	163	gene
10	932	structure	34	236	scale	57	160	genetic
11	900	analysis	35	229	time	58	159	centrality
12	880	base	36	227	clustering	59	157	flow
13	727	block	37	220	theory	60	156	classification
14	494	use	38	213	large	61	155	partition
15	430	datum	39	209	self	62	155	hierarchical
16	407	modularity	40	205	matrix	63	150	application
17	398	method	41	204	dynamic	64	148	slip
18	373	dynamics	42	204	identification	65	146	small
19	357	structural	43	197	modeling	66	146	design
20	317	approach	44	197	pattern	67	146	link
21	300	blockmodel	45	195	detect	68	145	web
22	294	information	46	194	local	69	144	organize
23	293	optimization	47	190	world	70	143	spectral
24	293	random						

keywords can be split into single words. Lemmatization was used in WoS2Pajek to deal with the "word-equivalence problem". Table 2.7 lists the frequency counts for keywords attached to works in the network **WKc**. Having "network" as the most frequent keyword is trivial. The next two items, "community" and "detection", suggest a problem with keywords containing two words. As a term relating to clustering, "cluster" is only in the sixth place.

Many of the other frequently used terms in Table 2.7 including model, graph, and structure are generic with limited value. Other keywords – complex, social, base, use, datum, method, approach, information, fault, scale, self, local, world, gene, genetic, flow, slip, small, and organize – convey less information. Either keywords are utterly useless for understanding of scientific citation or they have to be *examined with great care in clearly defined contexts*. To this end, we identify parts of the citation network by identifying islands (see [7]) of closely related works in them. For this, keywords become very useful for discerning the major interests of the works in an island as a focused substantive context. The same idea is clear also when we consider bibliographic coupling.

2.3   Analyses of the Citation Networks

Given our focus on citation networks, we consider ways of identifying and interpreting important parts of these networks. They include components for identifying important paths through these networks based in the ideas formulated in [18], used to examine the DNA development literature in [19], applied to the network centrality literature in [20], and extended in [7].

2.3.1   Components

Our analyses of the primary "clustering citation network" (**CiB**) features components, the identification of main paths through this literature, identifying islands (as clusters of related works) and bibliometric coupling. Our main use of components is for identifying networks useful for obtaining important paths and islands. The network, **CiB**, has 690 (weak) components. The largest have sizes 12702, 21, 20, 19, 17, 10, and 9. Here, we limit our analysis to the largest component, labeled **CiteMain**.

The presence of the reciprocal dyads identified in Figure 2.1 remains. To obtain an acyclic network, we applied the preprint transformation (see [7]) to **CiteMain**. The resulting network, **CiteMacy** (Cite, Main, acyclic), has 12,712 nodes and 81,972 arcs. The increase in the number of works is due to some of them appearing twice with one name starting with a = sign indicating the "preprint" version of a paper. We computed the SPC weights on its arcs [2]. The total flow is 1.625×10^{20}.

2.3.2   The CPM Path of the Main Citation Network

We start by identifying main paths. Figure 2.2 shows the critical path method (CPM) main path [2, 7] through the network clustering literature (in **CiteMacy**). At the bottom of this main path there are seven publications, all cited by an influential paper by Cartwright and Harary appearing in 1956. They are important foundational works for social network analysis. It continues with 22 publications from the blockmodeling literature encompassing both unsigned and signed networks. This is followed by an important transition in this main path marking a transition between the social networks field and the work of natural scientists on social networks. This 2000 publication is the last work from the area of social network analysis. It analyzed the Erdős collaboration graph. The connecting link features this production and one by Newman in 2001. Thereafter, the rest of the main path features work from the community detection literature through 2016. We expand further on this description when discussing key-route paths.

The branching at the top of the figure reflects the end of the search period we used. The top four papers cite a work by Fortunato and Hric appearing in 2016. Were a new search used to expand this main path, undoubtedly these most recent works would be cited and the main path would continue through some of them. We note that when the network centrality literature was analyzed in [7] a similar transition between fields was identified: social networks to physics to neuroscience.

2.3.3   Key-Route Paths

The CPM approach yields a single main path through the literature. A more nuanced image of this feature is obtained by identifying key-routes through a network. This method, known as the Taiwan approach, was developed in [22]. The algorithm has been generalized and included

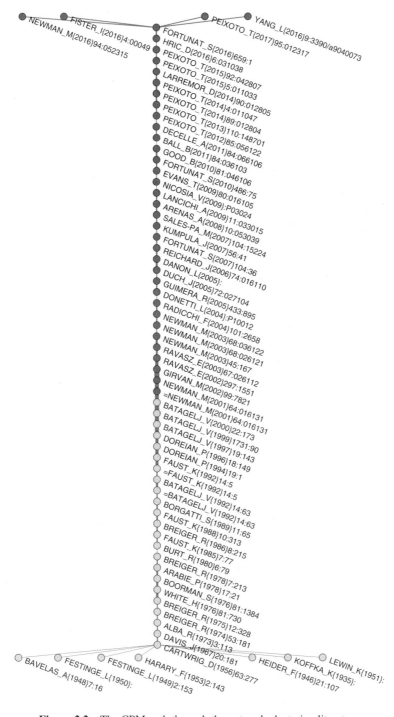

Figure 2.2   The CPM path through the network clustering literature.

in `Pajek`. The `Pajek` instruction for obtaining key-routes through 150 arcs with the largest weights is:

```
Network/Acyclic Network/Create (Sub)network/CPM/
    Global Search/Key-Route [1-150]
```

Figure 2.3 shows the results for this network clustering network. The starting and ending works are the same for both Figure 2.2 and Figure 2.3. However, between these ends, additional works are included to provide a more complex view of the evolution of the clustering field(s). The basic sequence between the social network and community detection literatures remains. Indeed, the transition point between these two literatures is a cut.

We divide our expanded discussion into two temporal periods.

2.3.3.1   The Period 1956–2000

Two papers by Cartwright and Harary and Davis formed the foundations for signed block-modeling. After these two papers, we would have expected to see the foundational paper for blockmodeling of Lorrain and White (appearing in1971), but it is not on the CPM main path nor on the key-routes. We account for this below in our discussion of Tables 2.8 and 2.9. Next comes a paper of Alba discussing cliques, a conceptual dead end even though it is much studied in the social networks area. This is followed by Breiger, who created the foundations for analyzing two-mode networks, a critically important development. The five papers involving Breiger, Boorman, Arabie, White, Levitt and Pattison, all important for creating the blockmodeling tradition, follow. Included is the work outlining the first algorithm, CONCOR, for blockmodeling and works with substantive interpretations of blockmodeling results involving White, Boorman, Breiger, Arabie, Levitt, and Pattison in the mid-to-late 1970s. Also appearing on the main path are papers on explanations of role structure theory in algebraic models involving Boorman and White and Breiger and Pattison. Burt proposed a rival algorithm for blockmodeling in 1976 which is not on the main path. A later paper from him, published in 1980, is on the main path.

A special issue of *Social Networks* devoted to blockmodeling appeared in 1992. Four papers from this issue are in the main path: two works by Batagelj, Doreian, Ferligoj introducing the direct approach to blockmodeling for structural and regular equivalence, and two papers by Faust and Wasserman (with one with Anderson) discussing the interpretation and evaluation of blockmodels and stochastic blockmodels. In 1994, Doreian, Batagelj and Ferligoj proposed generalized concepts of equivalence based on block types and corresponding criterion functions which provides an appropriate measure of fit of blockmodels to the empirical data.

Also on this main path is a paper by Doreian and Mrvar appearing in 1996 that used the generalized blockmodeling approach and applied it to signed networks and a paper by Batagelj (appearing in 1997) which provided a mathematical formalization of the generalized blockmodeling. The last two papers in the class of the social network contributions involve Batagelj, Mrvar and Zaveršnik, who proposed several clustering procedures for large networks and applied these algorithms to the Erdős collaboration graph. As noted above, this work is the bridge to the contributions of natural scientists, mostly working on community detection problems.

2.3.3.2   The Period 2001–2016

A paper by Newman, appearing in 2001, is the first production on the main path for works from the natural sciences. He presented a variety of statistical properties of scientific collaboration

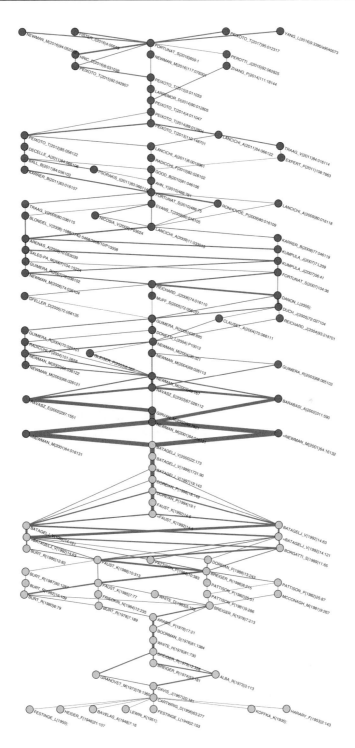

Figure 2.3   Key-route paths through the network clustering literature.

Table 2.8 List of works on CPM path (1), main paths (2), and island (3) – part 1

Label	Code	First author	Title	Journal
KOFFKA_K(1935):	12	Koffka, K	Principles of Gestalt Psychology	book
HEIDER_F(1946)21:107	12	Heider, F	Attitudes and cognitive organization	J PSYCHOL
BAVELAS_A(1948)7:16	12	Bavelas, A	A mathematical model for group structure	HUMAN ORG
FESTINGE_L(1949)2:153	12	Festinger, L	The analysis of sociograms using matrix algebra	HUMAN REL
FESTINGE_L(1950):	12	Festinger, L	Informal social communication	PSYCHO REV
LEWIN_K(1951):	12	Lewin, K	Field theory in social science	book
HARARY_F(1953)2:143	12	Harary, F	On the notion of balance of a signed graph	MICH MATH J
CARTWRIG_D(1956)63:277	123	Cartwright, D	Structural balance – a generalization of heider theory	PSYCHOL REV
HUBBELL_C(1965)28:377	3	Hubbell, CH	An input-output approach to clique identification	SOCIOMETRY
DAVIS_J(1967)20:181	123	Davis, JA	Clustering and structural balance in graphs	HUM RELAT
BOYD_J(1969)6:139	3	Boyd, JP	Algebra of group kinship	J MATH PSYCHOL
HARTIGAN_J(1972)67:123	3	Hartigan, JA	Direct clustering of a data matrix	J AM STAT ASSOC
ALBA_R(1973):113	123	Alba, RD	Graph-theoretic definition of a sociometric clique	J MATH SOCIOL
GRANOVET_M(1973)78:1360	23	Granovet.MS	The strength of weak ties	AM J SOCIOL
BREIGER_R(1974)53:181	123	Breiger, RL	Duality of persons and groups	SOC FORCES
BREIGER_R((1975)12:328	123	Breiger, RL	Algorithm for clustering relational data with applications to social network analysis and comparison with multidimensional-scaling	J MATH PSYCHOL
WHITE_H(1976)81:730	123	White, HC	Social-structure from multiple networks .1. Blockmodels of roles and positions	AM J SOCIOL
BOORMAN_S(1976)81:1384	123	Boorman, SA	Social-structure from multiple networks. 2. Role structures	AM J SOCIOL
BURT_R(1976)55:93	3	Burt, RS	Positions in networks	SOC FORCES
BURT_R(1977)56:106	3	Burt, RS	Positions in multiple network systems .1. General conception of stratification and prestige in a system of actors cast as a social topology	SOC FORCES
BURT_R(1977)56:551	3	Burt, RS	Positions in multiple network systems .2. Stratification and prestige among elite decision-makers in community of Altneustadt	SOC FORCES
ARABIE_P(1978)17:21	123	Arabie, P	Constructing blockmodels – how and why	J MATH PSYCHOL
SAILER_L(1978)1:73	3	Sailer, LD	Structural equivalence – meaning and definition, computation and application	SOC NETWORKS
BURT_R(1978)7:189	23	Burt, RS	Cohesion versus structural equivalence as a basis for network subgroups	SOCIOL METHOD RES
BREIGER_R(1978)7:213	123	Breiger, RL	Joint role structure of 2 communities elites	SOCIOL METHOD RES
SNYDER_D(1979)84:1096	3	Snyder, D	Structural position in the world system and economic-growth, 1955-1970 – multiple-network analysis of transnational interactions	AM J SOCIOL
BREIGER_R(1979)13:21	3	Breiger, RL	Toward an operational theory of community elite structures	QUAL QUANT
BREIGER_R(1979)42:262	3	Breiger, RL	Personae and social roles – network structure of personality-types in small-groups	SOC PSYCHOL
BURT_R(1980)6:79	123	Burt, RS	Models of network structure	ANNU REV SOCIOL
MCCONAGH_M(1981)9:267	23	Mcconaghy, MJ	The common role structure – improved block-modeling methods applied to 2 communities elites	SOCIOL METHOD RES
PATTISON_P(1981)9:286	23	Pattison, PE	A reply to Mcconaghy – equating the joint reduction with block-model common role structures	SOCIOL METHOD RES
BURT_R(1982)16:109	23	Burt, RS	Testing a structural model of perception – conformity and deviance with respect to journal norms in elite sociological methodology	QUAL QUANT

Table 2.9 List of works on CPM path (1), main paths (2), and island (3) – part 2

Label	Code	First author	Title	Journal
PATTISON_P(1982)25:51	23	Pattison, PE	A factorization procedure for finite-algebras	J MATH PSYCHOL
PATTISON_P(1982)25:87	23	Pattison, PE	The analysis of semigroups of multirelational systems	J MATH PSYCHOL
MANDEL_M(1983)48:376	3	Mandel, MJ	Local roles and social networks	AM SOCIOL REV
WHITE_D(1983)5:193	23	White, DR	Graph and semigroup homomorphisms on networks of relations	SOC NETWORKS
FRIEDKIN_N(1984)12:235	23	Friedkin, NE	Structural cohesion and equivalence explanations of social homogeneity	SOCIOL METHOD RES
DOREIAN_P(1985)36:28	3	Doreian, P	Structural equivalence in a journal network	J AM SOC INFORM SCI
FAUST_K(1985)7:77	123	Faust, K	Does structure find structure – a critique of Burt use of distance as a measure of structural equivalence	SOC NETWORKS
FIENBERG_S(1985)80:51	3	Fienberg, SE	Statistical-analysis of multiple sociometric relations	J AM STAT ASSOC
BREIGER_R(1986)8:215	123	Breiger, RL	Cumulated social roles – the duality of persons and their algebras	SOC NETWORKS
BURT_R(1987)92:1287	23	Burt, RS	Social contagion and innovation – cohesion versus structural equivalence	AM J SOCIOL
FAUST_K(1988)10:313	123	Faust, K	Comparison of methods for positional analysis – structural and general equivalences	SOC NETWORKS
DOREIAN_P(1988)13:243	23	Doreian, P	Equivalence in a social network	J MATH SOCIOL
PATTISON_P(1988)10:383	23	Pattison, PE	Network models – some comments on papers in this special issue	SOC NETWORKS
WINSHIP_C(1988)10:209	3	Winship, C	Thoughts about roles and relations – an old document revisited	SOC NETWORKS
BORGATTI_S(1989)11:65	123	Borgatti, SP	The class of all regular equivalences – algebraic structure and computation	SOC NETWORKS
IACOBUCC_D(1990)55:707	3	Iacobucci, D	Social networks with 2 sets of actors	PSYCHOMETRIKA
BURT_R(1990)12:83	23	Burt, Rs	Detecting role equivalence	SOC NETWORKS
BATAGELJ_V(1992)14:63	123	Batagelj, V	Direct and indirect methods for structural equivalence	SOC NETWORKS
BATAGELJ_V(1992)14:121	23	Batagelj, V	An optimizational approach to regular equivalence	SOC NETWORKS
ANDERSON_C(1992)14:137	3	Anderson, CJ	Building stochastic blockmodels	SOC NETWORKS
FAUST_K(1992)14:5	123	Faust, K	Blockmodels – interpretation and evaluation	SOC NETWORKS
DOREIAN_P(1994)19:1	123	Doreian, P	Partitioning networks based on generalized concepts of equivalence	J MATH SOCIOL
DOREIAN_P(1996)18:149	123	Doreian, P	A partitioning approach to structural balance	SOC NETWORKS
BATAGELJ_V(1997)19:143	123	Batagelj, V	Notes on blockmodeling	SOC NETWORKS
BATAGELJ_V(1999)1731:90	123	Batagelj, V	Partitioning approach to visualization of large graphs	LECT NOTES COMPUT SC
BATAGELJ_V(2000)22:173	123	Batagelj, V	Some analyses of Erdos collaboration graph	SOC NETWORKS
NEWMAN_M(2001)64:016131	123	Newman, MEJ	Scientific collaboration networks. I. Network construction and fundamental results	PHYS REV E
NEWMAN_M(2001)64:16132	23	Newman, MEJ	Scientific collaboration networks. II. Shortest paths, weighted networks, and centrality	PHYS REV E
GIRVAN_M(2002)99:7821	123	Girvan, M	Community structure in social and biological networks	P NATL ACAD SCI USA
RAVASZ_E(2002)297:1551	123	Ravasz, E	Hierarchical organization of modularity in metabolic networks	SCIENCE
BARABASI_A(2002)311:590	23	Barabasi, AL	Evolution of the social network of scientific collaborations	PHYSICA A
RAVASZ_E(2003)67:026112	123	Ravasz, E	Hierarchical organization in complex networks	PHYS REV E

networks. An important contribution for the development of the community detection approach is the paper of Girvan and Newman, also on the CPM main path, and the key-routes through this network. Here (and in some other papers not included in the main path but in the key-route paths and islands) they introduced the clustering coefficient. They also introduced the term community detection to avoid confusion with the clustering coefficient. We note that only recently, with the further development of stochastic blockmodels, did the social networks terminology get used again, albeit to a limited extent.

Next, two papers of Ravasz with her collaborators discuss the hierarchical organization in complex networks. Later, productions by Sales-Pardo and Arenas *et al.* also deal with this topic. Newman applied a variety of techniques and models to analyze complex networks and to examine the properties of highly clustered networks in 2003. In the same year, Newman and Park argued that social networks differ from most other types of networks. Next, four papers propose different algorithms for detecting network communities (Radicchi *et al.*, Donetti and Muñoz, Arenas, and Ball *et al.*). Guimera and Amaral in *Nature* analyzed complex metabolic networks. The first paper on the main path dealing with the statistical aspects of community detection by Reichardt and Bornholdt appeared in 2006. Fortunato and Barthelemy [14] found that modularity optimization may fail to identify smaller modules. Kumpula *et al.* then proposed an approach for dealing with this problem.

The following two papers (Lancichinetti *et al.* and Nicosia *et al.*) proposed an approach for detecting overlapping structures in complex networks. Evans and Lambiotte proposed clustering links of a network. The next paper on the main path is by Fortunato, a highly cited overview of community detection in networks. Good *et al.* studied the performance of modularity maximization. The first paper in the main path discussing stochastic blockmodels is by Decelle *et al.* This idea was developed further by Peixoto in several papers appearing between 2012 and 2014. Larremore *et al.* studied the community structure in bipartite networks. In 2015, Peixoto used a statistical approach to large network models to discern overlapping clusters. Similarly, Hric *et al.* developed a joint generative model for data and meta-data to attempt the prediction of missing nodes. Peixoto's terminology is becoming closer to the one used in social network analysis. The last paper in the main path by Fortunato and Hric is a user guide for community detection in networks.

2.3.3.3 Tables 2.8–2.12

These tables provide more details regarding the authors, works, and journals for the works in the CPM path, key-routes, and islands. They are also relevant for our discussion of islands in Section 2.7. The five tables form a single extended table that is separated only for pagination reasons. In these tables the labels of the works are given in the first column, the second column (code) describes in which analysis the work appeared (1 – CPM path, 2 – key-routes, and 3 – link island). The following columns give the first author of the work, the work's title, and the journal in which the work was published.

The items in Table 2.8 and all but the last six items in Table 2.9 come from the social networks literature. The earliest items set the foundations of, and inspiration for, the development of social network analysis. The foundational paper for blockmodeling was published in 1971 by Lorrain and White [24]. Its absence from both the CPM main and key-routes is due to it being mathematically "fierce" in its use of category theory. However, Breiger et al. (1975) [8],

Table 2.10 List of works on CPM path (1), main paths (2), and island (3) – part 3

Label	Code	First author	Title	Journal
NEWMAN_M(2003)45:167	123	Newman, MEJ	The structure and function of complex networks	SIAM REV
NEWMAN_M(2003)68:026121	123	Newman, MEJ	Properties of highly clustered networks	PHYS REV E
RIVES_A(2003)100:1128	3	Rives, AW	Modular organization of cellular networks	P NATL ACAD SCI USA
GUIMERA_R(2003)68:065103	23	Guimera, R	Self-similar community structure in a network of human interactions	PHYS REV E
HOLME_P(2003)19:532	3	Holme, P	Subnetwork hierarchies of biochemical pathways	BIOINFORMATICS
NEWMAN_M(2003)68:036122	123	Newman, MEJ	Why social networks are different from other types of networks	PHYS REV E
GLEISER_P(2003)6:565	23	Gleiser, PM	Community structure in jazz	ADV COMPLEX SYST
NEWMAN_M(2004)69:026113	23	Newman, MEJ	Finding and evaluating community structure in networks	PHYS REV E
NEWMAN_M(2004)38:321	23	Newman, MEJ	Detecting community structure in networks	EUR PHYS J B
REICHARD_J(2004)93:218701	23	Reichardt, J	Detecting fuzzy community structures in complex networks with a Potts model	PHYS REV LETT
ARENAS_A(2004)38:373	3	Arenas, A	Community analysis in social networks	EUR PHYS J B
CLAUSET_A(2004)70:066111	23	Clauset, A	Finding community structure in very large networks	PHYS REV E
RADICCHI_F(2004)101:2658	123	Radicchi, F	Defining and identifying communities in networks	P NATL ACAD SCI USA
DONETTI_L(2004):P10012	123	Donetti, L	Detecting network communities: a new systematic and efficient algorithm	J STAT MECH
GUIMERA_R(2004)70:025101	23	Guimera, R	Modularity from fluctuations in random graphs and complex networks	PHYS REV E
GUIMERA_R(2005)433:895	123	Guimera, R	Functional cartography of complex metabolic networks	NATURE
DUCH_J(2005)72:027104	123	Duch, J	Community detection in complex networks using extremal optimization	PHYS REV E
DANON_L(2005):	123	Danon, L	COSIN book	–
PALLA_G(2005)435:814	3	Palla, G	Uncovering the overlapping community structure of complex networks in nature and society	NATURE
MUFF_S(2005)72:056107	23	Muff, S	Local modularity measure for network clusterizations	PHYS REV E
GFELLER_D(2005)72:056135	23	Gfeller, D	Finding instabilities in the community structure of complex networks	PHYS REV E
GUIMERA_R(2005)102:7794	3	Guimera, R	The worldwide air transportation network: Anomalous centrality, community structure, and cities' global roles	P NATL ACAD SCI USA
NEWMAN_M(2006)103:8577	3	Newman, MEJ	Modularity and community structure in networks	P NATL ACAD SCI USA
REICHARD_J(2006)74:016110	123	Reichardt, J	Statistical mechanics of community detection	PHYS REV E
BOCCALET_S(2006)424:175	3	Boccaletti, S	Complex networks: Structure and dynamics	PHYS REP
NEWMAN_M(2006)74:036104	23	Newman, MEJ	Finding community structure in networks using the eigenvectors of matrices	PHYS REV E
FORTUNAT_S(2007)104:36	123	Fortunato, S	Resolution limit in community detection	P NATL ACAD SCI USA
KUMPULA_J(2007)56:41	123	Kumpula, JM	Limited resolution in complex network community detection with Potts model approach	EUR PHYS J B
KUMPULA_J(2007)7:L209	23	Kumpula, JM	Limited resolution and multiresolution methods in complex network community detection	FLUCT NOISE LETT
GUIMERA_R(2007)3:63	3	Guimera, R	Classes of complex networks defined by role-to-role connectivity profiles	NAT PHYS
ROSVALL_M(2007)104:7327	3	Rosvall, M	An information-theoretic framework for resolving community structure in complex networks	P NATL ACAD SCI USA
GUIMERA_R(2007)76:036102	23	Guimera, R	Module identification in bipartite and directed networks	PHYS REV E

Table 2.11 List of works on CPM path (1), main paths (2), and island (3) – part 4

Label	Code	First author	Title	Journal
SALES-PA_M(2007)104:15224	123	Sales-Pardo, M	Extracting the hierarchical organization of complex systems	P NATL ACAD SCI USA
ARENAS_A(2008)10:053039	123	Arenas, A	Analysis of the structure of complex networks at different resolution levels	NEW J PHYS
CLAUSET_A(2008)453:98	3	Clauset, A	Hierarchical structure and the prediction of missing links in networks	NATURE
KUMPULA_J(2008)78:026109	3	Kumpula, JM	Sequential algorithm for fast clique percolation	PHYS REV E
KARRER_B(2008)77:046119	23	Karrer, B	Robustness of community structure in networks	PHYS REV E
BLONDEL_V(2008):P10008	23	Blondel, VD	Fast unfolding of communities in large networks	J STAT MECH-THEORY E
LEUNG_I(2009)79:066107	3	Leung, IXY	Towards real-time community detection in large networks	PHYS REV E
LANCICHI_A(2009)11:033015	123	Lancichinetti, A	Detecting the overlapping and hierarchical community structure of complex networks	NEW J PHYS
LANCICHI_A(2009)80:016118	23	Lancichinetti, A	Benchmarks for testing community detection algorithms on directed and weighted graphs with overlapping communities	PHYS REV E
RONHOVDE_P(2009)80:016109	23	Ronhovde, P	Multiresolution community detection for megascale networks by information-based replica correlations	PHYS REV E
GOMEZ_S(2009)80:016114	3	Gomez, S	Analysis of community structure in networks of correlated data	PHYS REV E
TRAAG_V(2009)80:036115	23	Traag, VA	Community detection in networks with positive and negative links	PHYS REV E
NICOSIA_V(2009):P03024	123	Nicosia, V	Extending the definition of modularity to directed graphs with overlapping communities	J STAT MECH
EVANS_T(2009)80:016105	123	Evans, TS	Line graphs, link partitions, and overlapping communities	PHYS REV E
LANCICHI_A(2009)80:056117	3	Lancichinetti, A	Community detection algorithms: A comparative analysis	PHYS REV E
BARBER_M(2009)80:026129	3	Barber, MJ	Detecting network communities by propagating labels under constraints	PHYS REV E
FORTUNAT_S(2010)486:75	123	Fortunato, S	Community detection in graphs	PHYS REP
GOOD_B(2010)81:046106	123	Good, BH	Performance of modularity maximization in practical contexts	PHYS REV E
LANCICHI_A(2010)81:046110	3	Lancichinetti, A	Statistical significance of communities in networks	PHYS REV E
RADICCHI_F(2010)82:026102	23	Radicchi, F	Combinatorial approach to modularity	PHYS REV E
LANCICHI_A(2010)5:0011976	3	Lancichinetti, A	Characterizing the Community Structure of Complex Networks	PLOS ONE
AHN_Y(2010)466:761	23	Ahn, YY	Link communities reveal multiscale complexity in networks	NATURE
EVANS_T(2010)77:265	3	Evans, TS	Line graphs of weighted networks for overlapping communities	EUR PHYS J B
MUCHA_P(2010)328:876	3	Mucha, PJ	Community Structure in Time-Dependent, Multiscale, and Multiplex Networks	SCIENCE
GREGORY_S(2010)12:103018	3	Gregory, S	Finding overlapping communities in networks by label propagation	NEW J PHYS
KARRER_B(2011)83:016107	23	Karrer, B	Stochastic blockmodels and community structure in networks	PHYS REV E
EXPERT_P(2011)108:7663	23	Expert, P	Uncovering space-independent communities in spatial networks	P NATL ACAD SCI USA
PSORAKIS_I(2011)83:066114	23	Psorakis, I	Overlapping community detection using Bayesian non-negative matrix factorization	PHYS REV E
TRAAG_V(2011)84:016114	23	Traag, VA	Narrow scope for resolution-limit-free community detection	PHYS REV E
DECELLE_A(2011)107:065701	3	Decelle, A	Inference and Phase Transitions in the Detection of Modules in Sparse Networks	PHYS REV LETT
BALL_B(2011)84:036103	123	Ball, B	Efficient and principled method for detecting communities in networks	PHYS REV E
DECELLE_A(2011)84:066106	123	Decelle, A	Asymptotic analysis of the stochastic block model for modular networks and its algorithmic applications	PHYS REV E

Table 2.12 List of works on CPM path (1), main paths (2), and island (3) – part 5

Label	Code	First author	Title	Journal
LANCICHI_A(2011)6:0018961	23	Lancichinetti, A	Finding Statistically Significant Communities in Networks	PLOS ONE
LANCICHI_A(2011)84:066122	23	Lancichinetti, A	Limits of modularity maximization in community detection	PHYS REV E
NADAKUDI_R(2012)108:188701	3	Nadakuditi, RR	Graph Spectra and the Detectability of Community Structure in Networks	PHYS REV LETT
PEIXOTO_T(2012)85:056122	123	Peixoto, TP	Entropy of stochastic blockmodel ensembles	PHYS REV E
PEIXOTO_T(2013)110:148701	123	Peixoto, TP	Parsimonious Module Inference in Large Networks	PHYS REV LETT
GOPALAN_P(2013)110:14534	3	Gopalan, PK	Efficient discovery of overlapping communities in massive networks	P NATL ACAD SCI USA
PEIXOTO_T(2014)89:012804	123	Peixoto, TP	Efficient Monte Carlo and greedy heuristic for the inference of stochastic block models	PHYS REV E
PEIXOTO_T(2014)4:011047	123	Peixoto, TP	Hierarchical Block Structures and High-Resolution Model Selection in Large Networks	PHYS REV X
LARREMOR_D(2014)90:012805	123	Larremore, DB	Efficiently inferring community structure in bipartite networks	PHYS REV E
ZHANG_P(2014)111:18144	23	Zhang, P	Scalable detection of statistically significant communities and hierarchies, using message passing for modularity	P NATL ACAD SCI USA
KAWAMOTO_T(2015)91:012809	3	Kawamoto, T	Estimating the resolution limit of the map equation in community detection	PHYS REV E
ZHANG_X(2015)91:032803	3	Zhang, X	Identification of core-periphery structure in networks	PHYS REV E
PEIXOTO_T(2015)5:011033	123	Peixoto, TP	Model Selection and Hypothesis Testing for Large-Scale Network Models with Overlapping Groups	PHYS REV X
JIANG_J(2015)91:062805	3	Jiang, JQ	Stochastic block model and exploratory analysis in signed networks	PHYS REV E
PEIXOTO_T(2015)92:042807	23	Peixoto, TP	Inferring the mesoscale structure of layered, edge-valued, and time-varying networks	PHYS REV E
PEROTTI_J(2015)92:062825	23	Perotti, JI	Hierarchical mutual information for the comparison of hierarchical community structures in complex networks	PHYS REV E
ZHANG_P(2016)93:012303	3	Zhang, P	Community detection in networks with unequal groups	PHYS REV E
VALLES-C_T(2016)6:011036	3	Valles-Catala, T	Multilayer Stochastic Block Models Reveal the Multilayer Structure of Complex Networks	PHYS REV X
NEWMAN_M(2016)117:078301	23	Newman, MEJ	Estimating the Number of Communities in a Network	PHYS REV LETT
HRIC_D(2016)6:031038	123	Hric, D	Network Structure, Metadata, and the Prediction of Missing Nodes and Annotations	PHYS REV X
FORTUNAT_S(2016)659:1	123	Fortunato, S	Community detection in networks: A user guide	PHYS REP
NEWMAN_M(2016)94:052315	123	Newman, MEJ	Equivalence between modularity optimization and maximum likelihood methods for community detection	PHYS REV E
FISTER_I(2016)4:00049	123	Fister, I	Toward the Discovery of Citation Cartels in Citation Networks	FRONT PHYS
YANG_L(2016)9:3390/a9040073	123	Yang, LJ	Community Structure Detection for Directed Networks through Modularity Optimisation	ALGORITHMS
PEIXOTO_T(2017)95:012317	123	Peixoto, TP	Nonparametric Bayesian inference of the microcanonical stochastic block model	PHYS REV E

White *et al.* (1976) and Boorman *et al.* (1976) (the 16th through 18th items) provided the first algorithm for blockmodeling along with substantive interpretations of blockmodeling results. The next three papers in the table, by Burt, introduced a rival algorithm for blockmodeling, especially [9]. Other papers presented blockmodeling results, critiques of methods, and discussions of closely related topics, especially role structures.

The Heider (1946) paper (the second work in Table 2.8), along with the Harary (1953) (sixth), the Cartwright and Harary (1956) (eighth) and Davis (1967) (tenth) papers formed the foundations for the creation of signed blockmodeling by Doreian and Mrvar (1996) (the 23rd item in Table 2.9). The basic idea is located in the structural theorems in the papers of Cartwright and Harary, and in Davis, being coupled to the direct approach to blockmodeling [13].

Examining journals as venues for works is facilitated by considering the right-hand column of Tables 2.8–2.12. Many of the journals relating to blockmodeling in Table 2.8 are from the mainstream sociological literature. They include two from *The American Journal of Sociology* and four each from Social Forces and *Sociological Research and Methods*. The list of journals in Table 2.9 reveals a sharp transition, with *Social Networks* appearing 15 times. It appears just once in Table 2.8. It appears that (i) blockmodeling became more of a method for partitioning social networks with migration to a newer journal and (ii) the interest of sociologists in this research area diminished.

A similar pattern can be discerned for the subsequent community detection literature. The shift from "social" social networks to community detection development is marked by the appearance of works produced by Mark Newman and Michelle Girvan in *Physics Review E* and *PNAS* (the sixth, fifth and fourth items from the bottom in Table 2.9). In terms of a frequent venue, *Physics Review E* dominates with 44 works related to clustering appearing in its pages. It seems that *Physics Review E* plays the same "venue role" as *Social Networks* regarding clustering. However, it does so to a far larger research community having many more scientists and more journals (which are also larger in size).

There is a contrast between the list of journals in Table 2.6 and those listed in Tables 2.8–2.12. In the left panel of Table 2.6, the high ranking journals are *PNAS*, Nature, *Lecture Notes in Computer Science*, *Science*, *Physica A*, *Physics Review E*, and *Physics Review Letters*. In the right panel, the top four journals are *Lecture Notes in Computer Science*, *Physics Review E*, *Physica A* and *Social Networks*. In the main, the journals heading the lists in Table 2.6 largely vanish from the list in Tables 2.8–2.12. *Lecture Notes in Computer Science* does not appear, *Physica A* appears only once, Science twice, *Nature* thrice, and *Physics Review Letters* four times. Only *PNAS* and *Physics Review E* have works appearing with any regularity. It seems that community detection has a relatively narrow focus within the wider natural sciences literature – just as blockmodeling did in the earlier sociological literature. For researchers interested in the substantive meanings associated with partitioning with this literature, this raises interesting questions that are answered, in part, by examining islands in this literature.

2.3.4 *Positioning Sets of Selected Works in a Citation Network*

The original main path analyses produced figures like the one shown in Figure 2.2 with a single main path. A recent extension of this approach enables a researcher to determine main paths through a selected set of nodes (works) in a citation network. This can be used to position a given set of nodes in a citation network – they can either attach to the principal main path or form separate streams. This is illustrated with three examples involving valued networks, signed networks, and a geophysics network.

The basic idea is to select a set of works on specific topic. For the valued networks example, we focused on works authored by Žiberna and Nordlund extending blockmodeling for binary networks to valued networks. For the signed networks we selected papers by Doreian and Mrvar, who have written extensively on this topic. For the geophysics network we used works selected from the network discussed in Subsection 2.4.3. The new option determines, for each work from the set of given works, the corresponding main path passing through it. There are two possible outcomes. One is that the intersection of the principal main path and the obtained main path is non-empty. If so, then the selected work is related to those in the principal main path. This allows "branches" having ties to or from works in the principal main path. The second option is that the intersection is empty, implying that the selected work is focused on different issues.

The main path of Figure 2.2 is present in both Figures 2.4 and 2.5, which differ only by considering separately valued and signed networks. We consider first works on

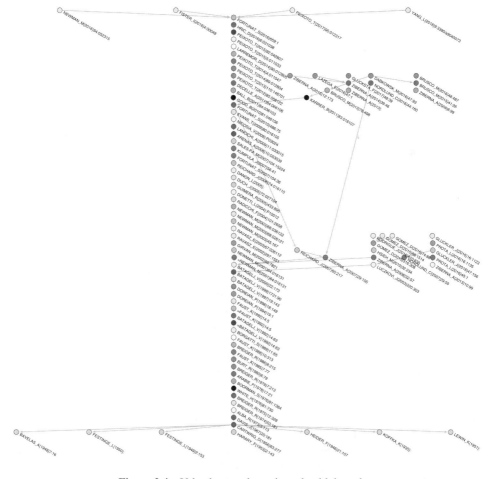

Figure 2.4 Valued networks main path with branches.

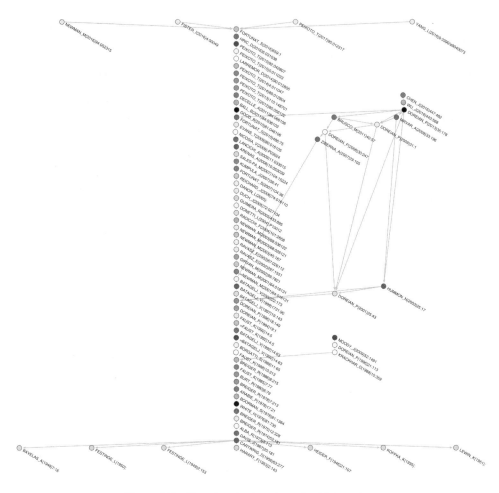

Figure 2.5 Signed network main path with branches.

valued networks, { ZIBERNA_A(2007)29:105, NORDLUND_C(2007)29:59, ZIBERNA_A (2008)32:57, ZIBERNA_A(2009)6:99, ZIBERNA_A(2013):, ZIBERNA_A (2013)10:99, ZIBERNA_A(2014)39:46, ZIBERNA_A(2016)12:137, NORDLUND_C(2016)44:160 }, as shown in Figure 2.4.

All the foundational papers are cited by a paper by Cartwright and Harary published in 1956. The next 22 works are all well known in the **social** network analysis with many in the blockmodeling tradition before a branch appears. It is a pair of branches with two ties to works by Batagelj regarding valued networks. One comes from a work of his Ljubljana colleague, Žiberna, and the other from a work involving Luczkovich. The former has a small main path with works considering valued networks. The second branch includes productions that appeared in a special issue of the *Journal of Economic Geography* focused on social networks distributed in geographic space. Blockmodeling ideas were mobilized in this special issue. The second branch, actually a double branch, involves the survey paper of Fortunato on community detection. There is a citation from that work to a work that cites one of works authored by Žiberna in one of the branches mentioned above. The other takes the form of a citation to the Fortunato work. The works on this branch involve works authored by social scientists regarding valued networks and algorithmic methods for partitioning networks. There is also a citation from a work involving Nordlund, a coauthor with Žiberna to another Žiberna production in the earlier branch.

We turn to consider the works on the signed network, { DOREIAN_P(1996)18:149, DOREIAN_P(1996)21:113,   DOREIAN_P(2001)25:43,   HUMMON_N(2003)25:17, DOREIAN_P(2009)31:1,   MRVAR_A(2009)33:196,   BRUSCO_M(2011)40:57,   DOR-EIAN_P(2013)35:178 }, as shown in Figure 2.5. The first branch appears with a citation to a 1986 paper authored by Breiger, firmly in the blockmodeling tradition, by Krackhardt. It appears far earlier than the first branches of Figure 2.5. This paper was cited by another production involving both Doreian and Krackhardt which in turn was cited by a paper involving Moody, a rising scholar in the social networks field. While this paper has been cited frequently, its content is not defined as being primarily within the domain defined by signed networks. This has been captured in Figure 2.5.

The next branch off the main path involves a book edited by Doreian and Stokman on the evolution of social networks. Ideas expressed there were picked up by Doreian and Krackhardt in a 2001 work, which provided compelling evidence against the widely accepted idea of signed networks tending towards balance. This was reinforced by a 2003 production by Hummon and Doreian.

Another branch of the main path is due to a citation from Žiberna to Batagelj, colleagues at the University of Ljubljana. This is connected through citations from productions involving Doreian, Mrvar, and Brusco on fitting signed blockmodels. There are ties from this group citing the 2001 and 2003 productions mentioned in the previous paragraph. There is also a tie from a work on partitioning signed two-mode into a work involving Fortunato late in the main path. It forms the last branch of Figure 2.5.

Note that in both examples a part of the principal main path is also a part of the main path through each work from the selected set of works. These parts combine in both cases in the complete principal main path.

The foregoing two examples of "attaching" branches to a main path were triggered by considering variants of networks studied in the blockmodeling literature. Now, we examine a completely different example. In Subsection 2.4.3, we examine link islands in the network partitioning literature. We selected some of the productions in this link island, { DIETERIC_J(1979)84:2161, CARLSON_J(1989)40:6470, CARLSON_J(1991) 44:6226, OLAMI_Z(1992)46:R1720, TURCOTTE_D(1999)62:1377 }, and repeated the above analyses. The results are shown in Figure 2.6, a simpler figure than for the two previous examples.

For this example, we start our discussion at the top of the main path. A 1999 work, one involving Turcotte, a very prominent contributor to this literature, cites two works off the "main" path. Each citation leads to a smaller and distinct main path. Both smaller main paths link back to the "main" path, albeit at different places. The one on the right of is a 1992 production that cites a production of the same year. The one on the left sent a tie to a production published in 1990. At the bottom of Figure 2.6, there is a publication on the main path that cites earlier foundational works for the geophysical literature. This is a common pattern for identified main paths.

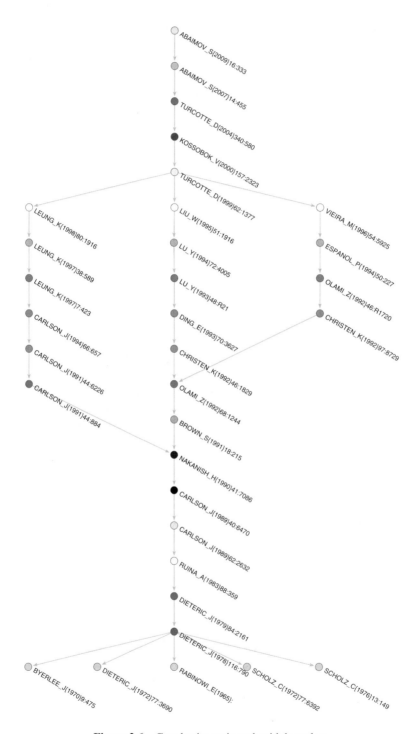

Figure 2.6   Geophysics main path with branches.

2.4   Link Islands in the Clustering Network Literature

A *link island* is a connected subnetwork having a higher internal cohesion than on the links to its neighbors. Identifying islands is a general and efficient approach for identifying locally "important" subnetworks in a given network. The details for doing this are described in [7] (Section 2.9). The method amounts to filtering networks to identify some manageable parts. In large networks, it is likely that many such islands will be identified. While islands are identified through the ties linking them, it is crucial to examine the substantive content of the islands. Just identifying topological features, while useful, is not enough. Islands are *coherent* with the coherence coming from substance and the kind of information contained in Tables 2.8–2.12.

Link islands were used extensively in [7] to examine the structure of scientific citation networks (Section 4.7), US patent networks (Section 5.6) and the US Supreme Court citation network (in Section 6.2). We use the same tools here to examine the clustering network as defined above. General `Pajek` instructions for doing this are contained in [7].

Figure 2.7 shows the ten link islands having sizes in the range [20,150] identified in the network clustering literature. Adopting George Orwell's phrase (from *Animal Farm*) "All animals are equal, but some animals are more equal than others" we change it to "All islands are interesting, but some islands are more interesting than others". It seems that the islands labeled 10, 7, 9 and 2 have the most interest value. The other islands have much smaller maximal weights, smaller diameter, and represent less important stories.

Island 10 is the largest of these islands having 150 works and a maximal weight of 0.5785. It has two clear parts separated by a cut. Island 7 is next in size with 74 works having a much lower maximal weight of 4.9611×10^{-18}. It also has two parts. The lower left part is centered on a single production while the upper right part appears to be centered on a set of inter-linked works. Both parts are highly centralized. Island 9 has 44 works with a maximal weight of 2.416×10^{-14}. Its structure suggests separate parts linked only through a cut. Island 2 has 33 nodes with a maximal weight of 2.462×10^{-19}. Apart from the presence of pendants linked to the main part of the island, there are no obvious sub-parts.

The obvious question is simple: What holds these islands together in terms of substance and content? We turn to examine this next.

2.4.1   Island 10: Community Detection and Blockmodeling

Figure 2.8 shows Island 10 in more detail with its works identified. The upper left part is exclusively in the community detection domain while the lower right part contains works from the social network literature. The clear cut is the last node of the latter literature as identified in Figure 2.3. This link island provides a more expanded view compared to the one in Figure 2.3 which was more expanded than Figure 2.2. Given our earlier detailed consideration of the main path and key-routes, little more needs to be written at this point. The additional works in Island 10 do provide a foundation for a more detailed examination of the transition between two fields and what is featured in the two parts of the network clustering literature.

On the blockmodeling side, the first authors involved with the most works in Island 10 are Burt (9), Batagelj (7), Breiger (6), Doreian (5), Faust (4), and Pattison (4). Many are co-authored productions involving some of these researchers. For those involved in community detection analyses, the first authors involved with the most works in the island are Newman (13), Peixoto (8), Lancichinetti (7), Guimera (6), and Arenas (4). In terms of indegree, the three most cited

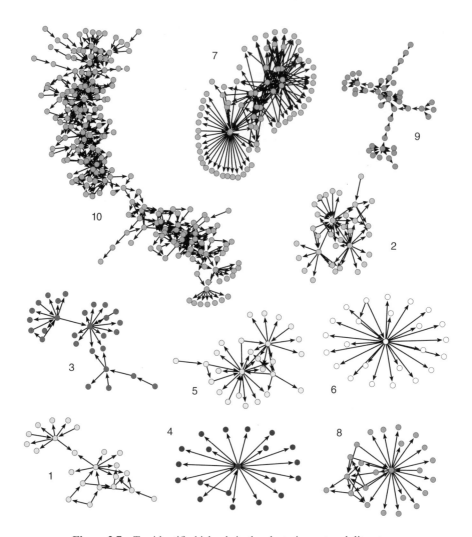

Figure 2.7   Ten identified islands in the clustering network literature.

items for the blockmodeling part of this island involve Arabie, Boorman, and Breiger. As noted above, all were involved in the foundational work for blockmodeling. Regarding community detection, the most cited researchers are Fortunato, Peixoto, and Newman. Their works appear to be either foundational or general surveys.

2.4.2   Island 7: Engineering Geology

The publication years for works in this island span 1965 through 2017. Studying this island leads to a caution regarding the boundary problem. Its works are present in the citation network through the term "sliding block analysis" used in this discipline. The earliest paper on the island

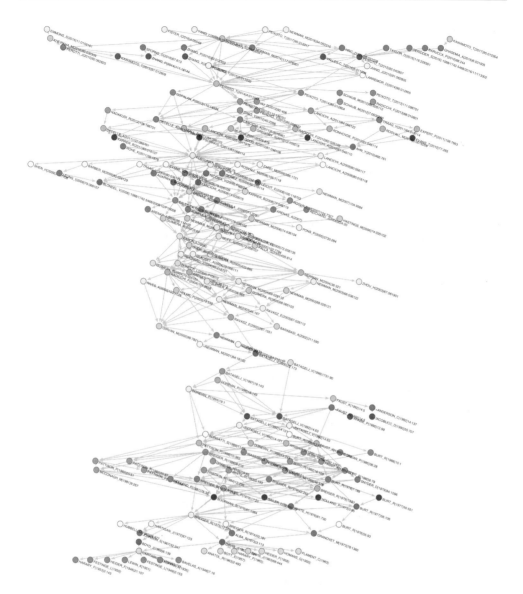

Figure 2.8   Island 10 community detection and blockmodeling.

appeared in *Geotechnique*. It contained a method for calculating the permanent displacements of soil slopes, embankments, and dams during seismic events. The model was recognized as having great value in studying earthquakes. Two papers appeared in *Geotechnique* about a decade later proposing another method. It also is valuable, especially for analyzing earth slopes and earthen dams. Indeed, there are a variety of methods for studying seismic events related to earthquakes and landslides. The works on this island are focused on methods for measuring seismic activity and predicting their consequences.

Some works in this link island stand out in terms of the number of citations they receive and make. One paper by Jibson (of the US Geological Survey) appeared in 2007 in *Engineering Geology*. Its citations topped both the indegree and outdegree values. With colleagues, six other papers involving this scientist are in this island. Also high on the outdegree listing are papers involving Stamatopoulos. The more recent works have as the primary focus, as reflected by keywords, predicting the dynamics of slip surfaces, saturated sands, slopes, and landslides. The methodological focus is clear also, with both multi-block models and sliding-block models playing a central role.

The final paper in this island used a large database of recorded ground motions to develop a predictive model of earth displacements. The empirical contexts for the body of work in this island are landslides and earthquakes linked through the impact of seismic events. The major journals for this line of work include Geotechnique, *Journal of Geotechnical and Geoenvironmental Engineering, Bulletin of Earthquake Engineering, Soils and Foundations, Engineering Geology* and *Soil Dynamics and Earthquake Engineering*. It is clearly part of a broader field of engineering geology, the application of geological knowledge to improve the design of engineering projects, their construction as well as their maintenance and operation, including the impact of seismic events.

2.4.3 Island 9: Geophysics

Island 9 is shown in Figure 2.9. Its works are focused on earthquake modeling. This part of the literature is in the clustering citation network because of the term "spring-block model". Again, this is another meaning of the term "block model". The works in Island 9 are part of the geophysics literature as evidenced by the journals where many of them appear. They include *Geophysical Research Letters, Physical Review Letters, Journal of Geophysics Research*, and *Physical Review A*.

One obvious question is simple to state: Why are Islands 7 and 9 not joined? Seismic events and earthquakes are features in both of them. Surely, they must be linked? After a closer inspection of the works in these two islands, there is a very simple answer to our question. The works in Island 9 are especially focused on *temporal* and *dynamic* issues in contrast to the works in Island 7, which entirely relate to *static* issues. The difference between Island 9 and Island 7 reveals a profound similarity between two completely distinct scientific fields. It seems there is a real difference between static and dynamic approaches to studying empirical phenomena. Surprisingly, it is present in *both* the natural and social sciences. For far too long, social network analyses ignored temporal issues. One set of approaches to dealing with the evolution of social networks appeared in the edited collection [12]. Subsequent contributions appeared in special issues of *Journal of Mathematical Sociology*. Since then, the focus on dynamic models of social networks has become far more extensive. It remains to be seen if the static and dynamic approaches of the works studying the seismic events of Islands 7 and 9 will be joined in geophysics and engineering geology.

2.4.4 Island 2: Electromagnetic Fields and their Impact on Humans

The papers appearing in this island deal with numerical methods for computations relating to electromagnetic fields. Their appearance in this island is due to the term "block model". In this

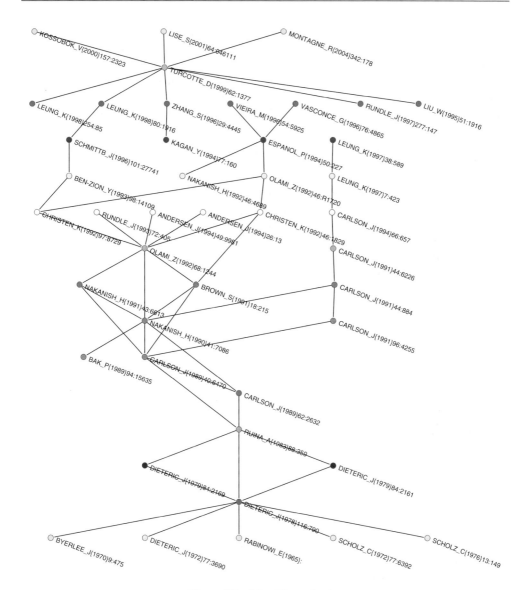

Figure 2.9   Island 9 geophysics.

different literature, "block models of people" use a limited number of cubical cells to predict the internal electromagnetic fields and specific absorption rate distributions inside human bodies. The earliest paper on this island appeared in 1968. Other early papers involving Hagmann and his colleagues are present also. A production by Massoudi appeared in 1985 with the words "limitations of the cubical block model" in the title. It has the highest indegree and the second highest outdegree. The main content concerns the meaning of a block model. One of the productions involving Hagmann (1986) has the highest outdegree. The Massoudi paper cites it, discussing the block model concept. Both papers appeared in *IEEE Transactions on Microwave Theory and Techniques*. Indeed, many of the works on this island appeared in this journal. They are strictly methodological.

A paper involving Zwamborn was published in 1991 (three papers with him as a co-author appear on this island). It has the second highest outdegree and appeared in the *Journal of the Optical Society of America A*. It concerned the computation of electromagnetic fields inside strongly inhomogeneous objects. The most recent paper on this island, appearing in 2002, was published in *Microwave and Optical Technical Letters*. It concerned resonant frequency calculation for inhomogeneous dielectric resonators. These papers are also methodological. If human bodies are inhomogeneous objects, there is continuity of this empirical focus.

In the analyses of bibliometric networks reported in [7] (Section 4.7.3), there was an "optical network line island". Many of the productions on this island involved journals published by the IEEE (Institute of Electrical and Electronics Engineers) and the Optical Society of America. The same appears to be the case for Island 2. The analysis in [7] examined the role of the institutional dominance of large professional organizations, something that appears relevant here also, especially when their interests are coupled.

2.4.5   Limitations and Extensions

Our brief examination of these four link islands implies some cautionary notes along with suggestions for further work.

- WoS is quite limited in the information it provides for individual works. Only half of the works on Island 2 had complete descriptions in WoS ($DC = 1$). This restriction is known already [7]. The problem was far less acute for the other islands. Clearly, different subject areas will have differing levels of this problem. These gaps in the information must be filled. One option is to extend the original WoS data with additional manually constructed descriptions for these works.
- The search terms used for extracting citation networks can be ambiguous. The search terms used here included block model*, blockmodel*, and block-model*. For those in the social networks sub-field, the term "blockmodel" is very well known. But, for the works in Island 2, "block model" means something quite different. The works in Islands 7 featured "sliding block analysis" while in Island 9 the core term was "spring-block model".
- Such differences in meaning for a search term can be discerned only through a careful examination of the identified literatures. Clearly, general terms have to be used to include as many potentially relevant works as possible. However, the results need to be considered carefully. We were surprised to learn of the other meanings for the term "block". No doubt researchers in geophysics and engineering geology would be surprised to find works from the social networks literature in their citation networks if searches were done using the term "block". The proposed approach enabled us to identify these other meanings and consider the maximal weight of corresponding islands, along with their importance.
- Examining temporal shifts in the keywords used in literature and the journals where works are published are important avenues of exploration for understanding the dynamics of scientific fields. The changes were most dramatic for the network clustering literature examined in Island 10.

• The structures of the islands shown in Figure 2.7 are quite different. An open problem is whether this has an impact on the production of knowledge and the social organization of scientific disciplines.

2.5 Authors

We consider, in more detail, the authors creating the papers in the network clustering literature. The network considered in this section is for works having complete information. We computed the networks **CiC**, **WAc**, **WKc**, and **WJc**. Their sizes are in Table 2.3.

The publication counts for authors are shown in Table 2.13 with a focus on the authors producing works in the core topic of this book. From **CiC**, it is straightforward to construct the counts of works by these authors. Authors with the largest number of papers about clustering networks are shown in Table 2.13. The large number of Chinese authors author names in the list may be an example of the "three Zhang, four Li" effect [29]. Lacking the resources to examine the relevant works to identify these authors, we proceed with a caution that some of the counts for these Chinese authors are not final.

The top ten entries in Table 2.13 come from the community detection area. Only four of the (first) authors listed in Table 2.13 work in the social networks literature. Their names are bolded. As all four are involved in collaborative work, the counts by single author names are limited as a summary of individual activity. The remaining works come from researchers in other disciplines, most of whom study community detection. The same caveat regarding collaborative production holds there also. Even so, these counts of works reflect accurately the far larger number of researchers and productions from the natural sciences, consistent with our results about the main path, key-routes, and Island 10.

We contend it may be more useful to examine productivity *inside* research groups and focus explicitly on collaboration. To this end, the idea of identifying cores in networks has value. A full description of k-cores and p-cores is provided in [7] (in Section 2.10.1). More importantly, for our purposes here, are P_S-cores also described in [7] (Section 4.10.1.3).

Table 2.13 Authors involved in the largest number of works

Rank	Frequency	Author	Rank	Frequency	Author	Rank	Frequency	Author
1	66	ZHANG_X	15	35	ZHANG_Z	28	26	ZHANG_H
2	57	WANG_Y	16	35	ZHANG_J	29	26	WANG_L
3	56	LIU_J	17	34	JIAO_L	30	26	TURCOTTE_D
4	51	WANG_X	18	33	ZHANG_S	31	26	**BORGATTI_S**
5	44	LI_J	19	32	WANG_S	32	26	**EVERETT_M**
6	42	WANG_H	20	31	**BATAGELJ_V**	33	26	WANG_C
7	41	LIU_Y	21	31	CHEN_H	34	24	LI_X
8	41	LI_Y	22	29	YANG_J	35	24	LI_L
9	40	NEWMAN_M	23	28	HANCOCK_E	36	24	LIU_X
10	39	WANG_J	24	28	WANG_W	37	23	LI_S
11	39	**DOREIAN_P**	25	27	CHEN_L	38	23	ZHOU_Y
12	38	CHEN_Y	26	26	LI_H	39	23	CHEN_X
13	36	ZHANG_Y	27	26	WU_J	40	23	LEE_J
14	35	WANG_Z						

2.5.1 Productivity Inside Research Groups

The network we use here is **Ct**, an undirected network obtained from $\mathbf{N}^T * \mathbf{N}$, where

$$\mathbf{N} = \mathrm{diag}(\frac{1}{\max(1, \mathrm{outdeg}(p))})\mathbf{WA}$$

with symmetrization [4].

A subset of nodes C is a P_S-core at some threshold iff for each of its nodes the sum of weights on links to other nodes from C is greater or equal to that threshold and C is the maximum such subset. Authors with the largest P_S-core values in **Ct** [6] are listed in Table 2.14. Again, bolding is used for researchers in the social network field. The number of researcher names from the social network side is now up to 14, still a minority. The values for authors equals the sum of all their fractional contributions to works with authors inside the core, a better measure than counts of publications bearing their names.

Figure 2.10 shows the links between author names with the size of nodes being proportional to their P_S-core value. For visual clarity, loops are removed. The names for researchers in the social network community are marked in boldface. The large top left P_S-core features researchers from the physical sciences with a clear central part. While Newman is connected to this P_S-core through one link, the size of his node is the largest. Several paths link other prominent researchers to this central part. They include one linking Peixoto to Fortunato to Lancichietti to Wang_J. There is also a path from Barabasi to Newman to Zhang_X. One surprise, at least for us, is the connection of Borgatti and Everett, having the strongest tie in Figure 2.11, to the central part of the P_S-core featuring natural scientists through their links with Boyd and his many links within this core. All three met, and worked, at the University of California at Irvine, an important center for social network analysis. This merits further attention.

Immediately to the right of this large P_S-core is a much smaller one involving Wasserman, Pattison and Breiger, who worked with each other on role systems and helped create the foundations for exponential random graph modeling of networks. Below this P_S-core is one centered

Table 2.14 Authors with the largest P_S-core values in **Ct**

Rank	P_S-core value	Author	Rank	P_S-core value	Author	Rank	P_S-core value	Author
1	21.0347	NEWMAN_M	15	6.0292	WANG_J	28	5.2589	**BRUSCO_M**
2	15.9653	**BORGATTI_S**	16	5.7500	PIZZUTI_C	29	5.2500	DIETERIC_J
3	15.9653	**EVERETT_M**	17	5.7014	STAMATOP_C	30	5.2483	LANCICHI_A
4	12.5000	**BURT_R**	18	5.6736	SUN_P	31	5.2483	FORTUNAT_S
5	12.5000	**DOREIAN_P**	19	5.6669	ZHANG_S	32	5.1111	**BOYD_J**
6	10.4722	PEIXOTO_T	20	5.6307	WANG_H	33	5.0633	WANG_X
7	10.1126	TURCOTTE_D	21	5.6307	LIU_J	34	5.0278	QIAN_X
8	8.7900	**FERLIGOJ_A**	22	5.5417	YANG_J	35	5.0208	**WASSERMA_S**
9	8.7900	**BATAGELJ_V**	23	5.5417	LESKOVEC_J	36	5.0000	OKAMOTO_H
10	6.5115	WANG_Y	24	5.5417	ZHANG_J	37	5.0000	JESSOP_A
11	6.4097	**PATTISON_P**	25	5.4432	HANCOCK_E	38	4.9881	BARABASI_A
12	6.4097	**BREIGER_R**	26	5.4432	ZHANG_Z	39	4.9775	**KRACKHAR_D**
13	6.2083	**MRVAR_A**	27	5.2589	**STEINLEY_D**	40	4.9112	ZHANG_H
14	6.0292	ZHANG_X						

The bolded authors published in journals in the social sciences.

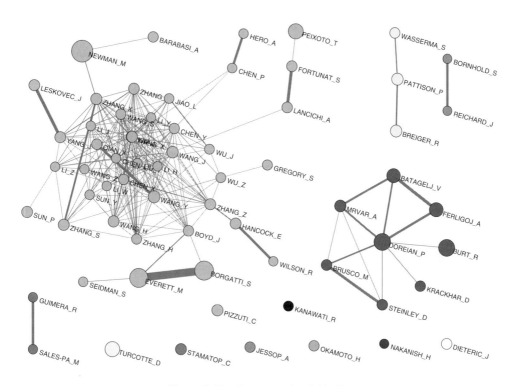

Figure 2.10 P_S-cores at level 4 in **Ct**.

on Doreian, who collaborated with all of the other researchers in this P_S-core. The links are strongest with Batagelj, Ferligoj and Mrvar. The next strongest tie is between him and Brusco – they worked on algorithms for blockmodeling along with Steinley. The strongest dyadic links in this core are between Batagelj and Ferligoj and between Brusco and Steinley.

We note two items: (i) many of the author names in Table 2.14 involve researchers participating in collaborative work (see below for more on this) and (ii) many of the names in this table have been mentioned in the above analyses, adding to the coherence of the results we report.

2.5.2   Collaboration

Collaboration is a critically important, and increasing, feature of modern scientific research. To examine this we use **Ct′**, an undirected network without loops obtained from $\mathbf{N}^T * \mathbf{N}'$, where

$$\mathbf{N}' = \text{diag}(\frac{1}{\max(1, \text{outdeg}(p) - 1)})\mathbf{WA},$$

through symmetrization and setting the diagonal values to 0 [10]. In the network **Ct** each work co-authored by an author contributes $\frac{1}{k^2}$ (k is the number of co-authors) to "self-collaboration" (value on the loop) of that author. The network **Ct′** describes the true collaboration with *others*.

Authors with the largest P_S-core values in **Ct′** are listed in Table 2.15 and presented in Figure 2.11. Heading the list are Borgatti and Everett, who have published together on

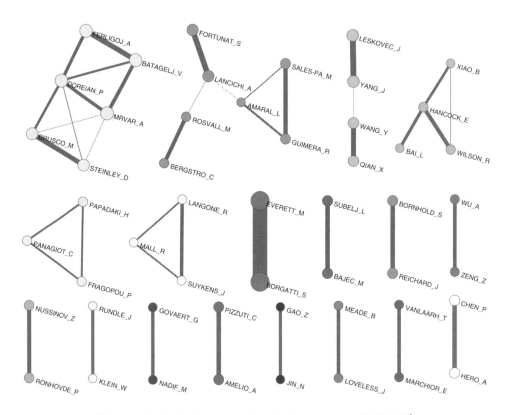

Figure 2.11   Links between authors in a P_S-core at level 3.5 in **Ct'**.

Table 2.15   Authors with the largest P_S-core values in **Ct'** s

Rank	Value	Author	Rank	Value	Author
1	15.8333	**BORGATTI_S**	15	5.0000	AMELIO_A
1	15.8333	**EVERETT_M**	15	5.0000	BAJEC_M
2	7.6667	**FERLIGOJ_A**	15	5.0000	SUBELJ_L
2	7.6667	**BATAGELJ_V**	15	5.0000	CHEN_P
2	7.6667	**MRVAR_A**	15	5.0000	PIZZUTI_C
2	7.6667	**DOREIAN_P**	15	5.0000	REICHARD_J
7	6.4333	**STEINLEY_D**	15	5.0000	BORNHOLD_S
7	6.4333	**BRUSCO_M**	23	4.8333	SALES-PA_M
9	6.3333	YANG_J	23	4.8333	GUIMERA_R
9	6.3333	LESKOVEC_J	23	4.5833	NUSSINOV_Z
11	6.0000	LANCICHI_A	23	4.5833	RONHOVDE_P
11	6.0000	FORTUNAT_S	27	4.3333	ROSVALL_M
13	5.3333	QIAN_X	27	4.3333	BERGSTRO_C
13	5.3333	WANG_Y	27	4.3333	WILSON_R
15	5.0000	HERO_A	37	4.3333	HANCOCK_E

The bolded authors published in journals in the social sciences.

blockmodeling for a long time. Next comes publications involving Ferligoj, Batagelj, Doreian, and Mrvar, who also have worked together for an extensive period. Both Steinley and Brusco, who have collaborated with Doreian, appear next, but they also worked together on clustering problems before publishing papers with Doreian on blockmodeling. It is interesting that the leading "nonsocial" authors from Table 2.14, Newman, Peixoto, and Turcotte, are missing in Table 2.15. The reasons are a combination of publishing of single author papers and publishing with many different co-authors.

Similar analyses have been performed for social networks as a whole in [7]. The citation network studied there was far larger as a more extensive literature was studied. Many of the above names appear also in the tables and figures of [7]. Comparing the two sets of analyses makes it clear that the role of these authors in this literature largely, but not completely, involves blockmodeling.

There is always a choice regarding which links are included for further examination of the structure of any studied network. Figure 2.11 shows the network when the threshold was set at 3.5. Necessarily, the results are more fragmented with 18 smaller link islands. In the middle of Figure 2.11 is the heavy Borgatti–Everett dyad having the highest value. The top left link island also features authors working on blockmodeling, consistent with our earlier results. The remaining items belong to the community detection literature.

2.5.3   Citations Among Authors Contributing to the Network Partitioning Literature

The network $\mathbf{Acite} = \mathbf{WAc}^T * \mathbf{CiC} * \mathbf{WAc}$ describes the citations among authors. The value of element $\mathbf{Acite}[u, v]$ is equal to the number of citations from works coauthored by u to works coauthored by v. While these numbers are inflated slightly when u and v collaborate, coauthorship is part of the citation structure. Collaboration matters greatly.

Link islands can be extracted from this network. The methods described in [7] require setting bounds for delineating islands. For this analysis they were [10, 50] with 16 islands identified for this network. They have quite different structures. Each can be examined but we focus on two of them as they pertain to community detection and (non-stochastic) blockmodeling.

The community detection island shown at the bottom of Figure 2.12 is large and massively centered on Newman. By far, works involving him are cited the most. Unsurprisingly to those in the field, a strong case can be made for him founding this research front both alone and with key collaborators. Fortunato is another highly cited author, most likely for his extensive and comprehensive summary of this research area. Note also that the terminal nodes (outdegree is 0) are the founders of complex networks approach Barabasi, Albert and Girvan.

The island containing publications about blockmodeling is smaller and is less centralized. The most central author is Doreian, but nowhere near to the extent of Newman. Moreover, there is more nuance in the structure with citations going *from* him to authors involved in creating the foundations of blockmodeling. Also, three distinct collaborative efforts are involved. One consists of works featuring him with Batagelj and Ferligoj on blockmodeling. One is with Mrvar on signed networks and the third involves his work with Brusco and Steinley on algorithms for partitioning networks. Citations go also to Borgatti and Everett without any corresponding reciprocating citations. Citations from Robins, Pattison, and Wasserman, all prominent in social networks, are reminders that this island is about blockmodeling. Were the focus on probabilistic

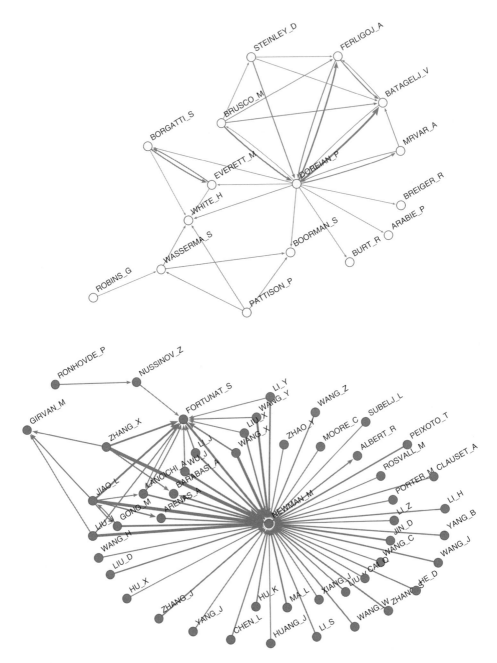

Figure 2.12 Citations among authors from two parts of the literature: community detection and blockmodeling.

approaches to studying social networks, especially exponential random graph models, that part of the network would expand greatly with the blockmodeling part disappearing. An expanded analysis of the whole social network literature, albeit for an earlier time period, is given in [7] (their Figure 4.17) and reinforces this point while showing links between these two areas of the literature.

No doubt researchers more familiar with the community detection literature, could paint a more nuanced picture for their part of the network clustering literature. One feature of the islands technique is the way it determines items more closely related among themselves compared to the connections from them to elsewhere in the network. While useful, an open problem is the examination of links between such islands. When coupled to the use of keywords and placed in a temporal framework, this will facilitate an examination of the movements of ideas within and between parts of citation and collaboration networks.

We could analyze also networks $n(\mathbf{WAc})^T * \mathbf{CiC} * n(\mathbf{WAc})$ (every citation has value 1 that is distributed among authors) and $n(\mathbf{WAc})^T * n(\mathbf{CiC}) * n(\mathbf{WAc})$ (every work has value 1 that is distributed among authors).

2.5.4   Citations Among Journals

There is a huge literature on citation relations between journals. It origins are found in the work of Garfield starting in the 1950s. Among his many contributions were establishing the Institute for Scientific Information (ISI) and the creation of the *Science Citation Index* (SCI) making extensive use of the aggregated journal-to-journal citation data provided annually by the Journal Citation Reports (JCR). See, for example, [15]. Also created was the *Social Science Citation Index* (SSCI). Much work has followed on mapping to structure these networks. A recent example is provided by Leydesdorf *et al.* examining structural shifts in journal-to-journal networks [22].

Our focus here has been on a sustained look at the citation network of works considering partitioning of networks. This can be extended to construct a journal-to-journal network for this literature only. Most likely, some of the works studied above will be found in the SCI. Others will be located in in the SSCI, with some overlap. Using only one of these data sources would be limiting and combining them would be difficult. Our case is somewhat special because of our interest in citations in the field of network clustering and not in general citations among journals. The task is one of counting the citation links between journals featuring works in this area.

2.5.4.1   Counting

To get information about citations among journals we compute the derived network

$$\mathbf{JJ} = \mathbf{WJ}^T * \mathbf{Ci} * \mathbf{WJ}$$

Its weights have the following meaning: $jj(i,j) =$ the number of citations from papers published in journal i to papers published in journal j, with attention confined to the network partitioning literature.

While this network can be searched for link islands, the results are limited due the different sizes of the journals involved. To obtain more useful results we applied the fractional approach described immediately below. Note that $n(\mathbf{WJ}) = \mathbf{WJ}$.

2.5.4.2 The Fractional Approach

In the fractional approach, we use the normalized citation network $n(\mathbf{Ci})$. The derived network is determined as follows:

$$\mathbf{JJf} = \mathbf{WJ}^T * n(\mathbf{Ci}) * \mathbf{WJ}$$

Its weights have the following meaning: $jjf(i,j) = fractional$ contribution of citations from papers published in journal i to papers published in journal j, again with attention restricted to the network partitioning literature.

There are 12 link islands in the [10, 50] range for the number of nodes (see Figure 2.13). Examining this figure more closely, the largest link island (top left) involves the journals where work on blockmodeling and community detection appeared. This island is considered in more detail below. The subject matter of the remaining islands contains surprises. Continuing to read across the top row of this figure, the primary subjects are dentistry and medical technologies, with and surgery, reconstructive surgery, and physical therapy as follow-up treatments to surgery. Dropping to the next row, reading from the left, the subjects are earthquakes and fluid mechanics, laser surgery in dentistry, petroleum engineering, and cardiovascular problems and treatments. Across the bottom row, the topics are archeology and antiquity studies, marine research and ship technology, linguistics (featuring only German language journals), and soil science.

This diversity of subject matter suggests a variety of issues. One is that the network partitioning literature is spread across far more disciplines than we anticipated. Of course, this could imply that the initial search was too broad. But if multiple disciples are involved, examining the journal-to-journal structures for these other disciplines has interest value. All of the islands are highly centralized, having either star-like or hierarchical structures. This is suggestive of another feature of the organization of scientific production at the journal level which merits further attention.

We label the largest island as the main island. It is presented in Figure 2.14. By far, most journals on this link island are from the physics-driven approach. Indeed, as shown at the bottom left, only a small number come from the social science approach to social networks. In part this reflects the institutional dominance of the natural sciences, especially physics. The only link from the physics literature to the social science literature is from *Physics Review E* to *Social Networks*. This is due to a link from a Newman paper in the former journal to a Batagelj paper in the latter literature, exactly the transition point between the blockmodeling literature and the community detection literature discussed in our analysis of main paths in Figure 2.2.

Figure 2.14 emphasizes its acyclic (hierarchical) structure with strong components. They are few in number. Only one is in the social science part of the network (lower left of the figure). It has *Social Networks* and *Journal of Mathematical Sociology*, both of which featured works on blockmodeling. The largest strong component has (*Physics Review E*, *Physics Review A*, *Physics Review Letters*, *Physica A*, *European Physical Journal B*, *Nature*, *Science* and *PNAS*). Note that the subgroup *Nature*, *Science*, and *PNAS* is linked back to the second strong component only with the arc between *PNAS* and *Physics Review E*.

Physics Review E, the primary journal for works on community detection and related topics, forms by two reciprocated dyads a strong component with *Physics Review Letters* and *Physica A*. The fractional tie from *Physica A* to *Physics Review E* is far stronger than the reverse tie. The fractional tie from *Physics Review E* to *Physics Review Letters* is stronger than the reverse tie, something meriting further attention. The third strong component also involves three journals

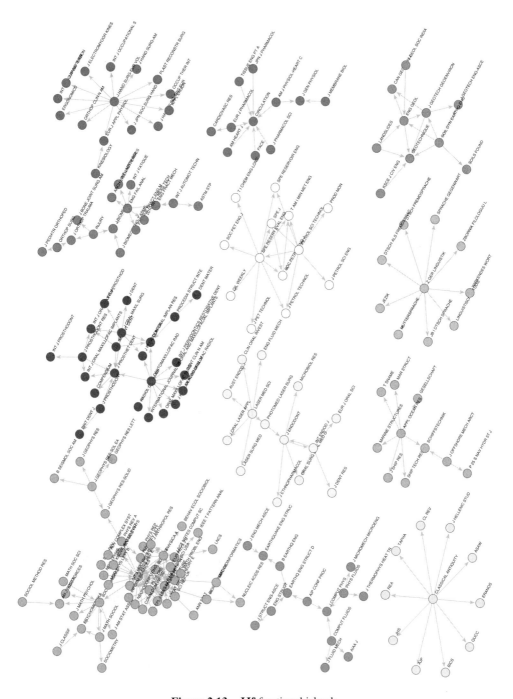

Figure 2.13 JJf fractional islands.

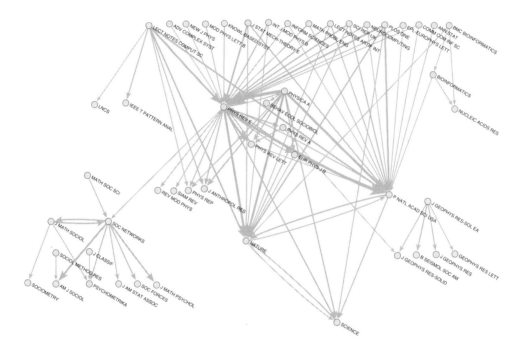

Figure 2.14 **JJf** main fractional island.

with two reciprocated dyads. *Nature* has reciprocated ties with both the *PNAS* and *Science*. All three journals are highly institutionalized within the natural sciences. Both *PNAS* and *Nature* are heavily cited, which reflects this institutional prominence. However, these ties are not reciprocated. It seems reasonable to assume that works in the other journals sent ties to these journals as a form of validation of their ideas. *Science* is relatively peripheral in Figure 2.14. This suggests that the works involving partitioning networks are not a central part of the overall scientific literature involving the natural sciences.

2.5.5 *Bibliographic Coupling*

Bibliographic coupling occurs when two works each cite a third work in their bibliographies. The idea was introduced by Kessler in 1963 [21] and has been used extensively since then. See Figure 2.15 where two citing works, *p* and *q*, are shown. Work *p* cites five works and *q* cites

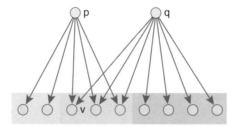

Figure 2.15   Bibliographic coupling.

seven works. The key idea is that there are three documents cited by both p and q. This suggests some content communality between p and q. It is thought that having more works citing pairs of prior works increases the likelihood of them sharing content. This is not unreasonable.

In `WoS2Pajek` the citation relation is p **Ci** q where work p cites work q. Therefore the *bibliographic coupling* network **biCo** can be determined as

$$\mathbf{biCo} = \mathbf{Ci} * \mathbf{Ci}^T$$

where $bico_{pq}$ = the number of works cited by both works p and $q = |\mathbf{Ci}(p) \cap \mathbf{Ci}(q)|$.

Bibliographic coupling weights are symmetric, $bico_{pq} = bico_{qp}$:

$$\mathbf{biCo}^T = (\mathbf{Ci} * \mathbf{Ci}^T)^T = \mathbf{Ci} * \mathbf{Ci}^T = \mathbf{biCo}$$

The pairs with the largest values involve works featuring reviews (or overviews of a field) and authors citing themselves. Review papers may require closer consideration when considering bibliographic coupling as they make many citations across wide areas.

Figure 2.16 shows the bibliographic coupling of works for links above a threshold of 25. There is one large set of such coupled works in a network along with three dyads and a triple of works. They feature productions involving physicists and computer scientists.

2.5.5.1   Fractional Bibliographic Coupling

Given the problems with works making many citations, especially with review works citing *many* works, we take a different approach. Necessarily, review papers cover a wide area (or multiple areas). That two works are cited in a broad review paper need not imply that they have common content. Ideally, it would be useful to separate specific contributions on *research fronts* from works looking back at what was done in general, but the literature contains both types of documents. We think a different strategy is required. Neutralizing the distorting impact of review documents suggests using normalized measures designed to control for this is useful (see [17]). We first consider:

$$\mathbf{biC} = n(\mathbf{Ci}) * \mathbf{Ci}^T$$

where $n(\mathbf{Ci}) = \mathbf{D} * \mathbf{Ci}$ and $\mathbf{D} = \mathrm{diag}(\frac{1}{\max(1,\mathrm{outdeg}(p))})$. $\mathbf{D}^T = \mathbf{D}$.

$$\mathbf{biC} = (\mathbf{D} * \mathbf{Ci}) * \mathbf{Ci}^T = \mathbf{D} * \mathbf{biCo}$$

$$\mathbf{biC}^T = (\mathbf{D} * \mathbf{biCo})^T = \mathbf{biCo}^T * \mathbf{D}^T = \mathbf{biCo} * \mathbf{D}$$

For $\mathbf{Ci}(p) \neq \emptyset$ and $\mathbf{Ci}(q) \neq \emptyset$ it holds (proportions)

$$\mathbf{biC}_{pq} = \frac{|\mathbf{Ci}(p) \cap \mathbf{Ci}(q)|}{|\mathbf{Ci}(p)|} \quad \text{and} \quad \mathbf{biC}_{qp} = \frac{|\mathbf{Ci}(p) \cap \mathbf{Ci}(q)|}{|\mathbf{Ci}(q)|} = \mathbf{biC}^T_{pq}$$

and $\mathbf{biC}_{pq} \in [0, 1]$. $\mathbf{biC}_{pq}$ is the proportion of its references that the work p shares with the work q.

Combining $\mathbf{biC}_{pq}$ and $\mathbf{biC}_{qp}$ we can construct different normalized measures such as

$$\mathbf{biCoa}_{pq} = \frac{1}{2}(\mathbf{biC}_{pq} + \mathbf{biC}_{qp}) \quad \text{Average}$$

$$\mathbf{biCom}_{pq} = \min(\mathbf{biC}_{pq}, \mathbf{biC}_{qp}) \quad \text{Minimum}$$

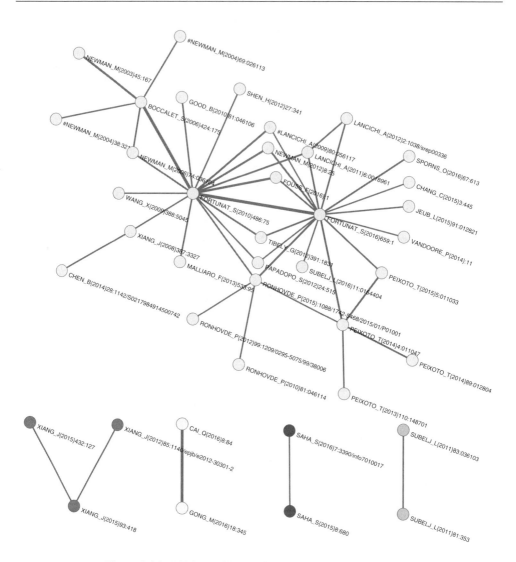

Figure 2.16 Bibliographic coupling above a threshold set at 25.

Other possible measures include geometric mean, the harmonic mean, and the Jaccard index. All these measures are symmetric. In the following we will use the Jaccard coefficient

$$\mathbf{biCoj}_{pq} = (\mathbf{biC}_{pq}^{-1} + \mathbf{biC}_{qp}^{-1} - 1)^{-1} = \frac{|\mathbf{Ci}(p) \cap \mathbf{Ci}(q)|}{|\mathbf{Ci}(p) \cup \mathbf{Ci}(q)|}$$

It is easy to verify that $biCoj_{pq} \in [0, 1]$ and $biCoj_{pq} = 1$ iff the works p and q are referencing the same works, $\mathbf{Ci}(p) = \mathbf{Ci}(q)$. To get a useful dissimilarity measure, use $dis = 1 - sim$ or

$dis = \frac{1}{sim} - 1$ or $dis = -\log sim$. For example

$$\mathbf{biCojD}_{pq} = 1 - \mathbf{biCoj}_{pq} = \frac{|\mathbf{Ci}(p) \oplus \mathbf{Ci}(q)|}{|\mathbf{Ci}(p) \cup \mathbf{Ci}(q)|} \quad \text{Jaccard distance}$$

which is the proportion of the number of distinct neighbors and all neighbors of works p and q in the citation network.

2.5.5.2 Jaccard Islands

We computed Jaccard similarity measures for the network CiteB and determined corresponding link islands having sizes in the range [5, 75]. The following table shows the distribution of the sizes of 133 islands that were identified.

Size	5	6	7	8	9	10	11	12	13	14	15	17	18	24	27
Number	33	16	11	17	12	8	4	2	2	3	1	4	2	1	1

Size	28	31	33	34	40	43	48	51	52	55	58	70	71	75
Number	1	2	1	1	1	1	1	1	2	1	1	1	1	1

We examine more closely a social networks Jaccard island (shown in Figure 2.17 with 70 works), a Jaccard island featuring works of physicists (in Figure 2.18 with 58 works), and three smaller Jaccard islands having 23, 22, and 18 works (see Figure 2.20).

The social networks Jaccard island is the largest such island. It has works spread over a variety of topics linked to partitioning social networks. There are many cuts linking these areas. One the top left of Figure 2.17, the works involve stochastic blockmodeling and exponential random graph models. The work by Sailer appeared in 1974 and is a cut connecting three sub-areas including the part just described. To the right of this cut are works involving the origins of blockmodeling. Below this cut are more works on classical blockmodeling. On the lower right of Figure 2.17 are works featuring discussions of the early algorithms for blockmodeling. At the bottom of the figure are more contemporary works on blockmodeling, including generalized blockmodeling. Many of these works were featured in Section 2.3.

The Jaccard island shown in Figure 2.18 features works by physicists regarding community detection and related methods for partitioning networks. It also has many cuts. Indeed, we suggest the presence of cuts is a feature of networks formed through bibliographic coupling links. In addition, it seems that bibliographic coupling is very useful for identifying different *sub-areas* of fields and how they are connected (see Figure 2.19).

It is straightforward to determine the citations received by works in these two Jaccard islands. The top numbers of received citations are shown in Table 2.18 where the relevant items from social network literature is on the left and those for the physicists are on the right.

Unsurprisingly, most of the works appearing in both columns have appeared earlier in our narrative. There are some clear differences between the two distributions. When the box-plots are drawn, the distribution for the social networks literature is far more skewed, with outliers present, than for the physicist part of the literature. Also the mean and median for these limited distributions are considerably higher in the physicist literature.

Figure 2.17   Bibliographic coupling in the social networks literature.

When the years of the publications are examined, another clear difference emerges. The range of years for the social networks part of the literature goes from 1950 to 1992. In contrast, the physicist works have dates ranging from 2002 to 2010. (The one document on the right in Table 2.18 that appeared in 1977 was written by an anthropologist. His data were latched upon by the physicists as useful data allowing the demonstration of community detection methods.) This reflects a clear difference between these two parts of the literature on clustering networks. One was developed over a longer period of time as a "leisurely" generation of methods, their application, and the generation of substantive results regarding the structure of *social* networks. It was merely one part of this literature that focused on many other issues regarding social networks The community detection literature exploded over a much shorter period of time with a focus on a clearly defined technical research issue.

It reflects also a difference in the social organization of science, something noted in [7]. Larger disciplines having more journals and a much longer institutionalized organization regarding professional organizations, as well as having more publication outlets, become far more visible.

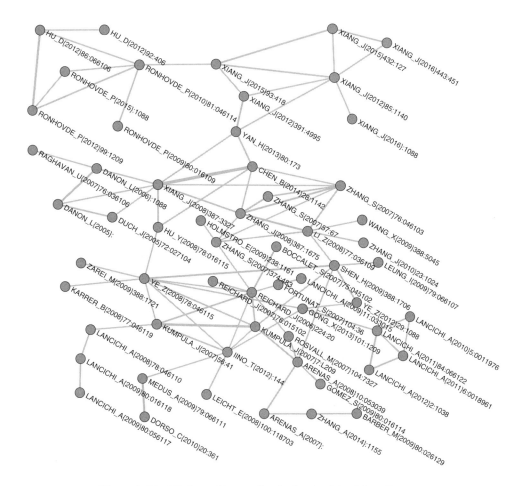

Figure 2.18   Bibliographic coupling in the physicist-driven literature.

We turn now to consider three smaller Jaccard islands. They are shown in Figure 2.20. The methods for determining citations are exactly the same as for the two largest islands (Table 2.16). These three smaller islands have works focused in three domains. The first deals with a part of the physicist and mathematical literature, the second with a part of the broader clustering literature, and the third with signed networks.

As indicated by the works in the left column of Table 2.17, the earliest work (by Erdős appearing in 1960 and at rank 10 of the column) set the foundations for the development of random graph theory. Another mathematical work appeared in 1985 (rank 6 in the column). There is an early social science work at rank 15 that attracted the attention of some physicists. A social networks text appears at rank 4 with sections on random graphs. The remaining works produced by physicists building on these ideas are concentrated between 1995 and 2001.

The top-ranked item in the second column of Table 2.17 appeared in *Psychometrika* in 1982. A companion paper by the same authors (Ferligoj and Batagelj) in the same journal appeared a year later. These works created the foundations for a distinctive approach to clustering relational

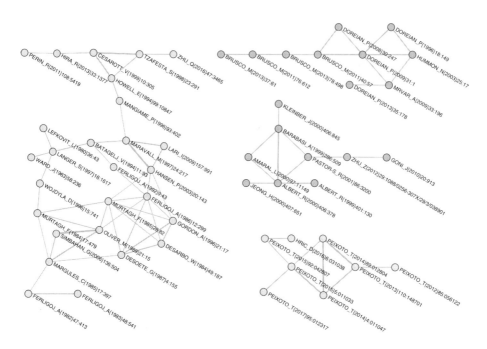

Figure 2.19   Bibliographic coupling – selected islands.

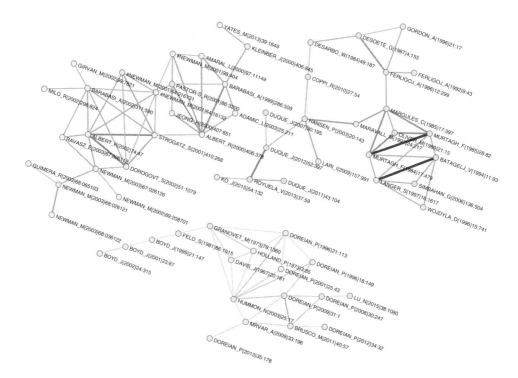

Figure 2.20   Bibliographic coupling for three smaller islands.

Table 2.16   Bibliographic coupling of the most cited works from the works of the two largest islands

Figure 2.17 (Social network literature)			Figure 2.18 (Physicist literature)		
Rank	Count	Work	Rank	Count	Work
1	58	LORRAIN_F(1971)1:49	1	45	GIRVAN_M(2002)99:7821
2	50	WHITE_H(1976)81:730	2	43	#NEWMAN_M(2004)69:026113
3	48	BREIGER_R(1975)12:328	3	40	CLAUSET_A(2004)70:066111
4	33	ARABIE_P(1978)17:21	4	38	DUCH_J(2005)72:027104
5	26	BOORMAN_S(1976)81:1384	5	36	GUIMERA_R(2005)433:895
6	24	SAILER_L(1978)1:73	6	35	#NEWMAN_M(2004)38:321
7	22	BURT_R(1976)55:93	7	34	RADICCHI_F(2004)101:2658
8	22	WHITE_D(1983)5:193	8	31	#DANON_L(2005):
9	15	NADEL_S(1957):	9	31	#ZACHARY_W(1977)33:452
10	14	HEIL_G(1976)21:26	10	27	FORTUNAT_S(2007)104:36
11	12	SAMPSON_S(1969):	11	25	ALBERT_R(2002)74:47
12	12	HOLLAND_P(1981)76:33	12	25	NEWMAN_M(2003)45:167
13	11	BURT_R(1983):	13	20	REICHARD_J(2006)74:016110
14	11	JOHNSON_S(1967)32:241	14	20	REICHARD_J(2004)93:218701
15	10	BURT_R(1982):	15	19	GUIMERA_R(2003)68:065103
16	10	HOMANS_G(1950):	16	19	NEWMAN_M(2006)103:8577
17	10	FAUST_K(1988)10:313	17	19	PALLA_G(2005)435:814
18	10	FREEMAN_L(1979)1:215	18	19	WU_F(2004)38:331
19	10	FIENBERG_S(1985)80:51	19	17	FLAKE_G(2002)35:66
20	9	BORGATTI_S(1989)11:65	20	17	#BLONDEL_V(2008):P10008
21	8	WHITE_H(1963):	21	17	BOCCALET_S(2006)424:175
22	8	BURT_R(1980)6:79	22	17	GLEISER_P(2003)6:565
23	8	BREIGER_R(1979)13:21	23	16	FORTUNAT_S(2010)486:75
24	8	BATAGELJ_V(1992)14:121	24	16	RAVASZ_E(2002)297:1551
25	7	MANDEL_M(1983)48:376	25	16	MEDUS_A(2005)358:593
26	7	KNOKE_D(1982):	26	16	#DONETTI_L(2004):P10012
27	7	DOREIAN_P(1988)13:243	27	15	NEWMAN_M(2006)74:036104
28	7	BREIGER_R(1978)7:213	28	13	BRANDES_U(2008)20:172
29	7	SNYDER_D(1979)84:1096	29	13	GUIMERA_R(2004)70:025101
30	7	HUBERT_L(1978)43:31	30	12	HOLME_P(2003)19:532

and attribute data that was picked up by others working on general clustering problems. The other works in this column came from researchers working on traditional clustering problems.

Most of the works appearing in the third column of Table 2.17 deal with signed networks. The top four ranked items set the foundations for a formal approach to structural balance. The conceptual approach came from Heider in 1946 and is ranked second. The top rank is for an initial formal statement by Cartwright and Harary in 1956 and extended by Davis in 1967. There are some items on blockmodeling that were picked up by Doreian and Mrvar in 1996 to create an algorithm for partitioning signed networks.

Bibliographic coupling the Most Frequent Keywords in Works of a Given Subnetwork
For the social networks island and the physicist island identified in Figures 2.17 and 2.18, the most frequent keywords in works of these islands were extracted. They are shown in Table 2.18.

Table 2.17 Bibliographic coupling in the three smaller islands

Rank	Physicist literature		Clustering literature		Signed networks	
1	23	WATTS_D(1998)393:440	21	FERLIGOJ_A(1982)47:413	13	CARTWRIG_D(1956)63:277
2	18	BARABASI_A(1999)286:509	11	LEFKOVIT_L(1980)36:43	12	HEIDER_F(1946)21:107
3	17	ALBERT_R(1999)401:130	10	PERRUCHE_C(1983)16:213	11	DAVIS_J(1967)20:181
4	15	WASSERMA_S(1994):	9	MURTAGH_F(1985)28:82	10	NEWCOMB_T(1961):
5	15	AMARAL_L(2000)97:11149	8	FERLIGOJ_A(1983)48:541	9	WHITE_H(1976)81:730
6	13	BOLLOBAS_B(1985):	6	GORDON_A(1996)21:17	8	HARARY_F(1965):
7	13	FALOUTSO_M(1999)29:251	4	DUQUE_J(2007)30:195	8	DOREIAN_P(1996)18:149
8	13	NEWMAN_M(2001)98:404	4	KIRKPATR_S(1983)220:671	7	DOREIAN_P(2005):
9	10	STROGATZ_S(2001)410:268	4	MACQUEEN_J(1967):281	7	HEIDER_F(1958):
10	10	ERDOS_P(1960)5:17	3	DESARBO_W(1984)49:187	6	BREIGER_R(1975)12:328
11	10	REDNER_S(1998)4:131	3	MARGULES_C(1985)17:397	6	HOMANS_G(1950):
12	9	JEONG_H(2000)407:651	3	HANSEN_P(2003)20:143	6	BATAGELJ_V(1998)21:47
13	9	ALBERT_R(2000)406:378	3	DUQUE_J(2011)43:104	5	BORGATTI_S(2002):
14	9	MOLLOY_M(1995)6:161	3	MARAVALL_M(1997)24:217	5	LORRAIN_F(1971)1:49
15	9	MILGRAM_S(1967)1:61	3	GAREY_M(1979):	5	WHITE_D(1983)5:193

We consider first the left column featuring the social networks part of the clustering literature. The top two keywords are social and network confirming the nature of the works in this island. The next two are solidly about blockmodeling which is based on conceptions of equivalence. Additional terms include role structural, relation, sociometric, position, regular (for a specific equivalence type), direct (for one approach to blockmodeling) and block. All of these terms are recognizable as relevant terms.

The word network also heads the list of keywords for the community detection literature. It is followed immediately by community. Again, the essence of the content of the island is identified. It is followed by complex, a term used far more by the physicists in the expression "complex networks". The term modularity is foundational for community detection. The presence of overlap as a keyword in this island reflects another difference between the two literatures with community detection authors being far more concerned with overlapping clusters. The presence of the keywords metabolic and biological provide a hint that the physicists study a broader set of networks than those working in social networks.

There are only seven keywords common to both lists: network, analysis, structure, graph, model, algorithm, and organization. Both areas are concerned with delineating structure, studying graphs, fitting models (albeit of different sorts), and mobilizing algorithms.

Co-citation is a concept with strong parallels with bibliographic coupling (see Small [25] and Marshakova-Shaikevich [28]). The focus is on the extent to which works are co-cited by later works. The basic intuition is that the more earlier works are cited, the higher the likelihood that they have common content. The *co-citation* network **coCi** can be determined as $\mathbf{coCi} = \mathbf{Ci}^T *$ $\mathbf{Ci}$. $coci_{pq}$ = the number of works citing both works p and q. $coci_{pq} = coci_{qp}$. The same kinds of analyses can be performed for co-citation. An example of doing this is in [7] regarding the Supreme Court. However, we do not pursue this here.

2.5.6   Linking Through a Jaccard Network

Bibliographic coupling networks are linking works to works. Let **S** be such a network. The derived network $\mathbf{WA}^T * \mathbf{S} * \mathbf{WA}$ links authors to authors through **S**. Again, the normalization

Table 2.18   The most frequent keywords of the two largest islands in the Jaccard bibliographic coupling network

Figure 2.17 (Social network literature)			Figure 2.18 (Physicist-driven literature)		
Rank	Count	Work	Rank	Count	Work
1	42	network	1	54	network
2	34	social	2	52	community
3	27	blockmodel	3	48	complex
4	24	equivalence	4	30	structure
5	23	analysis	5	30	modularity
6	17	structure	6	28	detection
7	17	role	7	19	algorithm
8	15	structural	8	18	graph
9	12	relation	9	17	metabolic
10	11	multiple	10	12	resolution
11	10	graph	11	12	model
12	10	datum	12	12	optimization
13	8	statistical	13	9	organization
14	7	model	14	8	detect
15	7	algorithm	15	8	cluster
16	7	sociometric	16	7	identification
17	7	position	17	6	dynamics
18	7	regular	18	6	analysis
19	6	relational	19	6	method
20	6	computation	20	5	use
21	6	two	21	5	base
22	5	organization	22	5	hierarchical
23	5	stochastic	23	4	overlap
24	5	approach	24	4	pott
25	5	direct	25	4	multi
26	4	block	26	4	maximization
27	4	similarity	27	4	world
28	4	group	28	4	information
29	4	application	29	4	biological
30	3	measure	30	4	limit

question must be addressed. Given different options, we selected the derived networks defined as:

$$\mathbf{C} = n(\mathbf{WA})^T * \mathbf{S} * n(\mathbf{WA})$$

It is easy to verify that:

- if $\mathbf{S}$ is symmetric, $\mathbf{S}^T = \mathbf{S}$, then $\mathbf{C}$ is also symmetric, $\mathbf{C}^T = \mathbf{C}$
- the total of weights of $\mathbf{S}$ is redistributed in $\mathbf{C}$:

$$\sum_{a \in L(C)} c(a) = \sum_{a \in L(S)} s(a)$$

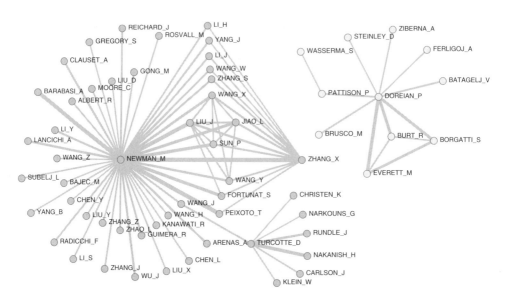

Figure 2.21   Authors Jaccard coupling – cut at level 11.

We applied this construction to combine the Jaccard network with networks **WA**, **WJ** and **WK**. We limited our analysis to networks with complete descriptions of works (**WAc**, **WJc**, **WKc**, **CiteC**).

As an example, Figure 2.21 presents a link-cut at level 11 in the authors Jaccard coupling network $\mathbf{ACoj} = n(\mathbf{WA})^T * \mathbf{biCoj} * n(\mathbf{WA})$. There are two disjoint parts to the figure. The smaller one on the right features authors active in the blockmodeling literature. It is centered on Doreian. The larger part on the left comes from the physics-driven literature and is centered on Newman. The result is very similar to the one shown in Figure 2.12. The social networks part is smaller in Figure 2.21 while the physics-driven part is larger with an additional part linked through Turcotte.

In Figure 2.22 a link-cut at level 1300 in the journals Jaccard coupling network $\mathbf{JCoj} = n(\mathbf{WJ})^T * \mathbf{biCoj} * n(\mathbf{WJ})$ is presented as another example. Because the links between journals have greater weights, a much larger link-cut is required. Overwhelmingly, the journals come from the physics-driven part of the literature. Three such journals are particularly prominent: *Physica A*, *Physics Review E* and *PLOS ONE*. The fourth prominent journal is *Lecture Notes Comput Science* from the computing science literature. Only two social science journals are present. *Journal of Mathematical Sociology* is linked to *Social Networks*, which is linked to only *Physica A*. Despite its name, *Social Network Analysis and Mining* is focused more on data mining in large networks, reflecting a computer science orientation.

In Figure 2.23 the main link-island in [1, 30] in the keywords Jaccard coupling network $\mathbf{KCoj} = n(\mathbf{WK})^T * \mathbf{biCoj} * n(\mathbf{WK})$ is presented. Table 2.18 presents separate lists of keywords in the social network literature and the physics-driven literature. Figure 2.23 shows how *some* of these keywords are linked. That the keywords network and community are the most prominent is not surprising given these keywords head the physics-driven literature list in Table 2.18. Having community and detection separated is problematic and reflects the problem of two-word

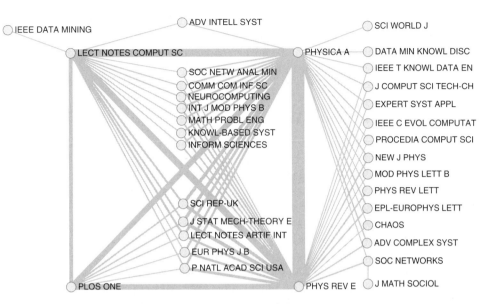

Figure 2.22 Journals Jaccard coupling – cut at level 1300.

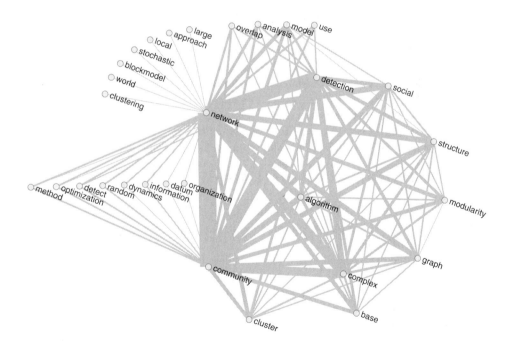

Figure 2.23 Keywords Jaccard coupling – main island.

keywords discussed earlier. Overall, 15 of the 30 keywords from the physics-driven literature are in Figure 2.23 while ten of the keywords from the social network literature are present (with seven common to both lists). Overall, the linkage of the keywords shown in Figure 2.23 seems more useful than the separate list in Table 2.18.

2.6   Summary and Future Work

We obtained citation data for the network clustering literature for a large citation network including both community detection and blockmodeling works through to February 22, 2017. The primary data source was the WoS. Details about recording, processing, and the resulting data sets were provided. In addition to having works as units, we included data on authors, journals, and keywords to generate some two-mode networks featuring works × authors, works × journals, and works × keywords. The boundary problem was discussed as was a treatment ensuring the studied citation network is acyclic.

Our results included descriptions of the most cited works and the most citing works as a preliminary delineation of the content of this research area. Lists of the most prominent journals where works in the network clustering literature appeared were created. In doing this, the importance of establishing network boundaries appropriately was discussed. The nature of keywords was discussed with a proviso that many cannot be taken at face value and using them to understand science must be done with great care.

Components of the studied network were identified with attention confined to the largest one. The CPM path through this component was identified. It revealed a clear transition from the social network part of the literature to the community detection part. The key-route paths revealed the same transition but with more works and a more nuanced view of it. Link islands, as clusters, were identified. There were ten of them. Detailed discussions were provided for four, including one with a clear distinction between the community detection and social networks literatures as being connected through a cut.

When attention was turned to considering authors, a listing of authors involved in the most works was provided. This was a limited result. To move beyond this, we examined productivity within research groups by using P_S-cores. A listing of authors having the largest P_S-core values was provided. To dig further into the contribution of authors, both co-authorship and collaboration were studied. This was extended to citations among the authors contributing to the network clustering literature, with close attention paid to the community detection and blockmodeling parts. Attention was paid to journal-to-journal networks for only the items identified in the network clustering literature.

Bibliographic coupling was considered and extended through fractional bibliographic coupling to use a better measure of the extent to which works are coupled. A total of 15 link islands were identified in the network of bibliometrically coupled documents. Again, attention was focused on two featuring, separately, the social network and physics-driven parts of this literature. Three more smaller link islands were examined, each with a clear sub-part of the literature. When keywords were examined in the context of link islands and bibliometric coupling, they were much more useful. Also, sub-areas were more clearly identified.

Together, these different ways of examining the network clustering literature provided a coherent and consistent understanding of its citation structure of works and the contributions of authors and journals. Future work will consider the other link islands in the citation network and those identified in the bibliometric coupling of works. Given the usefulness of bibliometric coupling, it is highly likely that the co-citation network will add additional insight to the coherence of this literature.

Acknowledgements

This work was supported in part by the Slovenian Research Agency (research programs P1-0294 and P5-0168, and research projects J1-9187 and J7-8279) and by Russian Academic Excellence Project '5-100'.

References

1. *histcomp* – (*comp*iled *Hist*oriography program), 2010. URL http://garfield.library.upenn.edu/histcomp/guide.html.
2. V. Batagelj. Efficient algorithms for citation network analysis. *CoRR*, cs.DL/0309023, 2003. URL http://arxiv.org/abs/cs.DL/0309023.
3. V. Batagelj. *WoS2Pajek: Manual for version 1.4*. IMFM, Ljubljana, Slovenia, 2016.
4. V. Batagelj and M. Cerinšek. On bibliographic networks. *Scientometrics*, 96(3):845–864, 2013.
5. V. Batagelj and A. Mrvar. *Pajek and Pajek-XXL, Program for Analysis and Visualization of Large Networks, Reference Manual, List of commands with short explanation*, 1996–2017. version 5.01.
6. V. Batagelj and M. Zaveršnik. Fast algorithms for determining (generalized) core groups in social networks. *Adv. Data Analysis and Classification*, 5(2):129–145, 2011.
7. V. Batagelj, P. Doreian, A. Ferligoj, and N. Kejžar. *Understanding Large Temporal Networks and Spatial Networks: Exploration, Pattern Searching, Visualization and Network Evolution*. Wiley Series in Computational and Quantitative Social Science Series. Wiley, 2014.
8. R. L. Breiger, S. A. Boorman, and P. Arabie. An algorithm for clustering relational data with applications to social network analysis and comparison with multidimensional scaling. *Journal of Mathematical Psychology*, 12(3):328–383, 1975.
9. R. S. Burt. Positions in networks. *Social Forces*, 55(1):93–122, 1976.
10. M. Cerinšek and V. Batagelj. Network analysis of zentralblatt MATH data. *Scientometrics*, 102(1):977–1001, 2015.
11. S. Cole and J. R. Cole. Visibility and the structural bases of awareness of scientific research. *American Sociological Review*, 33(3):397–413, 1968.
12. P. Doreian and F. N. E. Stokman. *Evolution of Social Networks*. Gordon & Breach, New York, 1997.
13. P. Doreian, V. Batagelj, and A. Ferligoj. *Generalized Blockmodeling*. Structural Analysis in the Social Sciences. Cambridge University Press, New York, 2005.
14. S. Fortunato and M. Barthélemy. Resolution limit in community detection. *PNAS* January 2, 104(1):36–41, 2007.
15. E. Garfield. Citation analysis as a tool in journal evaluation. *Science*, 178: 471–479, 1972.
16. E. Garfield, A. I. Pudovkin, and V. S. Istomin. Why do we need algorithmic historiography? *J. Am. Soc. Inf. Sci. Technol.*, 54(5):400–412, 2003.
17. M. Gauffriau, P. O. Larsen, I. Maye, A. Roulin-Perriard, and M. von Ins. Publication, cooperation and productivity measures in scientific research. *Scientometrics*, 73(2):175–214, 2007.
18. N. P. Hummon and P. Doreian. Connectivity in a citation network: The development of DNA theory. *Social Networks*, 13:39–63, 1989.
19. N. P. Hummon and P. Doreian. Computational methods for social network analysis. *Social Networks*, 12:273–288, 1990.
20. N. P. Hummon, P. Doreian, and L. C. Freeman. Analyzing the structure of the centrality-productivity literature. *Knowledge*, 11:460–481, 1990.
21. M. Kessler. Bibliographic coupling between scientific papers. *American Documentation*, 14(1):10–25, 1963.
22. L. Leydesdorff, C. S. Wagner, and L. Bornmann. Betweenness and diversity in journal citation networks as measures of interdisciplinarity—a tribute to eugene garfield. *Scientometrics*, 114:567–592, 2018.
23. J. S. Liu and L. Y. Y. Lu. An integrated approach for main path analysis: Development of the hirsch index as an example. *Journal of the American Society for Information Science and Technology*, 63(3):528–542, 2012.

24. F. Lorrain and H. C. White. Structural equivalence of individual in social networks. *Journal of Mathematical Sociology*, 1:49–80, 1971.
25. I. Marshakova. System of documentation connections based on references (sci). *Nauchno-Tekhnicheskaya Informatsiya Seriya*, 2(6):3–8, 1973.
26. W. D. Nooy, A. Mrvar, and V. Batagelj. *Exploratory Social Network Analysis with Pajek*. Cambridge University Press, New York, 2018.
27. R Development Core Team. *R: A Language and Environment for Statistical Computing*. R Foundation for Statistical Computing, Vienna, 2011.
28. H. Small. Co-citation in the scientific literature: A new measure of the relationship between two documents. *Journal of the American Society for Information Science*, 24(4):265–269, 1973.
29. Wikipedia. Chinese surname — Wikipedia, the free encyclopedia, 2013. URL https://en.wikipedia.org/wiki/Chinese_surname. [Online; accessed 22-July-2014].

3

Clustering Approaches to Networks

Vladimir Batagelj[1,2,3]
[1]IMFM, Ljubljana
[2]IAM, University of Primorska, Koper
[3]NRU, HSE, Moscow

3.1 Introduction

Clustering and classification are two related activities sometimes used as synonyms. In clustering, the goal is to identify in a given set of units, groups (clusters, classes) of (usually) similar units. In classification a given unit has to be assigned to the corresponding (predefined) group. These two activities are embedded in our language and are therefore basic for most of our daily tasks.

The earliest classification systems were taxonomies of animals and plants: Shen Nung, China, ~3000 BCE and Ebers Papyrus, Egypt, ~1500 BCE. A theoretical framework was proposed by Aristotle (384–322 BCE). The taxonomic systems were improved by Linnaeus (1707–1778), Darwin (1809–1882), DNA (1953), and PhyloCode (1998).

The first steps towards "numeric" clustering procedures were taken in the first half of 20th century by defining different (dis)similarity measures such as Czekanowski coefficient (1909), coefficient of racial likeness (Pearson, 1926), generalized distance (Mahalanobis, 1936), etc. Early methods were proposed inside biometrics and psychometrics by Driver and Kroeber (1932), Forbes (1933), Zubin (1938), Sturtevant (1939), etc. Kruskal's minimum spanning tree algorithm (1956) was predated by the unnoticed Borůvka (1926) and Jarnik (1930) algorithms.

The development of cluster analysis started in the 1950s and resulted in some fundamental books: Sokal and Sneath, *Principles of Numerical Taxonomy* (1963) [59]; Jardine and Sibson, *Mathematical Taxonomy* (1971) [41]; Benzécri, *L'analyse des données* (1973) [13]; Anderberg, *Cluster Analysis for Application* (1973) [2]; Hartigan, *Clustering algorithms* (1975) [38]; and later Jain and Dubes, *Algorithms for clustering data* (1988) [40]; Kaufman and Rousseeuw, *Finding Groups in Data: An Introduction to Cluster Analysis* (1990) [46]; and Gan, Ma, and Wu, *Data Clustering – Theory, Algorithms, and Applications* (2007) [32].

Advances in Network Clustering and Blockmodeling, First Edition.
Edited by Patrick Doreian, Vladimir Batagelj, and Anuška Ferligoj.
© 2020 John Wiley & Sons Ltd. Published 2020 by John Wiley & Sons Ltd.

Two streams of clustering research emerged – one inside pattern recognition and the other within data analysis. The data analytic stream was initially attached to the psychometric community until 1985 when the International Federation of Classification Socities (IFCS) was established with its own conference and journals, *Journal of Classification* (published by CSNA from 1984) and *Advances in Data Analysis and Classification* (from 2007). Conference proceedings are published in a Springer series *Studies in Classification, Data Analysis, and Knowledge Organization*. In the 1990s, the interests of the clustering community extended to data analysis and data science [39]. For details about the development of IFCS see [16]. At the turn of the millenium clustering was somehow absorbed also into data mining as one of its constituents. In social network analysis, the clustering problem is known as blockmodeling [26].

In this chapter we first present an optimization framework for a general clustering problem. In the second part we discuss the clustering of networks and clustering networks.

3.2   Clustering

In data analysis we usually follow the scheme

real or imaginary world		objects of analysis		data analysis
CONCEPTS		UNITS		DESCRIPTIONS
{X}	$\longleftrightarrow$	X	$\longleftrightarrow$	[X]
	formalization		operationalization	
{produced cars of type T}		car of type T		[seats=4, max-speed= …]

A *unit* $X \in \mathcal{U}$ is represented by a vector/*description* $X \equiv [X] = [x_1, x_2, ..., x_m]$ from the set $[\mathcal{U}]$ of all possible descriptions of units from *space* $\mathcal{U}$. $x_i = V_i(X)$ is the value of the ith of selected properties or *variables* on X. Variables can be measured on different *scales*: nominal, ordinal, interval, rational, absolute [58]. In concrete analysis, the *set of units* of our interset $U \subset \mathcal{U}$ is (usually) finite, $n = |U|$.

There exist other kinds of descriptions of units: symbolic objects [14, 24], lists of keywords from a text, chemical formulas, nodes in a given graph, and digital pictures, etc.

3.2.1   The Clustering Problem

We start with the formal setting of the clustering problem. We use the following notation: a nonempty subset of units $C, \emptyset \subset C \subseteq U$, is called a *cluster*. A set of clusters, $\mathbf{C} = \{C_i\}$, forms a *clustering*. Φ denotes the set of *feasible clusterings*. A *criterion function*, $P : \Phi \to \mathbb{R}_0^+$, evaluates the quality of a clustering.

With these notions, we can express the *clustering problem* $(\Phi, P, \min)$ as follows:
Determine the clustering, $\mathbf{C}^\star \in \Phi$, for which

$$P(\mathbf{C}^\star) = \min_{\mathbf{C} \in \Phi} P(\mathbf{C})$$

Since the set of units U is finite, the set of feasible clusterings is also finite. Therefore the set $Min(\Phi, P)$ of all solutions of the problem (optimal clusterings) is not empty. (In theory) the set

Min(Φ, P) can be determined by the complete search. We denote the value of criterion function for an optimal clustering by $\min(\Phi, P)$.

Generally, the clusters of clustering $\mathbf{C} = \{C_1, C_2, \dots, C_k\}$ need not to be pairwise disjoint; yet, clustering theory and practice mainly deal with clusterings which are the *partition*s of $\mathbf{U}$

$$\bigcup_{i=1}^{k} C_i = \mathbf{U} \quad \text{and} \quad i \neq j \Rightarrow C_i \cap C_j = \emptyset$$

Each partition determines an equivalence relation in $\mathbf{U}$, and vice versa. We denote the set of all partitions of $\mathbf{U}$ into k clusters (classes) by $P_k(\mathbf{U})$.

3.2.2   Criterion Functions

The criterion function is usually constructed as follows. When joining the individual units into a cluster C, we make a certain "error", and so create a certain "tension" among them – we denote this quantity by $p(C)$. A *simple* criterion function $P(\mathbf{C})$ combines these "partial/local errors" into a "global error". Usually, it takes one of two forms:

S. $P(\mathbf{C}) = \displaystyle\sum_{C \in \mathbf{C}} p(C)$, or

M. $P(\mathbf{C}) = \displaystyle\max_{C \in \mathbf{C}} p(C)$

Both can be unified and generalized in the following way: Let $(\mathbb{R}, \oplus, e, \leq)$ be an ordered abelian monoid then:

$\oplus$. $P(\mathbf{C}) = \displaystyle\bigoplus_{C \in \mathbf{C}} p(C)$

The *cluster error* $p(C)$ usually has the properties:

$$p(C) \geq 0 \quad \text{and} \quad \forall X \in \mathbf{U} : p(\{X\}) = 0$$

Continuing, we assume that these properties of $p(C)$ hold.

Often also,

$$p(C_1 \cup C_2) \geq p(C_1) \oplus p(C_2)$$

holds for disjoint clusters, $C_1 \cap C_2 = \emptyset$. In such a case, we have, for simple criterion functions, $\min(P_{k+1}\mathbf{U}), P) \leq \min(P_k(\mathbf{U}), P)$ – we fix the value of k and set $\Phi \subseteq P_k(\mathbf{U})$.

To express the cluster-error, $p(C)$, we define on the space of units $\mathcal{U}$ a *dissimilarity* $d : \mathcal{U} \times \mathcal{U} \to \mathbb{R}_0^+$ for which we require:

D1. $\forall X \in \mathcal{U} : d(X, X) = 0$

D2. *symmetric*: $\forall X, Y \in \mathcal{U} : d(X, Y) = d(Y, X)$

Usually the dissimilarity d is defined using another dissimilarity $\delta : [\mathcal{U}] \times [\mathcal{U}] \to \mathbb{R}_0^+$ defined on unit descriptions as

$$d(X, Y) = \delta([X], [Y])$$

The dissimilarity d is:

D3. *even*: $\forall X, Y \in \mathscr{U} : (d(X, Y) = 0 \Rightarrow \forall Z \in \mathscr{U} : d(X, Z) = d(Y, Z))$
D4. *definite*: $\forall X, Y \in \mathscr{U} : (d(X, Y) = 0 \Rightarrow X = Y)$
D5. *metric*: $\forall X, Y, Z \in \mathscr{U} : d(X, Y) \leq d(X, Z) + d(Z, Y)$ – the triangle inequality
D6. *ultrametric*: $\forall X, Y, Z \in \mathscr{U} : d(X, Y) \leq \max(d(X, Z), d(Z, Y))$
D7. *additive*, iff the Buneman's or four-point condition holds $\forall X, Y, U, V \in \mathscr{U}$:

$$d(X, Y) + d(U, V) \leq \max(d(X, U) + d(Y, V), d(X, V) + d(Y, U))$$

A dissimilarity d is a *distance* iff D4 and D5 hold. Since the description $[\ ] : U \rightarrow [U]$ does not need to be injective, d can be indefinite. Often, a weaker form of definiteness holds:

$$\forall X, Y \in \mathscr{U} : (d(X, Y) = 0 \Rightarrow [X] = [Y])$$

A dissimilarity d is selected according to the nature of the set of units descriptions $[\mathscr{U}]$ and our analytic goals. Many examples of dissimilarities can be found in [22].

3.2.2.1   Dissimilarities on $\mathbb{R}^m$

In the standard case, $X \in \mathbb{R}^m$, many different dissimilarities have been proposed. Some of them are presented in Table 3.1.

3.2.2.2   (Dis)similarities on $\mathbb{B}^m$

Let $\mathbb{B} = \{0, 1\}$. For binary vectors $X, Y \in \mathbb{B}^m$ we define $a = XY$, $b = X\overline{Y}$, $c = \overline{X}Y$, $d = \overline{XY}$. It holds $a + b + c + d = m$. The counters a; b; c; and d are used to define several resemblances – (dis)similarity measures on binary vectors. See Table 3.2.

In some cases, the definition can yield an indefinite expression $\frac{0}{0}$. To deal with this, we can restrict the use of the measure, or define the values also for indefinite cases. For example, we extend the values of Jaccard coefficient such that $s_4(X, X) = 1$. For Kulczynski coefficient, we preserve the relation $T = \frac{1}{s_4} - 1$ by

$$s_4 = \begin{cases} 1 & d = m \\ \dfrac{a}{a + b + c} & \text{otherwise} \end{cases} \qquad s_3^{-1} = T = \begin{cases} 0 & a = 0, \ d = m \\ \infty & a = 0, \ d < m \\ \dfrac{b + c}{a} & \text{otherwise} \end{cases}$$

We can transform a similarity s from $[1, 0]$ into dissimilarity d on $[0, 1]$ by $d = 1 - s$. For details see [8].

Table 3.1 Dissimilarities on $\mathbb{R}^m$

n	Measure	Definition	Range	Note
1	Euclidean	$\sqrt{\sum_{i=1}^{m}(x_i-y_i)^2}$	$[0,\infty)$	$M(2)$
2	Sq. Euclidean	$\sum_{i=1}^{m}(x_i-y_i)^2$	$[0,\infty)$	$M(2)^2$
3	Manhattan	$\sum_{i=1}^{m}|x_i-y_i|$	$[0,\infty)$	$M(1)$
4	rook	$\max_{i=1}^{m}|x_i-y_i|$	$[0,\infty)$	$M(\infty)$
5	Minkowski	$\sqrt[p]{\sum_{i=1}^{m}|x_i-y_i|^p}$	$[0,\infty)$	$M(p)$
6	Canberra	$\sum_{i=1}^{m}\frac{|x_i-y_i|}{|x_i+y_i|}$	$[0,\infty)$	
7	Heincke	$\sqrt{\sum_{i=1}^{m}(\frac{|x_i-y_i|}{|x_i+y_i|})^2}$	$[0,\infty)$	
8	Self-balanced	$\sum_{i=1}^{m}\frac{|x_i-y_i|}{\max(x_i,y_i)}$	$[0,\infty)$	
9	Lance-Williams	$\frac{\sum_{i=1}^{m}|x_i-y_i|}{\sum_{i=1}^{m}x_i+y_i}$	$[0,\infty)$	
10	Correlation c.	$\frac{\mathrm{cov}(X,Y)}{\sqrt{\mathrm{var}(X)\mathrm{var}(Y)}}$	$[1,-1]$	

3.2.2.3 Dissimilarities between Sets

Let $\mathscr{F}$ be a finite family of subsets of the finite set U; $A,B \in \mathscr{F}$ and let $A \oplus B = (A \setminus B) \cup (B \setminus A)$ denote the symmetric difference between A and B. The "standard" dissimilarity between sets is the *Hamming distance*:

$$d_H(A,B) := \mathrm{card}(A \oplus B)$$

Other dissimilarities, normalized to $[0,1]$, between sets are:

$$d_s(A,B) = \frac{\mathrm{card}(A \oplus B)}{\mathrm{card}(A)+\mathrm{card}(B)} \qquad d_u(A,B) = \frac{\mathrm{card}(A \oplus B)}{\mathrm{card}(A \cup B)} = 1 - \frac{\mathrm{card}(A \cap B)}{\mathrm{card}(A \cup B)}$$

$$d_m(A,B) = \frac{\max(\mathrm{card}(A \setminus B),\mathrm{card}(B \setminus A))}{\max(\mathrm{card}(A),\mathrm{card}(B))}$$

For all these dissimilarities, $d(A,B) = 0$ if $A = B = \emptyset$.

Table 3.2 (Dis)similarities on $\mathbb{B}^m$

n	Measure	Definition	Range
1	Russel and Rao (1940)	$\dfrac{a}{m}$	$[1,0]$
2	Kendall, Sokal-Michener (1958)	$\dfrac{a+d}{m}$	$[1,0]$
3	Kulczynski (1927), T^{-1}	$\dfrac{a}{b+c}$	$[\infty,0]$
4	Jaccard (1908)	$\dfrac{a}{a+b+c}$	$[1,0]$
5	Kulczynski	$\dfrac{1}{2}\left(\dfrac{a}{a+b}+\dfrac{a}{a+c}\right)$	$[1,0]$
6	Sokal & Sneath (1963), un_4	$\dfrac{1}{4}\left(\dfrac{a}{a+b}+\dfrac{a}{a+c}+\dfrac{d}{d+b}+\dfrac{d}{d+c}\right)$	$[1,0]$
7	Driver & Kroeber (1932)	$\dfrac{a}{\sqrt{(a+b)(a+c)}}$	$[1,0]$
8	Sokal & Sneath (1963), un_5	$\dfrac{ad}{\sqrt{(a+b)(a+c)(d+b)(d+c)}}$	$[1,0]$
9	Q_0	$\dfrac{bc}{ad}$	$[0,\infty]$
10	Yule (1927), Q	$\dfrac{ad-bc}{ad+bc}$	$[1,-1]$
11	Pearson, ϕ	$\dfrac{ad-bc}{\sqrt{(a+b)(a+c)(d+b)(d+c)}}$	$[1,-1]$
12	$-bc-$	$\dfrac{4bc}{m^2}$	$[0,1]$
13	Baroni-Urbani, Buser (1976), S^{**}	$\dfrac{a+\sqrt{ad}}{a+b+c+\sqrt{ad}}$	$[1,0]$
14	Braun-Blanquet (1932)	$\dfrac{a}{\max(a+b,a+c)}$	$[1,0]$
15	Simpson (1943)	$\dfrac{a}{\min(a+b,a+c)}$	$[1,0]$
16	Michael (1920)	$\dfrac{4(ad-bc)}{(a+d)^2+(b+c)^2}$	$[1,-1]$

3.2.2.4 Equivalent Resemblances

Resemblances r and s are *(order) equivalent*, $r \cong s$, iff they induce the same or reverse ordering in the set of unordered pairs of units, i.e., iff

$$\forall X, Y, U, V \in \mathscr{U} : (r(X, Y) < r(U, V)) \Leftrightarrow (s(X, Y) < s(U, V))$$

or

$$\forall X, Y, U, V \in \mathscr{U} : (r(X, Y) < r(U, V)) \Leftrightarrow (s(X, Y) > s(U, V)).$$

3.2.2.5 Transformations

Dissimilarities usually take values in the interval $[0, 1]$ or in the interval $[0, \infty]$. They can be transformed, one into the other, by mappings:

$$\frac{d}{1-d} : [0, 1] \to [0, \infty] \quad \text{and} \quad \frac{d}{1+d} : [0, \infty] \to [0, 1],$$

or, in the case $d_{max} < \infty$, by

$$\frac{d}{d_{max}} : [0, d_{max}] \to [0, 1].$$

To transform a distance d into another distance we often use the mappings:

$$\log(1 + d), \quad \min(1, d) \quad \text{and} \quad d^r, \ 0 < r < 1.$$

Not all resemblances are dissimilarities. For example, the correlation coefficient has the interval $[1, -1]$ as its range. We can transform any of these mappings to the interval $[0, 1]$ by mappings:

$$\frac{1}{2}(1 - d), \quad \sqrt{1 - d^2}, \quad 1 - |d|, \quad \text{etc.}$$

When applying these transformations to a measure d, we want all of the nice properties of d to be preserved. In this respect, the following theorems are useful.

Proposition 3.1 *Let d be a dissimilarity on $\mathcal{U}$ and let a mapping $f : d(\mathcal{U} \times \mathcal{U}) \to \mathbb{R}_0^+$ have the property $f(0) = 0$, then $d'(X, Y) = f(d(X, Y))$ is also a dissimilarity.*

Proposition 3.2 *Let d be a distance on $\mathcal{U}$ and let the mapping $f : d(\mathcal{U} \times \mathcal{U}) \to \mathbb{R}$ have the properties:*

(a) $f(x) = 0 \Leftrightarrow x = 0$,
(b) $x < y \Rightarrow f(x) < f(y)$,
(c) $f(x + y) \leq f(x) + f(y)$,

then $d'(X, Y) = f(d(X, Y))$ is also a distance and $d' \cong d$.

All concave functions have also the sub-additivity property (c). The following concave functions satisfy the last theorem:

(a) $f(x) = \alpha x, \ \alpha > 0$,   (b) $f(x) = \log(1 + x), \ x \geq 0$,

(c) $f(x) = \dfrac{x}{1 + x}, \ x \geq 0$,   (d) $f(x) = \min(1, x)$,

(e) $f(x) = x^\alpha, \ 0 < \alpha \leq 1$,   (f) $f(x) = \arcsin x, \ 0 \leq x \leq 1$.

Proposition 3.3 *Let $d : \mathcal{U} \times \mathcal{U} \to \mathbb{R}$ has the property $Di, \ i = 1, ..., 7$, then $f(d), \ f \in$ (a)-(f) also has this property.*

Proposition 3.4 [42] *For each (nonnegative) dissimilarity measure d there is a unique non-negative real number p, called a metric index, such that d^α is a metric for all $\alpha \leq p$, and d^α is not a metric for all $\alpha > p$.*

Therefore, if a dissimilarity d is not metric, it can be transformed into one by using the power transformation.

3.2.2.6 Problems with Dissimilarities

There is an issue when dealing with *mixed units* in which variables are measured on different types of scales. Two approaches are usually used:

- Convert them to a common type of measurement scale (see [2]).
- Compute selected dissimilarities on homogeneous parts and combine them. See for example Gower's dissimilarity [35].

In both cases, we have to consider the *fairness* of a dissimilarity in which all variables must contribute equally. A partial solution to this problem is to use the normalized variables. We can also consider the dependencies among variables, such as in the Mahalanobis distance [63].

3.2.3 Cluster-Error Function/Examples

Now we can define several types of cluster-error functions:

S. $p(C) = \sum_{X,Y \in C, X < Y} w(X) \cdot w(Y) \cdot d(X, Y)$

$\overline{S}$. $p(C) = \dfrac{1}{w(C)} \sum_{X,Y \in C, X < Y} w(X) \cdot w(Y) \cdot d(X, Y)$

where $w : \mathscr{U} \to \mathbb{R}^+$ is a *weight* of units, which is extended to clusters by:

$$w(\{X\}) = w(X), \quad X \in \mathscr{U}$$

$$w(C_1 \cup C_2) = w(C_1) + w(C_2), \quad C_1 \cap C_2 = \emptyset$$

Often, $w(X) = 1$ holds for each $X \in \mathscr{U}$. Then $w(C) = \mathrm{card}(C)$.

M. $p(C) = \max_{X,Y \in C} d(X, Y) = \mathrm{diam}(C)$ – diameter

T. $p(C) = \min_{T \text{ is a spanning tree over } C} \sum_{(X:Y) \in T} d(X, Y)$

We use the labels in front of the forms of (cluster-) criterion functions to denote *types* of criterion functions. For example:

SM. $P(\mathbf{C}) = \sum_{C \in \mathbf{C}} \max_{X,Y \in C} d(X, Y)$

It is easy to prove:

Proposition 3.5 *Let* $P \in \{\text{SS}, \overline{\text{SS}}, \text{SM}, \text{MS}, \overline{\text{MS}}, \text{MM}\}$, *then there exists an* $\alpha_k^P(\mathbf{U}) > 0$ *such that for each* $\mathbf{C} \in P_k(\mathbf{U})$ *the following holds:*

$$P(\mathbf{C}) \geq \alpha_k^P(\mathbf{U}) \cdot \max_{C \in \mathbf{C}} \max_{X,Y \in C} d(X, Y).$$

Note that this inequality can be written also as $P(\mathbf{C}) \geq \alpha_k^P(\mathbf{U}) \cdot \text{MM}(\mathbf{C})$.

The criterion function $P(\mathbf{C})$, based on a dissimilarity d, is *sensitive* iff for each feasible clustering $\mathbf{C}$ it holds

$$P(\mathbf{C}) = 0 \iff \forall C \in \mathbf{C} \ \forall X, Y \in C : d(X, Y) = 0$$

and it is α-*sensitive* iff there exists an $\alpha_k^P(\mathbf{U}) > 0$ such that for each $\mathbf{C} \in P_k(\mathbf{U})$:

$$P(\mathbf{C}) \geq \alpha_k^P(\mathbf{U}) \cdot \text{MM}(\mathbf{C})$$

Proposition 3.6 *Every* α-*sensitive criterion function is also sensitive.*

Proposition 3.5 can be re-expressed as:

Proposition 3.7 *The criterion functions* $\text{SS}, \overline{\text{SS}}, \text{SM}, \text{MS}, \overline{\text{MS}}, \text{MM}$ *are* α-*sensitive.*

Another form of a cluster-error function, one frequently used in practice, is based on the notion of a leader or representative of the cluster C:

R. $p(C) = \min_{L \in \mathbf{F}} \sum_{X \in C} w(X) \cdot d(X, L)$

where $\mathbf{F} \subseteq \mathscr{F}$ is the set of *representatives*. The element $\overline{C} \in \mathbf{F}$, which minimizes the right side expression, is called the *representative* of the cluster C. It is not always uniquely determined.

Proposition 3.8 *Let* $p(C)$ *be of type* R *then*

a) $p(C) + w(X) \cdot d(X, \overline{C \cup \{X\}}) \leq p(C \cup \{X\}), \quad X \notin C$
b) $p(C \setminus \{X\}) + w(X) \cdot d(X, \overline{C}) \leq p(C), \quad X \in C$

3.2.3.1 The Generalized Ward's Criterion Function

To obtain the *generalized Ward's clustering problem* we rely on the equality

$$p(C) = \sum_{X \in C} d_2^2(X, \overline{C}) = \frac{1}{2\text{card}(C)} \sum_{X,Y \in C} d_2^2(X, Y)$$

and replace the expression for $p(C)$ with

$$p(C) = \frac{1}{2w(C)} \sum_{X,Y \in C} w(X) \cdot w(Y) \cdot d(X, Y) = \overline{S}(C)$$

Note that d can be **any** dissimilarity on $\mathscr{U}$.

From this definition, we can easily derive the following equality: If $C_u \cap C_v = \emptyset$ then

$$w(C_u \cup C_v) \cdot p(C_u \cup C_v) = w(C_u) \cdot p(C_u) + w(C_v) \cdot p(C_v) + \sum_{X \in C_u, Y \in C_v} w(X) \cdot w(Y) \cdot d(X, Y)$$

In [5] it is shown also how to replace $\overline{C}$ by a generalized, possibly imaginary (with descriptions not neccessary in the same set as $\mathscr{U}$), central element in a way to preserve the properties characteristic for Ward's clustering problem.

Let $\mathscr{U}^*$ denote the space of units extended with generalized centers. The *generalized center* of cluster C is called an (abstract) element $\overline{C}$ for which the dissimilarity between it and any $U \in \mathscr{U}^*$ is determined by

$$d(U, \overline{C}) = d(\overline{C}, U) = \frac{1}{w(C)} \left(\sum_{X \in C} w(X) \cdot d(X, U) - p(C) \right)$$

When for all units $w(X) = 1$, the right side of the definition can be read as: the average dissimilarity between the unit/center U and cluster C diminished by the average radius of cluster C.

We have a suggestion: For each dissimilarity, find its metric index p and in the generalized Huygens theorem use d if $p \geq 1$, otherwise (if $p < 1$) use d^p.

For the generalized Ward's criterion function, the *generalized Huygens theorem* holds:

Proposition 3.9

$$I_T = I_W + I_B$$

where

$$I_T = p(\mathbf{U}) = \frac{1}{2w(\mathbf{U})} \sum_{X, Y \in U} w(X) \cdot w(Y) \cdot d(X, Y)$$

$$I_W = \sum_{C \in \mathbf{C}} p(C) \quad and \quad I_B = \sum_{C \in \mathbf{C}} w(C) \cdot d(\overline{C}, \overline{\mathbf{U}})$$

For a given set of units $\mathbf{U}$, the value of their "total inertia" I_T is fixed. Therefore when minimizing the "standard" criterion function (within inertia) I_W, we are also maximazing the function (between inertia) I_B – the traditional definition of clustering problem.

3.2.3.2   Other Criterion Functions

Several other types of criterion functions have been proposed in the literature. A very important class among them are the "statistical" criterion functions based on the assumption that the units are sampled from a mixture of the multivariate normal distributions [51].

The *modularity* criterion function from the complex networks approach [15, 55] is also of type S with cluster error

$$p(C) = \left(\frac{d(C)}{2m} \right)^2 - \frac{m(C)}{m}$$

where m is the number of links, $m(C) = |\{e \in L : ext(e) \subseteq C\}|$ and $d(C) = |\{e \in L : ext(e) \cap C \neq \emptyset\}|$; ext is a function assigning to a link $e(u, v)$ its end-nodes $\{u, v\}$.

   Not all clustering problems can be expressed by a simple criterion function. In some applications, a *general* criterion function of the form

$$P(\mathbf{C}) = \underset{(C_1, C_2) \in \mathbf{C} \times \mathbf{C}}{\bigoplus} q(C_1, C_2), \quad q(C_1, C_2) \geq 0$$

is needed. We use this in the optimizational approach to blockmodeling [26].

   In some problems, several criterion functions can be defined $(\Phi, P_1, P_2, \ldots, P_s)$ and the clustering problem is formulated as *multicriteria clustering* problem [30].

   Note that for a criterion function of type SS, we have a similar situation as in the generalized Huygens theorem:

Proposition 3.10

$$P_T = P_W + P_B$$

where, denoting $p(C, D) = \sum_{X \in C, Y \in D} d(X, Y),$

$$P_T = p(\mathbf{U}, \mathbf{U}), \quad P_W = \sum_{C \in \mathbf{C}} p(C, C) = SS(\mathbf{C}), \quad and \quad P_B = \sum_{\substack{C, D \in \mathbf{C} \\ C \neq D}} p(C, D)$$

3.2.3.3   Example: Partitioning a Generation of School Pupils into a Given Number of Classes

This is an example of nontraditional clustering problem in which the clusters are not characterized as "groups of similar units". We consider the problem of partitioning a generation of pupils into a given number of classes so that these classes will have (almost) the same number of pupils and, further, that they will have a structure as similar as possible. An appropriate criterion function is

$$P(\mathbf{C}) = \underset{\substack{\{C_1, C_2\} \in \mathbf{C} \times \mathbf{C} \\ \text{card}(C_1) \geq \text{card}(C_2)}}{\max} \; \underset{\substack{f : C_1 \to C_2 \\ f \text{ is surjective}}}{\min} \; \underset{X \in C_1}{\max} \; d(X, f(X))$$

where $d(X, Y)$ is a measure of dissimilarity between pupils X and Y.

3.2.4   The Complexity of the Clustering Problem

Because the set of feasible clusterings Φ is finite, it is tempting to think the clustering problem, (Φ, P), could be solved by a brute force approach inspecting all feasible clusterings. Unfortunately, the number of feasible clusterings grows dramatically with n. For example

$$\text{card}(P_k) = S(n, k) = \frac{1}{k!} \sum_{i=0}^{k-1} (-1)^i \binom{k}{i} (k - i)^n, \quad 0 < k \leq n$$

where $S(n, k)$ is a Stirling number of the second kind. For this reason, the brute force algorithm is only of theoretical interest with little relevance for the clustering problem in all empirical contexts when n is large.

We assume that readers are familiar with the basic notions of the theory of complexity of algorithms [33]. Although there are some types of clustering problems of polynomial complexity, for example (P_2, MM) and (P_k, ST), it seems that they are mainly NP-hard. Brucker [18] showed that ($\propto$ denotes the polynomial reducibility of problems [33]):

Theorem 3.11 *Let the criterion function*

$$P(\mathbf{C}) = \bigoplus_{C \in \mathbf{C}} p(C)$$

be α-sensitive, then for each problem $(P_k(\mathbf{U}), P)$ there exists a problem $(P_{k+1}(\mathbf{U}'), P)$, such that $(P_k(\mathbf{U}), P) \propto (P_{k+1}(\mathbf{U}'), P)$.

Theorem 3.12 *Let the criterion function P be sensitive then 3-COLOR $\propto (P_3, P)$.*

Note that, by Theorem 3.11, the clustering problems (P_k, MM), $k > 3$, are also NP-hard.

The complexity results for some types of clustering criterion functions are summarized in Table 3.3.

From these results, it follows (as it is believed) that no efficient (polynomial) exact algorithm exists for solving the clustering problem. Therefore, the procedures should be used that give "good" results, but not necessarily the best, in a reasonable time. In the following section we present some standard approaches for solving the clustering problem.

3.3   Approaches to Clustering

3.3.1   *Local Optimization*

Often, for a given optimization problem $(\Phi, P, \min)$, there exist rules relating to each element of the set Φ some elements of Φ. They are *local transformations*. The elements which can be obtained from a given element are called *neighbors* – local transformations determine the *neighborhood relation $S \subseteq \Phi \times \Phi$ in the set Φ. The *neighborhood* of element $X \in \Phi$ is called the

Table 3.3   The complexity of clustering problems

Polynomial	NP-hard	Note
(P_2, MM)	(P_3, MM)	Theorem 3.12
	(P_3, SM)	Theorem 3.12
	(P_2, SS)	MAX-CUT $\propto (P_2, \mathrm{SS})$
	$(P_2, \mathrm{S\bar{S}})$	$(P_2, \mathrm{SS}) \propto (P_2, \mathrm{S\bar{S}})$
	(P_2, MS)	PARTITION $\propto (P_2, \mathrm{MS})$
$(\mathbb{R}_2^m, \mathrm{S\bar{S}})$		
$(\mathbb{R}_k^1, \mathrm{S\bar{S}})$		
$(\mathbb{R}_k^1, \mathrm{SM})$		
$(\mathbb{R}_k^1, \mathrm{MM})$		

set $S(X) = \{Y : X\,S\,Y\}$. The element $X \in \Phi$ is a *local minimum* for the *neighborhood structure* (Φ, S) iff

$$\forall Y \in S(X) : P(X) \leq P(Y)$$

In the following, we assume that S is reflexive, $\forall X \in \Phi : X\,S\,X$.

The relation S is a basis of the *local optimization procedure*:

select X_0; $X := X_0$;
while $\exists Y \in S(X) : P(Y) < P(X)$ **do** $X := Y$;

which starting in an initial element $X_0 \in \Phi$ repeats moving to an element, in its neighborhood determined by local transformation, which creates a better value of the criterion function until no such element exists. To get a good solution, we repeat the procedure many times with random initial element X_0 and keep the best solution found.

3.3.1.1   Clustering Neighborhoods

Usually the neighborhood relation in local optimization clustering procedures over $P_k(U)$ is determined by the following two transformations:

- *transition*: clustering $\mathbf{C}'$ is obtained from $\mathbf{C}$ by moving a unit $X_s \in C_u$ from one cluster, C_u, to another, C_v,

$$\mathbf{C}' = (\mathbf{C} \setminus \{C_u, C_v\}) \cup \{C_u \setminus \{X_s\}, C_v \cup \{X_s\}\}$$

- *transposition*: clustering $\mathbf{C}'$ is obtained from $\mathbf{C}$ by interchanging two units, $X_p \in C_u$ and $X_q \in C_v$, from different clusters

$$\mathbf{C}' = (\mathbf{C} \setminus \{C_u, C_v\}) \cup \{(C_u \setminus \{X_p\}) \cup \{X_q\}, (C_v \setminus \{X_q\}) \cup \{X_p\}\}$$

The transpositions preserve the number of units in clusters. The local optimization based on transitions and/or transpositions is usually called the *relocation* method.

Using Proposition 3.8, we can prove the following important property of the minimal solutions of the clustering problem $(P_k, SR, \min)$:

Proposition 3.13 *In the locally, with respect to transitions, minimal clustering for the problem* $(P_k, SR, \min)$

$$\text{SR.} \qquad P(C) = \sum_{C \in \mathbf{C}} \sum_{X \in C} w(X) \cdot d(X, \overline{C})$$

each unit is assigned to the nearest representative: Let $\mathbf{C}^\bullet$ *be locally with respect to transitions minimal clustering then it holds:*

$$\forall C_u \in \mathbf{C}^\bullet \quad \forall X \in C_u \quad \forall C_v \in \mathbf{C}^\bullet \setminus \{C_u\} : d(X, \overline{C}_u) \leq d(X, \overline{C}_v)$$

Two basic implementation approaches are usually used: the *stored data* approach and the *stored dissimilarity matrix* approach.

If the constraints are not too stringent, the relocation method can be applied directly on Φ. Otherwise, we can transform, using the *penalty function method*, the problem to an equivalent non-constrained problem $(P_k, Q, \min)$ with $Q(\mathbf{C}) = P(\mathbf{C}) + \alpha K(\mathbf{C})$ where $\alpha > 0$ is a large constant and $K(\mathbf{C}) = 0$, for $\mathbf{C} \in \Phi$, and $K(\mathbf{C}) > 0$ otherwise.

There exist several improvements of the basic relocation algorithm, including simulated annealing and tabu search [1].

3.3.1.2   On Testing

The condition $P(\mathbf{C}') < P(\mathbf{C})$ is equivalent to $P(\mathbf{C}) - P(\mathbf{C}') > 0$. For the S criterion function,

$$\Delta P(\mathbf{C}, \mathbf{C}') = P(\mathbf{C}) - P(\mathbf{C}') = p(C_u) + p(C_v) - p(C'_u) - p(C'_v)$$

Some additional simplifications can be achieved by considering the relations between C_u and C'_u, and between C_v and C'_v.

We illustrate this using the generalized Ward's method. For this purpose, it is useful to introduce the quantity

$$a(C_u, C_v) = \sum_{X \in C_u, Y \in C_v} w(X) \cdot w(Y) \cdot d(X, Y)$$

Using $a(C_u, C_v)$, we can express $p(C)$ in the form $p(C) = \frac{a(C,C)}{2w(C)}$ and the equality mentioned in the introduction of the generalized Ward clustering problem: if $C_u \cap C_v = \emptyset$ then

$$w(C_u \cup C_v) \cdot p(C_u \cup C_v) = w(C_u) \cdot p(C_u) + w(C_v) \cdot p(C_v) + a(C_u, C_v)$$

We analyze the transition of a unit X_s from cluster C_u to cluster C_v. We have $C'_u = C_u \setminus \{X_s\}$, $C'_v = C_v \cup \{X_s\}$,

$$w(C_u) \cdot p(C_u) = w(C'_u) \cdot p(C'_u) + a(X_s, C'_u) = (w(C_u) - w(X_s)) \cdot p(C'_u) + a(X_s, C'_u)$$

as well as:

$$w(C'_v) \cdot p(C'_v) = w(C_v) \cdot p(C_v) + a(X_s, C_v)$$

From $d(X_s, X_s) = 0$, it follows $a(X_s, C_u) = a(X_s, C'_u)$. Therefore

$$p(C'_u) = \frac{w(C_u) \cdot p(C_u) - a(X_s, C_u)}{w(C_u) - w(X_s)} \qquad p(C'_v) = \frac{w(C_v) \cdot p(C_v) + a(X_s, C_v)}{w(C_v) + w(X_s)}$$

and, finally,

$$\Delta P(\mathbf{C}, \mathbf{C}') = p(C_u) + p(C_v) - p(C'_u) - p(C'_v) =$$
$$= \frac{w(X_s) \cdot p(C_v) - a(X_s, C_v)}{w(C_v) + w(X_s)} - \frac{w(X_s) \cdot p(C_u) - a(X_s, C_u)}{w(C_u) - w(X_s)}$$

In the case when d is the squared Euclidean distance, it is possible to derive also an expression for the corrections of centers [60].

3.3.2 Dynamic Programming

Suppose that $Min(\Phi_k, P) \neq \emptyset$, $k = 1, 2, \ldots$. Denoting $P^*(\mathbf{U}, k) = P(\mathbf{C}_k^*(\mathbf{U}))$ we can derive the generalized *Jensen equality* [10]:

$$
P^*(\mathbf{U}, k) = \begin{cases} p(\mathbf{U}) & \{\mathbf{U}\} \in \Phi_1 \\ \min_{\substack{\emptyset \subset C \subset \mathbf{U} \\ \exists C \in \Phi_{k-1}(\mathbf{U} \setminus C) : C \cup \{C\} \in \Phi_k(\mathbf{U})}} (P^*(\mathbf{U} \setminus C, k-1) \oplus p(C)) & k > 1 \end{cases}
$$

This is a *dynamic programming* (Bellman) equation which, for some special constrained problems, keeps the size of Φ_k small, and allows us to solve the clustering problem by the adapted Fisher's algorithm [10].

3.3.3 Hierarchical Methods

The set of feasible clusterings Φ determines the *feasibility predicate* $\underline{\Phi}(\mathbf{C}) \equiv \mathbf{C} \in \Phi$ defined on $\mathscr{P}(\mathscr{P}(\mathbf{U}) \setminus \{\emptyset\})$, and, conversely, $\Phi \equiv \{\mathbf{C} \in \mathscr{P}(\mathscr{P}(\mathbf{U}) \setminus \{\emptyset\}) : \underline{\Phi}(\mathbf{C})\}$.

In the set Φ, the relation of *clustering inclusion* $\sqsubseteq$ can be introduced by

$$
\mathbf{C}_1 \sqsubseteq \mathbf{C}_2 \equiv \forall C_1 \in \mathbf{C}_1, C_2 \in \mathbf{C}_2 : C_1 \cap C_2 \in \{\emptyset, C_1\}.
$$

The clustering $\mathbf{C}_1$ is a *refinement* of the clustering $\mathbf{C}_2$.

It is well known that $(P(\mathbf{U}), \sqsubseteq)$ is a partially ordered set (and also a semimodular lattice). Because any subset of partially ordered set is also partially ordered, we have: Let $\Phi \subseteq P(\mathbf{U})$ then $(\Phi, \sqsubseteq)$ is a partially ordered set.

The clustering inclusion determines two related relations (on Φ):

$$
\mathbf{C}_1 \sqsubset \mathbf{C}_2 \equiv \mathbf{C}_1 \sqsubseteq \mathbf{C}_2 \wedge \mathbf{C}_1 \neq \mathbf{C}_2 \quad \text{– a strict inclusion, and}
$$
$$
\mathbf{C}_1 \sqsubset\!\!\!\!\cdot\ \mathbf{C}_2 \equiv \mathbf{C}_1 \sqsubset \mathbf{C}_2 \wedge \neg \exists \mathbf{C} \in \Phi : (\mathbf{C}_1 \sqsubset \mathbf{C} \wedge \mathbf{C} \sqsubset \mathbf{C}_2) \quad \text{– a predecessor.}
$$

Part of the following text we presented already is in Section 9.3 of [11]. We include it here to make the text self-contained. We assume that the set of feasible clusterings $\Phi \subseteq P(\mathbf{U})$ satisfies the following conditions:

F1. $\mathbf{O} \equiv \{\{X\} : X \in \mathbf{U}\} \in \Phi$

F2. The feasibility predicate $\underline{\Phi}$ is *local* – it has the form $\underline{\Phi}(\mathbf{C}) = \wedge_{C \in \mathbf{C}} \varphi(C)$ where $\varphi(C)$ is a predicate defined on $\mathscr{P}(\mathbf{U}) \setminus \{\emptyset\}$ (clusters). The intuitive meaning of $\varphi(C)$ is that: $\varphi(C) \equiv$ the cluster C is "good". Therefore, the locality condition can be read as a "good" clustering $\mathbf{C} \in \Phi$ consists of "good" clusters.

F3. The predicate Φ has the property of *binary heredity* with respect to the *fusibility* predicate $\psi(C_1, C_2)$, i.e.,

$$
C_1 \cap C_2 = \emptyset \wedge \varphi(C_1) \wedge \varphi(C_2) \wedge \psi(C_1, C_2) \Rightarrow \varphi(C_1 \cup C_2)
$$

This condition means: in a "good" clustering, a fusion of two "fusible" clusters produces a "good" clustering.

F4. The predicate ψ is *compatible* with the clustering inclusion $\sqsubseteq$, i.e.,

$$\forall C_1, C_2 \in \Phi : (C_1 \sqsubset C_2 \wedge C_1 \setminus C_2 = \{C_1, C_2\} \Rightarrow \psi(C_1, C_2) \vee \psi(C_2, C_1))$$

F5. The *interpolation* property holds in Φ, i.e., $\forall C_1, C_2 \in \Phi$:

$$(C_1 \sqsubset C_2 \wedge \text{card}(C_1) > \text{card}(C_2) + 1 \Rightarrow \exists C \in \Phi : (C_1 \sqsubset C \wedge C \sqsubset C_2))$$

These conditions provide a framework within which the hierarchical methods can be applied also to constrained clustering problems $\Phi_k(\mathbf{U}) \subset P_k(\mathbf{U})$. In the ordinary problem, both predicates $\varphi(C)$ and $\psi(C_p, C_q)$ are always true – all conditions F1–F5 are satisfied.

3.3.3.1 Greedy Approximation

We call a *dissimilarity between clusters* a function $D : (C_1, C_2) \to \mathbb{R}_0^+$ which is symmetric, i.e., $D(C_1, C_2) = D(C_2, C_1)$

Let $(\mathbb{R}_0^+, \oplus, e, \leq)$ be an ordered abelian monoid. Then the criterion function $P(\mathbf{C}) = \bigoplus_{C \in \mathbf{C}} p(C)$, $\forall X \in \mathbf{U} : p(\{X\}) = 0$ is *compatible* with a dissimilarity D over Φ iff for all $C \subseteq \mathbf{U}$ holds:

$$\varphi(C) \wedge \text{card}(C) > 1 \Rightarrow p(C) = \min_{(C_1, C_2) : C_2 = C \setminus C_1 \wedge \psi(C_1, C_2)} (p(C_1) \oplus p(C_2) \oplus D(C_1, C_2))$$

Proposition 3.14 *An S criterion function is compatible with a dissimilarity D defined by*

$$D(C_p, C_q) = p(C_p \cup C_q) - p(C_p) - p(C_q)$$

In this case, let $\mathbf{C}' = \mathbf{C} \setminus \{C_p, C_q\} \cup \{C_p \cup C_q\}$ $C_p, C_q \in \mathbf{C}$ *then*

$$P(\mathbf{C}') = P(\mathbf{C}) + D(C_p, C_q)$$

Proposition 3.15 *Let P be compatible with D over Φ, with $\oplus$ distributes over min, and F1–F5 holding, then*

$$P(\mathbf{C}_k^*) = \min_{\mathbf{C} \in \Phi_k} P(\mathbf{C}) = \min_{\substack{C_1, C_2 \in \mathbf{C} \in \Phi_{k+1} \\ \psi(C_1, C_2)}} (P(\mathbf{C}) \oplus D(C_1, C_2))$$

The equality from Proposition 3.15 can be written also in the form

$$P(\mathbf{C}_k^*) = \min_{\mathbf{C} \in \Phi_{k+1}} (P(\mathbf{C}) \oplus \min_{\substack{C_1, C_2 \in \mathbf{C} \\ \psi(C_1, C_2)}} D(C_1, C_2))$$

from which we get the following "greedy" approximation:

$$P(\mathbf{C}_k^*) \approx P(\mathbf{C}_{k+1}^*) \oplus \min_{\substack{C_1, C_2 \in \mathbf{C}_{k+1}^* \\ \psi(C_1, C_2)}} D(C_1, C_2)$$

It is the basis for the agglomerative (binary) procedure for solving the clustering problem.

3.3.3.2   Agglomerative Methods

Here is the agglomerative procedure for solving the clustering problem:

1. $k := n; \mathbf{C}(k) := \{\{X\} : X \in \mathbf{U}\};$
2.   **while** $\exists C_i, C_j \in \mathbf{C}(k) : (i \neq j \wedge \psi(C_i, C_j))$ **repeat**
2.1. $(C_p, C_q) := \text{argmin}\{D(C_i, C_j) : i \neq j \wedge \psi(C_i, C_j)\};$
2.2. $C := C_p \cup C_q; k := k - 1;$
2.3. $\mathbf{C}(k) := \mathbf{C}(k+1) \setminus \{C_p, C_q\} \cup \{C\};$
2.4.   determine $D(C, C_s)$ for all $C_s \in \mathbf{C}(k)$
3. $m := k$

Note that, because it is based on an approximation, this procedure is not an exact procedure for solving the clustering problem.

For another, *probabilistic*, view on agglomerative methods, see [43].

Divisive methods work in the reverse direction. The problem here is how to efficiently find a good split (C_p, C_q) of a cluster C.

In derivations of between the clusters dissimilarity $D(C_u, C_v)$ for different "classical" agglomerative methods, we use the generalized Ward's cluster error function, $p(C)$, and the generalized centers [5]. We consider:

Minimal: $D^m(C_u, C_v) = \min\limits_{X \in C_u, Y \in C_v} d(X, Y)$

Maximal: $D^M(C_u, C_v) = \max\limits_{X \in C_u, Y \in C_v} d(X, Y)$

Average: $D^a(C_u, C_v) = \dfrac{1}{w(C_u)w(C_v)} \sum\limits_{X \in C_u, Y \in C_v} w(X) \cdot w(Y) \cdot d(X, Y)$

Gower-Bock: $D^G(C_u, C_v) = d(\overline{C}_u, \overline{C}_v) = D^a(C_u, C_v) - \dfrac{p(C_u)}{w(C_u)} - \dfrac{p(C_v)}{w(C_v)}$

Ward: $D^W(C_u, C_v) = \dfrac{w(C_u)w(C_v)}{w(C_u \cup C_v)} D^G(\overline{C}_u, \overline{C}_v)$

Inertia: $D^I(C_u, C_v) = p(C_u \cup C_v)$

Variance: $D^V(C_u, C_v) = var(C_u \cup C_v) = \dfrac{p(C_u \cup C_v)}{w(C_u \cup C_v)}$

Weighted increase of variance:

$$D^v(C_u, C_v) = var(C_u \cup C_v) - \frac{w(C_u) \cdot var(C_u) + w(C_v) \cdot var(C_v)}{w(C_u \cup C_v)} = \frac{D^W(C_u, C_v)}{w(C_u \cup C_v)}$$

For all of these measures, the *Lance-Williams-Jambu formula* holds:

$$D(C_p \cup C_q, C_s) = \alpha_1 D(C_p, C_s) + \alpha_2 D(C_q, C_s) + \beta D(C_p, C_q) +$$
$$+ \gamma |D(C_p, C_s) - D(C_q, C_s)| + \delta_1 v(C_p) + \delta_2 v(C_q) + \delta_3 v(C_s)$$

The coefficients $\alpha_1, \alpha_2, \beta, \gamma$ and δ are given in Table 3.4.

Table 3.4 Lance-Williams-Jambu coefficients

Method	α_1	α_2	β	γ	δ_t	$v(C_t)$
Minimum	$\dfrac{1}{2}$	$\dfrac{1}{2}$	0	$-\dfrac{1}{2}$	0	$-$
Maximum	$\dfrac{1}{2}$	$\dfrac{1}{2}$	0	$\dfrac{1}{2}$	0	$-$
Average	$\dfrac{w_p}{w_{pq}}$	$\dfrac{w_q}{w_{pq}}$	0	0	0	$-$
Gower-Bock	$\dfrac{w_p}{w_{pq}}$	$\dfrac{w_q}{w_{pq}}$	$-\dfrac{w_p w_q}{w_{pq}^2}$	0	0	$-$
Ward	$\dfrac{w_{ps}}{w_{pqs}}$	$\dfrac{w_{qs}}{w_{pqs}}$	$-\dfrac{w_s}{w_{pqs}}$	0	0	$-$
Inertia	$\dfrac{w_{ps}}{w_{pqs}}$	$\dfrac{w_{qs}}{w_{pqs}}$	$\dfrac{w_{pq}}{w_{pqs}}$	0	$-\dfrac{w_t}{w_{pqs}}$	$p(C_t)$
Variance	$\dfrac{w_{ps}^2}{w_{pqs}^2}$	$\dfrac{w_{qs}^2}{w_{pqs}^2}$	$\dfrac{w_{pq}^2}{w_{pqs}^2}$	0	$-\dfrac{w_t}{w_{pqs}^2}$	$p(C_t)$
Weighted increase of variance	$\dfrac{w_{ps}^2}{w_{pqs}^2}$	$\dfrac{w_{qs}^2}{w_{pqs}^2}$	$-\dfrac{w_s w_{pq}}{w_{pqs}^2}$	0	0	$-$

$$w_p = w(C_p),\ w_{pq} = w(C_p \cup C_q),\ w_{pqs} = w(C_p \cup C_q \cup C_s)$$

3.3.3.3 Hierarchies

The agglomerative clustering procedure produces a series of feasible clusterings $\mathbf{C}(n)$, $\mathbf{C}(n-1)$, $\dots$, $\mathbf{C}(m)$ with $\mathbf{C}(m) \in Max\Phi$ (maximal elements for $\sqsubseteq$). Their union $\mathcal{T} = \bigcup_{k=m}^{n} \mathbf{C}(k)$ is called a *hierarchy* and has the property:

$$\forall C_p, C_q \in \mathcal{T} : C_p \cap C_q \in \{\emptyset, C_p, C_q\}$$

The set inclusion $\subseteq$ is a *tree*, or a *hierarchical* order, on $\mathcal{T}$. The hierarchy $\mathcal{T}$ is *complete* iff $U \in \mathcal{T}$.

For $W \subseteq U$, we define the *smallest cluster* $C_{\mathcal{T}}(W)$ from $\mathcal{T}$ containing W as:

c1. $W \subseteq C_{\mathcal{T}}(W)$
c2. $\forall C \in \mathcal{T} : (W \subseteq C \Rightarrow C_{\mathcal{T}}(W) \subseteq C)$

$C_{\mathcal{T}}$ is a *closure* on $\mathcal{T}$ with a special property

$$Z \notin C_{\mathcal{T}}(\{X, Y\}) \Rightarrow C_{\mathcal{T}}(\{X, Y\}) \subset C_{\mathcal{T}}(\{X, Y, Z\}) = C_{\mathcal{T}}(\{X, Z\}) = C_{\mathcal{T}}(\{Y, Z\})$$

A mapping $h : \mathcal{T} \to \mathbb{R}_0^+$ is a *level function* on $\mathcal{T}$ iff

11. $\forall X \in \mathbf{U} : h(\{X\}) = 0$
12. $C_p \subseteq C_q \Rightarrow h(C_p) \le h(C_q)$

A simple example of level function is $h(C) = \text{card}(C) - 1$.

Every hierarchy/level function determines an ultrametric dissimilarity on $\mathbf{U}$

$$\delta(X, Y) = h(C_{\mathscr{T}}(\{X, Y\}))$$

The converse is true also (see [25]): Let d be an ultrametric on $\mathbf{U}$. Denote a closed ball in X with radius r with $\overline{B}(X, r) = \{Y \in \mathbf{U} : d(X, Y) \le r\}$. Then for any given set $A \subset \mathbb{R}^+$ the set

$$C(A) = \{\overline{B}(X, r) : X \in \mathbf{U}, r \in A\} \cup \{\{\mathbf{U}\}\} \cup \{\{X\} : X \in \mathbf{U}\}$$

is a complete hierarchy, and $h(C) = \text{diam}(C)$ is a level function.

The pair $(\mathscr{T}, h)$ is called a *dendrogram* or a *clustering tree* because it can be visualized as a tree.

Unfortunately, the function $h_D(C) = D(C_p, C_q)$, $C = C_p \cup C_q$ is not always a level function – for some Ds the *inversions*, $D(C_p, C_q) > D(C_p \cup C_q, C_s)$, are possible. Batagelj showed [4]:

Proposition 3.16 h_D *is a level function for the Lance-Williams procedure* $(\alpha_1, \alpha_2, \beta, \gamma)$ *iff:*

(i) $\gamma + \min(\alpha_1, \alpha_2) \ge 0$
(ii) $\alpha_1 + \alpha_2 \ge 0$
(iii) $\alpha_1 + \alpha_2 + \beta \ge 1$

The dissimilarity D has the *reducibility* property iff

$$D(C_p, C_q) \le t, \quad D(C_p, C_s) \ge t, \quad D(C_q, C_s) \ge t \quad \Rightarrow \quad D(C_p \cup C_q, C_s) \ge t$$

Proposition 3.17 [19] *If a dissimilarity D has the reducibility property then h_D is a level function.*

A very fast agglomerative clustering procedure exists for dissimilarities that have reducibility property [54].

3.3.4   Adding Hierarchical Methods

Suppose we have already built a clustering tree $\mathscr{T}$ over the set of units $\mathbf{U}$. To add a new unit X into the tree $\mathscr{T}$, we start in the root and branch while looking down. Assuming we have reached the node corresponding to cluster C, one obtained by joining sub-clusters C_p and C_q, $C = C_p \cup C_q$. There are three possibilities: (i) add X to C_p or (ii) adding X to C_q or (iii) forming a new cluster $\{X\}$. See Figure 3.1.

Consider again the "greedy approximation"

$$P(C_k^{\bullet}) = P(C_{k+1}^{\bullet}) + D(C_p, C_q)$$

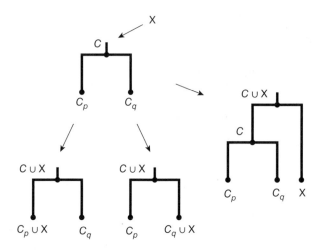

Figure 3.1   Adding hierarchical method.

where $D(C_p, C_q) = \min_{C_u, C_v \in \mathbf{C}^{\bullet}_{k+1}} D(C_u, C_v)$ and $\mathbf{C}^{\bullet}_i$ are greedy solutions. Since we wish to minimize the value of criterion function P, it follows from the greedy relation we need to select the case corresponding to the maximal among values $D(C_p \cup \{X\}, C_q)$, $D(C_q \cup \{X\}, C_p)$, and $D(C_p \cup C_q, \{X\})$.

This is a basis for the adding clustering method. We start with a tree on the first two units and then, successively, add to it the remaining units. The unit X is included within all clusters through which we branch it in a downward direction.

3.3.5   The Leaders Method

To support our intuition regarding further developments, we briefly describe a simple version of the dynamic clusters method – the *leaders* or k-means method [23, 38] which forms the basis of several recent "data-mining" and "big data" analytic methods. In the leaders method, the criterion function has the form SR. The basic scheme of the leaders method is simple:

    Select $\mathbf{C}_0$; $\mathbf{C} := \mathbf{C}_0$;
    repeat
        determine for each $C \in \mathbf{C}$ its leader $\overline{C}$;
        the new clustering $\mathbf{C}$ is obtained by assigning each unit to its nearest leader
    until leaders stabilize

To obtain a "good" solution and an impression of its quality, this procedure can be repeated using different (random) $\mathbf{C}_0$ partitions.

The dynamic clusters method is a generalization of the above scheme. We denote:

Λ	– set of *representatives*
$L \subseteq \Lambda$	– *representation*
Ψ	– set of *feasible representations*
$W : \Phi \times \Psi \to \mathbb{R}_0^+$	– *extended criterion function*
$G : \Phi \times \Psi \to \Psi$	– *representation function*
$F : \Phi \times \Psi \to \Phi$	– *clustering function*

and the following conditions must be satisfied:

W0. $P(\mathbf{C}) = \min_{L \in \Psi} W(\mathbf{C}, L)$

and the functions G and F tend to improve (diminish) the value of the extended criterion function, W:

W1. $W(\mathbf{C}, G(\mathbf{C}, L)) \leq W(\mathbf{C}, L)$
W2. $W(F(\mathbf{C}, L), L) \leq W(\mathbf{C}, L)$

then the *dynamic clusters method* (DCM) can be described by the scheme:

> select $\mathbf{C} := \mathbf{C}_0;$ $L := L_0;$
> **repeat**
> $L := G(\mathbf{C}, L);$
> $\mathbf{C} := F(\mathbf{C}, L)$
> **until** the clustering $\mathbf{C}$ stabilizes

To this scheme correspond the sequence $v_n = (\mathbf{C}_n, L_n), n \in \mathbb{N}$ determined by relations

$$L_{n+1} = G(\mathbf{C}_n, L_n) \quad \text{and} \quad \mathbf{C}_{n+1} = F(\mathbf{C}_n, L_{n+1})$$

and the sequence of values of the extended criterion function, $u_n = W(\mathbf{C}_n, L_n)$. We denote $u^* = P(\mathbf{C}^*)$. Then it holds:

Proposition 3.18 *For every* $n \in \mathbb{N}$, $u_{n+1} \leq u_n$, $u^* \leq u_n$, *and if for* $k > m$, $v_k = v_m$ *then* $\forall n \geq m :$ $u_n = u_m$.

Proposition 3.18 states that the sequence u_n is monotonically decreasing and bounded. Therefore, it is convergent. Note that the limit of u_n is not necessarily u^* – as the dynamic clusters method is a local optimization method.

Two types of sequences v_n are possible:

Type A: $\neg \exists k, m \in \mathbb{N}, k > m : v_k = v_m$

Type B: $\exists k, m \in \mathbb{N}, k > m : v_k = v_m$

Type B_0: Type B with $k = m + 1$

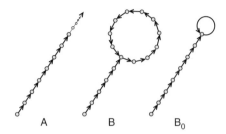

A   B B_0

For DCM to be a viable algorithm, the corresponding sequences, v_n, should be of type B. The DCM sequence (v_n) is of type B if

- sets Φ and Ψ are both finite, for example when selecting a representative of C among its members
- $\exists \delta > 0 : \forall n \in \mathbb{N} : (v_{n+1} \neq v_n \Rightarrow u_n - u_{n+1} > \delta)$

Because the sets **U**, and consequently Φ, are finite we expect a good dynamic clustering procedure to stabilize in a finite number of steps – it is of type of type B, as described above.

However, the conditions W0, W1 and W2 are not strong enough to ensure this. To compensate for the possibility that the set of representations, Ψ, is infinite, we include the additional requirement:

W3. $W(\mathbf{C}, G(\mathbf{C}, L)) = W(\mathbf{C}, L) \Rightarrow L = G(\mathbf{C}, L)$

With this requirement the "symmetry" between Φ and Ψ is destroyed. This can be reestablished with the requirement:

W4. $W(F(\mathbf{C}, L, L)) = W(\mathbf{C}, L) \Rightarrow \mathbf{C} = F(\mathbf{C}, L)$

Alas, it turns out that W4 often fails. For this reason, we avoid it henceforth.

Proposition 3.19 *If W3 holds and if there exists $m \in \mathbb{N}$ such that $u_{m+1} = u_m$, then also $L_{m+1} = L_m$.*

Usually, in the applications of the DCM, the clustering function takes the form $F : \Psi \to \Phi$. In this case, the condition W2 simplifies to $W(F(L), L) \leq W(\mathbf{C}, L)$, which can be expressed also as $F(L) \in \text{Min}_{\mathbf{C} \in \Phi} W(\mathbf{C}, L)$. For such *simple* clustering functions, it holds:

Proposition 3.20 *If the clustering function F is simple and if there exists $m \in \mathbb{N}$ such that $L_{m+1} = L_m$, then for every $n \geq m : v_n = v_m$.*

When G is *simple*, it has the form $G : \Phi \to \Psi$.

Proposition 3.21 *If W3 holds and the representation function G is simple then:*

a. $G(\mathbf{C}) = \arg\min_{L \in \Psi} W(\mathbf{C}, L)$

b. $\exists k, m \in \mathbb{N}, k > m \forall i \in \mathbb{N} : v_{k+i} = v_{m+i}$

c. $\exists m \in \mathbb{N} \forall n \geq m : u_n = u_m$

d. *if also F is simple then* $\exists m \in \mathbb{N} \forall n \geq m : v_n = v_m$

In the original dynamic clustering method [23], both of the functions F and G are simple – $F :$ $\Psi \to \Phi$ and $G : \Phi \to \Psi$.

If W3 also holds and the functions F and G are simple, then

> G0. $G(\mathbf{C}) = \operatorname{argmin}_{L \in \Psi} W(\mathbf{C}, L)$

and

> F0. $F(L) \in \operatorname{Min}_{\mathbf{C} \in \Phi} W(\mathbf{C}, L)$

In other words, given an extended criterion function W, the relations G0 and F0 define an appropriate pair of functions, G and F, such that the DCM stabilizes in a finite number of steps.

3.4   Clustering Graphs and Networks

When the set of units **U** consists of graphs (e.g. chemical molecules) we write about *clustering of graphs* (networks). For this purpose, we can use standard clustering approaches provided that we have an appropriate definition of dissimilarity between graphs.

The first approach is to define a vector description $[\mathbf{G}] = [g_1, g_2, \dots, g_m]$ of each graph $\mathbf{G}$, and then use some standard dissimilarity δ on $\mathbb{R}^m$ to compare these vectors $d(\mathbf{G}_1, \mathbf{G}_2) = \delta([\mathbf{G}_1], [\mathbf{G}_2])$. We can get $[\mathbf{G}]$, for example, by:

- **Invariants:** Compute the values of selected invariants (indices) on each graph [61].
- **Fragment counts:** Select a collection of subgraphs (fragments), for example triads, and count the number of appearences of each – a *fragments spectrum* [6, 57].

Let **Gph** be the set of all graphs. An *invariant* of a graph is a mapping $i : \mathbf{Gph} \to \mathbb{R}$ which is constant over isomorphic graphs

$$\mathbf{G} \approx \mathbf{H} \Rightarrow i(\mathbf{G}) = i(\mathbf{H})$$

The number of nodes, the number of arcs, the number of edges, maximum degree Δ, chromatic number χ, etc. are all graph invariants. Invariants have an important role in examining the isomorphism of two graphs. To prove that **G** is not isomorphic to **H** it is enough to find an invariant i such that $i(\mathbf{G}) \neq i(\mathbf{H})$.

Invariants on *families* of graphs are called *structural properties*: Let $\mathscr{F} \subseteq \mathbf{Gph}$ be a family of graphs. A property $i : \mathscr{F} \to \mathbb{R}$ is *structural* on $\mathscr{F}$ iff

$$\forall \mathbf{G}, \mathbf{H} \in \mathscr{F} : (\mathbf{G} \approx \mathbf{H} \Rightarrow i(\mathbf{G}) = i(\mathbf{H}))$$

A collection $\mathscr{I}$ of invariants/structural properties is *complete* iff

$$(\forall i \in \mathscr{I} : i(\mathbf{G}) = i(\mathbf{H})) \Rightarrow \mathbf{G} \approx \mathbf{H}$$

In most cases (families of graphs) there is no efficiently computable complete collection.

Different dissimilarities between strings are based on *transformations* including insert, delete, transpose [45, 50]. For binary trees, Robinson considered a dissimilarity based on the transformation of *neighbors exchange over an edge* (see Figure 3.2).

There is a natural generalization of this approach to graphs and other structured objects [6]: Let $\mathcal{T} = \{T_k\}$ be a set of *basic transformations* of units $T_k : \mathcal{U} \to \mathcal{U}$ and $v : \mathcal{T} \times \mathcal{U} \to \mathbb{R}^+$ a value or cost of transformation, which satisfies the conditions:

$$\forall T \in \mathcal{T} : (T : X \mapsto Y \Rightarrow \exists S \in \mathcal{T} : (S : Y \mapsto X \wedge v(T, X) = v(S, Y)))$$

and $v(\mathrm{id}, X) = 0$.

Suppose that for each pair $X, Y \in \mathcal{U}$, there exists a finite sequence $\tau = (T_1, T_2, \ldots, T_t)$ such that $\tau(X) = T_t \circ T_{t-1} \circ \ldots \circ T_1(X) = Y$. Then we can define:

$$d(X, Y) = \min_{\tau}(v(\tau(X)) : \tau(X) = Y)$$

where

$$v(\tau(X)) = \begin{cases} 0 & \tau = \mathrm{id} \\ v(\eta(T(X))) + v(T, X) & \tau = \eta \circ T \end{cases}$$

It is easy to verify that this dissimilarity, $d(X, Y)$, is a distance.

For example, see Figure 3.3. Using the transformations G1 and G2 we can transform any pair of connected simple graphs, one to the other. For triangulations of the plane on n nodes, S is such a transformation.

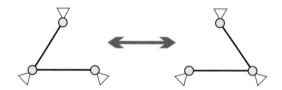

Figure 3.2   Neighbors exchange over an edge.

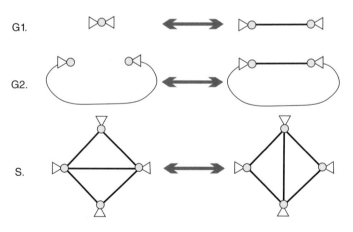

Figure 3.3   Examples of transformations.

3.5   Clustering in Graphs and Networks

Since in a graph $\mathbf{G} = (V, L)$ we have two kinds of objects, nodes and links, we can think about *clustering of nodes* and *clustering of links*. Usually we deal with clustering of nodes.

3.5.1   An Indirect Approach

Again, we use the standard clustering methods, provided that we have an appropriate definition of dissimilarity between nodes. The usual approach is to define a vector description $[v] = [t_1, t_2, \ldots, t_m]$ of each node $v \in V$, and then use some standard dissimilarity, δ, on $\mathbb{R}^m$ to compare these vectors $d(u, v) = \delta([u], [v])$.

We can assign to each node v different neighborhoods, such as $N(v) = \{u \in V : (v, u) \in L\}$, and other sets. In these cases, the dissimilarities between sets are used on them.

For a given graph $\mathbf{G} = (V, L)$, a property $t : V \to \mathbb{R}$ is *structural* iff for every automorphism φ of $\mathbf{G}$ it holds

$$\forall v \in V : t(v) = t(\varphi(v))$$

Examples of such properties are:

$t(v) =$ degree (the number of neighbors) of node v
$t(v) =$ the number of nodes at distance d from node v
$t(v) =$ the number of triads of type x at node v
$t(v) =$ the number of graphlets of type x at node v [57]

For a given graph $\mathbf{G} = (V, L)$, a *property of pairs of nodes* $q : V \times V \to \mathbb{R}$ is *structural* if for every automorphism φ of $\mathbf{G}$, it holds:

$$\forall u, v \in V : q(u, v) = q(\varphi(u), \varphi(v))$$

Some examples of structural properties of pairs of nodes are:

$q(u, v) =$ **if** $(u, v) \in L$ **then** 1 **else** 0;
$q(u, v) =$ number of common neighbors of units u and v;
$q(u, v) =$ length of the shortest path from u to v.

Using a selected property of pairs of nodes, q, we can describe each node u with a vector

$$[u] = [q(u, v_1), q(u, v_2), \ldots, q(u, v_n), q(v_1, u), \ldots, q(v_n, u)]$$

and we define the dissimilarity between nodes $u, v \in V$ as $d(u, v) = \delta([u], [v])$.

Corrected dissimilarities based on properties of pairs of nodes for measuring the similarity between nodes v_i and v_j $(p \geq 0)$ must be used [26] such as: The corrected Manhattan distance:

$$d_c(p)(v_i, v_j) = \sum_{\substack{s=1 \\ s \neq i,j}}^{n} (|q_{is} - q_{js}| + |q_{si} - q_{sj}|) + p \cdot (|q_{ii} - q_{jj}| + |q_{ij} - q_{ji}|)$$

The corrected Euclidean distance:

$$d_e(p)(v_i, v_j) = \sqrt{\sum_{\substack{s=1 \\ s \neq i,j}}^{n} ((q_{is} - q_{js})^2 + (q_{si} - q_{sj})^2) + p \cdot ((q_{ii} - q_{jj})^2 + (q_{ij} - q_{ji})^2)}$$

The corrected dissimilarities with $p = 1$ are usually used.

3.5.2   A Direct Approach: Blockmodeling

A partition $\mathbf{C} = \{C_i\}$ splits the set of links (arcs) $L \subseteq V \times V$ into *blocks* $B_{ij} = L \cap C_i \times C_j$ – a subgraph of arcs from cluster C_i to cluster C_j. In blockmodeling, analysts adopting this approach attempt to find partitions producing blocks of selected types (complete, empty, regular, etc.), while allowing for some "errors" in the form of links not consistent with the specified block types [26]. Usually, the relocation method is used for solving the corresponding optimization problems.

Regarding blockmodeling, the dissimilarity based criterion functions for the indirect approach usually use the concept of structural equivalence. While this may be too limited, we do not pursue this issue further.

3.5.3   Graph Theoretic Approaches

The basic decomposition of graphs is to (weakly) connected components (partition of nodes, and links) and to (weakly) biconnected components (partition of links). For both, very efficient algorithms exist [21]. For directed graphs, the fundamental decomposition results can be found in [20].

From a network $\mathbf{N} = (V, L, w)$ we can get, for a threshold, t, a *link-cut* – a subnetwork $\mathbf{N}(t) = (V, L_t, w)$ where $L_t = \{p \in L : w(p) \geq t\}$. From it, we can get a clustering $\mathbf{C}(t)$ with connected components as clusters. For different thresholds, these clusterings form a hierarchy. An elaborated version of cuts is provided with the *islands* approach [[11], Subsection 2.9.1]. Islands also form a hierarchy for a selected node property of a given network.

In the 1970s and 1980s, Matula studied different types of connectivities in graphs and the structures they induce [52]. In most cases the algorithms are too demanding to be used on larger graphs. A nice overview of connectivity algorithms can be found in Esfahanian [27]. The graph partitioning problem has also several technical applications supported by special algorithms [36, 44, 47].

3.6   Agglomerative Method for Relational Constraints

Suppose that the units are described by attribute data $a : \mathbf{U} \to [\mathbf{U}]$ and are related by a binary *relation* $R \subseteq \mathbf{U} \times \mathbf{U}$ that determines the *relational data* or *network* $(\mathbf{U}, R, a)$ [9].

We want to cluster the units according to some (dis)similarity of their descriptions, but also considering the relation R which imposes *constraints* on the set of feasible clusterings [9, 28, 29, 31], usually in the following form:

$\Phi(R) = \{\mathbf{C} \in P(\mathbf{U})$: each cluster $C \in \mathbf{C}$ induces a subgraph $(C, R \cap C \times C)$ in the graph $(\mathbf{U}, R)$
           of the required type of connectedness$\}$

and criterion function of type SR:

$$P(\mathbf{C}) = \sum_{C \in \mathbf{C}} p(C), \quad p(C) = \sum_{X \in C} d(X, T_C)$$

We can define different types of sets of feasible clusterings for the same relation R. Some examples of *types of relational constraints*, $\Phi^i(R)$, are [29]

Clusterings	Type of connectedness
$\Phi^1(R)$	Weakly connected units
$\Phi^2(R)$	Weakly connected units that contain at most one center
$\Phi^3(R)$	Strongly connected units
$\Phi^4(R)$	Clique
$\Phi^5(R)$	The existence of a trail containing all the units of the cluster

In a directed graph a *trail* is a walk in which all arcs are distinct.

The set $R(X) = \{Y : X R Y\}$ is a *set of successors* of unit $X \in \mathbf{U}$ and, for a cluster $C \subseteq \mathbf{U}$, $R(C) = \bigcup_{X \in C} R(X)$. A set of units, $L \subseteq C$, is a *center* of a cluster C in the clustering of type $\Phi^2(R)$ iff the subgraph induced by L is strongly connected and $R(L) \cap (C \setminus L) = \emptyset$.

The sets of feasible clusterings $\Phi^i(R)$ are linked as follows: $\Phi^4(R) \subseteq \Phi^3(R) \subseteq \Phi^2(R) \subseteq \Phi^1(R)$ and $\Phi^4(R) \subseteq \Phi^5(R) \subseteq \Phi^2(R)$. If the relation R is symmetric, then $\Phi^3(R) = \Phi^1(R)$. If the relation R is an equivalence relation, then $\Phi^4(R) = \Phi^1(R)$.

The corresponding fusibility predicates are:

$\psi^1(C_1, C_2) \equiv \exists X \in C_1 \exists Y \in C_2 : (XRY \vee YRX)$

$\psi^2(C_1, C_2) \equiv (\exists X \in L_1 \exists Y \in C_2 : XRY) \vee (\exists X \in C_1 \exists Y \in L_2 : YRX)$

$\psi^3(C_1, C_2) \equiv (\exists X \in C_1 \exists Y \in C_2 : XRY) \wedge (\exists X \in C_1 \exists Y \in C_2 : YRX)$

$\psi^4(C_1, C_2) \equiv \forall X \in C_1 \forall Y \in C_2 : (XRY \wedge YRX)$

$\psi^5(C_1, C_2) \equiv (\exists X \in T_1 \exists Y \in I_2 : XRY) \vee (\exists X \in I_1 \exists Y \in T_2 : YRX)$

where I denotes initial nodes in a cluster C and T denotes terminal nodes in a cluster C. For ψ^3 the property F5 fails.

We can use both hierarchical and local optimization methods for solving some types of problems with relational constraint [11, 28, 29]. Here, we present only the hierarchical method:

1. $k := n; \mathbf{C}(k) := \{\{X\} : X \in \mathbf{U}\};$
2.   **while** $\exists C_i, C_j \in \mathbf{C}(k) : (i \neq j \wedge \psi(C_i, C_j))$ **repeat**
2.1. $(C_p, C_q) := \text{argmin}\{D(C_i, C_j) : i \neq j \wedge \psi(C_i, C_j)\};$
2.2. $C := C_p \cup C_q; k := k - 1;$
2.3. $\mathbf{C}(k) := \mathbf{C}(k + 1) \setminus \{C_p, C_q\} \cup \{C\};$
2.4.   determine $D(C, C_s)$ for all $C_s \in \mathbf{C}(k);$
2.5.   adjust the relation R as required by the clustering type
3. $m := k$

To get clustering procedures, it is necessary to further elaborate the questions how to adjust the relation after joining two clusters and how to update the dissimilarity $D(C, C_s)$.

In Figures 3.4 and 3.6, four adjusting *strategies* are presented. They are compatible with the corresponding types of constraints: Φ^1 – tolerant, Φ^2 – leader, Φ^4 – strict, and Φ^5 – two.way. In Figure 3.5 an example of application of strategies is presented.

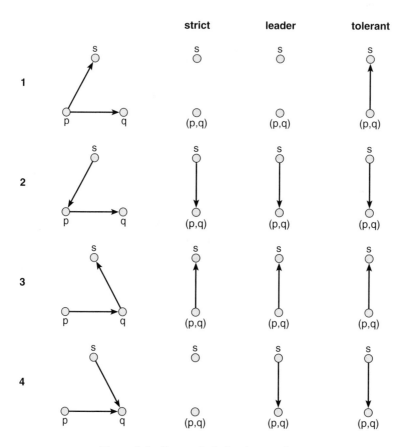

Figure 3.4 Types of relational constraints.

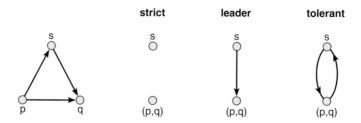

Figure 3.5 A composite example.

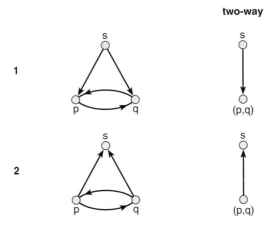

Figure 3.6 The two-way strategy.

The effects of these strategies can be described also as updates of the sets of successors $R(C)$:

Tolerant

$$R(C_r) = \{C_r\} \cup R(C_p) \cup R(C_q) \setminus \{C_p, C_q\}$$
$$R(C_s) = \{C_r\} \cup R(C_s) \setminus \{C_p, C_q\}, \quad \text{for } s \neq r \wedge \{C_p, C_q\} \cap R(C_s) \neq \emptyset$$

Strict

$$R(C_r) = \begin{cases} \{C_r\} \cup R(C_p) \cup R(C_q) \setminus \{C_p, C_q\}, & \text{for } C_q \, R \, C_p \\ \{C_r\} \cup R(C_s) \setminus \{C_p, C_q\}), & \text{otherwise} \end{cases}$$

$$R(C_s) = \begin{cases} \{C_s\} \cup R(C_s) \setminus \{C_p, C_q\}), & \text{for } \begin{array}{l} s \neq r \wedge (C_p \in R(C_s) \vee \\ C_q \in R(C_s) \wedge C_q R C_p) \end{array} \\ R(C_s) \setminus \{C_p, C_q\}), & \text{otherwise for } s \neq r \end{cases}$$

Leader

$$R(C_r) = \begin{cases} \{C_r\} \cup R(C_p) \cup R(C_q) \setminus \{C_p, C_q\}, & \text{for } C_q \, R \, C_p \\ \{C_r\} \cup R(C_s) \setminus \{C_p, C_q\}, & \text{otherwise} \end{cases}$$

$$R(C_s) = \begin{cases} \{C_s\} \cup R(C_s) \setminus \{C_p, C_q\}, & \text{for } s \neq r \wedge \{C_p.C_q\} \cap R(C_s) \neq \emptyset \\ R(C_s) \setminus \{C_p, C_q\}, & \text{otherwise for } s \neq r \end{cases}$$

Two-way

$$R(C_r) = \{C_r\} \cup (R(C_p) \cap R(C_q)) \setminus \{C_p, C_q\}$$

$$R(C_s) = \begin{cases} \{C_s\} \cup R(C_s) \setminus \{C_p, C_q\}, & \text{for } s \neq r \wedge \{C_p.C_q\} \subseteq R(C_s) \\ R(C_s) \setminus \{C_p, C_q\}, & \text{otherwise for } s \neq r \end{cases}$$

In the original approach [28, 29], a complete dissimilarity matrix is needed. To obtain fast algorithms that can be applied to large data sets we propose *considering only the dissimilarities between linked units*. For large data sets, we assume that the relation R is *sparse*.

For step 2.4, "determine $D(C, C_s)$ for all $C_s \in C(k)$" in the agglomerative procedure requires the adjustment of dissimilarities – computing the dissimilarities between a new cluster C and other remaining clusters. In the case of the relational constraints, we can limit the computation only to clusters that are related/linked to C.

This can be done efficiently in the following two ways:

- **A first approach**: we define a dissimilarity $D(S, T)$ between clusters S and T that allows quick updates (as in the Lance-Williams formula).
- **A second approach**: to each cluster we assign a representative and can efficiently compute a representative of merged clusters along with a dissimilarity between clusters in terms of their representatives.

The first approach was described already in [11]. Let $(\mathbf{U}, R)$, $R \subseteq \mathbf{U} \times \mathbf{U}$ be a graph and $\emptyset \subset S, T \subset \mathbf{U}$ and $S \cap T = \emptyset$. We call a *block* of relation R for S and T its part $R(S, T) = R \cap S \times T$. The *symmetric closure* of relation R we denote with $\hat{R} = R \cup R^{-1}$. It holds: $\hat{R}(S, T) = \hat{R}(T, S)$.

For all dissimilarities between clusters $D(S, T)$ we set:

$$D(\{s\}, \{t\}) = \begin{cases} d(s, t) & s \, \hat{R} \, t \\ \infty & \text{otherwise} \end{cases}$$

where d is a selected dissimilarity between units.

Minimum

$$D_{\min}(S, T) = \min_{(s,t) \in \hat{R}(S,T)} d(s, t)$$

$$D_{\min}(S, T_1 \cup T_2) = \min(D_{\min}(S, T_1), D_{\min}(S, T_2))$$

Maximum

$$D_{\max}(S, T) = \max_{(s,t) \in \hat{R}(S,T)} d(s, t)$$

$$D_{\max}(S, T_1 \cup T_2) = \max(D_{\max}(S, T_1), D_{\max}(S, T_2))$$

Average

$w : V \to \mathbb{R}$ – is a weight on units; for example $w(v) = 1$, for all $v \in \mathbf{U}$.

$$D_a(S, T) = \frac{1}{w(\hat{R}(S, T))} \sum_{(s,t) \in \hat{R}(S,T)} d(s, t)$$

$$w(\hat{R}(S, T_1 \cup T_2)) = w(\hat{R}(S, T_1)) + w(\hat{R}(S, T_2))$$

$$D_a(S, T_1 \cup T_2) = \frac{w(\hat{R}(S, T_1))}{w(\hat{R}(S, T_1 \cup T_2))} D_a(S, T_1) + \frac{w(\hat{R}(S, T_2))}{w(\hat{R}(S, T_1 \cup T_2))} D_a(S, T_2)$$

All three dissimilarities have the reducibility property. In this case, also the *nearest neighbor network* for a given network is preserved after joining the nearest clusters. This allows us to develop a very fast agglomerative hierarchical clustering procedure [54] and [11] (Subsection 9.3.5). It is available in the program `Pajek`. The same approach can be extended also to clustering of links of network [17] by transforming a given network into its line-graph in which the original links become new nodes.

For the second approach, we need the representatives of clusters and a dissimilarity between clusters that can be expressed in terms of representatives. For symbolic objects described by discrete distributions (histograms, barcharts) there exist some possibilities [12, 49].

3.6.1   Software Support

The first approach is implemented for weighted networks (weight is a dissimilarity) in `Pajek` – a program for analysis and visualization of large networks [56]. We also implemented it in R package cluRC [7]. An implementation in R of the second approach is still a work in progress.

3.7   Some Examples

To illustrate the hierarchical clustering with relational constraints, we use two examples:

- Clustering the US states according to the selected variables into geographically contiguous clusters.
- Clustering the authors from the network clustering literature (see Chapter 2) according to their citations into clusters with a single leaders group.

3.7.1   The US Geographical Data, 2016

From the site https://datausa.io/profile/geo/united-states/ we obtained the data about US states in 2016 for the following variables: *crime* – homicide deaths, *violent* – violent crimes, *smoking* – adult smoking prevalence, *drinking* – excessive drinking prevalence, *diabetes* – diabetes prevalence, *opioid* – opioid overdose death rate, and *income* – median household income.

In his book *The Stanford GraphBase* [48] Knuth provided a description of neighboring relation for the contiguous part of USA `contiguous-usa.dat` (without Alaska and Hawai). Because of missing data we removed also Washington DC.

We first applied the Ward's hierarchical clustering method using the squared Euclidean dissimilarity between units with standardized variables. On the basis of the corresponding dendrogram (see the left top part of Figure 3.7, we considered a clustering into five clusters:

$C_1 = \{AL, AR, LA, MS, NM, TN, SC\}$,
$C_2 = \{AZ, CA, DE, FL, GA, IL, IN, KS, MI, MO, NC, NV, NY, OH, OK, PA, TX\}$,
$C_3 = \{CO, IA, ID, ME, MN, MT, ND, NE, OR, SD, WY, RI, WI, WA, VT\}$,
$C_4 = \{CT, MA, MD, NH, NJ, UT, VA\}$,
$C_5 = \{KY, WV\}$.

The middle left part of Figure 3.7 shows the dissimilarity matrix reordered according to the obtained clustering.

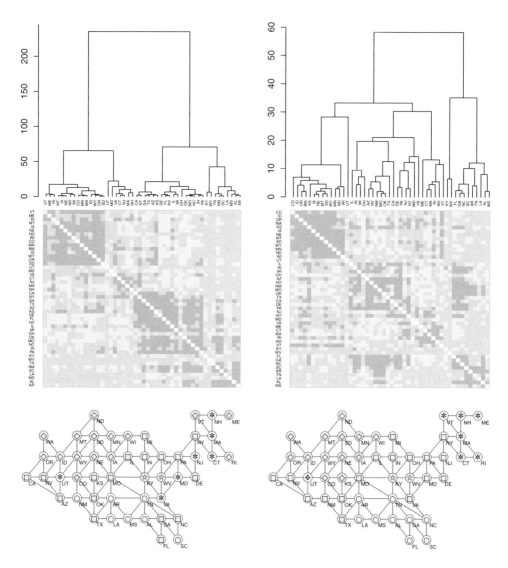

Figure 3.7   Ward clustering (left) and Maximum/Tolerant clustering (right).

To interpret the obtained clusters we produced Table 3.5 with averages of each variable over each cluster for raw and standardized units. The interpretation is left to the reader.

In the bottom left part of Figure 3.7, the obtained clustering/partition is represented with node colors on the network of neighboring US states. It is clear that the subnetworks induced by clusters are not all connected (forming contiguous regions). For example, the subnetwork induced by C_4 has four components $\{CT, MA, NH\}$, $\{NJ\}$, $\{MD, VA\}$ and $\{UT\}$.

Using hierarchical clustering with relational constraints with the Maximum/Tolerant strategy, we get a clustering that considers a given dissimilarity among units and produces clusters that

Table 3.5 Averages for Ward's clustering

	Crime	Violent	Smoking	Drinking	Diabetes	Opioid	Income
C_1	8.7857	496.45	0.2251	0.1447	0.1173	10.857	44631
C_2	5.9118	427.96	0.1826	0.1714	0.1048	13.853	53535
C_3	2.6333	239.99	0.1755	0.2023	0.0847	10.767	55908
C_4	3.8000	300.99	0.1521	0.1699	0.0903	23.657	69947
C_5	4.9000	273.02	0.2645	0.1195	0.1210	33.500	43727
All	4.9563	354.23	0.1856	0.1748	0.0989	14.700	54963
C_1	**1.5723**	**1.0924**	1.1363	−0.9927	1.2826	−0.4229	−1.1990
C_2	0.3923	0.5663	−0.0843	−0.1123	0.4134	−0.0932	−0.1657
C_3	**−0.9537**	**−0.8776**	−0.2887	**0.9094**	**−0.9924**	−0.4328	0.1097
C_4	−0.4747	−0.4090	**−0.9605**	−0.1617	−0.6005	0.9856	**1.7389**
C_5	−0.0231	−0.6239	**2.2668**	**−1.8260**	**1.5416**	**2.0687**	**−1.3039**

form contiguous regions. On the basis of the dendrogram in the right top part of Figure 3.7, we considered a clustering into six clusters:

$C_1 = \{AL, AR, FL, GA, LA, MS, NC, TN, SC\}$,
$C_2 = \{AZ, CA, DE, IL, IN, MD, MI, MO, NJ, NM, NV, NY, OH, OK, PA, VA, TX\}$,
$C_3 = \{CO, IA, ID, KS, MN, MT, ND, NE, OR, SD, WY, WI, WA\}$,
$C_4 = \{CT, MA, ME, NH, RI, VT\}$,
$C_5 = \{KY, WV\}$,
$C_6 = \{UT\}$.

The clusters of the obtained clustering/partition induce connected subnetworks, as expected. See the right bottom part of Figure 3.7 and Figure 3.8.

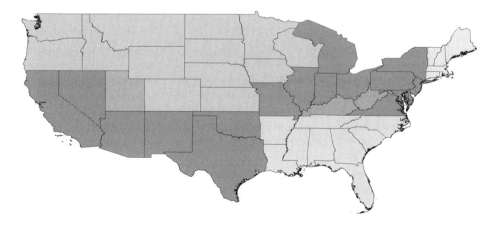

Figure 3.8 Maximum/Tolerant partition on the map.

Table 3.6   Averages for Maximum/Tolerant clustering

	Crime	Violent	Smoking	Drinking	Diabetes	Opioid	Income
C_1	8.1667	462.00	0.2140	0.1488	0.1160	10.788	46104
C_2	5.9701	425.91	0.1804	0.1719	0.1032	15.794	57054
C_3	2.8385	265.25	0.1765	0.2005	0.0852	7.408	55913
C_4	2.3833	234.31	0.1660	0.1932	0.0880	26.717	62751
C_5	4.9000	273.02	0.2645	0.1195	0.1210	33.500	43727
C_6	1.9000	204.72	0.0970	0.1210	0.0710	16.400	62518
All	4.9563	354.23	0.1856	0.1748	0.0989	14.700	54963
C_1	**1.3181**	**0.8278**	0.8162	−0.8584	1.1929	−0.4304	−1.0281
C_2	0.4165	0.5506	−0.1502	−0.0928	0.3026	0.1204	0.2427
C_3	−0.8695	−0.6836	−0.2620	**0.8523**	−0.9584	**−0.8024**	0.1103
C_4	−1.0564	−0.9212	−0.5625	0.6087	−0.7599	1.3223	**0.9039**
C_5	−0.0231	−0.6239	**2.2668**	**−1.8260**	**1.5416**	**2.0687**	**−1.3039**
C_6	**−1.2548**	**−1.1485**	**−2.5445**	−1.7764	−1.9456	0.1871	0.8767

The averages of each variable over these clusters for raw and standardized units are given in Table 3.6. The states of the first cluster C_1 have the highest rates of homicide deaths and violent crimes, high adult smoking levels and diabetes prevalence, and a low median household income. The states of the second cluster C_2 all have variables around the average; above average levels of homicide deaths and violent crimes. Typical for the cluster C_3 is the lowest opioid overdose death rate, the highest levels of excessive drinking and low crime rates. The states of the cluster C_4 have the highest income levels and a high level of excessive drinking and opioid death rate, but low levels of crime, smoking and diabetes. The two states in cluster C_5 have the lowest incomes and levels of excessive drinking, the highest values of smoking, diabetes and opioid death rates, and a lower crime rate. Utah, cluster C_6, has the lowest values of crime, smoking and diabetes, with very low levels of drinking, and a high income level.

3.7.2   Citations Among Authors from the Network Clustering Literature

We consider again the bibliometric data on the network clustering literature analyzed in Chapter 2. In Section 2.5.3, we analyzed the network **Acite** of citations among authors. Here, we analyze the normalized network of citations among authors $\mathbf{nAcite} = n(\mathbf{WAc})^T * n(\mathbf{CiteC}) * n(\mathbf{WAc})$. Every work has one point. They are distributed on arcs of the derived network. The weight $\mathbf{nAcite}[u, v]$ of the arc (u, v) is equal to the fractional share of works co-authored by u that are citing a work co-authored by v.

We removed loops (self-citations) and compute weighted indegrees. We look first at the largest weighted input degrees – the most cited authors, as presented in Table 3.7. The most cited ones are Mark Newman and Santo Fortunato by a wide margin. Quite high also are the most important researchers from the field of social network analysis, beginning with Ronald Burt.

In this example, we identified clusters such that the corresponding induced subnetworks are connected and contain a single center – type Φ^2. The **nAcite** weights are similarities, $s \in [\infty, 0]$.

Table 3.7 The most cited authors/fractional approach

i	w_i	Author	i	w_i	Author
1	329.8886	NEWMAN_M	26	19.7797	MALIK_J
2	155.4974	FORTUNAT_S	27	19.7317	ROSVALL_M
3	80.8228	GIRVAN_M	28	19.2631	VONLUXBU_U
4	51.6716	BARABASI_A	29	19.1634	BERGSTRO_C
5	45.1972	BURT_R	30	19.1422	BARTHELE_M
6	42.5944	ALBERT_R	31	18.6968	LEFEBVRE_E
7	39.6466	ZACHARY_W	32	18.6552	GUILLAUM_J
8	38.8163	LANCICHI_A	33	18.6261	DOREIAN_P
9	38.1660	CLAUSET_A	34	18.3258	KLEINBER_J
10	31.8938	SCHAEFFE_S	35	18.1618	BREIGER_R
11	31.7021	STROGATZ_S	36	17.4888	VICSEK_T
12	30.9933	FREEMAN_L	37	17.4204	BORGATTI_S
13	29.1247	WASSERMA_S	38	16.9268	PALLA_G
14	29.0661	MOORE_C	39	16.8126	OKADA_Y
15	26.1896	FAUST_K	40	16.7620	BOORMAN_S
16	24.8884	WATTS_D	41	15.8376	CHUNG_F
17	24.7421	WHITE_H	42	15.8216	GUIMERA_R
18	24.5679	NEWMARK_N	43	15.7187	RADICCHI_F
19	23.8077	BLONDEL_V	44	14.9995	CARLSON_J
20	23.0214	BATAGELJ_V	45	14.9914	EVERETT_M
21	22.6844	LAMBIOTT_R	46	14.6212	DUCH_J
22	22.5521	VANDONGE_S	47	14.5231	AMARAL_L
23	20.9136	ARENAS_A	48	14.4554	GRANOVET_M
24	19.8478	LESKOVEC_J	49	13.7216	DERENYI_I
25	19.8113	SHI_J	50	13.7216	FARKAS_I

To convert them to distances d, different transformations can be used, including $d = \frac{s_{max}}{s} - 1 \in [0, \infty]$ or $d = 1 - \frac{s}{s_{max}} \in [0, 1]$. We selected the second option with $s_{max} = 2.52$. On the obtained network, we applied, in `Pajek`, the hierarchical clustering with relational constraints procedure with the Maximum/Leader strategy and determined the partition of units into clusters of size at most 50. There are 257 such clusters. To reduce their number, we decided to consider only clusters with at least 20 units. There are 57 such clusters.

We extracted the corresponding subnetworks of citations among authors for visual inspection. Most of them are (double) star-like formed around the most prominent scientists in the field: Albert R + Barabási A, Bergstrom C + Rosvall M, Bezdek J, Blei D, Blondel V, Bonacich P + Kleinberg J, Breiger R, Burt R + Doreian P, Chung F + von Luxburg U, Clauset A, Dietrich J + Maede B, Fortunato S, Freeman L, Ghosh J, Girvan M, Goldberg D, Jaccard P, Jain A, Johnson D, Jordan M, Kaufman L, Knuth D, Leskovec J, Mac Queen J, Newman M, Newmark N, Okada Y, Palla G + Viscek T, Prescott W, Schaeffer S, Scott J, Sporus O, Stein C, Strehl A, Strogatz S, Van Dongen S, and some "cliques" of co-authors with attachments. We visually selected 12 clusters (Adamic L, Batagelj V + Ferligoj A, Bollobas B, Burt R + Doreian P, Faust K + Watts D, Fiedler M + Harary F, Granovetter M, Mizruchi M, Murtagh F, Nowicki K + Wasserman S, Robins G, Ward J, White H + Zachary W) with more interesting network structure for detailed inspection.

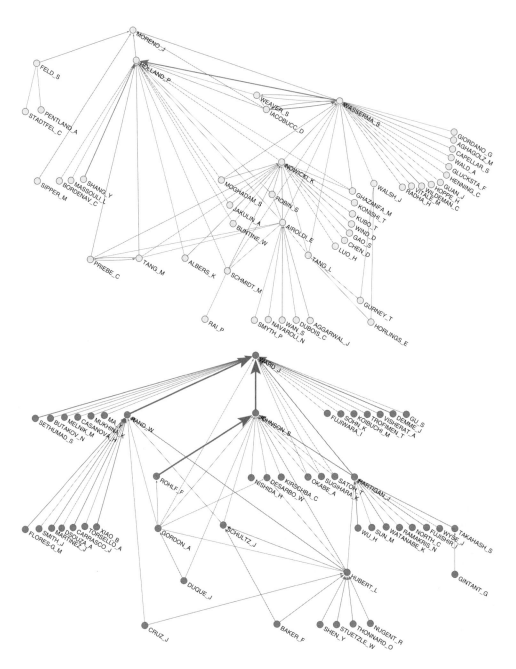

Figure 3.9   Subnetworks Wasserman and Ward.

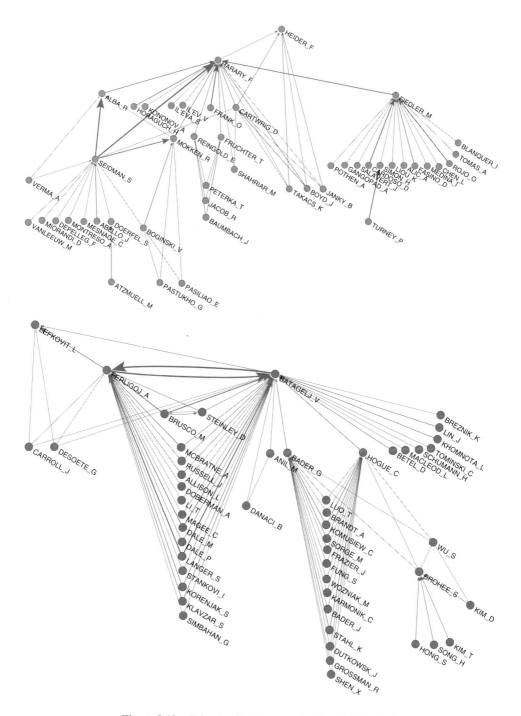

Figure 3.10   Subnetworks Harary and Batagelj + Ferligoj.

Most of the subnetworks of clusters for the Leader strategy have almost acyclic structure. This has to be considered also in their visualization. Because of space limitations, we present here only subnetworks induced by four among the selected clusters.

The central author in the first selected subnetwork, Figure 3.9 top, is Wasserman. He forms a strong component with Iacobucci and Weaver. The leader of this subnetwork is one of the founders of SNA (with sociometrics) Moreno [53]. Other important authors are Holland (with Leinhardt), the "father" of statistical approaches to SNA, Nowicki and Airoldi. The subnetwork features the statistical modeling of networks.

One of the most often used clustering methods is Ward's method [62]. Ward is the leader of the second subnetwork, Figure 3.9 bottom. It contains also other founders of clustering methods, Johnson and Rohlf, authors of fundamental books, Hartigan [38] and Gordon [34], and a theoretician, Hubert. The subnetwork is about cluster analysis.

The central author in the third subnetwork, Figure 3.10 top, is Harary, the author of the fundamental book on graph theory [37]. He is accompanied by other founders of graph-theoretic approaches to network analysis: Heider (signed networks), Alba (cliques), Cartwright (structure of directed networks), Seidman (cores), and Fiedler (eigen values/vectors).

Central to the fourth subnetwork, Figure 3.10 bottom, is a strong component with Batagelj and Ferligoj. They are citing the leader Lefkovitz. It contains also a strong component of authors Brusco and Steinley, working on efficient implementations of clustering algorithms, and several authors citing the paper [3] of Bader and Hogue describing the MCODE algorithm. The subnetwork is primarily about clustering with relational constraints.

3.8   Conclusion

In this chapter the "classical" approaches and results on clustering problem were presented and ways of adapting them for clustering of/in networks were shown. Most of the chapters in this monograph essentially propose different clustering criterion functions and some of them also describe new methods for obtaining the solutions. As already mentioned, most criterion functions are based on structural equivalence. One of the challenges for future research is to develop efficient algorithms for other types of equivalences for large networks.

Acknowledgements

This work is supported in part by the Slovenian Research Agency (research program P1-0294 and research projects J1-9187 and J7-8279) and by Russian Academic Excellence Project '5-100'.

References

1. E. Aarts and J. K. Lenstra, editors. *Local Search in Combinatorial Optimization*. John Wiley & Sons, Inc., New York, 1997.
2. M. R. Anderberg. *Cluster Analysis for Application*. Academic Press, New York, 1973.
3. G. D. Bader and C. W. V. Hogue. An automated method for finding molecular complexes in large protein interaction networks. *BMC Bioinformatics*, 4(2):1–27, 2003.
4. V. Batagelj. Note on ultrametric hierarchical clustering algorithms. *Psychometrika*, 46(3):351–352, 1981.

5. V. Batagelj. Generalized Ward and related clustering problems. In H. Bock, editor, *Classification and Related Methods of Data Analysis*, pages 67–74. North-Holland, Amsterdam, 1988.

6. V. Batagelj. Similarity measures between structured objects. In A. Graovac, editor, *MATH/CHEM/COMP 1988: Proceedings of an International Course and Conference on the Interfaces between Mathematics, Chemistry, and Computer Science, Dubrovnik, Yugoslavia, 20–25 June 1988*, Studies in Physical and Theoretical Chemistry, pages 25–40. Elsevier, 1989.

7. V. Batagelj. *clurc* – R package for clustering with relational constraint. 2017. URL https://github.com/bavla/cluRC.

8. V. Batagelj and M. Bren. Comparing resemblance measures. *Journal of Classification*, 12:73–90, 1995.

9. V. Batagelj and A. Ferligoj. Clustering relational data. In W. Gaul, O. Opitz, and M. Schader, editors, *Data Analysis*, Studies in Classification, Data Analysis, and Knowledge Organization, pages 3–15. Springer, Berlin, Heidelberg, 2000.

10. V. Batagelj, S. Korenjak-Černe, and S. Klavžar. Dynamic programming and convex clustering. *Algorithmica*, 11(2):93–103, 1994.

11. V. Batagelj, P. Doreian, A. Ferligoj, and N. Kejžar. *Understanding Large Temporal Networks and Spatial Networks: Exploration, Pattern Searching, Visualization and Network Evolution*. Wiley Series in Computational and Quantitative Social Science Series. Wiley, 2014.

12. V. Batagelj, N. Kejžar, and S. Korenjak-Černe. Clustering of modal valued symbolic data. *arXiv preprint arXiv:1507.06683*, 2015.

13. J. Benzécri and L. Bellier. *L'analyse des données: La Taxinomie*, volume 1 of *L'analyse des données*. Dunod, 1973.

14. L. Billard and E. Diday. *Symbolic Data Analysis: Conceptual Statistics and Data Mining*. Wiley Series in Computational Statistics. Wiley, 2012.

15. V. D. Blondel, J.-L. Guillaume, R. Lambiotte, and E. Lefebvre. Fast unfolding of communities in large networks. *Journal of Statistical Mechanics: Theory and Experiment*, 2008(10):P10008, 2008.

16. H.-H. Bock. A history of the international federation of classification societies. 2006. URL https://ifcs.boku.ac.at/site/lib/exe/fetch.php?media=pdfs:ifcs_history.pdf. This is a slightly modified, translated and updated version of Chapter 9 of H.-H. Bock, P. Ihm (eds.): *25 Jahre Gesellschaft für Klassifikation: Klassifikation und Datenanalyse im Wandel der Zeit*. Shaker Verlag, Aachen 2001, 184 pp.

17. J. Bodlaj and V. Batagelj. Hierarchical link clustering algorithm in networks. *Physical Review E*, 91(6):062814, 2015.

18. P. Brucker. On the complexity of clustering problems. In R. Henn, B. Korte, and W. Oettli, editors, *Optimization and Operations Research*, volume 157 of *Lecture Notes in Economics and Mathematical Systems*, pages 45–54. Springer, Berlin, Heidelberg, 1978.

19. M. Bruynooghe. Méthodes nouvelles en classification automatique des données taxinomiques nombreuses. *Statistique et Analyse des Données*, (3):24–42, 1977.

20. D. Cartwright and F. Harary. Structural balance: A generalization of Heider's theory. *Psychological Review*, 63:277–293, 1956.

21. T. H. Cormen, C. E. Leiserson, and R. L. Rivest. *Introduction To Algorithms*. MIT Press, Cambridge, 2nd edition, 2001.

22. M. M. Deza and E. Deza. *Encyclopedia of distances*. Springer, 2009.

23. E. Diday. *Optimisation en classification automatique, Tome 1, 2*. INRIA, Rocquencourt, 1979. (in French).

24. E. Diday and H. H. Bock. *Analysis of symbolic data: Exploratory methods for extracting statistical information from complex data*. Springer-Verlag, New York, 2000.

25. J. Dieudonné. *Foundations of modern analysis*. Academic Press, New York, 1960.

26. P. Doreian, V. Batagelj, and A. Ferligoj. *Generalized Blockmodeling*. Cambridge University Press, Cambridge, 2005.

27. A.-H. Esfahanian. On the evolution of connectivity algorithms. In L. W. Beineke and R. J. Wilson, editors, *Topics in structural graph theory*, volume 147 of *Encyclopedia of mathematics and its applications*. Cambridge University Press, New York, 2013.

28. A. Ferligoj and V. Batagelj. Clustering with relational constraint. *Psychometrika*, 47(4):413–426, 1982.

29. A. Ferligoj and V. Batagelj. Some types of clustering with relational constraints. *Psychometrika*, 48(4):541–552, 1983.

30. A. Ferligoj and V. Batagelj. Direct multicriteria clustering algorithms. *Journal of Classification*, 9:43–61, 1992.

31. A. Ferligoj and L. Kronegger. Clustering of attribute and/or relational data. *Metodološki vezki*, 6(2):135–153, 2009.

32. G. Gan, C. Ma, and J. Wu. *Data Clustering – Theory, Algorithms, and Applications*. SIAM, Philadelphia, 2007.

33. M. R. Garey and D. S. Johnson. *Computers and Intractability: A Guide to the Theory of NP-Completeness*. W. H. Freeman & Co., New York, 1979.

34. A. D. Gordon. *Classification*, 2nd Edition, volume 82 of *Monographs on Statistics and Applied Probability*. Chapman and Hall/CRC, Boca Raton, 1999.

35. J. Gower. A general coefficient of similarity and some of its properties. *Biometrics*, 27:857–874, 1971.

36. O. Grygorash, Y. Zhou, and Z. Jorgensen. Minimum spanning tree based clustering algorithms. In *18th IEEE International Conference on Tools with Artificial Intelligence,I CTAI'06*, pages 73–81. IEEE, 2006.

37. F. Harary. *Graph Theory*. Addison-Wesley, Reading, MA, 1969.

38. J. A. Hartigan. *Clustering algorithms*. Wiley-Interscience, New York, 1975.

39. C. Hayashi. Chikio Hayashi and Data Science – What is data science? *Student*, 2(1):44–51, 1997.

40. A. Jain and R. Dubes. *Algorithms for clustering data*. Prentice Hall, 1988.

41. N. Jardine, P. Jardine, and R. Sibson. *Mathematical Taxonomy*. Wiley Series in Probability and Mathematical Statistics. Wiley, 1971.

42. S. Joly and G. L. Calve. tude des puissances d'une distance. *Statistique et analyse des données*, 11(3):30–50, 1986.

43. S. D. Kamvar, D. Klein, and C. D. Manning. Interpreting and extending classical agglomerative clustering algorithms using a model-based approach. In *Proceedings of the Nineteenth International Conference on Machine Learning*, ICML '02, pages 283–290, San Francisco, CA, USA, 2002. Morgan Kaufmann Publishers Inc.

44. G. Karypis and V. Kumar. A fast and high quality multilevel scheme for partitioning irregular graphs. *SIAM Journal on Scientific Computing*, 20(1):359–392, 1998.

45. R. Kashyap and B. Oommen. A common basis for similarity measures involving two strings. *International Journal of Computer Mathematics*, 13(1):17–40, 1983.

46. L. Kaufman and P. J. Rousseeuw. *Finding Groups in Data: An Introduction to Cluster Analysis*. A Wiley-Interscience Publication. Wiley, 1990.

47. B. W. Kernighan and S. Lin. An efficient heuristic procedure for partitioning graphs. *The Bell System Technical Journal*, 49(2):291–307, 1970.

48. D. Knuth. *The Stanford GraphBase, A Platform for Combinatorial Computing*. ACM Press, New York, 1993.

49. S. Korenjak-Černe, N. Kejžar, and V. Batagelj. A weighted clustering of population pyramids for the world's countries, 1996, 2001, 2006. *Population Studies*, 69(1):105–120, 2015.

50. V. I. Levenshtein. Binary codes capable of correcting deletions, insertions, and reversals. *Doklady Akademii Nauk SSSR*, 163(4):845–848, 1965. English translation in Soviet Physics Doklady, 10(8):707–710, 1966.

51. F. H. C. Marriott. Optimization methods of cluster analysis. *Biometrika*, 69(2):417–421, 1982.

52. D. W. Matula. Graph theoretic techniques for cluster analysis algorithms. In J. Van Ryzin, editor, *Classification and clustering: Proceedings of an advanced seminar conducted by the Mathematics Research Center, the University of Wisconsin-Madison, May 3–5, 1976*, pages 95–130. Academic Press, 1977.

53. J. Moreno. *Who shall survive?: a new approach to the problem of Human Interrelations, Nervous and Mental Disease Monograph Series*, volume 58. Nervous and Mental Disease Publ., Washington, 1934.

54. F. Murtagh. *Multidimensional clustering algorithms*, volume 4. Physika Verlag, Vienna, 1985.

55. M. E. Newman and M. Girvan. Finding and evaluating community structure in networks. *Physical review E*, 69(2):026113, 2004.

56. W. D. Nooy, A. Mrvar, and V. Batagelj. *Exploratory Social Network Analysis with Pajek*, 3rd edition. Cambridge University Press, New York, 2018.

57. N. Pržulj. Biological network comparison using graphlet degree distribution. *Bioinformatics*, 23(2):e177–e183, 2007.

58. F. S. Roberts. *Discrete mathematical models, with applications to social, biological, and environmental problems*. Prentice-Hall, Englewood Cliffs, NJ, 1976.

59. R. R. Sokal and P. H. A. Sneath. *Principles of Numerical Taxonomy*. Books in Biology. W. H. Freeman, 1963.

60. H. Späth. *Cluster-Analyse-Algorithmen: zur Objektklassifizierung und Datenreduktion*. Datenverarbeitung: Oldenbourg. Oldenbourg R. Verlag GmbH, 1977.

61. R. Todeschini and V. Consonni. *Molecular Descriptors for Chemoinformatics*. John Wiley & Sons, New York, 2nd edition, 2009.

62. J. H. Ward. Hierarchical grouping to optimize an objective function. *Journal of the American Statistical Association*, 58(301):236–244, 1963.

63. Wikipedia. Mahalanobis distance. 2018. URL https://en.wikipedia.org/wiki/Mahalanobis_distance.

4

Different Approaches to Community Detection

Martin Rosvall[1], Jean-Charles Delvenne[2], Michael T. Schaub[3,4], and Renaud Lambiotte[4]

[1]Umeå University
[2]Université Catholique de Louvain
[3]Massachusetts Institute of Technology
[4]University of Oxford

This chapter is an extended version of The many facets of community detection in complex networks, Appl. Netw. Sci. 2: 4 (2017) by the same authors.

4.1 Introduction

A precise definition of what constitutes a community in networks has remained elusive. Consequently, network scientists have compared community detection algorithms on benchmark networks with a particular form of community structure and classified them based on the mathematical techniques they employ. However, this comparison can be misleading because apparent similarities in their mathematical machinery can disguise different reasons for why we would want to employ community detection in the first place. Here we provide a focused review of these different motivations that underpin community detection. This problem-driven classification is useful in applied network science, where it is important to select an appropriate algorithm for the given purpose. Moreover, highlighting the different approaches to community detection also delineates the many lines of research and points out open directions and avenues for future research.

While research related to community detection dates back to the 1970s in mathematical sociology and circuit design [21, 46], Newman's and Girvan's work on modularity in complex

Advances in Network Clustering and Blockmodeling, First Edition.
Edited by Patrick Doreian, Vladimir Batagelj, and Anuška Ferligoj.
© 2020 John Wiley & Sons Ltd. Published 2020 by John Wiley & Sons Ltd.

systems just over ten years ago revitalized the field of community detection, making it one of the main pillars of network science research [53, 54]. The promise of community detection, that we can gain a deeper understanding of a system by discerning important structural patterns within a network, has spurred a huge number of studies in network science. However, it has become abundantly clear by now that this problem has no canonical solution. In fact, even a general definition of what constitutes a community is still lacking. The reasons for this are not only grounded in the computational difficulties of tackling community detection. Rather, various research areas view community detection from different perspectives, illustrated by the lack of a consistent terminology: "network clustering", "graph partitioning", "community", "block" or "module detection" all carry slightly different connotations. This jargon barrier creates confusion, as readers and authors have different preconceptions and intuitive notions are not made explicit.

We argue that community detection should not be considered as a well-defined problem, but rather as an umbrella term with many facets. These facets emerge from different goals and motivations for what it is about the network that we want to understand or achieve, and lead to different perspectives on how to formulate the problem of community detection. It is critically important to be aware of these underlying motivations when selecting and comparing community detection methods. Thus, rather than an in-depth discussion of the technical details of different algorithmic implementations [16, 26, 28, 47, 52, 58, 72, 82], here we focus on the conceptual differences between different perspectives on community detection.

By providing a problem-driven classification, however, we do *not* argue that the different perspectives are unrelated. In fact, in some situations, different mathematical problem formulations can lead to similar algorithms and methods, and the different perspectives can offer valuable insights. For example, for undirected networks, optimizing the objective function modularity [54], initially proposed from a clustering perspective, can be interpreted as optimizing both a particular stochastic block model [50] and an auto-correlation measure of a particular diffusion process on the networks [20]. In other situations, however, such relationships disappear.

While some perspectives arguably are more principled than others, we do not assert that there is a particular perspective that is *a priori* better suited for any given network. In fact, as in data clustering [31], no one method can consistently perform the best on all kinds of networks [59]. Community detection is an unsupervised learning task that is blind to a researcher's intent with the analysis. Accordingly, to understand a particular method's usefulness, we must take the researcher's interest in the communities into context [80].

In the following, we unfold different aims underpinning community detection – in a relaxed form that includes assortative as well as disassortative group structures with dense and sparse internal connections, respectively – and discuss how the resulting problem perspectives relate to various applications. We focus on four broad perspectives that have served as motivation for community detection in the literature: (i) the cut-based perspective minimizes a constraint such as the number of links between groups of nodes, (ii) the clustering perspective maximizes internal density in groups of nodes, (iii) the stochastic block model perspective identifies groups of nodes in which nodes are stochastically equivalent, and (iv) the dynamical perspective identifies groups of nodes in which flows stay for a relatively long time such that they form building blocks of dynamics on networks (see Figure 4.1). While this categorization is not unique, we believe that it can help clarify concepts about community detection and serve as a guide to determining the appropriate method for a particular purpose.

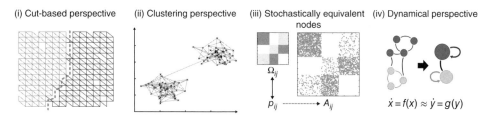

(i) Cut-based perspective   (ii) Clustering perspective   (iii) Stochastically equivalent nodes   (iv) Dynamical perspective

Figure 4.1 Schematic of four different approaches to community detection. (i) The cut-based perspective aims at minimizing the number of links between groups of nodes, independently of their intrinsic structure. (ii) The clustering perspective produces groups of densely connected nodes. (iii) The stochastic equivalence perspective looks for groups in which nodes are stochastically equivalent, typically inferred through a generative statistical network model. (iv) The dynamical perspective focuses on the impact of communities on dynamical processes and searches for dynamically relevant coarse-grained descriptions.

4.2 Minimizing Constraint Violations: the Cut-based Perspective

An early network partitioning application was circuit layout and design [3, 26]. This application spurred development of the now classical Kernighan–Lin algorithm [39] and the work by Donath and Hoffmann [21, 22], who were among the first to suggest the use of eigenvectors for network partitioning. For example, we might be confronted with a network that describes the signal flows between different components of a circuit. To design the circuit in an efficient way, our goal is now to partition the network into a fixed number of approximately equally sized groups for balanced load with a small number of edges between those groups for minimal communication overhead. The edges that run between the groups are commonly denoted as the cut. To design the most efficient circuit, our aim is thus to minimize this cut with more or less balanced groups.

To make this more precise, let us consider one specific variant of this scheme, known as *ratio cut* [32]. Let us denote the adjacency matrix of an undirected network $\mathcal{N}$ with n nodes by A, where $A_{uv} = 1$ if there is a connection from node u to node v, and $A_{uv} = 0$ if there is no connection. We can now write the problem of optimizing the ratio cut for a bipartition of all nodes $\mathcal{V}$ into two communities $\mathcal{V}_1$ and $\mathcal{V}_2 = \mathcal{V} \setminus \mathcal{V}_1$ as follows [32, 79]:

$$\min_{\mathcal{V}_1} \text{RatioCut}(\mathcal{V}_1, \mathcal{V}_2) := \min_{\mathcal{V}_u} \sum_u \frac{\text{cut}(\mathcal{V}_u, \mathcal{V} \setminus \mathcal{V}_u)}{|\mathcal{V}_u|}, \tag{4.1}$$

where $\text{cut}(\mathcal{V}_1, \mathcal{V}_2) := \sum_{u \in \mathcal{V}_1, v \in \mathcal{V}_2} (A_{uv} + A_{vu})/2$ is the sum of the possibly weighted edges between the two vertex sets $\mathcal{V}_1, \mathcal{V}_2$. Related problem formulations also occur in the context of parallel computations and load scheduling [64, 77], where approximately equally sized portions of work are to be sent to different processors, while keeping the dependencies between those tasks minimal. Further applications include scientific computing [64, 77], where partitioning algorithms divide the coordinate meshes when discretizing and solving partial differential equations. Image segmentation problems may also be phrased in terms of cut-based measures [76, 79].

Investigating these types of problems has led to many important contributions to partitioning networks, in particular in relation to spectral methods. The connection between spectral algorithms and cut-based problem formulations arises naturally by considering relaxations of the original, combinatorially hard discrete optimization problems, such as Equation (4.2), or other related objective functions such as the average or normalized cuts. This can be best seen when rewriting the above optimization problem as follows:

$$\min_f \quad f^T L f \tag{4.2}$$

$$\text{subject to} \quad f \perp \mathbf{1} \quad \| f \| = \sqrt{n} \tag{4.3}$$

$$\text{where} \quad f_u := \begin{cases} -\sqrt{|\mathcal{V}_2|/|\mathcal{V}_1|} & \text{if } u \in \mathcal{V}_1 \\ \sqrt{|\mathcal{V}_1|/|\mathcal{V}_2|} & \text{if } u \in \mathcal{V}_2 \end{cases} \tag{4.4}$$

Here the Laplacian matrix of the network has been defined as $L = D - A$, where D is the diagonal degree matrix with $D_{uu} = \sum_v A_{uv}$. Fiedler realized already in the 1970s that the second smallest eigenvalue of the Laplacian matrix is associated with the connectivity of the network, and that the associated eigenvector thus can be used to compute spectral bi-partitions [24, 25]. Such spectral ideas led to many influential algorithms and methods; see, for example, von Luxburg [79] for a tutorial on spectral algorithms.

In this cut-based problem formulation, there is no specification as to how the identified groups in the partition should be connected internally. While the implicit constraint is that the groups must not split into groups with an even smaller cut, there is no specification that the groups of nodes should be densely connected internally. Indeed, the type of networks considered in the context of cut-based partitions are often of a mesh- or grid-like form, for which several guarantees can be given in terms of the quality of the partitions obtained by spectral algorithms [77]. While such non-dense groupings emerging from the analysis of non-clique structures [73] can also be dynamically relevant (see section 4.5), they are likely missed when employing a community notion that focuses on finding dense groupings, as discussed next.

4.3   Maximizing Internal Density: the Clustering Perspective

A different motivation for community detection arises in the context of data clustering. We use the term clustering, which can have many definitions, in the following sense: for a set of given data points in a possibly high-dimensional space, the goal is to partition the points into a number of groups such that points within a group are close to or similar to each other in some sense, and points in different groups are more distant from each other. To achieve this goal, one often constructs a proximity or similarity network between the points and tries to group together nodes that are closer to each other than they are to the rest of the network. This approach results in a form of community detection problem where the closeness between nodes is described by the presence and weight of the edges between them.

Although minimizing the cut size and maximizing the internal number of links are closely related, there are differences pertaining to the typical constraints and search space associated with these objective functions. First, when employing a clustering perspective, there is normally no *a priori* information about the number of groups we are looking for. Second, we do not necessarily require the groups to be balanced in any way; rather we would like to find an optimal split into densely knit groups irrespective of their relative sizes.

Unsurprisingly, finding an optimal clustering is a computationally difficult problem. Further, as Kleinberg has shown [40], there are no clustering algorithms that satisfy a certain set of intuitive properties we might require from a clustering algorithm in continuous spaces. Similar problems also arise in the discrete setting for clustering of networks [13].

Nevertheless, there exists a large number of methods that follow a clustering-like paradigm and separate the nodes of a network into cohesive groups of nodes, often by optimizing a quality function. An important clustering metric in this context is the so-called conductance [4, 37, 41, 78]. Optimizing the global conductance was introduced as a way to produce a global bi-partition similarly to the two-way ratio-cut. However, this quantity has been successfully employed more recently as a local quality function to find localized clusters around one or more seed nodes. The local conductance of a set of nodes $\mathcal{V}_q \subset \mathcal{V}$ can be written

$$\phi(\mathcal{V}_q) := \frac{\sum_{u \in \mathcal{V}_q, v \notin \mathcal{V}_q} A_{uv}}{\min\{\mathrm{vol}(\mathcal{V}_q), \mathrm{vol}(\mathcal{V} - \mathcal{V}_q)\}}, \tag{4.5}$$

where $\mathrm{vol}(\mathcal{V}_q) := \sum_{u \in \mathcal{V}_q} \sum_v A_{uv}$ is the total degree of the nodes in set $\mathcal{V}_q$, commonly called its volume in analogy with geometric objects. Interestingly, it has been shown that, in specific contexts, the conductance can be a good predictor of some latent group structures in real-world applications [84].

Moreover, a local perspective on community detection has two appealing properties. First, the definition of a cluster does not depend on the global network structure but only on the relative local density. Second, only a portion of a network needs to be accessed, which is advantageous if there are computational constraints in using large networks, or we are only interested in a particular subsystem. In such cases, we would like to avoid having to apply a method to the whole network in order to find, for example, the cluster containing a particular node in the network.

The Newman–Girvan modularity [53, 54] is arguably one of the most common clustering measures used in the literature and was originally proposed from the clustering perspective discussed here. It is a global quality function and aims to find the community structure of the network as a whole. Given a partition $\mathbf{C} = \{\mathcal{V}_1, \dots, \mathcal{V}_k\}$ of a network into k groups, the modularity of $\mathbf{C}$ can be written as:

$$Q(\mathbf{C}) := \frac{1}{2m} \sum_{q=1}^{k} \sum_{u,v \in \mathcal{V}_q} \left[A_{uv} - \frac{d_u d_v}{2m} \right], \tag{4.6}$$

where $d_u = \sum_v A_{uv}$ is the degree of node u and $2m = \sum_u d_u$ is the total weight of all edges in the network. By optimizing the modularity measure over the space of all partitions, one aims to identify groups of nodes that are more densely connected to each other than one would expect from a statistical null model of the network. This statistical null model is commonly chosen to be the configuration model with preserved degree sequence.

However, a by-product of this choice of a global null-model is the tendency of modularity to balance the size of the groups in terms of their total connectivity. While different variants of modularity aim to account for this effect [26], it means modularity can be interpreted as a trade-off between a cut-based measure and an entropy [20]. Modularity is typically optimized with spectral or greedy algorithms [11, 26, 51]. While there are problems with modularity, such as its resolution limit [27] and other spurious effects [27, 29, 30, 44], the general idea has triggered researchers to develop a plethora of algorithms that follow a similar strategy [26]. Several works have addressed some of the shortcomings, by incorporating a resolution parameter, for example, or by explicitly accounting for the density inside each group [14, 15]. In practice, however, less seems to beat more and the original formulation of modularity remains the most widely used.

4.4   Identifying Structural Equivalence: the Stochastic Block Model Perspective

By grouping *similar* nodes that link to similar nodes within communities, we constrain ourselves to finding *assortative* group structure [28]. While we may also have hierarchical clusters with clusters of clusters, etc., such an assortative structural organization is too restrictive if we want to define groups based on more general connectivity patterns that include disassortative communities with weaker interactions within rather than between communities.

In social network analysis, a common goal is to identify nodes within a network that serve a similar structural role in terms of their connectivity profile. Accordingly, nodes are similar if they share the same kind of connection patterns to other nodes [46]. This idea is captured in concepts such as *regular equivalence*, which states that nodes are regularly equivalent if they are equally related to equivalent others [23, 33]. The first algorithms for identifying groups of "approximately equivalent" nodes were deterministic and permuted adjacency matrices to reveal block structures in so-called block models [6, 81].

A relaxation of regular equivalence is *stochastic equivalence* [34], where nodes are equivalent if they connect to equivalent nodes with equal probability. The stochastic formulation generalizes observations and forms generative models, which can be used for prediction. Because of this advantage over non-stochastic formulations, we focus on stochastic equivalence.

One of the most popular techniques to model and detect stochastically equivalent relationships in network data is to use stochastic block models (SBMs) [34, 56] and associated inference techniques. These models have their roots in the social networks literature [5, 34], and provide a flexible framework for modeling block structures within a network. When considering block models, we are interested in identifying node groups such that nodes within a community connect to nodes in other communities in an "equivalent way" [28].

Consider a network composed of n nodes divided into k classes. The standard SBM is defined by a set of node class labels and the affinity matrix Ω. More precisely, the link probability between two nodes u, v belonging to class c_u and c_v is given by:

$$p_{uv} := \mathbb{P}(A_{uv}) = \Omega_{c_u c_v}.$$

Under an SBM, nodes within the same class share the same probability of connecting to nodes of another class. This is the mathematical formulation of having stochastically equivalent nodes within each class. Finding the latent groups of nodes in a network now amounts to inferring the model parameters that provide the best fit for the observed network. That is, find the SBM with the highest likelihood of generating the data.

The standard SBM assumes that the expected degree of each node is a Poisson binomial random variable, a binomial random variable with possibly non-identical success probabilities in each trial. Because inferring the most likely SBM typically results in grouping nodes based on their degree in empirical networks with broad degree distributions, it can be advantageous to include a degree-correction into the model. In the degree corrected SBM [38], the probability p_{uv} that a link will appear between two nodes u, v depends both on their class labels c_u, c_v and their respective degree parameters d_i, d_j (each entry A_{ij} might be a Bernoulli or a Poisson random variable such as in [38]):

$$p_{uv} \sim d_u d_v \Omega_{c_u c_v}.$$

Thus, while edges in real-world networks tend to be correlated with effects such as triadic closure [26], by construction edges are conditionally independent random variables in SBMs.

Moreover, most common SBMs are defined for unweighted networks or networks with integer weights by modeling the network as a multi-graph. Though generalizations are available [2, 61], this is still a less studied area.

In contrast to the notions of community considered above, with stochastic equivalence we are no longer interested in maximizing some internal density or minimizing a cut. To see this, consider a bipartite network that from a cut- or density-based perspective contains no communities. From the stochastic equivalence perspective, however, we would say that this network contains two groups because nodes in each set only connect to nodes in the other set. When adopting an SBM to detect such structural organization of the links, we explicitly adopt a statistical model for the networks. The network is essentially an instance of an ensemble of possible networks generated from such a model.[1]

This model-based approach comes with several advantages. First, by defining the model, we effectively declare what is signal and what is noise in the data under the SBM. We can thus provide a statistical assessment of the observed data with, for example, p-values under the SBM. In other words, we can identify patterns that cannot be reasonably explained from density fluctuations of edges inherent to any realization of the model. Second, we are able, for example, to generate new networks from our model with a similar group structure, or predict missing edges and impute data. Third, we can make strong statements about the detectability of groups within a network. For example, precise criteria specify when any algorithm can recover the planted group structure for a network created by an SBM [18, 49]. By fitting an SBM to an observed adjacency matrix, it is possible to recover such a planted group structure down to its theoretical limit [48, 49]. These criteria apply to networks generated with SBMs and not real networks in general, in which case we do not know what kind of process created the network [59]. It is nevertheless a remarkable result since it highlights the fact that there are networks with undetectable block patterns.

Moreover, this model-based approach also offers ways to estimate the number of communities from the data by some form of model selection, including hypothesis testing [10], spectral techniques [42, 70], the minimum description length principle [60], or Bayesian inference [83] (see Chapter 11).

Finally, the generative nature of SBMs also makes them well suited for constructing benchmark networks. As a consequence, many benchmark networks proposed in the literature, such as the commonly used Lancichinetti–Fortunato-Radicchi (LFR) benchmarks [45], are specific types of SBMs. Results on these benchmark networks should therefore be taken for what they are: the ability to recover the underlying group structure of specific types of SBM-generated networks. For example, sparse networks without any underlying group structure still can contain meaningful dynamical building blocks.

4.5 Identifying Coarse-grained Descriptions: the Dynamical Perspective

Let us now consider a fourth alternative motivation for community detection, focusing on the processes that take place on the network. All notions of community outlined above are effectively structural in the sense that they are mainly concerned with the composition of

---

[1] This ensemble assumption is also reflected in the modularity formalism, where the observed network is compared to a null model.

the network itself or its representation as an adjacency matrix. However, in many cases one of the main reasons to apply tools from network science is to understand the *behavior* of a system. While the topology of a system puts constraints on the dynamics that can take place on the network, the network topology alone cannot explain the system behavior. For example, instead of finding a coarse-grained description of the adjacency matrix, we might be interested in finding a coarse-grained description of the dynamics acting on top of the network with multi-step paths beyond the nearest neighbors.

Take air traffic as an example. An airline network, with weighted links connecting cities according to the number of flights between them, can offer some interesting insights about air traffic. For instance, in the US air traffic network based on the number of flying passengers, Las Vegas and Atlanta form two major hubs. However, if we focus instead on the passenger flows based on actual multi-leg itineraries, the two cities show very different behaviors: Las Vegas is a tourist destination and typically the final destination of itineraries, whereas Atlanta is often a transfer hub to other final destinations [62, 69]. Thus, these airports play dynamically quite different roles in the network. Focusing on interconnection patterns alone can give an incomplete picture if we are interested in the dynamical behavior of a system, for which additional dynamical information should be taken into account. Conversely, a concentration of edges with high impact on the dynamics may arise just from a statistical fluctuation, if the network is seen as a realization of a particular random network model. In this way, structural and dynamical approaches can offer complementing information.

In general, however, they are blocks of nodes with different identities that trap the flow or channel it in specific directions. That is, they form reduced models of the dynamics where blocks of nodes are aggregated to single meta nodes with similar *dynamical function* with respect to the rest of the network. In this view, the goal of community detection is to find effective coarse-grained system descriptions of how the dynamics take place on the network structure.

To induce multi-step paths and couple also non-neighboring nodes, the dynamical approach to community detection has primarily focused on modeling the dynamics with Markovian diffusion processes [19, 43, 65], though the work of topological scales and synchronization share the same common ground [7]. Interestingly, for simple diffusion dynamics such as a random walk on an undirected network, which is essentially determined by the spectral properties of the network's Laplacian matrix, this perspective is tightly connected to the clustering perspective discussed in section 4.3. This is because the presence of densely knit groups within the network can introduce a time-scale separation in the diffusion dynamics: a random walker traversing the network will initially be trapped for a significant time inside a community corresponding to the fast time-scale, before it can escape and explore the larger network corresponding to a slower time-scale. However, this connection between link density and dynamical behavior breaks down for directed networks, even for a simple diffusion process [43, 65, 70]. This apparent relationship breaks down completely when focusing on longer pathways, possibly with memory effects in the dynamics [69, 71].

A dynamical perspective is useful especially in applications in which the network itself is well defined, but the emergent dynamics are hard to grasp. For instance, consider the nervous system of the roundworm *C. elegans*, for which there exists a distinct network. A basic generative network model, such as a Barabasi–Albert network or an SBM, might be too simple to capture the complex architecture of the network, and sampling alternative networks from such a model will not create valid alternative roundworm connectomes. Indeed, some more complicated network generative models have been proposed to model the structure of the network [55], and may be used to assess the significance of individual patterns compared to the background of the assumed model. However, if we are interested instead in assessing the dynamical implications of the evolutionary conserved network structure, it may be fruitful to engineer differences in the actual network and investigate how they affect the dynamical flows in the system. For instance, one can replicate experimental node ablations *in silico* and assess their dynamical impact [8].

In the dynamical perspective, we are typically interested in how short-term dynamics integrate into long-term behavior of the system and seek a coarse-grained description of the dynamics occurring on a given network. That is, the network itself represents the true structure, save for empirical imperfections. Therefore, in the dynamical perspective, model selection is in general not about comparing competing models [60, 83] but about comparing coarse-grained descriptions of the dynamics on resampled realizations of the observed network with, for example, the bootstrap [66] or cross-validation [63]. Nevertheless, it is possible to formulate generative statistical models for empirically observed pathways [62]. However, whereas the generative approach in, for example, [62] explicitly models the underlying state space of trajectories, we may simply be interested in effectively compressing the long-term behavior of the system [63].

Two methods that exploit the long-term dynamics of the system by identifying communities with long flow persistence are the Markov stability [20] and the map equation [65]. Whereas the Markov stability detailed in Chapter 12 takes a statistical approach and favors communities inside which a random walker is more likely to remain after a given time t than expected at infinite time, the map equation reveals modular regularities by compressing the dynamics. It is an information-theoretic approach that uses the duality between compressing data and finding regularities in the data [65, 75]. It measures the quality of communities by how much they can compress a modular description of the dynamics. The shorter description, the more detected regularities, such that the shortest description captures the most regularities. Given module assignments $\mathbf{C}$ of all nodes in the network, the map equation measures the description length $L(\mathbf{C})$ of a random walker that moves within and between modules from node to node by following the links between the nodes [68]:

$$L(\mathbf{C}) = q_\curvearrowright H(\mathcal{Q}) + \sum_{q=1}^{k} p_\circlearrowright^q H(\mathcal{P}^q) \tag{4.7}$$

Here the entropy $H(\mathcal{Q})$ measures the average per-step description length of movements between modules derived from module-enter rates $\mathcal{Q}$ of all k modules and $H(\mathcal{P}^q)$ measures the average per-step description length of movements within module q derived from node-visit and module-exit rates $\mathcal{P}^q$. The description lengths are weighted by their rate of use, $q_\curvearrowright$ and $p_\circlearrowright^q$, respectively. The visit rates can be obtained by first calculating the PageRank of links and nodes or directly from the data if they represent flow themselves. In any case, finding the optimal partition of the network by assigning each node to one or more modules corresponds to testing different node assignments and picking the one that minimizes the map equation. This simple formulation allows for straightforward generalizations to coarse-grained hierarchical [67] descriptions of dynamics in memory [69] and multilayer [17] networks.

As the air traffic example above illustrates, it can be crucial to go beyond standard network abstractions and consider memory and higher-order effects in multi-step pathways to better understand system behavior. For example, higher-order abstractions, such as memory and multilayer networks, provide principled means to reveal highly overlapping modular organization in complex systems: link clustering [1] and clique percolation [57] methods can be interpreted as trying to account for second-order Markov dynamics (see Supplementary Note 3 in [69]).

Compared to some of the other perspectives, the dynamical viewpoint has received somewhat less attention and has been confined mainly to diffusion dynamics (see Chapter 12). A key challenge is to extend this perspective to other types of dynamics and link it more formally to approaches of model order reduction considered in control theory. In light of the recently growing interest in the control of complex systems, this could help us better understanding complex systems.

4.6   Discussion

Community detection can be viewed through a range of different lenses. Rather than looking at community detection as a generic tool that is supposed to work in a generic context, considering the application in mind is important when choosing between or comparing different methods. Each of the perspectives outlined above has its own particularities, which may or may not be suitable for the problem of interest.

We emphasize the different perspectives in the following example. Given a real-world network generated by a possibly complex random assignment of edges, we assume that we are interested in some particular dynamics taking place on this network, such as epidemic spreading. We also assume that the network is structured such that the dynamics exhibit a time-scale separation. If, for instance, we want to coarse-grain an epidemic and identify critical links that should be controlled to confine the epidemic, then it does not matter whether or not random fluctuations generated the modules that induce the time-scale separation. In any case, these modules will be relevant for the dynamics.

Assume now that the same network encodes interdependency of tasks in a load-scheduling problem. In such a circumstance, a cut-based approach will find a relevant community structure, in that it will allow an optimally balanced assignment of tasks to processors that minimizes communication between processors. These communities may be different from the ones attached to the epidemic-spreading example.

If we instead assume that the links represent friendships, we may want to identify densely knit groups irrespective of their relative sizes. Accordingly, taking the clustering perspective and maximizing the internal density can give yet another set of communities.

In these three cases, we considered a single realization of the network with the goal of extracting useful information about its structure, independently of the possible mechanisms that generated it.

Let us finally consider the same network from a stochastic equivalence perspective, and assume for simplicity that the network is a particular realization of an Erdős–Rényi network. In this case, an approach based on the SBM is expected to declare that there is no significant pattern to be found here at all, as the encountered structural variations can already be explained by random fluctuations rather than by hidden class labels. Thus, communities in the SBM picture are defined via the latent variables within the statistical model of the network structure, and not via their impact on the behavior of the system. In this way, different motivations for community detection can find different answers even for the very same network.

To illustrate that different motivations can give different answers for the same network, we use an example from [65]. The directed, weighted network is formed as a ring of rings such that each internal ring captures flows for a relatively long time despite the stronger links between the rings (see Figure 4.2). For example, a random walker takes on average three steps within a ring highlighted as a cluster in Figure 4.2a before exiting. In contrast, a random walker takes on average only 2.4 steps within a cluster in Figure 4.2b. A method that seeks to coarse-grain the dynamics will therefore identify the flow modules in Figure 4.2a rather than the clusters with high internal density in Figure 4.2b. For example, the modular description quantified by the map equation is almost twice as efficient with the flow modules as it is with the clusters with high internal density. The opposite is true for a method that highlights structural regularity and high internal density: the modularity score is twice as large for the clustering in Figure 4.2b. While this example only illustrates the fundamental difference between two methods applied to a schematic network, methods from different perspectives will give different answers for real networks as well [36].

In addition to the differences *between* these perspectives, there are also variations *within* each perspective. For instance, distinct plausible generative models such as the standard SBM or the degree-corrected SBM will, for a given network, lead to different inferred community structure. Similar variations exist in the dynamical paradigm as well: distinct natural assumptions for the dynamics, such as dynamics with or without memory, uniform across nodes or edges, etc., applied to a given network will lead to different partitions. Also different balancing criteria (see section 4.2) or different concepts of high internal density (see section 4.3) will be valid in different contexts.

In fact, some of the internal variations make the perspectives overlap in particular scenarios. For instance, one can compare all the algorithms on simple, undirected LFR benchmark networks [45]. However, the LFR benchmark clearly imposes a density-based notion of communities. Similarly, for simple undirected networks, optimizing modularity corresponds to the inference of a particular SBM [50] or may be reinterpreted as a diffusion process on a network [20]. Nevertheless, this overlap of concepts, typically present in unweighted, undirected networks, is only partial, and breaks down, for example, in directed networks or for more complex dynamics.

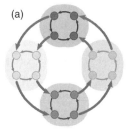 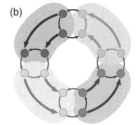

(a)

(b)

Modularity	$Q = 0.25$	Modularity	$Q = 0.50$
Map equation	$L = 2.67$ bits/step	Map equation	$L = 4.13$ bits/step

Figure 4.2 Communities that highlight different aspects of networks. Identifying coarse-graining flows in groups, here illustrated by the map equation, and densely connected groups, here illustrated by modularity, highlights different aspects of structure in directed and weighted networks. Each shaded area represents a cluster in two alternative clusterings of a schematic network. (a) The clustering as optimized by the map equation (minimum L). (b) The clustering as optimized by modularity (maximum Q). The thicker links have double the weight of the thinner links. Example from [65].

4.7   Conclusions

In summary, no general purpose algorithm will ever serve all applications or data types [59] because each perspective emphasizes a particular core aspect: a cut-based method provides good separation of balanced groups, a clustering method provides strong cohesiveness of groups with high internal density, stochastic block models provide strong similarity of nodes inside a group in terms of their connectivity profiles, and methods that view communities as dynamical building blocks aim to provide node groups that influence or are influenced by some dynamics in the same way. As more and more diverse types of data are collected, leading to ever more complex network structures, including directed [47], temporal [35, 74], multi-layer or multiplex networks [12], the differences between the perspectives presented here will become even more striking – the same network might have multiple valid partitions depending on the question about the network we are interested in. We might moreover not only be interested in partitioning the nodes, but also in partitioning edges [1], or even motifs [9]. Rather than striving to find a "best" community-detection algorithm for a better understanding of complex networks, we argue for a more careful treatment of what network aspects we seek to understand when applying community detection.

Acknowledgements

We thank Aaron Clauset, Leto Peel, and Daniel Larremore for fruitful discussions. MR was supported by the Swedish Research Council grant 2016-00796. MTS, JCD, and RL acknowledge support from FRS-FNRS, the Belgian Network DYSCO (Dynamical Systems, Control and Optimisation) funded by the Interuniversity Attraction Poles Programme initiated by the Belgian State Science Policy Office, and the ARC (Action de Recherche Concerte) on Mining and Optimization of Big Data Models funded by the Wallonia-Brussels Federation. MTS received funding from the European Union's Horizon 2020 research and innovation programme under the Marie Sklodowska-Curie grant agreement No 702410.

References

1. Y.-Y. Ahn, J. P. Bagrow, and S. Lehmann. Link communities reveal multiscale complexity in networks. *Nature*, 466(7307):761–764, June 2010. doi: 10.1038/nature09182.
2. C. Aicher, A. Z. Jacobs, and A. Clauset. Learning latent block structure in weighted networks. *Journal of Complex Networks*, page cnu026, 2014.
3. C. J. Alpert and A. B. Kahng. Recent directions in netlist partitioning: a survey. *Integration, the VLSI journal*, 19(1):1–81, 1995.
4. R. Andersen, F. Chung, and K. Lang. Local graph partitioning using pagerank vectors. In *2006 47th Annual IEEE Symposium on Foundations of Computer Science (FOCS'06)*, pages 475–486. IEEE, 2006.
5. C. J. Anderson, S. Wasserman, and K. Faust. Building stochastic blockmodels. *Social Networks*, 14(1):137–161, 1992.
6. P. Arabie, S. A. Boorman, and P. R. Levitt. Constructing blockmodels: How and why. *Journal of Mathematical Psychology*, 17(1):21–63, 1978.
7. A. Arenas, A. Díaz-Guilera, and C. J. Pérez-Vicente. Synchronization reveals topological scales in complex networks. *Physical Review Letters*, 96(11):114102, Mar. 2006.
8. K. A. Bacik, M. T. Schaub, M. Beguerisse-Díaz, Y. N. Billeh, and M. Barahona. Flow-based network analysis of the Caenorhabditis elegans connectome. *PLoS Comput Biol*, 12(8):1–27, 08 2016.

9. A. R. Benson, D. F. Gleich, and J. Leskovec. Higher-order organization of complex networks. *Science*, 353(6295):163–166, 2016. ISSN 0036-8075.

10. P. J. Bickel and P. Sarkar. Hypothesis testing for automated community detection in networks. *Journal of the Royal Statistical Society: Series B (Statistical Methodology)*, 78(1):253–273, 2016.

11. V. D. Blondel, J.-L. Guillaume, R. Lambiotte, and E. Lefebvre. Fast unfolding of communities in large networks. *Journal of Statistical Mechanics: Theory and Experiment*, 2008(10):P10008, 2008.

12. S. Boccaletti, G. Bianconi, R. Criado, C. I. Del Genio, J. Gómez-Gardeñes, M. Romance, I. Sendiña-Nadal, Z. Wang, and M. Zanin. The structure and dynamics of multilayer networks. *Physics Reports*, 544(1):1–122, 2014.

13. A. Browet, J. Hendrickx, and A. Sarlette. Incompatibility boundaries for properties of community partitions. *IEEE Transactions on Network Science and Engineering*, 83–99, 2017.

14. M. Chen, K. Kuzmin, and B. K. Szymanski. Community detection via maximization of modularity and its variants. *IEEE Transactions on Computational Social Systems*, 1(1):46–65, 2014.

15. M. Chen, T. Nguyen, and B. K. Szymanski. A new metric for quality of network community structure. *arXiv:1507.04308*, 2015.

16. M. Coscia, F. Giannotti, and D. Pedreschi. A classification for community discovery methods in complex networks. *Statistical Analysis and Data Mining*, 4(5):512–546, 2011.

17. M. De Domenico, A. Lancichinetti, A. Arenas, and M. Rosvall. Identifying modular flows on multilayer networks reveals highly overlapping organization in interconnected systems. *Physical Review X*, 5(1):011027, 2015.

18. A. Decelle, F. Krzakala, C. Moore, and L. Zdeborová. Inference and phase transitions in the detection of modules in sparse networks. *Physical Review Letters*, 107:065701, Aug. 2011.

19. J.-C. Delvenne, S. N. Yaliraki, and M. Barahona. Stability of graph communities across time scales. *Proceedings of the National Academy of Sciences*, 107(29):12755–12760, 2010.

20. J.-C. Delvenne, M. T. Schaub, S. N. Yaliraki, and M. Barahona. The stability of a graph partition: A dynamics-based framework for community detection. In *Dynamics On and Of Complex Networks, Volume 2*, pages 221–242. Springer, 2013.

21. W. E. Donath and A. J. Hoffman. Algorithms for partitioning of graphs and computer logic based on eigenvectors of connection matrices. *IBM Technical Disclosure Bulletin*, 15(3):938–944, 1972.

22. W. E. Donath and A. J. Hoffman. Lower bounds for the partitioning of graphs. *IBM Journal of Research and Development*, 17(5):420–425, 1973.

23. M. G. Everett and S. P. Borgatti. Regular equivalence: General theory. *Journal of Mathematical Sociology*, 19(1):29–52, 1994.

24. M. Fiedler. Algebraic connectivity of graphs. *Czechoslovak Mathematical Journal*, 23(2):298–305, 1973.

25. M. Fiedler. A property of eigenvectors of nonnegative symmetric matrices and its application to graph theory. *Czechoslovak Mathematical Journal*, 25(4):619–633, 1975.

26. S. Fortunato. Community detection in graphs. *Physics Reports*, 486(3): 75–174, 2010.

27. S. Fortunato and M. Barthélemy. Resolution limit in community detection. *Proceedings of the National Academy of Sciences*, 104(1):36–41, 2007.

28. S. Fortunato and D. Hric. Community detection in networks: A user guide. *Physics Reports*, 659:1–44, 2016.

29. B. H. Good, Y.-A. de Montjoye, and A. Clauset. Performance of modularity maximization in practical contexts. *Physical Review E*, 81(4):046106, Apr. 2010.

30. R. Guimera, M. Sales-Pardo, and L. A. N. Amaral. Modularity from fluctuations in random graphs and complex networks. *Physical Review E*, 70(2):025101, 2004.

31. I. Guyon, U. Von Luxburg, and R. C. Williamson. Clustering: Science or art. In *NIPS 2009 Workshop on Clustering Theory*, pages 1–11, 2009.

32. L. Hagen and A. B. Kahng. New spectral methods for ratio cut partitioning and clustering. *IEEE Transactions on Computer-aided Design of Integrated Circuits and Systems*, 11(9):1074–1085, 1992.

33. R. A. Hanneman and M. Riddle. *Introduction to social network methods*. University of California Riverside, 2005.

34. P. W. Holland, K. B. Laskey, and S. Leinhardt. Stochastic blockmodels: First steps. *Social Networks*, 5(2):109–137, 1983.

35. P. Holme and J. Saramäki. Temporal networks. *Physics Reports*, 519(3): 97–125, 2012.

36. D. Hric, R. K. Darst, and S. Fortunato. Community detection in networks: Structural communities versus ground truth. *Physical Review E*, 90(6):062805, 2014.

37. R. Kannan, S. Vempala, and A. Vetta. On clusterings: Good, bad and spectral. *Journal of the ACM*, 51(3):497–515, 2004.

38. B. Karrer and M. E. Newman. Stochastic blockmodels and community structure in networks. *Physical Review E*, 83(1):016107, 2011.
39. B. W. Kernighan and S. Lin. An efficient heuristic procedure for partitioning graphs. *Bell System Technical Journal*, 49(2):291–307, 1970.
40. J. Kleinberg. An impossibility theorem for clustering. *NIPS'02 Proceedings of the 15th International Conference on Neural Information Processing Systems*, 463–470, 2003.
41. K. Kloster and D. F. Gleich. Heat kernel based community detection. In *Proceedings of the 20th ACM SIGKDD International Conference on Knowledge Discovery and Data Mining*, pages 1386–1395. ACM, 2014.
42. F. Krzakala, C. Moore, E. Mossel, J. Neeman, A. Sly, L. Zdeborová, and P. Zhang. Spectral redemption in clustering sparse networks. *Proceedings of the National Academy of Sciences*, 110(52):20935–20940, 2013.
43. R. Lambiotte, J.-C. Delvenne, and M. Barahona. Random walks, Markov processes and the multiscale modular organization of complex networks. *IEEE Transactions on Network Science and Engineering*, 1(2):76–90, 2014.
44. A. Lancichinetti and S. Fortunato. Limits of modularity maximization in community detection. *Physical Review E*, 84:066122, Dec. 2011.
45. A. Lancichinetti, S. Fortunato, and F. Radicchi. Benchmark graphs for testing community detection algorithms. *Physical Review E*, 78(4):046110, Oct. 2008.
46. F. Lorrain and H. C. White. Structural equivalence of individuals in social networks. *Journal of Mathematical Sociology*, 1(1):49–80, 1971.
47. F. D. Malliaros and M. Vazirgiannis. Clustering and community detection in directed networks: A survey. *Physics Reports*, 533(4):95–142, 2013.
48. L. Massoulié. Community detection thresholds and the weak Ramanujan property. In *Proceedings of the 46th Annual ACM Symposium on Theory of Computing*, pages 694–703. ACM, 2014.
49. E. Mossel, J. Neeman, and A. Sly. A proof of the block model threshold conjecture. *Combinatorica*, 38(3):665–708.
50. M. Newman. Equivalence between modularity optimization and maximum likelihood methods for community detection. *Physical Review E*, 94(5): 052315, 2016.
51. M. E. Newman. Finding community structure in networks using the eigenvectors of matrices. *Physical Review E*, 74(3):036104, 2006.
52. M. E. Newman. Communities, modules and large-scale structure in networks. *Nature Physics*, 8(1):25–31, 2012.
53. M. E. J. Newman. Modularity and community structure in networks. *Proceedings of the National Academy of Sciences*, 103(23):8577–8582, 2006.
54. M. E. J. Newman and M. Girvan. Finding and evaluating community structure in networks. *Physical Review E*, 69(2):026113, Feb 2004.
55. V. Nicosia, P. E. Vértes, W. R. Schafer, V. Latora, and E. T. Bullmore. Phase transition in the economically modeled growth of a cellular nervous system. *Proceedings of the National Academy of Sciences*, 110 (19):7880–7885, 2013.
56. K. Nowicki and T. A. B. Snijders. Estimation and prediction for stochastic blockstructures. *Journal of the American Statistical Association*, 96(455):1077–1087, 2001.
57. G. Palla, I. Derényi, I. Farkas, and T. Vicsek. Uncovering the overlapping community structure of complex networks in nature and society. *Nature*, 435(7043):814–818, 2005.
58. S. Parthasarathy, Y. Ruan, and V. Satuluri. Community discovery in social networks: Applications, methods and emerging trends. In *Social network data analytics*, pages 79–113. Springer, 2011.
59. L. Peel, D. B. Larremore, and A. Clauset. The ground truth about metadata and community detection in networks. *Science Advances*, 3(5): e1602548, 2017.
60. T. P. Peixoto. Parsimonious module inference in large networks. *Physical Review Letters*, 110(14):148701, 2013.
61. T. P. Peixoto. Inferring the mesoscale structure of layered, edge-valued, and time-varying networks. *Physical Review E*, 92(4):042807, 2015.
62. T. P. Peixoto and M. Rosvall. Modelling sequences and temporal networks with dynamic community structures. *Nature Communications*, 8(1):582, 2017.
63. C. Persson, L. Bohlin, D. Edler, and M. Rosvall. Maps of sparse Markov chains efficiently reveal community structure in network flows with memory. *arXiv:1606.08328*, 2016.
64. A. Pothen. Graph partitioning algorithms with applications to scientific computing. In *Parallel Numerical Algorithms*, pages 323–368. Springer, 1997.
65. M. Rosvall and C. T. Bergstrom. Maps of random walks on complex networks reveal community structure. *Proceedings of the National Academy of Sciences*, 105(4):1118–1123, 2008.
66. M. Rosvall and C. T. Bergstrom. Mapping change in large networks. *PloS One*, 5(1):e8694, 2010.

67. M. Rosvall and C. T. Bergstrom. Multilevel compression of random walks on networks reveals hierarchical organization in large integrated systems. *PloS One*, 6(4):e18209, 2011.

68. M. Rosvall, D. Axelsson, and C. Bergstrom. The map equation. *The European Physical Journal Special Topics*, 178(1):13–23, 2009.

69. M. Rosvall, A. V. Esquivel, A. Lancichinetti, J. D. West, and R. Lambiotte. Memory in network flows and its effects on spreading dynamics and community detection. *Nature Communications*, 5, 4630 2014.

70. A. Saade, F. Krzakala, and L. Zdeborová. Spectral clustering of graphs with the bethe hessian. In *Advances in Neural Information Processing Systems 27 (Proceedings of Neural Information Processing Systems)*, Z. Ghahramani, M. Welling, C. Cortes, N.D. Lawrence and K.Q. Weinberger (eds), pages 406–414, 2014.

71. V. Salnikov, M. T. Schaub, and R. Lambiotte. Using higher-order Markov models to reveal flow-based communities in networks. *Scientific Reports*, 6:23194, 3 2016.

72. S. E. Schaeffer. Graph clustering. *Computer Science Review*, 1(1):27–64, 2007.

73. M. T. Schaub, J.-C. Delvenne, S. N. Yaliraki, and M. Barahona. Markov dynamics as a zooming lens for multiscale community detection: non clique-like communities and the field-of-view limit. *PloS One*, 7(2):e32210, 2012.

74. V. Sekara, A. Stopczynski, and S. Lehmann. Fundamental structures of dynamic social networks. *Proceedings of the National Academy of Sciences*, 113(36):9977–9982, 2016.

75. C. Shannon. A mathematical theory of communication. *Bell System Technical Journal*, 27:379–423, 1948.

76. J. Shi and J. Malik. Normalized cuts and image segmentation. *IEEE Transactions on Pattern Analysis and Machine Intelligence*, 22(8):888–905, 2000.

77. D. A. Spielman and S.-H. Teng. Spectral partitioning works: Planar graphs and finite element meshes. In *Proceedings of the 37th Annual Symposium on Foundations of Computer Science*, pages 96–105. IEEE, 1996.

78. D. A. Spielman and S.-H. Teng. A local clustering algorithm for massive graphs and its application to nearly linear time graph partitioning. *SIAM Journal on Computing*, 42(1):1–26, 2013.

79. U. Von Luxburg. A tutorial on spectral clustering. *Statistics and Computing*, 17(4):395–416, 2007.

80. U. Von Luxburg, R. C. Williamson, and I. Guyon. Clustering: Science or art? In *JMLR Workshop and Conference Proceedings: ICML Unsupervised and Transfer Learning*, volume 27, pages 65–80, 2012.

81. H. C. White, S. A. Boorman, and R. L. Breiger. Social structure from multiple networks. I. blockmodels of roles and positions. *American Journal of Sociology*, 81(4):730–780, 1976.

82. J. Xie, S. Kelley, and B. K. Szymanski. Overlapping community detection in networks: The state-of-the-art and comparative study. *ACM Computing Surveys*, 45(4):43, 2013.

83. X. Yan. Bayesian model selection of stochastic block models. In *IEEE/ACM International Conference on Advances in Social Networks Analysis and Mining (ASONAM)*, pages 323–328. IEEE, 2016.

84. J. Yang and J. Leskovec. Defining and evaluating network communities based on ground-truth. *Knowledge and Information Systems*, 42(1): 181–213, 2015.

5

Label Propagation for Clustering

Lovro Šubelj

University of Ljubljana, Faculty of Computer and Information Science, Ljubljana, Slovenia

Label propagation is a heuristic method initially proposed for community detection in networks [26, 50], while the method can be adopted also for other types of network clustering and partitioning [5, 28, 39, 62]. Among all the approaches and techniques described in this book, label propagation is neither the most accurate nor the most robust method. It is, however, without doubt one of the simplest and fastest clustering methods. Label propagation can be implemented with a few lines of programming code and applied to networks with hundreds of millions of nodes and edges on a standard computer, which is true only for a handful of other methods in the literature.

In this chapter, we present the basic framework of label propagation, review different advances and extensions of the original method, and highlight its equivalences with other approaches. We show how label propagation can be used effectively for large-scale community detection, graph partitioning, and identification of structurally equivalent nodes and other network structures. We conclude the chapter with a summary of label propagation methods and suggestions for future research.

5.1 Label Propagation Method

The label propagation method was introduced by Raghavan *et al.* [50] for detecting non-overlapping communities in large networks. There exist multiple interpretations of network communities [23, 54], as described in Chapter 4. For instance, a community can be seen as a densely connected group, or cluster, of nodes that is only loosely connected to the rest of the network, which is also the perspective that we adopt here.

For the sake of simplicity, we describe the basic label propagation framework for the case of detecting communities in simple undirected networks. Consider a network with n nodes and let

Advances in Network Clustering and Blockmodeling, First Edition.
Edited by Patrick Doreian, Vladimir Batagelj, and Anuška Ferligoj.
© 2020 John Wiley & Sons Ltd. Published 2020 by John Wiley & Sons Ltd.

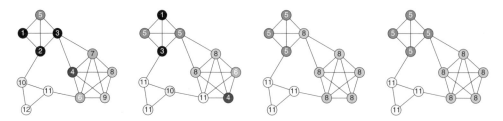

Figure 5.1 Label propagation in a small network with three communities. The labels and shades of the nodes represent community assignments at different iterations of the label propagation method.

Neighbors$_i$ denote the set of neighbors of node $i \in \{1, \dots, n\}$. Furthermore, let g_i be the group assignment or community label of node i which we would like to infer. The label propagation method then proceeds as follows. Initially, the nodes are put into separate groups by assigning a unique label to each node as $g_i = i$. Then, the labels are propagated between the nodes until an equilibrium is reached. At every iteration of label propagation, each node i adopts the label shared by most of its neighbors, *Neighbors$_i$*. Hence,

$$g_i = \underset{g}{\mathrm{argmax}}\, |\{j \in \textit{Neighbors}_i | g_j = g\}|. \tag{5.1}$$

Due to having numerous edges within communities, relative to the number of edges towards the rest of the network, the nodes of a community form a consensus on some label after only a couple of iterations of label propagation. More precisely, in the first few iterations, the labels form small groups in dense regions of the network, which then expand until they reach the borders of communities. Thus, when the propagation converges, meaning that Equation (5.1) holds for all of the nodes and the labels no longer change, connected groups of nodes sharing the same label are classified as communities. Figure 5.1 demonstrates the label propagation method on a small network, where it correctly identifies the three communities in just three iterations. In fact, due to the extremely fast structural inference of label propagation, the estimated number of iterations in a network with a billion edges is about 100 [59].

Label propagation is not limited to simple networks having, at most, one edge between each pair of nodes. Let A be the adjacency matrix of a network, where A_{ij} is the number of edges between nodes i and j, and A_{ii} is the number of self-edges or loops on node i. The label propagation rule in Equation (5.1) can be written as

$$g_i = \underset{g}{\mathrm{argmax}} \sum_j A_{ij}\delta(g_j, g), \tag{5.2}$$

where δ is the Kronecker delta operator that equals one when its arguments are the same and zero otherwise. Furthermore, in weighted or valued networks, the label propagation rule becomes

$$g_i = \underset{g}{\mathrm{argmax}} \sum_j W_{ij}\delta(g_j, g), \tag{5.3}$$

where W_{ij} is the sum of weights on the edges between nodes i and j, and W_{ii} is the sum of weights on the loops on node i. Label propagation can also be adopted for multipartite and other types of networks, which is presented in Section 5.4. However, there seems to be no obvious extension of label propagation to networks with directed arcs, since propagating the labels exclusively in the direction of arcs enables the exchange of labels only between mutually reachable nodes.

Figure 5.2   Resolution of ties between the maximal labels of the central nodes of the networks. The labels and shades of the nodes represent their current community assignments.

5.1.1   Resolution of Label Ties

At each step of label propagation, a node adopts the label shared by most of its neighbors denoted by the maximal label. There can be multiple maximal labels as shown on the left-hand side of Figure 5.2. In that case, the node chooses one maximal label uniformly at random [50]. Note, however, that the propagation might never converge, especially when there are many nodes with multiple maximal labels in their neighborhoods. This is because their labels could constantly change and label convergence would never be reached. The problem is particularly apparent in networks of collaborations between the authors of scientific papers, where a single author often collaborates with others in different research communities.

The simplest solution is always to select the smallest or the largest maximal label according to some predefined ordering [18], which has obvious drawbacks. Leung *et al.* [35] proposed a seemingly elegant solution to include also the concerned node's label itself into the maximal label consideration in Equation (5.2). This is equivalent to adding a loop on each node in a network. Nevertheless, the label inclusion strategy might actually create ties when there is only one maximal label in a node's neighborhood, which happens in the case of the central node of the network in the middle of Figure 5.2.

Most label propagation algorithms implement the label retention strategy introduced by Barber and Clark [5]. When there are multiple maximal labels in a node's neighborhood, and one of these labels is the current label of the node, the node retains its label. Otherwise, a random maximal label is selected to be the new node label. The main difference to the label inclusion strategy is that the current label of a node is considered only when there actually exist multiple maximal labels in its neighborhood. For example, the network on the right-hand side of Figure 5.2 is at equilibrium under the label retention strategy.

Random resolution of label ties represents the first of two sources of randomness in the label propagation method, hindering its robustness and consequently also the stability of the identified communities. The second is the random order of label propagation.

5.1.2   Order of Label Propagation

The discussion above assumed that, at every iteration of label propagation, all nodes update their labels simultaneously. This is called synchronous propagation [50]. The authors of the original method noticed that synchronous propagation can lead to oscillations of some labels in certain networks. Consider a bipartite or two-mode network with two types of nodes and edges only between the nodes of different type. Assume that, at some iteration of label propagation, the nodes of each type share the same label as in the example on the left-hand side of Figure 5.3. Then, at the next iteration, the labels of the nodes would merely switch and start to oscillate

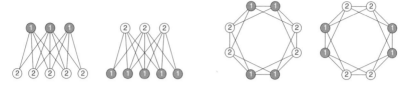

Figure 5.3 Label oscillations in bipartite and non-bipartite networks. The labels and shades of the nodes represent community assignments at two consecutive iterations of the label propagation method.

between two equivalent label configurations. For instance, such behavior occurs in networks with star-like communities consisting of one or few central hub nodes that are connected to many peripheral nodes, while the peripheral nodes themselves are not directly connected. Note that label oscillations are not limited to bipartite or nearly bipartite networks [18], as seen in the example on the right-hand side of Figure 5.3.

For this reason, most label propagation algorithms implement asynchronous propagation [50]. At every iteration of label propagation, the labels of the nodes are no longer updated all together, but sequentially in some random order, which is different for each iteration. This is in contrast to synchronous propagation, which always considers the labels from the previous iteration. Due to the random order of label updates, asynchronous propagation successfully breaks the cyclic oscillations of labels in Figure 5.3.

It must be stressed that asynchronous propagation with random tie resolution makes the label propagation method very unstable. In the case of the famous Zachary karate club network [76], the method identifies more than 500 different community structures [65], although the network consists of only 34 nodes. Asynchronous propagation applied to large online social networks and web graphs can wrongly also produce a giant community occupying the majority of the nodes in a network [35].

5.1.3   Label Equilibrium Criterium

Raghavan *et al.* [50] defined the convergence of label propagation as the state of label equilibrium when Equation (5.1) is satisfied for every node in a network. Let k_i denote the number of neighbors of node i and let k_i^g be the number of neighbors that share label g. The label propagation rule in Equation (5.1) can be rewritten as

$$g_i = \operatorname*{argmax}_g k_i^g. \tag{5.4}$$

The label equilibrium criterium thus requires that, for every node i, the following must hold

$$\forall g \ : \ k_i^{g_i} \geq k_i^g. \tag{5.5}$$

In other words, all nodes must be labeled with the maximal labels in their neighborhoods.

This criterion is similar, but not equivalent, to the definition of a strong community [49]. Strong communities require that every node has strictly more neighbors in its own community than in all other communities together, whereas at the label equilibrium every node has at least as many neighbors in its own community than in any other community.

An alternative approach is to define the convergence of label propagation as the state when the labels no longer change [5]. Equation (5.5) obviously holds for every node in a network and the label equilibrium is reached. Note, however, that this criterion must necessarily be combined with an appropriate label tie resolution strategy in order to ensure convergence when there are multiple maximal labels in the neighborhoods of nodes.

5.1.4   Algorithm and Complexity

As mentioned in the introduction, the label propagation method can be implemented with a few lines of programming code. Algorithm 5.1 shows the pseudocode of the basic asynchronous propagation framework defining the convergence of label propagation as the state of no label change and implements the retention strategy for label tie resolution.

Algorithm 5.1

```
label propagation {
   for each node i ∈ {1,…,n} {
      initialize node label gᵢ with i;
   }
   until node labels change repeat {
      for each node i ∈ {1,…,n} in random order {
         compute labels {g} that maximize kᵢᵍ = ∑ⱼAᵢⱼδ(gⱼ,g);
         if gᵢ ∉ {g} update gᵢ with random label from {g};
      }
   }
   report connected components induced by node labels;
}
```

When the state of label equilibrium is reached, groups of nodes sharing the same label are classified as communities. These can, in general, be disconnected, which happens when a node propagates its label to two or more disconnected nodes, but is itself relabeled in the later iterations of label propagation. Since connectedness is a fundamental property of network communities [23], groups of nodes with the same label are split into connected groups of nodes at the end of label propagation. Reported communities are thus connected components of the subnetworks induced by different node labels.

The label propagation method exhibits near-linear time complexity in the number of edges of a network denoted with m [35, 50]. At every iteration of label propagation, the label of node i can be updated with a sweep through its neighborhood which has complexity $\mathcal{O}(k_i)$, where k_i is the degree of node i. Since $\sum_i k_i = 2m$, the complexity of an entire iteration of label propagation is $\mathcal{O}(m)$. A random order or permutation of nodes before each iteration of asynchronous propagation can be computed in $\mathcal{O}(n)$ time, while the division into connected groups of nodes at the end of label propagation can be implemented with a simple network traversal, which has complexity $\mathcal{O}(n + m)$.

The overall time complexity of label propagation is therefore $\mathcal{O}(cn + cm)$, where c is the number of iterations before convergence. In the case of networks with a clear community structure, label propagation commonly converges in no more than ten iterations. Still, the number of iterations increases with the size of a network, as can be seen in Figure 5.4.

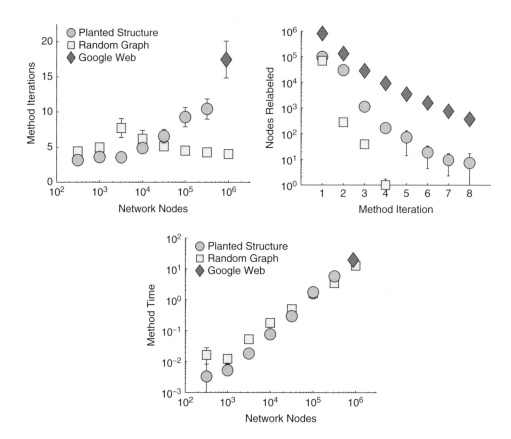

Figure 5.4   The number of iterations of label propagation, the number of relabeled nodes at the first eight iterations, and the running time in seconds. The markers are averages over 25 runs of the label propagation method, while the error bars show standard deviation.

Šubelj and Bajec [59] estimated the number of iterations of asynchronous label propagation from a large number of empirical networks, obtaining $c \approx 1.03m^{0.23}$. The time complexity of label propagation is thus approximately $\mathcal{O}(m^{1.2})$, which makes the method applicable to networks with up to hundreds of millions of nodes and edges on a standard desktop computer as long as the network fits into its main memory.

The left-hand side of Figure 5.4 shows the number of iterations of the label propagation framework in Algorithm 5.1 in artificial networks with planted community structure [33], Erdős–Rényi random graphs [22], and a part of the Google web graph [34] available at KONECT.[1] The web graph consists of $875,713$ nodes and $5,105,039$ edges, while the sizes of random graphs and artificial networks can be seen in Figure 5.4. In random graphs having no structure, label propagation correctly classifies all nodes into a single group in about five iterations, regardless of the size of a graph. Yet, the number of iterations increases with the size

---

[1] http://konect.uni-koblenz.de

in artificial networks with community structure, while the estimated number of iterations in a network with a billion edges is 113 [59].

Most nodes acquire their final label after the first few iterations of label propagation. The middle of Figure 5.4 shows the number of nodes that update their label at a particular iteration for the Google web graph, artificial networks having a planted community structure and random graphs with 10^5 nodes. The number of relabeled nodes drops exponentially with the number of iterations (logarithmic scales are used). For example, the percentages of relabeled nodes of the web graph after the first five iterations are 90.7%, 14.9%, 3.2%, 1.1%, and 0.4%, respectively. Furthermore, the algorithm running time is only 19.5 seconds, as shown on the right-hand side of Figure 5.4.

5.2   Label Propagation as Optimization

Here, we discuss the objective function of the label propagation method to shed light on label propagation as an optimization method.

At every iteration of label propagation each node adopts the most common label in its neighborhood. Therefore, label propagation can be seen as a local optimization method seeking to maximize the number of neighbors with the same label or, equivalently, minimize the number of neighbors with different labels. From the perspective of node i, the label propagation rule in Equation (5.2) assigns its group label g_i to maximize $\sum_j A_{ij}\delta(g_i, g_j)$, where A is the adjacency matrix of the network. Hence, the objective function maximized by the basic label propagation method is

$$\mathcal{F}(\{g\}) = \sum_{ij} A_{ij}\delta(g_i, g_j), \tag{5.6}$$

where $\{g\}$ is the group labeling of network nodes [5, 65]. Notice that $\mathcal{F}$ is non-negative and has the optimum of $2m$, where m is the number of edges in a network.

Equation (5.6) has a trivial optimal solution of labeling all nodes in a network with the same label, corresponding to putting all nodes into one group. Equation (5.2) then holds for every node and $\mathcal{F} = 2m$. However, starting with each node in its own group by assigning them unique labels when $\mathcal{F} = 0$, the label propagation process usually is trapped in a local optimum. For networks having a clear community structure, this corresponds to nodes of each community being labeled with the same label when $\mathcal{F} = 2m - 2m'$, where m' is the number of edges between communities. For example, the value of $\mathcal{F}$ for the community structure revealed on the right-hand side of Figure 5.1 is $46 - 8 = 38$.

Network community structure is only a local optimum of the label propagation process, whereas the global optimal solution corresponds to a trivial, undesirable, labeling. Thus, directly optimizing the objective function of label propagation with some other optimization method trying to escape a local optimum might not yield a favorable outcome. Furthermore, a network can have also many local optima that imply considerably different community structures. As already mentioned in Section 5.1.2, label propagation identifies more than 500 different structures in the Zachary karate club network [76] with 34 nodes and more than 10^5 in the *Saccharomyces cerevisiae* protein interaction network [31] with 2111 nodes [65]. Raghavan *et al.* [50] suggested aggregating labelings from multiple runs of label propagation. However, this can fragment a network into very small communities [65]. A more suitable method for combining different labelings of label propagation is consensus clustering [24, 32, 78], but this comes with increased time complexity.

The above perspective on label propagation as an optimization method results from the following equivalence. Tibély and Kertész [65] have shown that the label propagation in Equation (5.2) is equivalent to a ferromagnetic Potts model [48, 70]. The q-state Potts model is a generalization of the Ising model as a system of interacting spins on a lattice, with each spin pointing to one of q equally spaced directions. Consider the so-called standard q-state Potts model on a network placing a spin on each node [51]. Let σ_i denote the spin on node i which can be in one of q possible states, where q is set equal to the number of nodes in a network n. The zero-temperature kinetics of the model are defined as follows. One starts with each spin in its own state as $\sigma_i = i$ and then iteratively aligns the spins to the states of their neighbors as in the label propagation process. The ground state is ferromagnetic with all spins in the same state, while the dynamics can also get trapped at a metastable state with more than one spin state. The Hamiltonian of the model can be written as

$$\mathcal{H}(\{\sigma\}) = - \sum_{ij} A_{ij} \delta(\sigma_i, \sigma_j), \tag{5.7}$$

where $\{\sigma\}$ are the states of spins on network nodes. By setting $\sigma_i = g_i$, minimizing the described Potts model Hamiltonian $\mathcal{H}$ in Equation (5.7) is equivalent to maximizing the objective function of the label propagation method $\mathcal{F}$ in Equation (5.6).

As is almost any other clustering method, the label propagation method is nondeterministic and can produce different outcomes on different runs. Therefore, throughout the chapter, we report the results obtained over multiple runs of the method.

5.3   Advances of Label Propagation

Section 5.1 presented the basic label propagation method and discussed details of its implementation. Section 5.2 clarified the objective function of label propagation. In this section, we review different advances of the original method, addressing some of the weaknesses identified in the previous sections. Section 5.3.1 shows how to redefine the method's objective function by imposing constraints to use label propagation as a general optimization framework. Section 5.3.2 demonstrates different heuristic approaches changing the method's objective function implicitly by adjusting the propagation strength of individual nodes. This promotes the propagation of labels from certain desirable nodes or, equivalently, suppresses the propagation from the remaining nodes. Finally, Section 5.3.3 discusses different empirically motivated techniques to improve the overall performance of the method.

Unless explicitly stated otherwise, the above advances are presented for the case of non-overlapping community detection in simple undirected networks. Nevertheless, Section 5.4 presents extensions of label propagation to other types of networks such as multipartite, multilayer, and signed networks. Furthermore, in Section 5.5, we show how label propagation can be adopted to detect alternative types of groups such as overlapping or hierarchical communities and groups of nodes that are similarly connected to the rest of the network by structurally equivalent nodes as in Chapter 6. Note that different approaches and techniques described in Sections 5.3–5.5 can be combined. The advances of the basic label propagation method described in this section can be used directly with the extensions to other types of groups and networks described in the next sections.

5.3.1 Label Propagation Under Constraints

As shown in Section 5.2, the objective function of label propagation has a trivial optimal solution of assigning all nodes to a single group. A standard approach for eliminating such undesirable solutions is to add constraints to the objective function of the method. Let $\mathcal{H}$ be the objective function of label propagation expressed in the form of the ferromagnetic Potts model Hamiltonian as in Equation (5.7). The modified objective function minimized by label propagation under constraints is $\mathcal{H} + \lambda \mathcal{G}$, where $\mathcal{G}$ represents a penalty term with imposed constraints with λ being a regularization parameter weighing the penalty term $\mathcal{G}$ against the original objective function $\mathcal{H}$.

Barber and Clark [5] proposed a penalty term $\mathcal{G}_1$ borrowed from the graph partitioning literature requiring that nodes are divided into smaller groups of the same size:

$$\mathcal{G}_1(\{g\}) = \sum_g n_g^2, \tag{5.8}$$

where $n_g = \sum_i \delta(g_i, g)$ is the number of nodes in group g, g_i is the group label of node i, and $n = \sum_g n_g$ is the number of nodes in a network. The penalty term $\mathcal{G}_1$ has the minimum of n when all nodes are in their own groups and the maximum of n^2 when all nodes are in a single group, which effectively guards against the undesirable trivial solution. The modified objective function $\mathcal{H}_1 = \mathcal{H} + \lambda_1 \mathcal{G}_1$ can be written as

$$\mathcal{H}_1(\{g\}) = -\sum_{ij} (A_{ij} - \lambda_1) \delta(g_i, g_j), \tag{5.9}$$

where A is the adjacency matrix of a network. Equation (5.9) is known as the constant Potts model [67] and is equivalent to a specific version of the stochastic block model [77], while the regularization parameter λ_1 can be interpreted as the threshold between the density of edges within and between different groups. The label propagation rule in Equations (5.2) and (5.4) for the modified objective function $\mathcal{H}_1$ is

$$g_i = \underset{g}{\operatorname{argmax}} \sum_j (A_{ij} - \lambda_1) \delta(g_j, g)$$

$$= \underset{g}{\operatorname{argmax}} \, k_i^g - \lambda_1 n_g, \tag{5.10}$$

where $k_i^g = \sum_j A_{ij} \delta(g_j, g)$ is the number of neighbors of node i in group g. Equation (5.10) can be efficiently implemented with Algorithm 5.1 by updating n_g.

An alternative penalty term $\mathcal{G}_2$, which has been popular in the community detection literature, requires nodes being divided into groups having the same total degree [5]:

$$\mathcal{G}_2(\{g\}) = \sum_g k_g^2, \tag{5.11}$$

where $k_g = \sum_i k_i \delta(g_i, g)$ is the sum of degrees of nodes in group g and k_i is the degree of node i. The penalty term $\mathcal{G}_2$ is again minimized when all nodes are in their own groups and maximized when all nodes are in a single group, avoiding the trivial solution. The modified objective function $\mathcal{H}_2 = \mathcal{H} + \lambda_2 \mathcal{G}_2$ can be written as

$$\mathcal{H}_2(\{g\}) = - \sum_{ij} (A_{ij} - \lambda_2 k_i k_j) \delta(g_i, g_j), \tag{5.12}$$

while the corresponding label propagation rule is

$$g_i = \operatorname*{argmax}_g \sum_j (A_{ij} - \lambda_2 k_i k_j) \delta(g_j, g)$$

$$= \operatorname*{argmax}_g k_i^g - \lambda_2 k_i k_g + \lambda_2 k_i^2 \delta(g_i, g). \tag{5.13}$$

Equation (5.13) can be efficiently implemented with Algorithm 5.1 by updating k_g.

Equation (5.12) is a special case of the Potts model investigated by Reichardt and Bornholdt [51] and is a generalization of a popular quality function in community detection named modularity [45]. The modularity $\mathcal{Q}$ measures the number of edges within network communities against the expected number of edges in a random graph with the same degree sequence [46]. Formally,

$$\mathcal{Q}(\{g\}) = \frac{1}{2m} \sum_{ij} \left(A_{ij} - \frac{k_i k_j}{2m} \right) \delta(g_i, g_j). \tag{5.14}$$

Notice that setting $\lambda_2 = 1/2m$ in Equation (5.12) yields $\mathcal{H}_2 = -2m\mathcal{Q}$ [5].

Label propagation under the constraints of Equation (5.13) can be employed for maximizing the modularity $\mathcal{Q}$. Note, however, that the method might easily get trapped at a local optimum, not corresponding to very high $\mathcal{Q}$. For example, the average $\mathcal{Q}$ over 25 runs for the Google web graph from Figure 5.4 is 0.763. In contrast, the unconstrained label propagation gives a value of 0.801. For this reason, label propagation under constraints is usually combined with a multistep greedy agglomerative algorithm [55], one driving the method away from a local optimum. Using such an optimization framework, Liu and Murata [38] revealed community structures with the highest values of $\mathcal{Q}$ ever reported for some commonly analyzed empirical networks. Han et al. [28] recently adapted the same framework also for another popular quality function called map equation [53].

The third variant of label propagation under constraints [13] is based on the absolute Potts model [52] with the modified objective function $\mathcal{H}_3 = \mathcal{H} + \lambda_3 \mathcal{G}_3$ written as

$$\mathcal{H}_3(\{g\}) = - \sum_{ij} (A_{ij}(\lambda_3 + 1) - \lambda_3) \delta(g_i, g_j). \tag{5.15}$$

By setting $\lambda_1 = \lambda_3/(\lambda_3 + 1)$ in Equation (5.9), one derives $\mathcal{H}_1 = \mathcal{H}_3/(\lambda_3 + 1)$, implying the method is in fact equivalent to the constant Potts model [67].

5.3.2 Label Propagation with Preferences

Leung et al. [35] have shown that adjusting the propagation strength of individual nodes can improve the performance of the label propagation method in certain networks. Let p_i be the propagation strength associated with node i called the node preference. Incorporating the node preferences p_i into the basic label propagation rule in Equation (5.2) gives

$$g_i = \operatorname*{argmax}_{g} \sum_{j} p_j A_{ij} \delta(g_j, g), \qquad (5.16)$$

while the method objective function in Equation (5.7) becomes

$$\mathcal{H}_p(\{g\}) = -\sum_{ij} p_i p_j A_{ij} \delta(g_i, g_j). \qquad (5.17)$$

In contrast to Section 5.3.1, these node preferences impose constraints on the objective function only implicitly by either promoting or suppressing the propagation of labels from certain desirable nodes, as shown in the examples below.

An obvious choice is to set the node preferences equal to the degrees of the nodes as $p_i = k_i$ [35]. For instance, this improves the performance of community detection in networks with high degree nodes in the center of each community. Šubelj and Bajec [57, 59] proposed estimating the most central nodes of each community or group during the label propagation process using a random walk diffusion. Consider a random walker utilized on a network limited to the nodes of group g_i and let p_i be the probability that the walker visits node i. The probabilities p_i are high for the most central nodes of group g_i and low for the nodes on the border. It holds that

$$p_i = \sum_{j} \frac{p_j}{k_j^{g_j}} A_{ij} \delta(g_i, g_j), \qquad (5.18)$$

where $k_i^{g_i} = \sum_j A_{ij} \delta(g_i, g_j)$ is the number of neighbors of node i in its group g_i. Clearly $p_i = k_i^{g_i}$ is the solution of Equation (5.18), but initializing the probabilities as $p_i = 1$ and updating their values according to Equation (5.18) only when the nodes change their groups g_i gives a different result. This mimics the actual propagation of labels occurring in a random order and keeps the node probabilities p_i synchronized with the node groups g_i. Equation (5.18) can be efficiently implemented in Algorithm 5.1 by updating $k_i^{g_i}$.

Label propagation with node preferences defined in Equation (5.18) is called defensive propagation [59] as it restrains the propagation of labels to preserve a larger number of groups by increasing the propagation strength of their central nodes or, equivalently, decreasing the propagation strength of their border nodes. Another strategy is to increase the propagation strength of the border nodes, which results in a more rapid expansion of groups and a smaller number of larger groups. This is called offensive propagation [59] with the label propagation rule written as

$$g_i = \operatorname*{argmax}_{g} \sum_{j} (1 - p_j) A_{ij} \delta(g_j, g). \qquad (5.19)$$

The left-hand side of Figure 5.5 demonstrates the defensive and offensive label propagation methods in an artificial network with two planted communities that are only loosely separated. While defensive propagation correctly identifies the communities planted in the network, offensive propagation spreads the labels beyond the borders of the communities and reveals no structure in this network. The right-hand side of Figure 5.5 compares the methods also on a graph partitioning problem. The methods are applied to a triangular grid with four edges

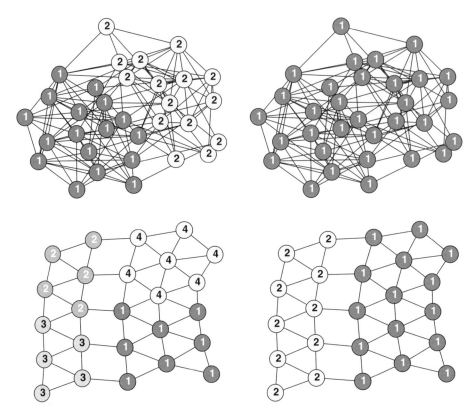

Figure 5.5 Comparison of defensive and offensive label propagation in artificial networks with a planted community structure and a triangular grid with four missing edges. The labels and shades of the nodes represent communities or groups identified by the two methods.

Table 5.1 Degeneracy diagrams of the label propagation methods displaying the non-degenerate ranges of the revealed groups (thick lines), while the percentages show the fraction of nodes in the tiny groups (left) and in the largest group (right). The values are averages over 25 runs of the methods

Method	European roads		Wikipedia users	
Standard propagation	61.5% ⊢————┼———⊣	0.9%	5.8% ⊢┼——┼——	67.6%
Defensive propagation	53.6% ⊢———┼——⊣	0.9%	6.6% ⊢┼———————┼—⊣	16.8%
Offensive propagation	7.1% ⊢┼—————┼⊣	8.5%	4.3% ⊢┼┼——	79.7%

removed, which makes a division into two groups the only sensible partition. In contrast, offensive propagation correctly partitions the grid into two groups, whereas defensive propagation overly restrains the spread of labels and recovers four groups.

Table 5.1 further compares the defensive and offensive label propagation methods on the European road network [58] with 1174 nodes and a network of user interactions on

Wikipedia [43] with 126,514 nodes. Both networks are available at KONECT. Degeneracy diagrams in Table 5.1 show the non-degenerate or effective ranges of the revealed groups that span the fraction of nodes not covered by the tiny groups with three nodes or less, or the largest group [64] (left and right percentages, respectively). Ideally, the thick lines in Table 5.1 would span from left to right. Due to the sparse grid-like structure of the road network, defensive propagation partitions 53.6% of the nodes into tiny groups, which is not a useful result. This can be avoided by using offensive propagation, where this percentage equals 7.1%. However, in the case of much denser Wikipedia network, offensive propagation returns one giant group occupying 79.7% of the nodes, thus defensive propagation with 16.8% is preferred. Note that the crucial difference between these two networks requiring the use of different methods is their density. A generally applicable approach is first to use defensive propagation and then iteratively refine the revealed groups with offensive propagation [57, 59], in this order. For example, such an approach reveals a partition of the road network with 7.9% of the nodes in the tiny groups and 6.4% of the nodes in the largest group on average.

An alternative definition of defensive and offensive label propagation is to replace the random walk diffusion in Equation (5.18) with the eigenvector centrality [14] defined as

$$p_i = \kappa_{g_i}^{-1} \sum_j p_j A_{ij} \delta(g_i, g_j), \tag{5.20}$$

where κ_{g_i} is a normalizing constant equal to the leading eigenvalue of the adjacency matrix A reduced to the nodes in group g_i. Zhang *et al.* [77] have shown that defensive label propagation with the eigenvector centrality for the node preferences is equivalent to the maximum likelihood estimation of a stochastic block model with Gaussian weights on the edges. This relates the label propagation method with yet another popular approach in the literature that is more thoroughly described in Chapter 11.

5.3.3   Method Stability and Complexity

Here, we discuss different techniques to improve the performance of the label propagation method by either increasing its stability or reducing its complexity.

One of the main sources of instability of the method is the random order of label updates in asynchronous propagation [35, 50]. Recall that the primary reason for this is to break cyclic oscillations of labels in synchronous propagation as it occurs in Figure 5.3. Li *et al.* [36] also proposed to use synchronous propagation even though this can lead to oscillations of labels, but to break the oscillations by making the label propagation rule in Equation (5.2) probabilistic. The probability that the node i with group label g_i updates its label to g is defined as

$$P_i(g) \propto \delta(g_i, g) + \sum_j A_{ij} \delta(g_j, g). \tag{5.21}$$

Although this successfully eliminates the oscillations of labels in Figure 5.3, probabilistic label propagation can make the method even more unstable. It must be stressed that this instability represents a major issue, especially in very large networks.

Cordasco and Gargano [17, 18] proposed a more elegant solution called semi-synchronous label propagation based on node coloring. A coloring of network nodes is an assignment of colors to nodes such that no two connected nodes share the same color [44]. Notice that if two nodes are not connected their labels do not directly depend on one another in Equation (5.2) and can therefore be updated simultaneously using synchronous propagation. Given a coloring of the network, semi-synchronous propagation traverses different colors in a random order as in asynchronous propagation. In contrast, the labels of the nodes with the same color are updated simultaneously as in synchronous propagation. For instance, coloring each node with a different color is equivalent to asynchronous propagation, while a simple greedy algorithm can find a coloring with at most $\Delta + 1$ colors, where Δ is the maximum degree in a network. In contrast to synchronous and asynchronous propagation, the convergence of semi-synchronous propagation can be formally proven.

Šubelj and Bajec [58, 61] observed empirically that updating the labels of the nodes in some fixed order drives the label propagation process towards similar solutions as setting the node preferences in Equation (5.16) higher (lower) for the nodes that appear earlier (later) in the order and then updating their labels in a random order as in asynchronous propagation. The node preferences can thus be used as node balancers to counteract the randomness introduced by asynchronous propagation. Let t_i be a normalized position of the node i in some random order, which is set to $1/n$ for the first node, $2/n$ for the second node and so on, where n is the number of nodes in a network. The value t_i represents the time at which the label of node i is updated. Balanced label propagation sets the node preferences using a logistic function as

$$g_i = \underset{g}{\operatorname{argmax}} \sum_j \frac{1}{1 + e^{-\gamma(2t_j-1)}} A_{ij}\delta(g_j, g), \qquad (5.22)$$

where γ is a parameter of the method. For $\gamma = 0$, Equation (5.22) is equivalent to the standard label propagation rule in Equation (5.2), while $\gamma > 0$ makes the method more stable, but this increases its time complexity. In practice, one must therefore decide on a compromise between the method stability and its time complexity.

The method stability is tightly knit with its performance. Figure 5.6 compares community detection of the label propagation methods in artificial networks with four planted communities [25]. Community structure is controlled by a mixing parameter μ that represents the fraction of nodes' neighbors in their own community. For example, the left-hand side of Figure 5.6 shows realizations of networks for $\mu = 0.1$ and 0.4. Performance of the methods is measured with the normalized mutual information [23], where higher is better (see [23] for the exact definition). As seen in the right-hand side of Figure 5.6, balanced label propagation combined with the defensive node preferences in Equation (5.18) performs best in these networks, when $\gamma = 1$.

Another prominent approach for improving community detection of the label propagation methods is consensus clustering [24, 32, 78]. One first applies the method to a given network multiple times and constructs a weighted consensus graph, where weights represent the number of times two nodes are classified into the same community. Note that only edges with weights above a given threshold are kept. The entire process is then repeated on the consensus graph until the revealed communities no longer change. For example, the left-hand side of Figure 5.7 shows two realizations of groups obtained with the standard label propagation method in Equation (5.2) in artificial networks for $\mu = 0.33$. Although these do not exactly coincide with the planted communities, label propagation in the corresponding consensus graph recovers the correct community structure as demonstrated on the right-hand side of Figure 5.7. For another example, Figure 5.8 shows the largest connected component of the European road network from Table 5.1 and the largest groups revealed by the offensive label propagation method in Equation (5.19) with 25 runs of consensus clustering.

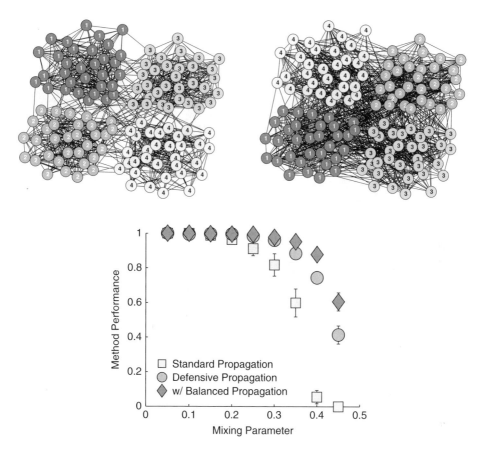

Figure 5.6 Performance of the label propagation methods in artificial networks with planted community structure represented by the labels and shades of the nodes. The markers are averages over 25 runs of the methods, while the error bars show standard errors.

Note, however, that consensus clustering can substantially increase the method's computational time. Other work has thus considered different hybrid approaches to improve the stability of community detection of the label propagation methods, where community structure revealed by one method is refined by another [57, 59], possibly proceeding iteratively or incrementally [19, 35]. For instance, label propagation under constraints [28, 38] has traditionally been combined with a multistep greedy agglomeration [55].

In the remaining sections, we also briefly discuss different approaches to reduce the complexity of the label propagation method. Although the time complexity is already nearly linear $\mathcal{O}(m^{1.2})$, where m is the number of edges in a network [59], one can still further improve the computational time. As shown in Figure 5.4, the number of nodes that update their label at a particular iteration of label propagation drops exponentially with the number of iterations. Thus, after a couple of iterations, most nodes have already acquired their final label and no longer need to be updated. For instance, one can selectively update only the labels of those nodes for which the fraction of neighbors sharing the same label is below a certain threshold [35], which

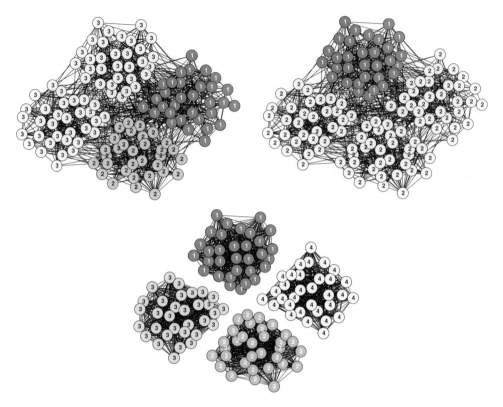

Figure 5.7 Label propagation in artificial networks with planted community structure and the corresponding consensus graph. The labels and shades of the nodes represent communities identified by the label propagation method.

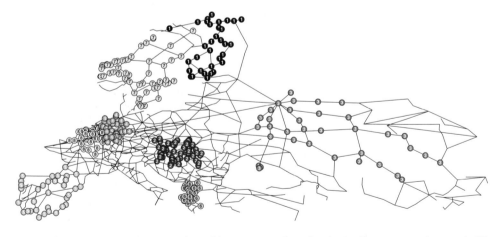

Figure 5.8 Offensive label propagation with consensus clustering in the European road network. The labels and shades of the nodes represent the largest eight groups identified by the method.

can make the method truly linear $\mathscr{O}(m)$. Xie and Szymanski [72] further formalized this idea using the concept of active and passive nodes. A node is said to be passive if updating would not change its label. Otherwise, the node is active. The labels are therefore propagated only between the active nodes until all nodes become passive.

Due to its algorithmic simplicity, the label propagation method is easily parallelizable, especially with synchronous or semi-synchronous propagation mentioned above. The method is thus suitable for application in distributed computing environments such as Spark[2] [16] or Hadoop[3] [47] and on parallel architectures [56]. In this way, label propagation has been successfully used on billion-node networks [16, 69].

5.4   Extensions to Other Networks

Throughout the chapter, we have assumed that the label propagation method is applied to simple undirected networks. Nevertheless, the method can easily be extended to networks with multiple edges between the nodes as in Equation (5.2) and networks with weights on the edges as in Equation (5.3). This holds also for the different advances of the propagation methods presented in Section 5.3. In contrast, there seems to be no straightforward extension to networks with directed arcs. The reason for this is that propagating the labels exclusively in the direction of arcs enables exchange of labels only between mutually reachable nodes forming a strongly connected component. Since any directed network is a directed acyclic graph on its strongly connected components, the labels can propagate between the nodes of different strongly connected components merely in one direction. Therefore, one usually disregards the directions of arcs when applying the label propagation method to directed networks except in the case when most arcs are reciprocal.

The method can be extended to signed networks with positive and negative edges between the nodes, as in the approach of Doreian and Mrvar [20]. In order to partition the network in such a way that positive edges mostly appear within the groups and negative edges between the groups, one assigns some fixed positive (negative) weight to positive (negative) edges and then applies the standard label propagation method for weighted networks in Equation (5.3). According to the objective function in Equation (5.7), the method thus simultaneously tries to maximize the number of positive edges within the groups and the number of negative edges between the groups. Still, this does not ensure that the nodes connected by a negative edge are necessarily assigned to different groups, but merely restricts the propagation of labels along the negative edges [1].

Table 5.2 shows the standard and signed label propagation methods applied to the Wikipedia web of trust network [43] available at KONECT. The network consists of $138,587$ nodes connected by $629,689$ positive edges and $110,417$ negative edges. Standard label propagation ignoring the signs of edges reveals one giant group occupying 89.0% of the nodes on average. Most positive edges are thus obviously within the groups, but the same also holds for negative edges. Signed label propagation with positive and negative weights on the edges reduces the size of the largest group to 60.6% of the nodes on average. Most positive edges remain within the groups, while more than half of negative edges are between the groups. Note that the method

---

[2] http://spark.apache.org
[3] http://hadoop.apache.org

Table 5.2 Comparison of the label propagation methods on the signed Wikipedia web of trust network. The values are averages over 25 runs of the methods, while $\mathcal{H}$ is defined in Equation (5.7)

Method	+ Edges within	− Edges between	Hamiltonian $\mathcal{H}$
Standard propagation	96.6%	6.7%	−528185.8
Signed propagation	90.9%	56.7%	−535065.2
With equal weights	75.6%	81.8%	−460413.1

assigns weights 1 and −1 to positive and negative edges. Since only 12.0% of the edges in the network are negative, this actually puts more emphasis on the positive edges. To circumvent the latter, one can assign equal total weight to positive and negative edges by using weights $1/m_p$ and $-1/m_n$, where m_p and m_n are the numbers of positive and negative edges. Signed label propagation with equal total weights returns a larger number of groups with 43.2% of the nodes in the largest group, and about the same fraction of positive edges within the groups and negative edges between the groups. For further discussion on partitioning signed networks see Chapter 8.

Any label propagation method can also be used on bipartite networks with two types of nodes and edges only between the nodes of different type as on the left-hand side of Figure 5.9. For instance, Barber and Clark [5] adopted the label propagation methods under constraints

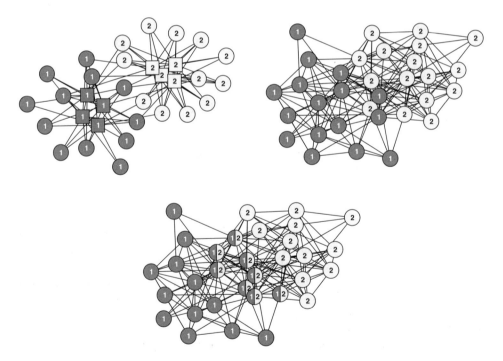

Figure 5.9 Non-overlapping and overlapping label propagation in artificial networks with planted community structure. The labels and shades of the nodes represent communities identified by different methods, while the types of nodes of the bipartite network are shown with distinct symbols.

to optimize bipartite modularity [4]. Liu and Murata [39, 40] proposed a proper extension of the label propagation framework to bipartite networks. This is a special case of semi-synchronous propagation with node coloring discussed in Section 5.3.3. Recall that semi-synchronous propagation updates the labels of the nodes with the same color synchronously, while different colors are traversed asynchronously. In bipartite networks, the types of the nodes can be taken for their colors, thus the method alternates between the nodes of each type, while the propagation of labels always occurs synchronously. The same principle can be extended also to multipartite networks, where again the nodes of the same type are assigned the same color. However, in multirelational or multilayer networks [11], one can separately consider the nodes of different layers, but the propagation of labels within each layer requires asynchronicity for the method to converge.

5.5   Alternative Types of Network Structures

The label propagation method was originally designed to detect non-overlapping communities in networks [35, 50]. In the following, we show how the method can be extended to more diverse network structures. We consider extensions to overlapping groups of nodes, groups of nodes at multiple resolutions that form a nested hierarchy, and groups of structurally equivalent nodes. Note that, in contrast to the extensions to other types of networks in Section 5.4, this increases the time complexity of the method derived in Section 5.1.4. As shown in the following, the time complexity increases by a factor depending on the type of groups considered.

5.5.1   *Overlapping Groups of Nodes*

Extension of the label propagation method to overlapping groups of nodes is relatively straightforward [26, 71]. Instead of assigning a single group label g_i to node i as the standard label propagation method in Equation (5.2), multiple labels are assigned to each node. Let ρ_i be the group function of node i where $\rho_i(g)$ represents how strongly the node is affiliated to group g. In particular, the node belongs to groups g for which $\rho_i(g) > 0$, while its group affiliations are normalized to one as $\sum_g \rho_i(g) = 1$. At the beginning of label propagation, each node is put into its own group by setting $\rho_i(i) = 1$. Then, at every iteration, each node adopts the group labels of its neighbors. The affiliation $\rho_i(g)$ of node i to group g is computed as the average affiliation of its neighbors. Hence,

$$\rho_i(g) = \sum_j \frac{\rho_j(g)}{k_i} A_{ij}, \qquad (5.23)$$

where A is the network adjacency matrix and k_i is the degree of node i. Equation (5.23) can be combined also with an inflation operator raising $\rho_i(g)$ to some exponent [74]. Obviously, the groups can now overlap as the nodes can belong to multiple groups. For example, the right-hand side of Figure 5.9 demonstrates the non-overlapping and overlapping label propagation methods in an artificial network with two planted overlapping communities.

Notice, however, that the label propagation rule in Equation (5.23) inevitably leads to every node in a network belonging to all groups. It is therefore necessary to limit the number of groups a single node can belong to. Gregory [26] proposed that, after each iteration of label propagation, the group affiliations $\rho_i(g)$ below $1/v$ are set to zero and renormalized, where v is a method parameter. Since $\sum_g \rho_i(g) = 1$ for every node, the nodes can thus belong to at most v groups. The parameter v can be difficult to determine if a network consists of overlapping and non-overlapping groups. Wu *et al.* [71] suggested replacing the parameter v by a node-dependent threshold ρ to keep node i affiliated to group g as long as

$$\frac{\rho_i(g)}{\max_g \rho_i(g)} \geq \rho. \tag{5.24}$$

The time complexity of the described overlapping label propagation method is $\mathcal{O}(cmv)$, where c is the number of iterations of label propagation, m is the number of edges in a network, and v is the maximum number of groups a single node belongs to. The method is implemented by a popular community detection algorithm COPRA[4] [26].

It is also possible to detect overlapping groups of nodes by using the standard non-overlapping label propagation method. Xie and Szymanski [73, 75] proposed associating a memory with each node to store group labels from previous iterations. Running the label propagation for c iterations assigns c labels to each node's memory. The probability of observing label g in the memory of node i or, equivalently, the number of occurrences of g in the memory of i can then be interpreted as the group affiliation $\rho_i(g)$ as defined above. Note that label propagation with node memory splits the label propagation rule in Equation (5.2) into two steps. Each neighbor j of the considered node i first propagates a random label from its memory, with the label g being selected with probability $\rho_j(g)$, while node i then adds the most frequently propagated label to its memory. The time complexity of the method is $\mathcal{O}(cm)$, where c is a small constant set to say 25. The method is implemented by another popular community detection algorithm SLPA[5] [75] and its successor SpeakEasy[6] [24].

DEMON[7] [19] is a well-known community detection algorithm that also uses non-overlapping label propagation to detect overlapping groups. Instead of assigning a memory to each node as above, this label propagation method is separately applied to the subnetworks reduced to the neighborhoods of the nodes. All of the resulting groups that are, in general, overlapping are then merged together.

5.5.2 Hierarchy of Groups of Nodes

Label propagation can be applied in a hierarchical manner in order to reveal a nested hierarchy of groups of nodes [35, 37, 59, 62]. The bottom level of such a hierarchy represents groups of nodes. The next level represents groups of groups of nodes and so on. Cutting the hierarchy at different levels results in groups of nodes at multiple resolutions. For example, Figure 5.10 demonstrates the hierarchical label propagation method in artificial networks with two levels of planted community structure. Let $G_1, G_2, \ldots$ denote the groups revealed by the basic label propagation method in Equation (5.2), which represent the bottom level of the group hierarchy. One then constructs a meta-network, where nodes correspond to different groups G_i and an edge

---

[4] http://gregory.org/research/networks/software/copra.html
[5] http://sites.google.com/site/communitydetectionslpa
[6] http://www.cs.rpi.edu/~szymansk/SpeakEasy
[7] http://www.michelecoscia.com/?page_id=42

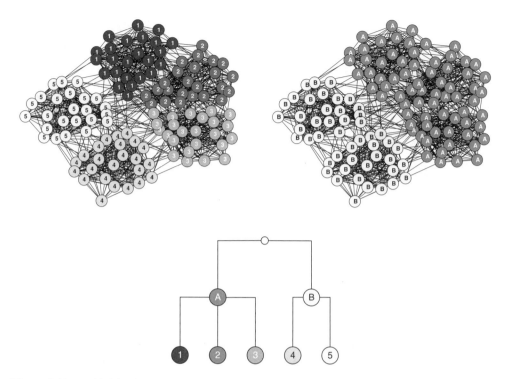

Figure 5.10 Artificial networks with two levels of planted community structure and the correspond-ing group hierarchy. The labels and shades of the nodes represent communities identified by the label propagation method.

is put between the groups G_i and G_j if their nodes are connected in the original network. The weight of the edge is set to the number of edges between the groups G_i and G_j in the original network. Similarly, a loop is added to each group G_i with a weight equal to the number of edges within the group G_i in the original network. Finally, one applies the weighted label propagation method in Equation (5.3) to the constructed meta-network to reveal groups of groups G_i. These constitute the next level of the group hierarchy. The entire process of such bottom-up group agglomeration is repeated iteratively until a single group is recovered, which is the root of the hierarchy. Note that label propagation with group agglomeration is algorithmically equivalent to the famous Louvain modularity optimization method [10, 66].

Figure 5.11 shows the meta-networks of the largest connected components of the Google web graph from Figure 5.4 with 875, 713 nodes and the Pennsylvania road network [34] with 1, 087, 562 nodes. Both networks are available at KONECT. The meta-networks were revealed by the hierarchical label propagation method with two and three steps of group agglomeration, and consist of 564 and 235 nodes, respectively. Notice that, although the networks are reduced to less than a thousandth of their original size, the group agglomeration process preserves a dense central core of the web graph and a sparse homogeneous topology of the road network [9].

Bottom-up group agglomeration can be effectively combined with top-down group refine-ment [62, 63]. Let $G_1, G_2, \ldots$ be the groups revealed at some step of the group agglomeration.

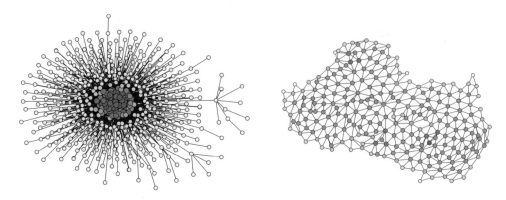

Figure 5.11 The meta-networks of the Google web graph and the Pennsylvania road network identified by the hierarchical label propagation method. The shades of the nodes are proportional to their corrected clustering coefficient [6], where darker (lighter) means higher (lower).

Prior to the construction of the meta-network, one separately applies the label propagation method to the subnetworks of the original network limited to the nodes of groups G_i. As this process repeats recursively until a single group is recovered, a sub-hierarchy of groups is revealed for each group G_i. Bottom-up agglomeration with top-down refinement enables the identification of a very detailed hierarchy of groups present in a network [24, 62]. One can also further control the resolution of groups by adjusting the weights on the loops in the meta-network [27]. The time complexity of the described hierarchical label propagation method is $\mathcal{O}(cmh)$, where c is the number of iterations and m the number of edges as before, while h is the number of levels of the group hierarchy.

5.5.3 Structural Equivalence Groups

The Different label propagation methods presented so far can be used to reveal connected and cohesive groups of nodes in a network. This includes detection of densely connected communities and graph partitioning as demonstrated in Figure 5.5. However, the methods cannot be adopted for detection of any kind of disconnected groups of nodes. Therefore, possibly the most interesting extension of the label propagation method is to find groups of structurally equivalent nodes [42, 60–62]. Informally, two nodes are said to be structurally equivalent if they are connected to the same other nodes in the network and thus have the same common neighbors [21, 41], whereas the nodes themselves may be connected or not. We here consider a relaxed definition of structural equivalence in which nodes can have only the majority of their neighbors in common. For example, the left-hand side of Figure 5.12 shows an artificial network with two planted communities of nodes labeled with 2 and 4, and two groups of structurally equivalent nodes labeled with 1 and 3 that form a bipartite structure. The former are also called assortative groups, while the latter are referred to as disassortative groups [23].

Let k_i denote the degree of node i and k_{ij} the number of common neighbors of nodes i and j. Hence, $k_i = \sum_j A_{ij}$ and $k_{ij} = \sum_k A_{ik}A_{kj}$, where A is the network adjacency matrix.

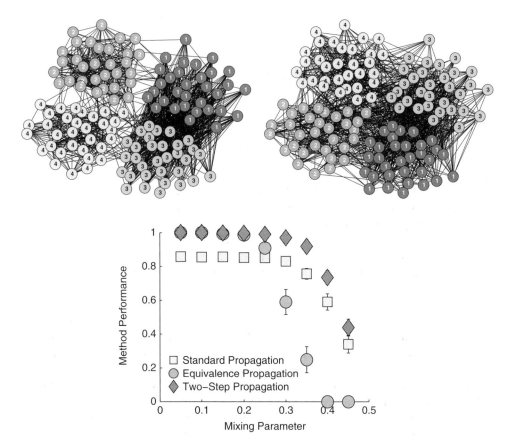

Figure 5.12 Performance of the label propagation methods in artificial networks with planted communities and structural equivalence groups represented by the labels and shades of the nodes. The markers are averages over 25 runs of the methods, while the error bars show standard errors.

Xie and Szymanski [72] modified the label propagation rule in Equation (5.2) as

$$g_i = \arg\max_g \sum_j (1 + k_{ij}) A_{ij} \delta(g_j, g), \qquad (5.25)$$

which increases the strength of propagation between structurally equivalent nodes. Notice that Equation (5.25) is in fact equivalent to simultaneously propagating the labels between the neighboring nodes as standard and also through their common neighbors represented by the term k_{ij}. Yet, the labels are propagated merely between connected nodes, thus the method can still reveal only connected groups of nodes.

Šubelj and Bajec [61, 62] proposed a proper extension of the label propagation method for structural equivalence that separately propagates the labels between the neighboring nodes and through nodes' common neighbors. Let τ_g be a parameter of group g that is set close to one for connected groups and close to zero for structural equivalence groups. The label propagation rule for general groups of nodes is then written as

$$g_i = \underset{g}{\mathrm{argmax}} \left(\tau_g \sum_j A_{ij}\delta(g_j, g) + (1 - \tau_g) \sum_{kj \neq i} \frac{1}{k_k - 1} A_{ik}A_{kj}\delta(g_j, g) \right). \quad (5.26)$$

The left-hand sum propagates the labels between the neighboring nodes i and j, while the right-hand sum propagates the labels between the nodes i and j through their common neighbors k. The degree k_k in the denominator ensures that the number of terms in both sums is proportional to k_i. By setting all group parameters in Equation (5.26) as $\tau_g = 1$, one retrieves the standard label propagation method in Equation (5.2) that can detect connected groups of nodes like communities, while setting $\tau_g \approx 0$, the method can detect structural equivalence groups. In the case when a community consists of structurally equivalent nodes as in a clique of nodes, any of the two methods can be used. In practice, the group parameters τ_g can be inferred from the network structure or estimated during the label propagation process [61, 62]. However, this can make the method very unstable. For this reason, we propose a much simpler approach.

Applying the standard label propagation method to the network on the left-hand side of Figure 5.12 reveals three groups of nodes, since both structural equivalence groups are detected as a single group of nodes. In general, configurations of connected structural equivalence groups are merged together by the method. One can, however, employ this behavior to detect structural equivalence groups using a two-step approach with top-down group refinement introduced before [62, 63]. The first step reveals connected groups of nodes using the standard label propagation method by setting $\tau_g = 1$ in Equation (5.26). This includes communities and configurations of connected structural equivalence groups. In the second step, one separately tries to refine each group from the first step using the structural equivalence label propagation method by setting $\tau_g = 0$ in Equation (5.26). While communities are still detected as a single group of nodes, configurations of structural equivalence groups are now further partitioned into separate structural equivalence groups.

The right-hand side of Figure 5.12 compares group detection of the label propagation methods in artificial networks with four groups discussed above [61]. Network structure is controlled by a mixing parameter μ that represents the fraction of edges that comply with the group structure, while the examples on the left-hand side of Figure 5.12 show realizations of networks for $\mu = 0.1$ and 0.4. Performance of the methods is measured with the normalized mutual information [23], where higher is better. As already mentioned, standard label propagation combines the two structural equivalence groups into a single group. Yet, label propagation for structural equivalence can reveal all four groups, but only when these are clearly defined in the network structure. Finally, the two-step approach performs best in these networks, and can accurately detect communities and structural equivalence groups as long as the latter can first be identified as a single connected group of nodes.

In Section 5.4 we argued that standard label propagation cannot be easily extended to directed networks. In contrast, label propagation for structural equivalence can in fact be adopted for detection of specific groups of nodes in directed networks. For instance, consider a network of citations between scientific papers. Let A be the network adjacency matrix where A_{ij} represents an arc from node i to node j meaning that paper i cites paper j. One might be interested in revealing groups of papers that cite the same other papers, which is known as cocitation [8, 15].

The label propagation rule for cocitation is

$$g_i = \operatorname*{argmax}_{g} \sum_{kj \neq i} A_{ik} A_{jk} \delta(g_j, g), \tag{5.27}$$

which propagates the labels between papers i and j through their common citations k. An alternative concept is bibliographic coupling [30], which refers to groups of papers that are cited by the same other papers. The label propagation rule for bibliographic coupling is

$$g_i = \operatorname*{argmax}_{g} \sum_{kj \neq i} A_{ki} A_{kj} \delta(g_j, g). \tag{5.28}$$

As an example, we constructed a citation network of 26,038 papers published in *Physical Review E*[8] between the years 2001 and 2015. This also includes 13 references of this chapter namely references [4, 5, 33, 35, 45, 46, 50–52, 55, 59, 66, 67]. Twelve of these focus on topics in network community detection and graph partitioning, whereas [46] discusses random graph models. We first ignore the directions of citations and apply the standard label propagation method in Equation (5.2) with 25 runs of consensus clustering introduced in Section 5.3.3. The method reveals 3033 groups of papers. The largest group consists of 1276 papers on network structure and dynamics including [46] with the most frequent terms in the titles of the papers being "network", "scale-free", "complex", "epidemic", "percolation", "random", "small-world" and "social". The remaining references mentioned above are all included in the fourth largest group with 189 other papers on network community detection. The left-hand side of Figure 5.13 shows a word cloud generated from the titles of these papers displaying the most frequently appearing terms in an aesthetically pleasing way.[9] These are "community", "network", "detection", "modularity", "structure", "complex", "finding" and "clustering".

We next consider the directions of citations by employing the cocitation label propagation method in Equation (5.27) that is again combined with 25 runs of consensus clustering. The method reveals 1016 cocitation groups with 2427 papers in the largest group. The latter consists

Figure 5.13 Word clouds demonstrating two of the largest groups of nodes revealed by different label propagation methods in the *Physical Review E* paper citation network. These show the most frequently appearing terms in the titles of the corresponding papers.

---

[8] http://journals.aps.org/pre
[9] https://www.jasondavies.com/wordcloud

of papers on various topics in network science including all the thirteen references from above. The right-hand side of Figure 5.13 shows a word cloud generated from the titles of these papers, where the most frequent terms are "network", "scale-free", "complex", "synchronization", "community", "random", "small-world" and "oscillators".

As shown in Section 5.1.4, the time complexity of a single iteration of the standard label propagation method is $\mathcal{O}(m) = \mathcal{O}(\langle k \rangle n)$, where n and m are the number of nodes and edges in a network, and $\langle k \rangle = \sum_i k_i / n$ is the average node degree. Since the structural equivalence methods presented above propagate the labels also between the nodes two steps apart, the time complexity of a single iteration becomes $\mathcal{O}(\langle k^2 \rangle n)$, where $\langle k^2 \rangle = \sum_i k_i^2 / n$ is the average node degree squared. The total time complexity of the methods is therefore $\mathcal{O}(c \langle k^2 \rangle n)$, where c is the number of iterations of label propagation.

5.6 Applications of Label Propagation

The label propagation methods are most commonly used for clustering and partitioning large networks with the main goal being network abstraction. In this section, we briefly review selected other applications of label propagation.

People You May Know is an important feature of the Facebook social service providing recommendations for future friendship ties between its users. Most friendship recommendations are of type "friend-of-friend", meaning that the users are suggested other users two hops away in the Facebook social graph [2]. Due to the immense size of the graph, it is distributed among multiple physical machines thus each machine stores some local part of the graph consisting only of a subset of users. When a friendship recommendation has to be made for a given user, it is desired that the users two steps away in the graph reside at the same machine as the concerned user, in order to minimize the communication between the machines. As reported in 2013 [68], the users are effectively partitioned among machines using a variant of label propagation under constraints presented in Section 5.3.1.

A related application is a compression of very large web graphs and online social networks to enable their analysis on a single machine [12]. Most compression algorithms rely on a given ordering of network nodes such that the edges are mainly between the nodes that are close in the ordering. In the case of web graphs, one can order the nodes representing web pages lexicographically by their URL, whereas no equivalent approach exists for social networks. Boldi *et al.* [13] adopted the label propagation method in Equation (5.15) to compute the ordering of network nodes iteratively starting from a random one. Using such a setting, the authors reported a major improvement in compression with respect to other known techniques. Most social networks and web graphs can be compressed to just a couple of bits per edge, while still allowing for an efficient network traversal. For instance, this compression approach was in fact used to reveal the four degrees of separation between the active users of Facebook in 2011 [3].

5.7 Summary and Outlook

In this chapter we have presented the basic label propagation method for network clustering and partitioning, together with its numerous variants and advances, extensions to different types of networks and clusterings, and selected large-scale applications. Due to the high popularity of label propagation in the literature, our review here is by no means complete. In particular, we have focused primarily on the results reported in the physics and computer science literature. However, the very same approach is also commonly used in the social networks literature [7, 21], where it is known under the name relocation algorithm or simply as a local greedy optimization. The label propagation method and the relocation algorithm thus provide a sort of common ground between two diverging factions of network science in the natural and social science literature [29].

As stated already in the introduction, label propagation is neither the most accurate nor the most robust clustering method. Yet it is a very fast and versatile method that can readily be applied to the largest networks and easily adopted for a particular application. It should be used as the first choice for gaining a preliminary insight into the structure of a network before trying out more sophisticated and expensive methods. In the case of very large online social networks and web graphs, the label propagation method is in fact often the only choice. Future research should therefore focus more on specific applications of label propagation in large networks, where the use of simple and efficient methods is unavoidable, and less on new *ad hoc* modifications of the original method, since there are already many of these.

References

1. J.-P. Attal and M. Malek. A new label propagation with dams. In *Proceedings of the International Conference on Advances in Social Networks Analysis and Mining*, pages 1292–1299, Paris, France, 2015.
2. L. Backstrom and J. Leskovec. Supervised random walks: Predicting and recommending links in social networks. In *Proceedings of the ACM International Conference on Web Search and Data Mining*, pages 635–644, Hong Kong, China, 2011.
3. L. Backstrom, P. Boldi, M. Rosa, J. Ugander, and S. Vigna. Four degrees of separation. In *Proceedings of the ACM International Conference on Web Science*, pages 45–54, Evanston, IL, USA, 2012.
4. M. J. Barber. Modularity and community detection in bipartite networks. *Physical Review E*, 76(6):066102, 2007.
5. M. J. Barber and J. W. Clark. Detecting network communities by propagating labels under constraints. *Physical Review E*, 80(2):026129, 2009.
6. V. Batagelj. Corrected overlap weight and clustering coefficient. In *Proceedings of the INSNA International Social Network Conference*, pages 16–17, Newport Beach, CA, USA, 2016.
7. V. Batagelj and A. Ferligoj. Clustering relational data. In *Data Analysis*, pages 3–15. Springer, Berlin, 2000.
8. V. Batagelj, P. Doreian, A. Ferligoj, and N. Kejžar. *Understanding Large Temporal Networks and Spatial Networks*. Wiley, Chichester, 2014.
9. N. Blagus, L. Šubelj, and M. Bajec. Self-similar scaling of density in complex real-world networks. *Physica A: Statistical Mechanics and its Applications*, 391(8):2794–2802, 2012.
10. V. D. Blondel, J.-L. Guillaume, R. Lambiotte, and E. Lefebvre. Fast unfolding of communities in large networks. *Journal of Statistical Mechanics: Theory and Experiment*, P10008, 2008.
11. S. Boccaletti, G. Bianconi, R. Criado, C. I. del Genio, J. Gómez-Gardeñes, M. Romance, I. Sendiña-Nadal, Z. Wang, and M. Zanin. The structure and dynamics of multilayer networks. *Physics Reports*, 544(1):1–122, 2014.
12. P. Boldi and S. Vigna. The WebGraph framework I: Compression techniques. In *Proceedings of the International Conference on World Wide Web*, pages 595–601, New York, NY, USA, 2004.
13. P. Boldi, M. Rosa, M. Santini, and S. Vigna. Layered label propagation: A multiresolution coordinate-free ordering for compressing social networks. In *Proceedings of the International World Wide Web Conference*, pages 587–596, Hyderabad, India, 2011.
14. P. Bonacich. Power and centrality: A family of measures. *American Journal of Sociology*, 92(5):1170–1182, 1987.
15. K. W. Boyack and R. Klavans. Co-citation analysis, bibliographic coupling, and direct citation: Which citation approach represents the research front most accurately? *Journal of the American Society for Information Science and Technology*, 61(12):2389–2404, 2010.

16. N. Buzun, A. Korshunov, V. Avanesov, I. Filonenko, I. Kozlov, D. Turdakov, and H. Kim. EgoLP: Fast and distributed community detection in billion-node social networks. In *Proceedings of the IEEE International Conference on Data Mining Workshop*, pages 533–540, Shenzhen, China, 2014.

17. G. Cordasco and L. Gargano. Community detection via semi-synchronous label propagation algorithms. In *Proceedings of the IMSAA Workshop on Business Applications of Social Network Analysis*, pages 1–8, Bangalore, India, 2010.

18. G. Cordasco and L. Gargano. Label propagation algorithm: A semi–synchronous approach. *International Journal of Social Network Mining*, 1(1):3–26, 2011.

19. M. Coscia, G. Rossetti, F. Giannotti, and D. Pedreschi. DEMON: A local-first discovery method for overlapping communities. In *Proceedings of the ACM SIGKDD International Conference on Knowledge Discovery and Data Mining*, pages 615–623, Beijing, China, 2012.

20. P. Doreian and A. Mrvar. A partitioning approach to structural balance. *Social Networks*, 18(2):149–168, 1996.

21. P. Doreian, V. Batagelj, and A. Ferligoj. *Generalized Blockmodeling*. Cambridge University Press, Cambridge, 2005.

22. P. Erdős and A. Rényi. On random graphs I. *Publicationes Mathematicae Debrecen*, 6:290–297, 1959.

23. S. Fortunato and D. Hric. Community detection in networks: A user guide. *Physics Reports*, 659:1–44, 2016.

24. C. Gaiteri, M. Chen, B. Szymanski, K. Kuzmin, J. Xie, C. Lee, T. Blanche, E. C. Neto, S.-C. Huang, T. Grabowski, T. Madhyastha, and V. Komashko. Identifying robust communities and multi-community nodes by combining top-down and bottom-up approaches to clustering. *Scientific Reports*, 5: 16361, 2015.

25. M. Girvan and M. E. J. Newman. Community structure in social and biological networks. *Proceedings of the National Academy of Sciences of the USA*, 99(12):7821–7826, 2002.

26. S. Gregory. Finding overlapping communities in networks by label propagation. *New Journal of Physics*, 12(10):103018, 2010.

27. J. Han, W. Li, and W. Deng. Multi-resolution community detection in massive networks. *Scientific Reports*, 6:38998, 2016.

28. J. Han, W. Li, Z. Su, L. Zhao, and W. Deng. Community detection by label propagation with compression of flow. *European Physical Journal B*, 89 (12):1–11, 2016.

29. C. A. Hidalgo. Disconnected, fragmented, or united? A trans-disciplinary review of network science. *Applied Network Science*, 1:6, 2016.

30. B. Jarneving. Bibliographic coupling and its application to research-front and other core documents. *Journal of Infometrics*, 1(4):287–307, 2007.

31. H. Jeong, S. P. Mason, A.-L. Barabási, and Z. N. Oltvai. Lethality and centrality of protein networks. *Nature*, 411(6833):41–42, 2001.

32. A. Lancichinetti and S. Fortunato. Consensus clustering in complex networks. *Scientific Reports*, 2:336, 2012.

33. A. Lancichinetti, S. Fortunato, and F. Radicchi. Benchmark graphs for testing community detection algorithms. *Physical Review E*, 78(4):046110, 2008.

34. J. Leskovec, K. J. Lang, A. Dasgupta, and M. W. Mahoney. Community structure in large networks: Natural cluster sizes and the absence of large well-defined clusters. *Internet Mathematics*, 6(1):29–123, 2009.

35. I. X. Y. Leung, P. Hui, P. Liò, and J. Crowcroft. Towards real-time community detection in large networks. *Physical Review E*, 79(6):066107, 2009.

36. S. Li, H. Lou, W. Jiang, and J. Tang. Detecting community structure via synchronous label propagation. *Neurocomputing*, 151(3):1063–1075, 2015.

37. W. Li, C. Huang, M. Wang, and X. Chen. Stepping community detection algorithm based on label propagation and similarity. *Physica A: Statistical Mechanics and its Applications*, 472:145–155, 2017.

38. X. Liu and T. Murata. Advanced modularity-specialized label propagation algorithm for detecting communities in networks. *Physica A: Statistical Mechanics and its Applications*, 389(7):1493–1500, 2009.

39. X. Liu and T. Murata. Community detection in large-scale bipartite networks. In *Proceedings of the IEEE/WIC/ACM International Joint Conference on Web Intelligence and Intelligent Agent Technology*, pages 50–57, Milan, Italy, 2009.

40. X. Liu and T. Murata. How does label propagation algorithm work in bipartite networks. In *Proceedings of the IEEE/WIC/ACM International Joint Conference on Web Intelligence and Intelligent Agent Technology*, pages 5–8, Milan, Italy, 2009.

41. F. Lorrain and H. C. White. Structural equivalence of individuals in social networks. *Journal of Mathematical Sociology*, 1(1):49–80, 1971.

42. H. Lou, S. Li, and Y. Zhao. Detecting community structure using label propagation with weighted coherent neighborhood propinquity. *Physica A: Statistical Mechanics and its Applications*, 392(14):3095–3105, 2013.
43. S. Maniu, T. Abdessalem, and B. Cautis. Casting a web of trust over Wikipedia: An Interaction-based approach. In *Proceedings of the International Conference on World Wide Web*, pages 87–88, New York, NY, USA, 2011.
44. M. E. J. Newman. *Networks: An Introduction*. Oxford University Press, Oxford, 2010.
45. M. E. J. Newman and M. Girvan. Finding and evaluating community structure in networks. *Physical Review E*, 69(2):026113, 2004.
46. M. E. J. Newman, S. H. Strogatz, and D. J. Watts. Random graphs with arbitrary degree distributions and their applications. *Physical Review E*, 64(2):026118, 2001.
47. M. Ovelgönne. Distributed community detection in web-scale networks. In *Proceedings of the International Conference on Advances in Social Networks Analysis and Mining*, pages 66–73, Niagara, Canada, 2013.
48. R. B. Potts. Some generalized order-disorder transformations. *Mathematical Proceedings of the Cambridge Philosophical Society*, 48(1):106–109, 1952.
49. F. Radicchi, C. Castellano, F. Cecconi, V. Loreto, and D. Parisi. Defining and identifying communities in networks. *Proceedings of the National Academy of Sciences of the USA*, 101(9):2658–2663, 2004.
50. U. N. Raghavan, R. Albert, and S. Kumara. Near linear time algorithm to detect community structures in large-scale networks. *Physical Review E*, 76(3):036106, 2007.
51. J. Reichardt and S. Bornholdt. Statistical mechanics of community detection. *Physical Review E*, 74(1):016110, 2006.
52. P. Ronhovde and Z. Nussinov. Local resolution-limit-free Potts model for community detection. *Physical Review E*, 81(4):046114, 2010.
53. M. Rosvall and C. T. Bergstrom. Maps of random walks on complex networks reveal community structure. *Proceedings of the National Academy of Sciences of the USA*, 105(4):1118–1123, 2008.
54. M. T. Schaub, J.-C. Delvenne, M. Rosvall, and R. Lambiotte. The many facets of community detection in complex networks. *Applied Network Science*, 2:4, 2017.
55. P. Schuetz and A. Caflisch. Efficient modularity optimization by multistep greedy algorithm and vertex mover refinement. *Physical Review E*, 77(4): 046112, 2008.
56. J. Soman and A. Narang. Fast community detection algorithm with GPUs and multicore architectures. In *Proceedings of the IEEE International Parallel Distributed Processing Symposium*, pages 568–579, Anchorage, AK, USA, 2011.
57. L. Šubelj and M. Bajec. Unfolding network communities by combining defensive and offensive label propagation. In *Proceedings of the ECML PKDD Workshop on the Analysis of Complex Networks*, pages 87–104, Barcelona, Spain, 2010.
58. L. Šubelj and M. Bajec. Robust network community detection using balanced propagation. *European Physical Journal B*, 81(3):353–362, 2011.
59. L. Šubelj and M. Bajec. Unfolding communities in large complex networks: Combining defensive and offensive label propagation for core extraction. *Physical Review E*, 83(3):036103, 2011.
60. L. Šubelj and M. Bajec. Generalized network community detection. In *Proceedings of the ECML PKDD Workshop on Finding Patterns of Human Behaviors in Network and Mobility Data*, pages 66–84, Athens, Greece, 2011.
61. L. Šubelj and M. Bajec. Ubiquitousness of link-density and link-pattern communities in real-world networks. *European Physical Journal B*, 85(1): 32, 2012.
62. L. Šubelj and M. Bajec. Group detection in complex networks: An algorithm and comparison of the state of the art. *Physica A: Statistical Mechanics and its Applications*, 397:144–156, 2014.
63. L. Šubelj and M. Bajec. Network group discovery by hierarchical label propagation. In *Proceedings of the European Social Networks Conference*, page 284, Barcelona, Spain, 2014.
64. L. Šubelj, N. J. Van Eck, and L. Waltman. Clustering scientific publications based on citation relations: A systematic comparison of different methods. *PLoS One*, 11(4):e0154404, 2016.
65. G. Tibély and J. Kertész. On the equivalence of the label propagation method of community detection and a Potts model approach. *Physica A: Statistical Mechanics and its Applications*, 387(19-20):4982–4984, 2008.
66. V. A. Traag. Faster unfolding of communities: Speeding up the Louvain algorithm. *Physical Review E*, 92(3):032801, 2015.
67. V. A. Traag, P. Van Dooren, and Y. Nesterov. Narrow scope for resolution-limit-free community detection. *Physical Review E*, 84(1): 016114, 2011.
68. J. Ugander and L. Backstrom. Balanced label propagation for partitioning massive graphs. In *Proceedings of the ACM International Conference on Web Search and Data Mining*, pages 507–516, Rome, Italy, 2013.

69. L. Wang, Y. Xiao, B. Shao, and H. Wang. How to partition a billion-node graph. In *Proceedings of the IEEE International Conference on Data Engineering*, pages 568–579, Chicago, IL, USA, 2014.

70. F. Y. Wu. The Potts model. *Reviews of Modern Physics*, 54(1):235–268, 1982.

71. Z.-H. Wu, Y.-F. Lin, S. Gregory, H.-Y. Wan, and S.-F. Tian. Balanced multi-label propagation for overlapping community detection in social networks. *Journal of Computer Science and Technology*, 27(3):468–479, 2012.

72. J. Xie and B. K. Szymanski. Community detection using a neighborhood strength driven label propagation algorithm. In *Proceedings of the IEEE International Workshop on Network Science*, pages 188–195, West Point, NY, USA, 2011.

73. J. Xie and B. K. Szymanski. Towards linear time overlapping community detection in social networks. In *Proceedings of the Pacific-Asia Conference on Knowledge Discovery and Data Mining*, pages 25–36, Kuala Lumpur, Malaysia, 2012.

74. J. Xie and B. K. Szymanski. LabelRank: A stabilized label propagation algorithm for community detection in networks. In *Proceedings of the IEEE International Workshop on Network Science*, pages 138–143, West Point, NY, USA, 2013.

75. J. Xie, B. K. Szymanski, and X. Liu. SLPA: Uncovering overlapping communities in social networks via a speaker-listener interaction dynamic process. In *Proceedings of the ICDM Workshop on Data Mining Technologies for Computational Collective Intelligence*, pages 344–349, Vancouver, Canada, 2011.

76. W. W. Zachary. An information flow model for conflict and fission in small groups. *Journal of Anthropological Research*, 33(4):452–473, 1977.

77. J. Zhang, T. Chen, and J. Hu. On the relationship between Gaussian stochastic blockmodels and label propagation algorithms. *Journal of Statistical Mechanics: Theory and Experiment*, P03009, 2015.

78. L. Zong-Wen, L. Jian-Ping, Y. Fan, and A. Petropulu. Detecting community structure using label propagation with consensus weight in complex network. *Chinese Physics B*, 23(9):098902, 2014.

6

Blockmodeling of Valued Networks

Carl Nordlund[1,2] and Aleš Žiberna[3]

[1]The Institute for Analytical Sociology, Linköping University, Sweden
[2]Center for Network Science, Central European University, Hungary
[3]Faculty of Social Sciences, University of Ljubljana

6.1   Introduction

In the wide variety of networks that connect and make up our worlds, relations not only either exist or not, but often carry a weight. Friendships can be ranked, interactions can be timed, and economic exchanges can be valued, details that provide us with a deeper, nuanced, and higher-resolution understanding of such networks than is provided by the mere existence of ties.

As with most methods and heuristics in network analysis, approaches for clustering[1] or blockmodeling are primarily geared to binary data [23, p. 25; 25]. For the set of indirect methods that do work with valued networks when determining clusters of equivalent actors, valued data still pose a dilemma when interpreting possible blockmodels derived from such approaches. Since the ideal blocks[2] of generalized blockmodeling (as well as density-based structural blockmodeling) are specified in terms of binary ties, these are not readily comparable with the intra- and inter-block patterns of valued relations. Valued networks are therefore often dichotomized, either prior to identifying equivalent sets of actors or when determining patterns within and between clusters using a statistically, theoretically or arbitrarily determined network-wide threshold.

---

[1] Following the terminology in this book, the terms "cluster" and "clustering" as used in this chapter refer to subsets of, and the procedure of identifying, actors that fulfill some meaningful definition of equivalence (also referred to as a "position" in some literature).
[2] In this chapter, "block" refers to the non-overlapping submatrices in a blockmodel as delineated by one or two clusters (i.e. subsets of actors; see the previous footnote).

Advances in Network Clustering and Blockmodeling, First Edition.
Edited by Patrick Doreian, Vladimir Batagelj, and Anuška Ferligoj.
© 2020 John Wiley & Sons Ltd. Published 2020 by John Wiley & Sons Ltd.

Whereas dichotomization can be theoretically motivated in some cases, *particularly when it comes to ranked/ordinal data*, dichotomization constitutes an inevitable reduction of the resolution and detailed level of valued data. In addition, the setting of a global cut-off to distinguish prominent and non-prominent valued ties is inherently problematic when actors in a valued network have different relational capacities [4, 41]. Good examples of networks in which actors have different relational capacities are trade networks. For instance, whereas the USD 59.4 billion worth of commodities that went from Germany to Austria in 2010 represents half of Austria's total imports, this very same flow only corresponds to 8% of Germany's total exports.[3] With unequal relational capacities such as between Austria and Germany, where the local perception of what constitutes a prominent tie differs widely, the use of absolute (network-wide) dichotomization thresholds show us, at best, just one possible way of looking at the data. In contrast to the trade flow example, an assumption of equal relational capacities would be more reasonable when, for instance, tracking the interaction times of school children during a common 45-minute break.

Besides dichotomization, other types of transformation can be useful for analyzing valued networks, both for identifying clusters and discerning how the block patterns of valued ties relate to the set of ideal binary blocks used in blockmodeling. A growing number of direct and indirect blockmodeling heuristics is also designed explicitly for valued networks, approaches that often are combined with, or based on, various transformations.

This chapter provides an overview of approaches, classical as well as recent innovations, to the clustering/blockmodeling of valued networks. This chapter focuses exclusively on one-mode, single-relational, non-signed,[4] possibly directional and possibly with self-ties, valued networks and the particularities that apply to the clustering/blockmodeling of such networks. We thus do not delve deeper into the properties of networks and methods that are equally relevant to binary networks.[5] We also do not cover stochastic blockmodeling, although versions of it for valued networks exist [e.g. 1].

In addition to the various direct and indirect methods applicable to valued networks, we describe various transformations found useful when clustering valued networks. The technicalities of each method and transformation are described, and we also discuss the particular variety of equivalence a respective method is geared to capture. We exemplify how different heuristics yield different results using two datasets, embodying different properties of valued networks: the EIES friendship data[6] at time point 2 [28] and intra-European commodity trade [41].

The next section describes the different types of valued network data that exist. This is followed by a description of various transformations that can be applied to such data prior to clustering methods being used. An overview of clustering methods and heuristics for valued networks follows, divided into indirect and direct approaches. An example section comes next

---

[3] Taken from the EU/EFTA trade example in the second part of this chapter, we occasionally use this data in our theoretical section to exemplify the effects of transformations.

[4] Some methods described in this chapter are also appropriate for signed networks (e.g. homogeneity blockmodeling [25]).

[5] Such as hierarchical clustering methods and their different varieties: although useful for grouping actors based on indirect measures of equivalence, such tools are equally useful for binary networks as well. That said, we do demonstrate that different hierarchical clustering methods yield different partitions in our first EIES data example.

[6] EIES data obtained from https://sites.google.com/site/ucinetsoftware/datasets/freemanseiesdata

in which we demonstrate various clustering/blockmodeling techniques on our two example networks. A concluding section rounds off the chapter.

6.2   Valued Data Types

As with statistical variables generally, tie values[7] can be measured on different scales. In this chapter, only networks where ties can be treated as measured on at least an interval scale are considered. We remain relatively liberal in terms of what can be treated as interval scales, as is often the practice in the social sciences. Therefore, tie values on 1–5 rating scales (Likert-type items) and similar are deemed acceptable.

Below, we list some of the more commonly used types of valued networks based on the meaning and/or origin of ties:

- A subjective judgment of relationships using different rating scales (e.g. a 5-point scale with descriptions at the extremes only, a 5-point scale with a description of each category, line measurement ...). The quality of measurement using different scales has been discussed elsewhere [e.g. 27, 30]. These rating scales share certain features:
  - The measurement is subjective (based on the perception of the responding actor).
  - At least the scale extremes are defined and are the same for all actors. In principle at least, all actors can also select the extreme option. This also means the definition of a "strong" tie is given from the outside (but interpreted by the respondent). Examples of such networks include, among others, all kinds of perceived social support networks, advice networks [e.g. 48] and friendship networks [28].
- Direct measurements of different flows or the frequency of some kind of interactions. Examples of such networks are email exchanges between people [47, 54], citations among journals [3], and trade among countries [41]. While these measurements may lack a pre-specified maximum, natural limits may exist such as when mapping minutes of school kids, interaction during a lunch break of limited duration. Typically, however, the perception of what constitutes a strong tie in such measurements varies among actors.

One distinction we concentrate on in this chapter is whether actors have equal or unequal relational capacities [41]. That is, whether all actors, at least in theory or principle, have similar capabilities for creating/maintaining ties of a certain strength and/or having a similar upper ceiling with respect to the sum of such tie strengths. This is very closely related to the question of how to determine which ties can be deemed important, strong, prominent etc. Can we determine this uniformly for the whole network, for parts of the network, or should this be done per actor or even at the dyadic level?

In more practical terms, when doing blockmodeling on networks where at least empirically different relational capabilities exist (e.g. a notable variance in valued in- and out-degrees), two interrelated questions must be addressed prior to the analysis:

- Does the perception of what constitutes a strong tie differ significantly among actors? Put differently: do relational capacities differ among actors?

---

[7] Also known as tie weights, particularly in the network science literature.

- If so, do we want such differences in relational capacities to influence our results? Are such would-be differences part of what we are looking for, or do we want to somehow discount such differences in our analysis?

Depending on the response to such questions, a variety of transformations and metrics exist that, with different variations in the specific conceptualization of equivalence, can be used in the context of valued blockmodeling.

6.3 Transformations

Prior to blockmodeling valued networks, relational data are often transformed in various ways to either adapt them to methods and heuristics that are primarily used for binary data or to assist a method in assessing a specific kind of tie prominence. The most trivial transformation is that of dichotomization: using a network-wide absolute threshold, the valued network is binarized to which conventional heuristics and methods apply. Treating dichotomization separately in what follows, other types of transformations have been used to cluster/blockmodel valued data. These can be separated into element-wise transformations where the individual ties are transformed independently of each other, or what we may call structural transformations where the transformation of a valued tie in some way depends on the properties of other valued relations in the network.

Transformations of valued networks can be combined and applied at different stages of a clustering/blockmodeling analysis of valued networks. Describing the various steps and procedural pathways for going from the raw valued network to clusters of actors, we propose the schematic in Figure 6.1, mainly to show how the different steps described later in the chapter

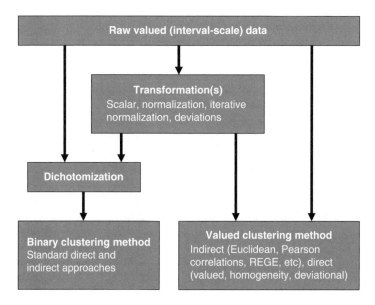

Figure 6.1 Procedures for clustering valued data.

come together. In Section 6.6 we provide some guidelines (also in the form of a flow chart in Figure 6.2) for selecting suitable approaches (including transformation).

We thus identify four main procedures for identifying clusters in valued networks:

- Raw valued data → dichotomization → binary clustering method → clusters
- Raw valued data → (non-dichotomous) transformation → dichotomization → binary clustering method → clusters
- Raw valued data → (non-dichotomous) transformation → valued clustering method → clusters
- Raw valued data → valued clustering method → clusters

Once the clusters are defined and a blockmodel is created, the common final step (not included above) is to interpret the blockmodel and the inter- and intra-block patterns of ties. Such block and blockmodel interpretations can be based on the original valued data, the would-be transformed data prior to clustering, or an additional post-clustering transformation.

The choice of transformations to apply to valued data before and possibly after clustering might at times seem like a somewhat arbitrary decision, often based on what yields the most intuitive, plausible, and interpretable results. Yet each transformation can be seen as the operationalization of a particular interpretation of prominence in the context of the valued network being studied. An example of this interplay between different transformations and final interpretations is found in Breiger's study [15] of intra-OECD trade. When applying the CONCOR algorithm (see below) to the same valued network of trade flows, but testing two different pre-CONCOR transformations – a dichotomization that keeps the quintile of the largest ties, and an iterative row-column normalization procedure – Breiger demonstrates how final interpretations of a dataset are intrinsically tied to the transformation methods that are chosen. We show a similar effect in our EFTA trade example.

In what follows, we look at various transformations that have been applied in the valued blockmodeling context. Starting off with the element-wise scaling transformations and dichotomization, we then consider normalization-based transformations and other structural transformations.

6.3.1   Scaling Transformations

Apart from absolute-value dichotomization, various element-wise scaling transformations can be applied prior to an analysis, transformations that thus retain the rank order of the valued ties. An example is the world-system-inspired study of Mahutga [34] in which he analyzes international trade flows of various commodity bundles for five points in time between 1965 and 2000, i.e. valued networks whose unequal relational capacities among countries (e.g. as given by the skewness of the tie value distributions) is a characteristic feature of the network. Determining regular equivalence using the REGE algorithm (see below), Mahutga applied a log-10 transformation of the commodity flow matrix prior to the algorithm that "maintains the relative differences between countries while aiding the algorithm" [34, p. 1884]. Such transformations are appropriate when the skewness or some other property of the tie value distribution is problematic in terms of the blockmodeling method later used, similar to, for example, using a log transform prior to regression analysis.

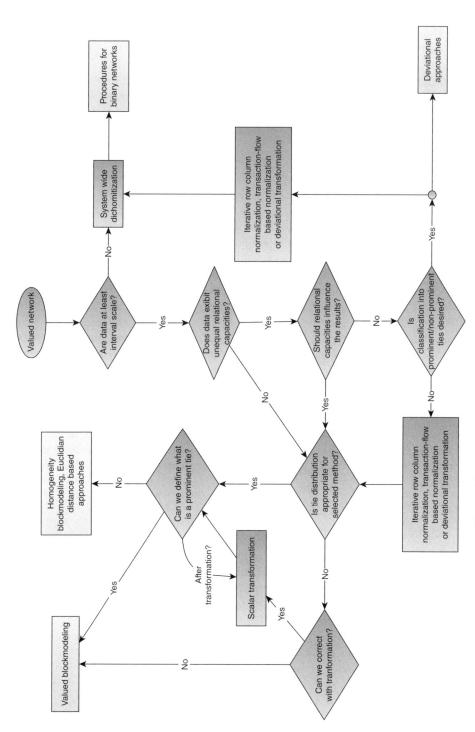

Figure 6.2 A "control chart" for choosing a suitable approach.

6.3.2   Dichotomization

The most common transformation of valued data is to dichotomize them. Irrespective of when the dichotomization is performed (see Figure 6.1), its feasibility hinges on the substantive research question and the properties of the valued ties, particularly their distribution. For ranked/ordinal data, where each value represents a particular strength, e.g. the quality or intensity of a social tie [e.g. 28, 51], dichotomization is used to filter out that particular strength of a tie. This is, we argue, the only case where dichotomization may be preferable to using valued data. If the range of continuous values is constrained, e.g. when mapping playtime among school kids during a 45-minute lunch break, it could be feasible to distinguish prominent (1) and not-so-prominent (0) pair-wise interaction on the basis of a system-wide absolute threshold. When theory offers no guidance on the suitability of such thresholds, distributions/histograms of tie values and valued degrees can be useful for determining possible thresholds. One can also try different dichotomization thresholds on the same valued network and compare the findings: as was done, for example in [22].

Rather than choosing a given absolute threshold for dichotomizing valued data, this can also be decided on the basis of total flow values. For instance, having sorted the valued ties in a network, one option is to retain a certain share of the largest values [15, p. 361], i.e. pruning away the corresponding share of the smallest values.

Dichotomization is not only directly applicable to raw valued data but can be, and often is, used as a secondary transformation prior to clustering. For instance, whereas normalization can be used to transform a valued network (see below), the normalized version of the network may have to be subsequently dichotomized prior to finding clusters using binary-oriented methods [see, e.g., 40]. As normalization and similar structural transformations are typically applied to counter skewness in valued degree distributions, i.e. as an attempt to counter unequal relational capacities among actors, choosing a dichotomization cut-off for the normalized version of the network should be easier than with the raw valued network.

6.3.3   Normalization Procedures

Normalization is a structural transformation that rescales the individual dyads relative to the values of other ties. It can be applied either prior to identifying clusters to assist a particular clustering heuristic [e.g. 3, 15] or in the interpretation of blocks [e.g. 40]. Normalization can either be done on the basis of individual actors and their immediate ego-networks or for the entire network as a whole. The former type of normalization can be done with respect to either inbound (column-wise normalization) or outbound ties (row-wise normalization). Normalization is applicable to all types of valued networks, whether directional or symmetric, and can include self-ties.

The most common type is marginal (or sum) normalization where each value is rescaled such that the sum of all values equals a given constant (typically unity). A full-matrix marginal normalization thus implies a rescaling of all relational values so that they sum up to unity. Although this does not change the relative differences between values and thus retains would-be unequal relational capacities, full-matrix marginal normalization is a useful transformation for comparing two or more networks as it rescales the values of each network into a common standardized value range. For instance, when comparing pupil interactions during a 45-minute break with those of a 1-hour break, full-matrix normalizations of both could allow for an easier comparison of the two.

Row- and column-wise marginal normalization are transformations where, respectively, the out- and in-degrees for each actor sum up to unity.[8] In the above example on Germany's exports to Austria, a row-wise marginal normalization would thus transform this dyad to 0.078 whereas a column-wise marginal normalization would transform it to 0.504. In these directional normalizations, the prominence of a tie is thus determined on a directional per-actor basis, indicating the share of a tie value with respect to the total (in- or outbound) tie values. A marginal normalization on rows (columns) is calculated by dividing each valued relation with the total valued outdegree (indegree). If an actor has a zero out- or in-degree, its corresponding row- or column-normalized values are in principle undefined, but often coded as 0.

For directional valued networks, assuming a lack of theoretically compelling reasons to only look at outbound or inbound ties, another possibility is to combine the two marginal-normalized matrices for rows and columns. In [40], given particular clusterings, both the row- and the column-wise normalized matrices were analyzed simultaneously to determine occurrences of regular blocks. Using a standardized threshold value, the row-regularity of blocks was identified by looking at the row-wise normalized matrix while the corresponding column-regularity was identified using the column-wise ditto. An alternative is to combine the row- and column-normalized matrices into a common matrix. Adding the elements of the row- and column-wise marginal-normalized matrices produces a matrix in which the cell values, ranging from 0 to 2, indicate the total relative significance of a valued flow for both actors. In the context of international trade, this composite metric has been used as a measure of the economic importance of individual flows [21, p. 182]. In the case of the directional trade flow from Germany to Austria (see above), this flow would then be transformed to 0.582.

Common alternatives to the marginal type of normalization are normalizations with respect to the mean, maximum, Z score, and standard deviation, each applicable to the whole network as well as the ego-centric normalizations with respect to in- and outbound ties, with or without self-ties. Of these, the mean variety has an advantage in that it works in row or column normalizations where actors might lack out- or inbound ties, whereas the marginal-, standard deviation- and Z-score-based normalizations yield invalid results when an actors corresponding in- or outdegree is zero.

6.3.4   Iterative Row-column Normalization

An extension of the above normalizations is to use the iterative row-column normalization approach [see, e.g., 46, p. 267; 37, 42, 43]. Repeatedly normalizing the valued data by rows, columns, rows, columns etc. will for most non-trivial data lead to a convergence where the sums of both the row and column vectors, respectively, sum up to unity (for marginal-normalization) or zero (for mean-normalization). The latter approach was used as a pre-clustering transformation by Breiger [15] and Baker [3]: whereas Breiger prepared his intra-OECD trade data prior to applying the CONCOR algorithm, Baker used this iterative approach prior to the Euclidean-based indirect measure of structural equivalence (see the next section).

---

[8] For symmetric (non-directional) valued networks, the row-wise marginal-normalized matrix is the transpose of the column-wise one.

The choice between the marginal- and mean-type iterative normalization is not arbitrary; instead, they are distinctly different transformations. The choice should primarily be made on substantive grounds. While the results of marginal row-column normalization can be thought of as values equivalent to raw ones in a hypothetical case where all units have exactly the same relational capacity, the meaning of "equivalent" for mean-based normalization is harder to assess. In addition to the mean and marginal iterative normalizations, similar iterative convergences occur when conducting row and column normalizations based on standard deviations (dividing each element in the row/column by the standard deviation of that row/column) or Z-scores (subtracting the mean from each element, subsequently dividing by the standard deviation). However, contrary to what is the case for the mean and marginal varieties where it is irrelevant whether the iteration starts with the rows or a column, that is not the case for the standard deviation and Z-scores iterative approaches.[9]

6.3.5 Transaction-flow and Deviational Transformations

The dilemma of unequal relational capacities precedes contemporary network analysis and so did the suggested solutions for distinguishing the prominent ties of a valued network from those that could be deemed less prominent. Much of this work was in the context of international trade and the much-debated transaction flow model and Relative Acceptance index of Savage and Deutsch [45; also see 2, 29, 31]. Savage and Deutsch aimed to create a null (quasi-independence) model of flow value probabilities that accounts for the individual distributions and magnitudes of both imports and exports of each country.[10] Contrasting these probabilities with actually occurring flows, the dyadic Relative Acceptance metric could be used, they argued, to identify interesting relations and segments of a trade flow network.

The Relative Acceptance metric was used as an indicator of dyadic prominence in several studies [e.g. 14, 21, 31; cf. 38]. A modification of the index was proposed by Goodman [29] where the null model of expected flows was based on the empirical topology: rather than assuming a null model distribution among all possible alters, Goodman's alternative restricted null distributions to the existing topology. In the context of international relations, Brams found that, although the default Savage–Deutsch null model might be feasible in the context of international trade, the Goodman variation and its assumption of a fixed topology was more useful when looking at the size and distribution of inter-country diplomatic missions [e.g. 14, p. 883, note 14].

While the Savage–Deutsch transformation has (to our knowledge) not been applied in blockmodeling, the deviational approach to valued blockmodeling suggested by Nordlund [41] shares the same principal objective of transforming valued ties with respect to the relative share of in- and outflows of each actor. Similar to the Relative Acceptance transformation, the deviational approach is based on a model of expected flows on the assumption of no interactions. However,

---

[9] The mathematical properties of Z-score-based iterative normalization is analyzed in [42], describing and exemplifying the procedure by beginning with Z-score-normalization on rows. Their example transformation is done by starting with columns: if starting with rows instead, following their own description, the transformed values differ quite substantially from those obtained when starting the iterated normalizations with respect to columns.

[10] This model could also be used to compute the expected values of cell values, which could then be used for normalization (e.g. dividing the actual value with the expected).

instead of transforming (through iteration) each valued dyad into a singular value, the deviational approach analytically arrives at, for each valued network, two separate matrices capturing deviations from expected values for, respectively, outbound (RD) and inbound (CD) flows. The formulas are given and described in [41, p. 163].

As the deviational transformation produces two matrices rather than one, it is not immediately useful for conventional direct and indirect clustering methods. Instead, Nordlund [41] proposes both indirect and direct clustering approaches where both RD and CD are simultaneously used. Apart from measuring the degree of equivalence and identifying subsets of equivalent actors, heuristics for interpreting the resulting blocks based on both of these deviational matrices are also suggested. These will be addressed in subsequent sections.

The deviations for corresponding cells in RD and CD are in many cases quite similar. In the previous Germany-to-Austria trade example, this flow deviates positively for both, albeit somewhat differently: from Germany's perspective, this flow is 83% higher than expected ($rd_{DEU,AUT} = 0.825$), whereas this import flow to Austria is 120% higher than expected ($cd_{DEU,AUT} = 1.199$). In this example on intra-EU/EFTA trade (see below), there are also examples of differences in the signs of the deviations. Whereas the trade flow from France to Germany is 13% higher than expected from France's point of view, it is 1.4% lower than expected from Germany's point of view.

6.4   Indirect Clustering Approaches

Capturing different specific notions of equivalence, indirect approaches and algorithms estimate the degree of such equivalences for each possible pair of actors in the network. Having established such measures of equivalence, suitable partitions of equivalent actors can then be identified using cluster analysis, such as hierarchical clustering, where the number of equivalent classes is either determined theoretically or guided by a suitable procedure for choosing an optimal number of clusters (see [36] for an overview of such procedures). Whereas this general procedure of indirect clustering methods is the same for binary and valued networks, below we focus on common indirect dis(similarities) applicable to valued networks – two for structural equivalence and the REGE algorithms for regular equivalence. We also discuss specific issues concerning valued networks when interpreting the resulting blockmodels derived from such indirect methods.

6.4.1   Structural Equivalence: Indirect Metrics

For structural equivalence, defined as when two actors have the same relations (and lack thereof) with their alters [33], the operationalizations of indirect measures are quite intuitive. In a symmetric network represented as a sociomatrix, the row or column vectors for each pair of actors are compared with each other, whereas both the row and column vectors are compared for directional networks.[11]

---

[11] For such comparisons of actors, profiles, there are different ways to cater for would-be self-ties [see, e.g., [53], p. 367]. Regardless of which metrics is used, it is important to pay special attention to self-ties (if present) and ties between the two units being evaluated for equivalence by using corrected measures [8]. For example, in a directed network with

Although there are many measures of (dis)similarity that are useful for valued networks (e.g. Tanimoto, cosine, Manhattan etc. [see 25, p. 181ff]), the most widely used indirect metrics of structural equivalence seem to be Pearson correlations and Euclidean distances. These indirect approaches thus transform the original sociomatrix into, respectively, either a correlation matrix (i.e. where a value approaching unity implies equivalence) or a distance matrix (i.e. where a value approaching zero implies equivalence).

Whereas both Pearson correlations and Euclidean distances are used to capture notions of the structural equivalence of actor-pairs in valued networks, they represent two slightly different takes on what structural equivalence means. Comparing the two metrics [e.g. 52, p. 374], the Euclidean distance (also called the "social distance" [see 19, p. 95]) metric is sensitive to differences in means and variances of the tie values of two actors [26], while the correlation metric caters for such differences. From this, the correlation-based indirect measure is oriented to identifying similar profiles of alter ties [e.g. 52, p. 375], with the Euclidean metric of structural equivalence instead capturing similarities with respect to tie strengths. This is exemplified in [41] where blockmodels stemming from, respectively, the correlation- and Euclidean-based indirect metrics are compared for Baker's citation data [3]. Whereas the Euclidean-based metric of structural equivalence produces clusters of actors with similar gross (in- plus out-) degrees, the clusters arising from the correlation-based indirect metric contained actors with wider variations in (unequal) relational capacities.

6.4.2   The CONCOR Algorithm

Building on the above indirect measure of structural equivalence, the CONCOR algorithm is an iterative algorithm that builds on the phenomenon of the convergence of iterated correlations. As an independent co-discovery of Breiger and Schwartz [16, 46], the phenomenon of convergence of iterated correlations had previously been observed by McQuitty [35].

Starting with the correlation-based indirect measure of structural equivalence (see above), the CONCOR algorithm repeats correlating the correlation matrix. With few exceptions, this procedure converges to a matrix that contains only positive and negative unity, these corresponding to partitioning the network into two, arguably structurally equivalent, clusters. The CONCOR algorithm can then be applied to each of these subsets, further partitioning the network in a hierarchical fashion.

As the CONCOR algorithm is based on correlations, it works for both binary and valued networks. This implies it is sensitive to differences in the means and variances of the valued relations of actors, i.e. there is an emphasis on similarities in the patterns rather than the strengths of ties. However, although the CONCOR algorithm has seen extensive use [see 16, 39, 50, see 32, for an extensive example], it has some well-known problems. First, CONCOR always produces a dual partition. For networks containing three very accentuated and similar clusters, the CONCOR algorithm always yields a split into two parts. Second, the choice of clusters for subsequent CONCOR splits can be arbitrary. In [50], the choice of which clusters to split was based on the size of the remaining clusters, implicitly assuming that the sets of structurally equivalent clusters would be similar in size. Third, the CONCOR algorithm can be quite sensitive to small

---

self-ties, if units i and j are to be structurally equivalent, tie $i \rightarrow j$ should be equivalent to tie $j \rightarrow i$ and self-tie of i to self-tie of j.

variations in the input data. This was demonstrated in Breiger's study [15] of OECD trade data: applying the CONCOR algorithm to both raw trade data and means-based iterative row-column normalized data (see the previous section), he found significant results in the partitions suggested by and derived from the CONCOR algorithm.

6.4.3   Deviational Structural Equivalence: Indirect Approach

Whereas the indirect approaches described above are applied to individual[12] matrices, these matrices can, as shown in the previous section, be subject to various types of transformations prior to clustering. However, as the deviational transformation yields two transformed versions of the original valued network, the standard indirect methods do not apply. Adapting the standard correlation-based formula to cater for both the RD and CD deviational matrices (see Section 3.5), Nordlund [41] proposed an indirect measure of deviational structural equivalence, where row and column vectors were analyzed by looking at, respectively, the row-based (RD) and column-based (CD) deviations [41, p. 165].

Once calculated for each pair of actors, this indirect measure of deviational structural equivalence is then subject to hierarchical clustering or a similar method for finding equivalent actors. Both RD and CD are used in the subsequent interpretation (i.e. the identification of complete and null blocks) of the resulting blockmodel (see below).

6.4.4   Regular Equivalence: The REGE Algorithms

While structural equivalence implies having the same patterns of ties to individual alter actors, regular equivalence implies having the same patterns of ties to actors that are themselves equivalent actors (White and Reitz 1983). Unlike how structural blockmodeling examines a blockmodel for complete (filled with ties) and null (void of ties) ideal blocks, regular blockmodeling replaces the former with ideal regular blocks, meaning blocks where each row and column, respectively, contain at least one tie. Formulated by Sailer [44], regular equivalence was further developed by White and Reitz [53]. White and Reitz [53] also proposed the REGE algorithm as an indirect method for measuring regular equivalence between pairs of actors. Through an iterative point-scoring procedure, the algorithm measures how well the relations of each pair of actors match each other in terms of both the strength and equivalence of the actors on the other end [53; 52, p. 479ff; 11, 13, 57]. A limitation of the REGE algorithms is that they only search for *maximal* regular equivalence.[13]

The notion of regular equivalence and the definition of a regular block is specified for binary networks. For formal definitions for valued networks, see [57]. Although the REGE algorithm

---

[12] With the caveat of multiplex networks: the classical indirect measures of equivalence presented here also have implementations for networks with multiple layers of relations. For such situations, the resulting measures of equivalence thus represent a composite measure of equivalence for each layer. Examples of such applications are the classical series of world-system studies [34, 39, 49, 50], where multiple types of international relations and commodity classes were simultaneously analyzed when identifying structurally and regularly equivalent countries.

[13] Maximal regular equivalence is the regular equivalence where the classes are the largest possible. For binary networks, this also means that REGE algorithms are appropriate for undirected connected networks (no isolates), as then all units represent one regular equivalence class.

applies to both binary and valued data, it has been shown to have some shortcomings, particularly when it comes to valued networks. Due to the workings of the matching function in REGE, a few similar ties of large magnitudes could very well dwarf several similar ties of lower magnitudes [see 11, 13]. The REGE algorithm thus tends to emphasize the strengths, rather than the patterns, of ties when estimating the degree of regular equivalence between pairs of actors. As noted by Žiberna [57], more than one variant of the REGE algorithm exists.

6.4.5   Indirect Approaches: Finding Clusters, Interpreting Blocks

As the above-mentioned indirect methods result in dyadic measures of equivalence, the definition and interpretation of equivalence depend on the chosen metric and would-be pre-cluster transformations. The partition of the correlation and distance matrices derived from, respectively, correlation- and Euclidean-based indirect measures also depends on the chosen clustering method. Although agglomerative hierarchical clustering is often used to delineate clusters of equivalent actors, the choice of the agglomeration method, i.e. single-link, complete-link, weighted/unweighted average, Ward etc., also affects the resulting partition. In addition, the choice of selecting a suitable number of clusters remains, a choice that can be informed by theoretical considerations, a review of different partitions, or procedures or indices for determining the number of clusters, or a combination of these. These issues concerning indirect-based partitioning apply equally to binary as well as valued networks and are thus not discussed here further.

Once a clustering and a corresponding blockmodel have been established, the subsequent step is typically to interpret the resulting blocks. For binary networks (including pre-cluster dichotomized valued networks), such interpretations are fairly straightforward, i.e. by comparing how the patterns of binary ties compare to the set of ideal blocks for structural, regular, or generalized equivalence. The comparison between valued empirical blocks and such ideal block types is more complicated where valued networks are involved. Žiberna [55, p. 108] specified ideal blocks for valued networks, which are only applicable in cases where raw values are compared (e.g. in terms of indirect approaches, if Euclidian distance-based matrices are used and not correlation or deviational based ones). When using homogeneity blockmodeling and structural equivalence, this means the blocks are represented and interpreted based on block means (or some other central values) of original and/or transformed data (if a transformation was used prior to blockmodeling). Using block means is also common and very suitable for Euclidean distance-based approaches (for valued or binary networks).

To match empirical valued blocks with the set of ideal binary blocks, post-cluster dichotomization can be employed. This can work well if the notion of equivalence includes magnitudinal differences of tie strengths, e.g. how the indirect Euclidean measure of structural equivalence (and perhaps also the REGE algorithm) works. If the chosen indirect method does take unequal relational capacities into account, e.g. such as the correlation-based metric, post-cluster dichotomization using a system-wide threshold of the raw valued data will emphasize the strength of ties when identifying various ideal block types. However, if a pre-cluster transformation for countering unequal relational capacities is applied, a post-cluster dichotomization of these transformed values could be used to identify prominent tie patterns for comparison with the ideal binary blocks.

To identify regular blocks in valued networks, Nordlund [40] proposes a post-cluster heuristic. Given a determined partition into regularly equivalent subsets, the raw valued data is normalized into, respectively, both a row- and column-normalized matrix (i.e. RN and CN above). Regular blocks are then identified by checking for row regularity in the RN matrix and column regularity in CN. Although determining tie prominence on the basis of both actors in a dyad, the heuristic involves dichotomization in this step as a standardized threshold is used to distinguish prominent and non-prominent ties.

Finally, for indirect deviational blockmodeling, Nordlund [41, p. 166] proposes that prominent ties be identified by looking at the signs and magnitude of deviations in both RD and CD. As the deviation for a valued tie may differ between RD and CD, i.e. depending on the actor in the dyad, the approach could result in valued ties being categorized as neither prominent nor non-prominent, but non-determined. Density block images based on this approach could thus result in blocks where the densities are expressed as ranges rather than precise values.

6.5   Direct Approaches

Direct approaches search for a partition based on criterion function that directly analyze network data, as opposed to indirect approaches that work on some (dis)similarity metrics computed from such raw data. Here we focus on generalized blockmodeling, especially the approaches for valued networks, including the deviational direct approach.

6.5.1   Generalized Blockmodeling

Generalized blockmodeling (for binary networks) has already been broadly discussed in the book *Generalized Blockmodeling* [25] and numerous other papers [e.g. 5, 7–10]. Generalized blockmodeling is a direct approach, meaning that a criterion function measures inconsistencies between an empirical partitioned network and one (or more) possible ideal blockmodel(s). In generalized blockmodeling, equivalence is defined by a set of allowed block types, which can either be set globally for the whole network or individually for each block. A block forms part of the network/matrix that represents ties from one cluster to another (these two can also be the same, i.e. blocks can also represent ties within a cluster). For each block, out of a set of allowed block types (which might also be only one), the block type with the minimum inconsistency vis-à-vis the empirical block in the same position is selected. The value of the criterion function for the whole network and a partition is then the sum of these block inconsistencies for all blocks.

The partition is optimized (by testing a large number of different partitions) so that the value of the criterion function is as low as possible. In the original as well as most other versions of generalized blockmodeling, relocation algorithms are used to optimize the criterion function [10, 25, 55, 60], although other attempts also exist [17, 18, 20]. For such relocation algorithms, the number of clusters must be set in advance.

There are several approaches to generalized blockmodeling. Each approach to generalized blockmodeling has its own logic concerning how the inconsistencies between empirical and ideal blocks are computed. Furthermore, each also defines a set of possible allowed block types. For binary, valued, and homogeneity blockmodeling (see below), these and their respective methods for calculating inconsistencies are presented in [55, Tables 2 and 3]. Implicit blockmodeling [6, 56, 58] is not discussed here as we find that other approaches are usually more appropriate, while approaches for signed networks [24] are discussed in Chapter 8. The deviational direct approach [41] is discussed in a later section. In this chapter, we only present the main ideas entailed in the selected blockmodeling approaches.

Binary blockmodeling only treats ties as either prominent (present) or non-prominent (not present). The inconsistencies are essentially computed as the number of times a tie is present where it should not be (based on ideal blocks) or vice versa. However, as ideal blocks are specified in terms of binary ties, the approach of computing the inconsistencies between the observed and ideal ties cannot directly apply to valued networks unless the valued ties have first been dichotomized appropriately. The question of whether it is appropriate to apply binary (generalized) blockmodeling to a valued network is thus often a question of whether some dichotomization transformation is appropriate (see Section 6.3).

6.5.2 Generalized Blockmodeling of Valued Networks

While valued blockmodeling is in essence similar to its binary counterpart in terms of "classifying" ties as either prominent or not, it also takes the value of the tie into account in such classifications. Therefore, the inconsistencies are computed as the sum of deviations from either 0 (when according to an ideal block a tie should not be present) or from a threshold we call m that determines, when a tie is considered prominent, if the tie should be present. Of course, when a tie should be present, values exceeding m do not cause inconsistencies. In the classical setting, a tie lower than $m/2$ is considered closer to being non-prominent and one higher than $m/2$ is considered closer to being prominent. Obviously, m is a parameter that must be selected by the researcher based on what can be considered a prominent tie. While some suggestions for how to select the value of the m parameter based on empirical data were given by Žiberna [55, 56], it is noted that valued blockmodeling is probably not appropriate when m cannot be selected based on what constitutes a prominent tie. Examples of how m was selected based on which ties we want to be treated as relevant can be found in the example section in the EIES friendship data and (after normalization) in the EFTA trade data.

Homogeneity blockmodeling is quite different because it does not "separate" ties into prominent or non-prominent, but simply strives to have selected values be as homogenous as possible. These selected values could be all tie values in a block (as in the case of complete blocks), values of some function over rows or columns (as in the case of f-(row/column) regular blocks) or some other set of values. Therefore, the inconsistencies are measured by some measure of variability. Sum of squared deviations from the mean and sum of absolute deviations from the median were suggested by Žiberna [55] and are used within this approach. The sum of squares approach is in the case of structural equivalence very similar to using average within-block variance suggested by Borgatti and Everett [12] as measure of blockmodel fit. The main advantage of using a measure not normalized by the size of a block (sum of squares) compared to using the normalized one (variance) is that the "errors" in small and large blocks have equal weight. The advantage of the sum of squares approach over the absolute deviations one is that mean is taken as a representative value (which is usually desired), while the advantage of the absolute deviations approach is that it is less sensitive to extreme values. For that reason, the sum of squares approach is usually the "default" option, leaving the absolute deviations approach to be used when we wish to reduce the effect of more extreme values like in one case seen in the EFTA trade example. Yet they often produce similar results.

One of the primary advantages of homogeneity blockmodeling is that there is no need to specify any parameters in advance (like the m parameter or the dichotomization threshold), while a disadvantage is that the compatibility of some block types is questionable. Originally, the approach also suffered from the so-called null block problem, caused by the fact that the null block in homogeneity blockmodeling is technically speaking a special case of the complete block (i.e. where all tie values are "homogenous" to 0) and thus only "identified" in real networks if the null block was without error. Yet this problem was solved later [59] by so-called restricted blocks where the value from which the deviations are computed is restricted to be no less than some pre-specified value. These restricted blocks (complete, f-regular, etc.) can be used instead of the classical blocks where we would like to identify null blocks (including imperfect ones). The characterization of ideal blocks and formulas for computing block inconsistencies for binary, valued and homogeneity blockmodeling are presented in [59, Tables 1, 2, and 3].

6.5.3   Deviational Generalized Blockmodeling

Just like in generalized blockmodeling described above, deviational generalized blockmodeling [41] searches for an optimal blockmodel with ideal blocks reducing a penalty function containing the total number of inconsistencies between ideal and empirical blocks. Allowing for the same set of ideal binary blocks as in (binary) generalized blockmodeling, the differences lie in how the various penalty scores are calculated and how the ideal blocks are identified. Beginning with the deviational transformation resulting in the row- and column-based deviational matrices (RD and CD, see Section 3.5), these two matrices are then dichotomized into binary versions (RB and CB). The default (recommended) dichotomization here is simply based on the sign of the deviations in RD and CD, but one can also apply either a one-sided (positive) or two-sided (positive and negative) threshold to find prominent and non-prominent ties. In the case of a two-sided threshold, non-determined ties can also appear [see 41, p. 164]. The penalties for different ideal blocks are calculated by looking simultaneously at both of these binary matrices (RB and CB) [see 41, p. 165]. Apart from the standard set of ideal blocks as found in generalized blockmodeling [e.g. 25, p. 187], complete and null blocks exist in both a strong and weak variety, depending on whether the criteria for the respective block is fulfilled in either, or both, deviational matrices.

Since deviational blockmodeling allows for ties to be neither prominent nor prominent, but also non-determined/contradictory, the approach also provides measures of interpretational uncertainty for each ideal block type [see 41, p. 165]. Given that such ties could mean that the sign of the deviation in RD and CD differs, these uncertainties are thus kept out from the actual penalty function. For generalized deviational blockmodeling, it is suggested that the sign of deviations is used when converting from RD/CD to RB/CB, finding the optimal blockmodel(s), and then increasing a two-sided deviational threshold until the total uncertainty score for the blockmodel increases. We can then state that a particular blockmodel is certain up to a specific two-sided deviational threshold, thus turning the inbound cutoff parameter into an outbound indicator of overall interpretational certainty for the blockmodel [41, p. 164].

6.6   On the Selection of Suitable Approaches

No study is the same-research questions, notions of equivalence, relevant aspects, and, of course, datasets vary between studies, making it difficult to provide the analyst with analytical schematics and rules of universal relevance. With the particular dilemmas pertaining to valued networks and the plethora of transformations and metrics available, this is even more the case with blockmodeling and clustering of valued networks. That said, we offer some guidelines we believe are generally applicable.

Whereas indirect approaches aim to capture notions of equivalence independently of how the resulting blockmodels appear, direct approaches identify suitable partitions and clusters on the patterns of ties of the final product, i.e. the blockmodel. Although effective search algorithms exist for finding partitions with low penalty scores, direct approaches are typically very time-consuming [25, p. 134], especially for large networks. Although such direct search algorithms should be run multiple times using different random starting partitions, they could end up with less-than-optimal solutions, i.e. where the criteria/penalty function reaches a local minimum. This problem is confounded the bigger the network is. Accordingly, direct approaches are mostly useful for relatively small networks while the indirect methods are better suited to dealing with large networks. What constitutes a small network and what a large network of course depends on computing power. We thus suggest using direct approaches when they can be estimated within a reasonable time. However, if we want the result to match a certain pre-specified blockmodel (e.g. hierarchical model, cohesive groups, core-periphery), generalized blockmodeling must be used (unless specialized approaches exist).

For blockmodeling valued networks, the choice between clustering methods and possible transformations must be based on the interpretation of the tie strengths and actors. If tie values have an equal interpretation across actors, such as when measuring playtime among pupils during a 45-minute lunch break, our general recommendation is to use the direct approaches for valued networks or distance-based measures (e.g. Euclidean) together with indirect approaches.

However, when actors seem to have unequal relational capacities, for instance as revealed by examining the distributions of tie values and/or valued degrees, another question must be addressed: should such relational capacities influence the result or not? If we perceive such differences as part of what defines equivalence, we once again recommend classical direct approaches and distance-based indirect measures applied directly to the raw valued data. If we wish to discount for such unequal relational capacities, we recommend the deviational direct approach and the correlation-based indirect measure or the application of a suitable transformation as described above prior to applying a more classical direct or indirect approach. For instance, in our EFTA trade example, we used iterative row-column marginal normalization and on this normalized network used "classical" direct approaches, more specifically valued and homogeneity blockmodeling.

To provide general guidance regarding the choice of suitable clustering approaches for valued networks, we propose the control chart shown in Figure 6.2, which comes with two caveats. First, the chart and its recommendations are only approximate and each decision should be taken after also considering other factors, at least the aim of blockmodeling in the specific case and the exact meaning of the data. Second, the figure in most cases does not provide an exact decision, e.g. the exact transformations are not specified and the choice between direct and indirect methods remains open. We have given guidelines on these and other aspects above or in particular sections of this chapter.

In the examples that follow, we test different approaches on two different valued networks. As is evident in these examples, two or more approaches can be equally "good", all depending on whether their respective clustering methods and pre- and post-cluster transformations capture what we intend to capture in our analysis.

6.7 Examples

In this section, we analyze two example networks: data on EIES friendship in time period 2 [28] and intra-EU/EFTA commodity trade in 2010 [41]. Whereas the trade data contain actors with unequal relational capacities, the Likert scale of the EIES friendship data depicts a network in which the actors, at least in theory, have equal capacities for specifying their outbound tie values and where each value on the scale corresponds to a specified degree of acquaintance.

We also exploit the simplicity of the first example (in the sense that no normalizations are used) to allow us to present its matrix partitioned into a larger number of clusters, on one hand using them to justify our selection of the number of clusters and, on the other, using them to allow readers to decide whether they agree with both our suggestions and the number of clusters based on scree plots.

6.7.1 EIES Friendship Data at Time 2

The first dataset on which we demonstrate the use of valued blockmodeling techniques is Freeman's EIES[14] acquaintance network at time point 2 [28]. The data were gathered in 1978 among researchers working on social network analysis. They were first introduced to a system for computer conferencing. Three networks were collected, two of which are acquaintance networks, one gathered at the time of introducing the system and one (which we are using here) after using the system for 7 months. In addition, the number of messages sent using the system was recorded (which we do not use). As mentioned, here we only use the data at the second time point for acquaintance data for 32 researchers. The acquaintance data were gathered using a survey where the participants answered using a 5-point scale coded as:

- 0 – a person who is unknown to me (or no reply)
- 1 – a person I've heard of, but not met
- 2 – a person I've met
- 3 – a friend
- 4 – a close personal friend

The Freemans [28, p. 367] warn that "[s]ince most of the participants were American speakers of English, it is difficult to determine whether their use of the term 'friend' refers to an affect-based tie. Americans tend to use that term to describe anyone from the most superficial acquaintance to a trusted lifelong intimate". The coding is, of course, important for interpreting the results, but also for selecting various parameter values such as the threshold for dichotomizing networks (for use with approaches for binary networks) or the m parameter of valued generalized blockmodeling.

---

[14] EIES data obtained from https://sites.google.com/site/ucinetsoftware/datasets/freemanseiesdata.

This may be considered a classical set of social science network data where the ties represent social relations that are usually measured with a questionnaire. It is also an example of a dataset where at least theoretically the relational capacities are equal for all respondents. Of course, one could argue that these do at least partly differ by respondents. However, these differences are typically not huge and, secondly, are not deterministically pre-determined (like in, for example, trade flow data, where a small country simply cannot trade as much as a big country). Therefore, in such cases the use of a normalizing technique (or methods directly suitable for networks with different relational capacities) is neither needed nor desired.

Such data can therefore be directly analyzed using methods for blockmodeling valued networks, i.e. without initial value transformations, and depending on the analyst's interest, the use of a network-wide dichotomization threshold could indeed be appropriate. One should note that, strictly speaking, the measurement scale used here is ordinal. However, as methods for blockmodeling valued networks assume at least an interval measurement scale, we treat these data here as interval scale data when applying generalized valued blockmodeling. Treating ordinal scales as an interval is often better than alternative solutions [61].

The classical approach to dealing with such data is to dichotomize the data. If we are explicitly interested in values above a specific threshold, for instance "friend" and above in the EIES data, dichotomization is indeed reasonable. Here the use of ties with a value equal to or greater than "friend" (3) seems natural, although other values might also be reasonable for certain purposes. Another reason for using this threshold here is that using a higher threshold (e.g. treating only the "close personal" friend values as ties) would be too stringent and result in just 51 ties, yielding a network density of 0.051 and an average in-out degree of 1.6.

For small (relatively dense) friendship networks, the structural equivalence model is appropriate.[15] Moreover, since the network is small, direct approaches are the most suitable.[16] Therefore, binary generalized blockmodeling according to structural equivalence was used on these data.

One question in need of answer is how to choose the number of clusters. We chose the number of clusters after reviewing the results for partitions into different numbers of clusters and after checking the scree diagram of criteria values at different numbers of clusters, all shown in Figure 6.3. Based on the scree diagram, four clusters seem appropriate (the curve "breaks" at 4, where subsequent additional clusters do not decrease the criterion function as much) and looking at the partitioned matrices confirms this result, as a five-cluster solution does not seem to increase the clarity of the block much. The four clusters that are obtained may be described as:

- Cluster 1: The most "popular" singleton cluster (i.e. containing a singular actor), with reciprocal ties to all actors in cluster 3.
- Cluster 2: A large cohesive group that also has relatively numerous ties (especially outgoing) to units from other groups.
- Cluster 3: A smaller cohesive group with reciprocal ties to the most popular unit.
- Cluster 4: A peripheral cluster – the least internally connected cluster, mainly connected to the singleton cluster 1.

---

[15] In addition, in this chapter we focus on issues connected to using valued data and therefore, as a rule, are sticking to the simplest equivalence that makes sense usually structural equivalence.

[16] In general, direct approaches are suggested unless the size of network is too large to be analyzed in a reasonable time with these approaches.

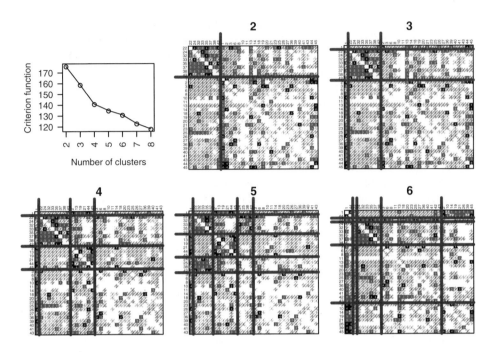

Figure 6.3   Binary generalized blockmodeling.

The number above the matrix plot indicates the number of clusters. In matrices, the valued network is plotted. The ties that were converted to 0s in the dichotomization are represented by shaded cells, while the ones converted to 1s are represented by shaded completely filled cells.

Binary blockmodeling applies if we are strictly interested in ties above a certain level, e.g. "friends or more" above. Yet if we are more interested in acquaintanceship in general, we could use valued blockmodeling on the assumption that the data are at least on an interval scale (i.e. each step between Likert-scale responses corresponds to the same degree or depth of acquaintanceship, for this purpose, the m parameter must be set. On one hand, m should be set to a value that means the two units are strongly connected, or at the maximal range of the scale. Both of these rules would suggest 4 is a suitable value. However, a more precise (and perhaps complicated) rule is that m should be set so that tie values closer to m than to 0 can be considered prominent, which compared to binary blockmodeling suggests using double the threshold that is used for dichotomization. This rule leads to the selection of $m = 5$, which is a more appropriate value although it means no tie can achieve this criterion[17] [55]. However, the advantage is that ties 3 and 4 (i.e. "a friend" and "a close personal friend") will be considered prominent and ties 0, 1, and 2 (i.e. a person they have only met or even less) will be considered non-prominent.

The results are presented in a similar fashion as for the binary blockmodeling in Figure 6.4. In this case, the scree diagram suggests three clusters (or at least eight) and, based on the review of the matrices partitioned into several clusters, this seems reasonable. However, after also inspecting the image matrices (Figure 6.5) we would opt for four or five clusters. We usually omit the

---

[17] It also follows from this that complete blocks will have an inconsistency of at least the number of ties in the block.

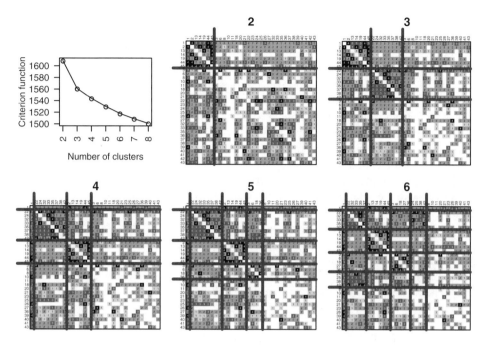

Figure 6.4 Valued generalized blockmodeling with $m = 5$. The number above the matrix plot indicates the number of clusters.

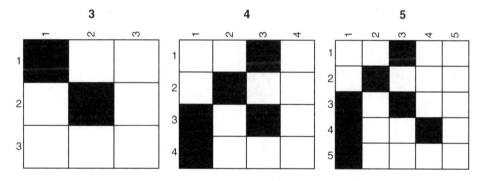

Figure 6.5 Image matrices for valued generalized blockmodeling with $m = 5$. The black squares indicate complete blocks and the white squares indicate null blocks. The number above the matrix plot indicates the number of clusters.

presentation of the image matrices to save space, yet here they were the reason that four or five clusters are suggested. While in the three-cluster solution it seems that no clusters are connected, the image matrices for the four- and five-cluster partitions show the first cluster (the single-most connected individual) has ties to cluster 3 and from all clusters but cluster 2 (and itself). Here we focus on four clusters for comparability with the results of the binary blockmodeling, although the five-cluster solution does find a new small cohesive group from the units from the peripheral cluster. In the case of four clusters, the interpretation of the clusters remains the same as for the binary blockmodeling. The membership of clusters is also very similar, only two units from a smaller cohesive group and a peripheral cluster change places (units 14 and 21). Unit 14 has all its prominent ties to or from the small cohesive cluster valued at 4 and most of its binary non-ties valued at 2 (almost relevant), which causes it to move into this cluster. On the other hand, unit 21 has all its binary ties to or from this cluster valued at 3, which was not strong enough evidence in the valued case.

Other approaches were also tested on this dataset. These included the indirect approach using corrected Euclidian distance [8] with $p = 2$ with Ward and complete hierarchical clustering and sum of squares homogeneity blockmodeling according to structural equivalence [55]. As the correlation measure and various transformations (including deviation based) are theoretically unsuitable, we do not present their results here. The four-cluster results for these selected methods together with the selected partitions that were previously presented are shown together in Figure 6.6.

The first partition was obtained using binary blockmodeling and is the same as in Figure 6.3, but here it is drawn on the valued network without any modifications, showing that the possible impression of clarity (based on Figure 6.3) might be false. Comparing this partition to the valued blockmodeling one with $m = 5$ (the next figure), we see they are indeed very similar. This is expected because in both approaches values 0, 1, and 2 fit better in the null block and values 3 and 4 in the complete block. The only difference between these partitions is seen in the binary partition, where unit 14 is in the fourth cluster, while unit 21 is in the third, while these units are exchanged in the valued $m = 5$ partition. The reason for the difference should be searched for in the diagonal block of the third cluster, which in both solutions is classified as complete.

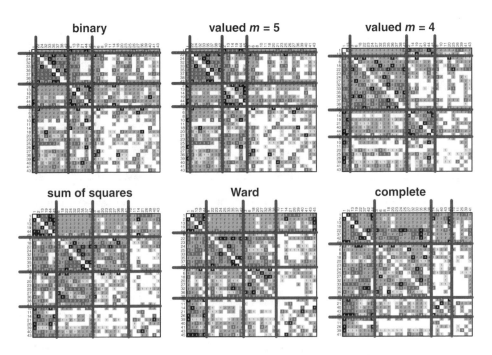

Figure 6.6 Four-cluster partitions by selected approaches.

Unit 21 has five 3s and four 2s as ties to or from other units of the third cluster. This makes it indifferent to these ties being in the null (fourth cluster) or complete (third cluster) block in both approaches. On the other hand, unit 14 has four 4s, five 2s, and one unity tie as ties to and from units from the second cluster. Given that there are more ties below 2.5 than above, this fits better than the null block in the binary approach. However, the sum of distances to 5 is 23, which is less than the sum of distances to 0, which is 27. This makes these ties better fitted for the complete blocks in the valued $m = 5$ approach. The third figure also captures a valued blockmodeling partition, although here the value of $m = 4$ is used. This means that now value 2 ("a person I've met") is as close to complete as it is to the null block. Consequently, complete blocks become larger.

In the bottom row, approaches based on sum of squares or (corrected) Euclidian distance are presented. All these tend to cluster units together with exactly the same tie values and thus the blocks in these partitions are much more homogenous. This is especially evident in the blocks from the first to the second and third cluster, which contain almost only 2s in these solutions (especially homogeneity sum of squares and Ward's solution). In particular, the results of sum of squares blockmodeling and Ward's hierarchical clustering are very similar, which is expected because both try to minimize sum of squared deviations from the mean, for which the homogeneity generalized blockmodeling approach is more successful. For example, based on this we can see that units of cluster 1 are friends (or more) with each other and some of them with selected units from cluster 4, who have met practically everyone. On the other hand, units from cluster 3 have very mixed relationships with each other, although they all have met or are even friends with people from clusters 1 and 2 and have often not even heard of (and have at most met) people from cluster 4. The results of the complete linkage hierarchical clustering are included here mainly to demonstrate that the choice of the hierarchical clustering approach has a big effect on the final result.

To conclude, in this example we find valued blockmodeling (with $m = 5$) the most appropriate method, although binary blockmodeling (with threshold 2.5) comes close behind. Both methods try to find groups that have either a friendly or non-existent relationship with each other, which we find more important than the homogeneity of the ties in this example. The advantage of valued blockmodeling is that it considers how far ties are from these "ideal" (friends, non-existent) states.

Sum of squares blockmodeling can, on the other hand, provide additional insight by trying to find as homogenous blocks as possible, which would be relevant if we were also interested in other levels of the continuum and not only in whether they are closer to being friends or not having a relationship.

6.7.2   Commodity Trade Within EU/EFTA 2010

This example consists of the total bilateral commodity trade between the 30 countries within EU and EFTA for the year 2010 [41]. This network is almost a complete graph, having a topological density of 99.8%. Stated in millions of US dollars, tie values range from 1 to 103,434, with a mean of 3954, a median of 483, and a standard deviation of 10,749.

If binary blockmodeling were to be used on this almost-complete graph, some kind of dichotomization would be necessary. On such a rich valued dataset, we think this should really be avoided.

As is almost a characteristic feature of international trade data, the actors (countries) in this network have vastly different relational capacities. Even if we assume that countries are equally integrated into the global economy, their economies are hugely different in magnitude, resulting in widely different perceptions of what constitutes an important trade flow for each country.

As seen in the previous section on transformations, there are several ways to take the values into account. One way is to simply use the raw data: as argued by Smith and White [49], the normalization of trade data could introduce artifacts for small flows, and the magnitudinal differences between countries constitute an important aspect of the trading system at large. While we agree with this, suitably transformed data (before or within the clustering procedure) can reveal different aspects of the data. In this example, we demonstrate which kinds of results and insights can be obtained by following both paths and explore several options along both paths. However, in all cases we focus only on structural equivalence. As our emphasis here is on transformations, not many clustering techniques will be applied. We begin by analyzing the raw data.

6.7.2.1 Analyzing Raw and Scaled Data

Since there are no theoretically given threshold for values to be considered as relevant, homogeneity generalized blockmodeling or equivalent direct approaches are the most appropriate. Due to the, in principle, superiority of direct approaches, we concentrate more on them. Therefore, we first use both homogeneity approaches, that is, sum of squares blockmodeling and absolute deviations blockmodeling, on the raw data.

Figure 6.7 shows the results of the sum of squares approach. Based on the scree diagram and a review of several solutions, we deem the four-cluster solution the most appropriate. However, regardless of the number of clusters a core-periphery structure is revealed whereby the core

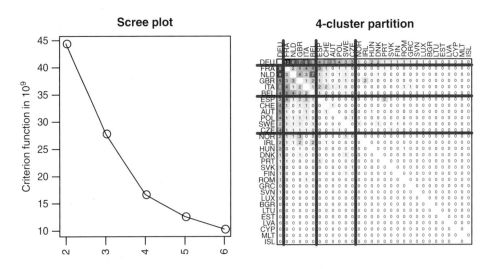

Figure 6.7 Results of sum of squares blockmodeling.

countries are those with larger trade. In the four-cluster solution, the clusters are as follows, named using the traditional trimodality used in world-system analysis:

- Dominant: Germany, having the strongest ties to practically all countries, especially to those in the other core cluster.
- Core: France, Netherlands, Great Britain, Italy, Belgium. These are strongly linked to Germany and with each other. The Semi-periphery also trades with them.
- Semi-periphery: Spain, Switzerland, Austria, Poland, Sweden, Czech Republic. These have very weak ties with each other, somewhat stronger ties to the Core and the strongest ties to Germany.
- Periphery: All remaining countries. Very weak ties to most countries, with some exceptions (especially with Germany).

As these results and especially Figure 6.8 show, the cluster membership is practically determined by the combined trade volume (the valued degree is computed as the sum of all incoming and outgoing ties).

One could rightly argue the clustering obtained above is heavily influenced by the different relational capacities of the countries. This could indeed constitute the main aspect of the structural equivalence we wish to capture, and the above methods seem adequate for capturing such differences. However, the relative influence of such differences when determining structurally equivalent actors can be reduced by applying a suitable scaling transformation. This is what Mahutga [34] did with his trade flow data when he applied a log-transformation prior to clustering. For our EU/EFTA trade flow example, Figure 6.9 shows that log-transformations indeed bring the distribution of the tie values much closer to a normal distribution. The requirement for this transformation is that all values are positive. In our example, we have two exceptions to this because both trade flows between Iceland and Cyprus are 0. To avoid the problem of an undefined logarithm, we added 1 to all tie values. If these values were to be lower than 1, the logarithm would be negative, although this does not pose a problem for homogeneity and similar approaches that can handle negative tie values. The smaller this value compared to what the smallest non-zero value in the network will be (in our case, we made it the same as the smallest

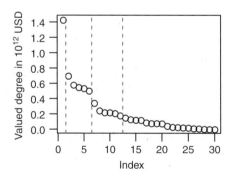

Figure 6.8 Valued gross degree (in- plus outdegree) of countries ordered by a four-cluster sum of squares solution. Dotted lines indicate cluster borders. The results are very similar if we use absolute deviations blockmodeling, although the size of some clusters changes (mainly a few countries move up in the hierarchy).

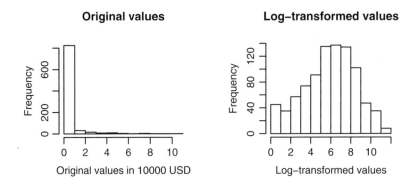

Figure 6.9   The distribution of the original and log-transformed tie values.

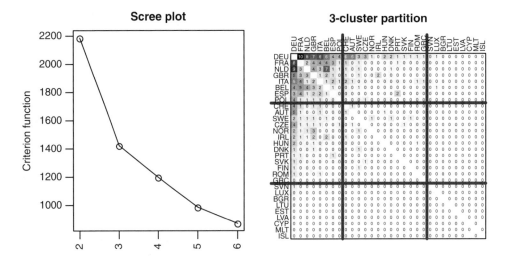

Figure 6.10   Results of sum of squares blockmodeling on the log-transformed network.

non-zero value), the bigger the distinction between the zero ties and the other (non-zero) tie values. The results of this analysis are presented in Figure 6.10. We can see that the general core-periphery structure remains, only Germany loses its special cluster and the core groups generally become larger. Consequently, judging by the scree plot, a three-cluster partition now seems the most appropriate (i.e. without a singleton Germany cluster).

6.7.2.2   Analyzing Iteratively Row-column Normalized Data

To counter the unequal relational capacities within this network, we first used iterated row-column marginal normalization. Here the results show strong evidence of trade gravity effects, i.e. where the magnitude of trade between countries is inversely related to the spatial distance between them. As this normalization is prone to producing relatively extreme values for countries trading small volumes, we use absolute deviations blockmodeling on the transformed network. The results are presented in Figures 6.11 and 6.12. Whereas the scree

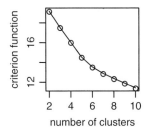

Figure 6.11 Scree diagram for the absolute deviations approach on an iterated row-column marginal normalized network.

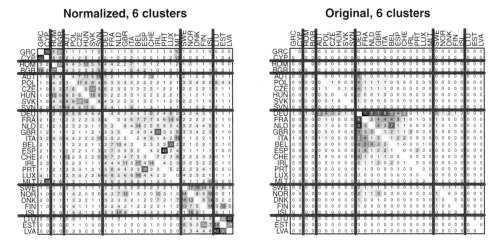

Figure 6.12 Six-cluster partition obtained using absolute deviations on an iterated row-column marginal-normalized network depicted on a normalized and original network.

diagram (Figure 6.11) does not suggest a suitable number of clusters, we found the six-cluster partition to be both substantially interesting and representing a suitable balance between details and aggregation.

In this solution, we have two clusters consisting of two neighboring countries, which are listed below the first. The remaining clusters represent specific sub-regions in Europe. These are:

- Greece and Cyprus: Greece is not exactly a neighbor of Cyprus, but it is its nearest neighbor in the dataset and, more importantly, the majority ethnic group on Cyprus is Greek.
- Romania and Bulgaria: The reason for them comprising a separate cluster is not only the strong ties between each other, but also that they are strongly connected to both the Greece/Cyprus cluster and Central-Eastern Europe.
- Central-Eastern Europe: Austria, Poland, Czech Republic, Hungary, Slovakia, and Slovenia.
- Central-Western Europe[18] (countries not listed elsewhere).

---

[18] This cluster is split into more seafaring and more inland-oriented countries in the eight-cluster solution.

- Northern Europe (Scandinavia): Sweden, Norway, Denmark, Finland, and Iceland.
- Baltic countries: Lithuania, Estonia, and Latvia.

On such an iteratively-normalized network, we could also specify which ties represent a relevant tie. Assuming a uniform distribution of tie values, the expected tie value here is 1/(number of units -1), which in our case is 1/29. With this, the use of valued blockmodeling is appropriate where m is set to twice that value, which is 2/29. The results are quite different; however, the regional trade-gravity effects are still obvious. The difference occurs as in the valued blockmodeling approach, values over m only cause inconsistencies in the null blocks and do not increase the fit in the complete blocks. The similarity of ties (apart from being close to 0 or m) is also not important. Fewer clusters are optimal in this case (three or four, with the former being more pronounced). The results for the three- and four-cluster solutions (on normalized and original networks) are presented in Figure 6.13. We selected the four-cluster solution due to its easier interpretability as the partition represents South-Eastern Europe, Central-Eastern Europe, Western Europe, and Northern Europe + Baltic.

6.7.2.3 Indirect Approaches

In this example, where possible we are focusing on the direct approaches. However, using indirect approaches brings the advantage that some distance measures can be used which do not have their equivalents in direct approaches. Two of these distance measures are:

- Corrected (Squared) Euclidian distance based on separate row and column (sum) normalization, where the distance is computed among rows for row profiles and among columns for column profiles.
- Using correlation (or actually (1–correlation)/2) as a distance measure.

The dendrogram for the first measure (Euclidian distance on separate row and column normalized networks) using Ward's hierarchical clustering together with the four-cluster solution[19] on both the original and separate row and separate column marginal-normalized networks is presented in Figure 6.14. The clusters in this partition represent Central-Eastern Europe, Western Europe, Northern Europe (including the Baltics), and South-Eastern Europe (including Slovakia).

The results for the second measure (correlations), again using Ward's hierarchical clustering, are shown in Figure 6.15. This procedure also captures trade gravity effects and regional trade concentrations. The five-cluster solution produces the following clusters:

- Western Europe
- Central-Eastern Europe
- Northern Europe and Ireland
- Greece, Cyprus, and Malta
- Baltic.

---

[19] Based on the dendrogram, the appropriate number of clusters is not very clear. The four-cluster solution is the last before singleton clusters emerge and is one of the more suitable ones based on the dendrogram.

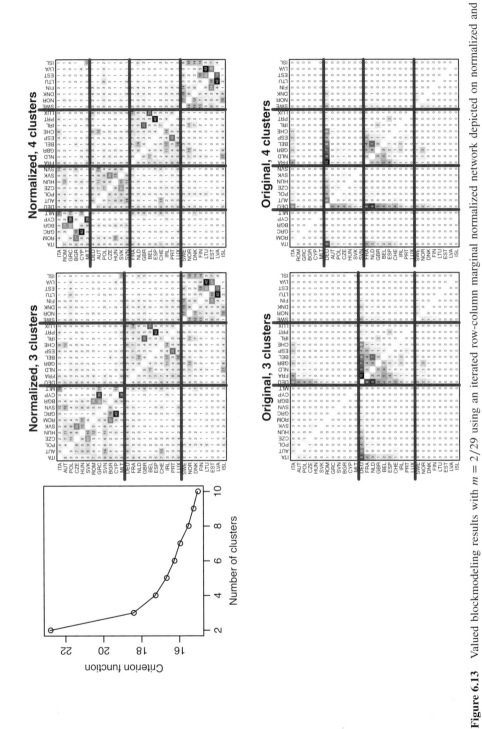

Figure 6.13 Valued blockmodeling results with $m = 2/29$ using an iterated row-column marginal normalized network depicted on normalized and original data (network).

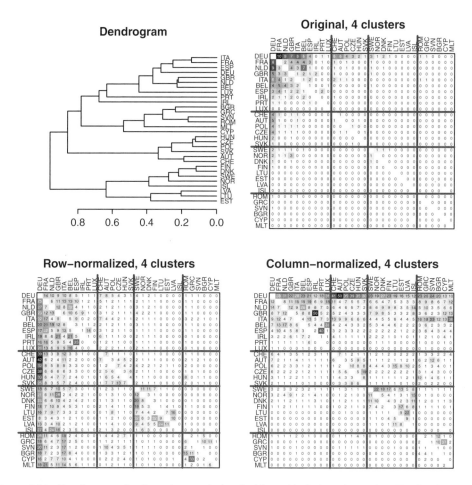

Figure 6.14 Dendrogram plus four-cluster solution for hierarchical clustering (using Ward's criteria) on Euclidian distance on separate row and column normalized networks depicted on original, row-normalized, and column-normalized data.

Using these indirect approaches admittedly has a certain appeal over using either generalized blockmodeling on iteratively row-column marginal-normalized networks or using a deviational approach: the transformations here are much simpler or not required at all (e.g. when using the correlation-based indirect metric), thus making it easier to motivate their use. Of course, generalized blockmodeling's main advantage is its versatility, particularly its ability to use other equivalences/block types and to use pre-specified blockmodels.

The final approach we try here is the indirect measure of deviational structural equivalence. This dataset was initially analyzed to demonstrate deviational blockmodeling [41], including its indirect approach. We use the same measures of deviational structural equivalence (correlations on RD and CD matrices) as suggested in [41] but apply Ward-type hierarchical clustering instead of the weighted-average version used in [41, p. 174]. In addition, rather than using an incremental two-sided deviational threshold (post-clustering), we identify prominent and non-prominent ties based on sign similarity in RD and CD. Here, non-determined ties are thus those where the

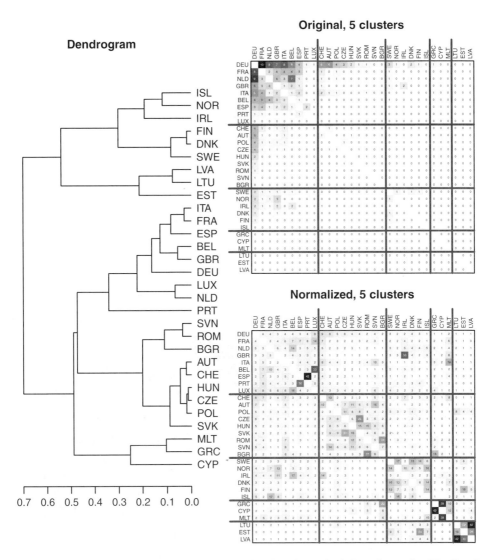

Figure 6.15 Dendrogram and the five-cluster solution for hierarchical clustering (using Ward's criteria) on correlations on the original network (depicted on original and iteratively row-column normalized networks).

corresponding values in RD and CD have different signs. The four-cluster partition is given in Figure 6.16 below.

Despite using the same deviational coefficients as those in [41, p. 174], the partition we obtain in Figure 6.16 differs somewhat from that of the original study. The differences are due to the particular linkage method used in each study: whereas Nordlund used weighted-average clustering, Figure 6.16 is obtained using Ward's hierarchical clustering. Thus, like for the case of binary networks (as well as any application of hierarchical clustering), the resulting partitions

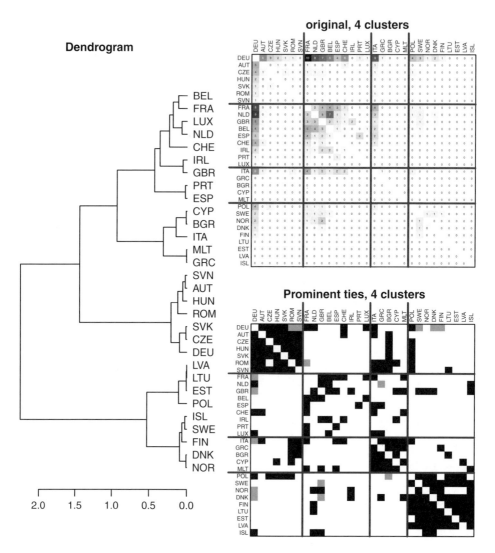

Figure 6.16 Dendrogram and four-cluster solution for Ward's hierarchical clustering on deviational correlations on the original network. Top blockmodel: original values; bottom blockmodel: prominent (black), non-prominent (white), and contradictory ties (gray).

are not only due to the choice of picking a specific indirect measure of equivalence and possible preceding transformations, but the results also depend on the particular clustering methods used.

6.7.2.4 Summary

This section demonstrates how trade flow data, in which units have vastly different relational capacities, can be viewed from at least two different angles. The first uses raw (non-normalized) or scaled networks whereby a core-periphery structure emerges, with partitions that are almost solely defined by the valued degree of the units, i.e. a strength-oriented focus. The other view is that based on somehow transformed data, where we try to cluster units not on the similarity of their raw ties but the ties where we somehow try to compensate for their differential relational capacity. When following this route, which we could call a pattern-oriented focus, a clear structure of cohesive groups emerges, indicating a preference for regional trade. While both views may be relevant, given that the second one is not solely influenced by degrees, we could say that this one in some way better harnesses the richness of the relational data. Of course, much of this had already been pointed out by Nordlund [41]. Yet here we have shown that other approaches (especially when using appropriate transformations) can also be used to produce similar results as those emerging from the deviational approach.

Whereas this example primarily applied direct methods, we also approached the data using various indirect correlation-based measures of structural equivalence, testing different pre-clustering transformations. When discounting for unequal relational capacities, the indirect approaches also seem capable of capturing patterns of regional trade. However, while the indirect deviational measure seems adequate for capturing such effects, we demonstrate how the choice of the clustering heuristic can, at least in part, affect the obtained partitions. Thus, even though a transformation might assist in capturing structural equivalence that discounts for unequal relational capacities among actors, the indirect blockmodeling of valued networks is – similar to their binary – siblings equally influenced by the particular clustering heuristic we use.

6.8   Conclusion

As suggested in the multitude of available approaches, the clustering and blockmodeling of network data is anything but a trivial exercise. Not only do many different definitions of equivalence abound, but there are also multiple heuristics and approaches for identifying and measuring equivalences and the classification of actors according to shared notions of equivalence. Of the two broad approaches for doing this, i.e. direct and indirect, the former has evolved into a general analytical framework – generalized blockmodeling – that allows different types of equivalences to be captured. The indirect approaches, constituting various approaches and algorithms for capturing pair-wise measures of different types of equivalences, typically imply additional choices and dilemmas when it comes to the actual clustering of such pair-wise indirect measures.

Whereas the above applies to the clustering and blockmodeling of networks generally, the existence of valued ties implies additional dilemmas for the analyst. Indeed, providing richer, higher-resolution insights than their binary counterparts, the comparison of ties in a valued network is not necessarily as straightforward as is the case with binary networks. To this end, several possible clustering methods for valued data and a range of transformations can be applied when clustering valued networks. However, prior to this, the analyst has to understand how the different transformations and clustering heuristics, direct and indirect, carry different ideas of equivalence that stretch beyond their binary counterparts. The core question that must be addressed when clustering valued networks is what the tie values actually mean in the context of equivalence. Are the tie values directly comparable across the network or do actors have different capacities when it comes to tie values? Is the sought-after notion of equivalence related to such differences or should such unequal relational capacities be discounted for?

This chapter has addressed various clustering approaches that apply to valued networks. Transformations play an important role here in both preparing valued networks for use with existing direct and indirect approaches as well as interpreting the resulting blockmodels. The most rudimentary of these is the dichotomization of valued data into binary data, an approach we find inherently problematic for multiple reasons. Even if we assume equal relational capacities among actors and are able to set a system-wide threshold to distinguish prominent and non-prominent ties, it will inevitably reduce the resolution of the data.

Scaling transformations can be used to adjust the overall distribution of the tie values; however, they do not account for unequal relational capacities. Iterative row column normalization can be used to in a sense adjust for unequal relational capacities as it makes the in, out, and overall (valued) degrees of all actors equal. If classification into prominent and non-prominent ties is needed and, importantly, it makes sense in the data and research context, system-wide threshold dichotomization can be employed. Other transformations explicitly aim to identify prominent ties in valued networks where actors have unequal relational capacities. We examine transaction flow models and a novel deviational transformation, two transformations where the prominence of ties is determined based on the different relational capacities and degrees of actors in a dyad.

We also examine indirect measures of equivalence that are applicable to valued networks, exemplifying this by testing various approaches, transformations, and measures on two datasets – the EIES friendship network and commodity trade among countries in the EU/EFTA region. For structural equivalence, we note (as has previously been noted) that the two most common indirect measures – Euclidean distances and correlation-based – operationalize different takes on structural equivalence. Whereas the Euclidean distance-based indirect measure takes tie values into account in its measure of structural equivalence, the correlation-based metric is more inclined to capture the profiles, rather than the strengths, of the ties.

Something similar can be said for generalized blockmodeling approaches to valued networks [55] and the deviational blockmodeling of Nordlund [41]. While the former focuses on actual tie values, the latter concentrates on whether these values are lower or higher than expected (based on a no interactions model) and therefore patterns of ties. As such, these approaches offer different views on the network and thus complement each other.

For both direct and indirect approaches, the approaches that focus more on tie strengths can often offer similar analysis as those focusing on profiles/deviations, the data are suitably transformed prior to analysis. However, such transformation is not always trivial or even possible.

Our general recommendations are:

- One should avoid the dichotomization of valued networks (and application of binary approaches) whenever possible, unless the tie values cannot be considered to be measured on at least an approximately interval scale.
- The approach used should be selected while considering what the ties or tie values mean and the properties of the blockmodeling solution we are looking for:
  - If all units in the network have equal relational capacities (like friendship data), approaches focusing on tie strengths should be selected.
  - If all units do not have equal capacities, we must decide whether we want such differences in relational capacities to influence the results or not.
    * If we do, once again approaches focusing on tie strengths should be selected.
    * If we do not want different relational capacities to affect the results, then correlation or deviational based approaches should be selected. In some cases, using strength-focused

> approaches after suitable transformations is also appropriate, while in others finding suitable transformations is too problematic.

- If equivalent direct and indirect approaches exist, direct approaches are preferred as long as the network is not too large to be analyzed by these approaches in a reasonable time. However, some indirect approaches (e.g. correlation-based ones) do not have an exact equivalent in direct approaches.
- When generalized equivalence or pre-specified blockmodels are needed, generalized block-modeling approaches must be used.

While this chapter has aimed to shed some light on contemporary approaches for blockmodeling valued networks, several pertinent unsettled questions should be noted. One question is how the deviations of actual ties from the expected ones based on a model of no interactions should be measured. Second, whereas our focus in this chapter was on structural equivalence, we find it important to also explore how different transformations can be used in conjunction with other types of equivalences. Third, related to both of these, we argue the overarching question for blockmodeling of valued networks concerns the triadic interdependence between, respectively, pre-clustering transformations, notions of equivalence, and substantive interpretations. As each step in blockmodeling a valued network offers so many distinct choices – whether the valued data should be transformed, which transformation to use, different notions of equivalence, various methods to identify such, methods and choices related to the clustering of actors, etc. – these choices must be informed, guided, and motivated by the specific theoretical and substantive underpinnings at hand. Thus, although we propose a general guideline for choosing methods depending on the dataset and the research questions, the context is, as always, crucial.

Acknowledgements

The authors acknowledge financial support from the Slovenian Research Agency (research core funding No. P5-0168 and project "Blockmodeling multilevel and temporal networks" No. J7-8279) and Budapest Közép-Európai Egyetem Alapitvány (CEU BPF), with additional support from the Institute for Analytical Sociology, Linköping University.

References

1. C. Aicher, A. Z. Jacobs, and A. Clauset. Learning latent block structure in weighted networks. *Journal of Complex Networks*, 3(2):221–248, 2015.
2. H. R. Alker. An IBM 709 program for gross analysis of transaction flows. *Behavioral Science*, 7:498–499, 1962.
3. D. R. Baker. A Structural Analysis of Social Work Journal Network: 1985–1986. *Journal of Social Service Research*, 15(3-4):153–168, 1992.
4. A. Barrat, M. Barthelemy, R. Pastor-Satorras, and A. Vespignani. The architecture of complex weighted networks. *Proceedings of the National Academy of Sciences*, 101(11):3747–3752, 2004.
5. V. Batagelj. Notes on blockmodeling. *Social Networks*, 19(2):143–155, 1997.
6. V. Batagelj and A. Ferligoj. Clustering relational data. In *Data Analysis*, pages 3–15. Springer, 2000.
7. V. Batagelj, P. Doreian, and A. Ferligoj. An optimizational approach to regular equivalence. *Social Networks*, 14(1-2):121–135, 1992.
8. V. Batagelj, A. Ferligoj, and P. Doreian. Direct and indirect methods for structural equivalence. *Social Networks*, 14(1-2):63–90, 1992.

9. V. Batagelj, A. Ferligoj, and P. Doreian. Fitting pre-specified blockmodels. In *Data Science, Classification, and Related Methods*, pages 199–206. Springer, 1998.

10. V. Batagelj, A. Mrvar, A. Ferligoj, and P. Doreian. Generalized blockmodeling with Pajek. *Metodoloski Zvezki*, 1 (2):455, 2004.

11. S. Borgatti and M. Everett. Regular Equivalence: Algebraic Structure and Computation. In *Proceedings of the Networks and Measurement Conference*. University of Irvine, California, Irvine, 1991.

12. S. P. Borgatti and M. G. Everett. Regular blockmodels of multiway, multimode matrices. *Social Networks*, 14(1):91–120, 1992.

13. S. P. Borgatti and M. G. Everett. Two algorithms for computing regular equivalence. *Social Networks*, 15(4):361–376, 1993.

14. S. J. Brams. Transaction flows in the international system. *American Political Science Review*, 60(4):880–898, 1966.

15. R. Breiger. Structures of economic interdependence among nations. *Continuities in structural inquiry*, pages 353–380, 1981.

16. R. L. Breiger, S. A. Boorman, and P. Arabie. An algorithm for clustering relational data with applications to social network analysis and comparison with multidimensional scaling. *Journal of Mathematical Psychology*, 12(3):328–383, 1975.

17. M. Brusco and D. Steinley. A variable neighborhood search method for generalized blockmodeling of two-mode binary matrices. *Journal of Mathematical Psychology*, 51(5):325–338, 2007.

18. M. Brusco and D. Steinley. A tabu-search heuristic for deterministic two-mode blockmodeling of binary network matrices. *Psychometrika*, 76 (4):612–633, 2011.

19. R. S. Burt. Positions in networks. *Social Forces*, 55(1):93–122, 1976.

20. J. Chan, S. Lam, and C. Hayes. Increasing the scalability of the fitting of generalised block models for social networks. In *IJCAI Proceedings of the International Joint Conference on Artificial Intelligence*, volume 22, page 1218, 2011.

21. J. I. Dominguez. Mice that do not roar: Some aspects of international politics in the world's peripheries. *International Organization*, 25(2): 175–208, 1971.

22. P. Doreian. A note on the detection of cliques in valued graphs. *Sociometry*, 32(2): 237–242, 1969.

23. P. Doreian. Actor network utilities and network evolution. *Social Networks*, 28(2):137–164, 2006.

24. P. Doreian and A. Mrvar. Partitioning signed social networks. *Social Networks*, 31(1):1–11, 2009.

25. P. Doreian, V. Batagelj, and A. Ferligoj. *Generalized Blockmodeling*. Cambridge University Press, New York, 2005.

26. K. Faust and A. K. Romney. Does structure find structure? A critique of Burt's use of distance as a measure of structural equivalence. *Social Networks*, 7(1):77–103, 1985.

27. A. Ferligoj and V. Hlebec. Evaluation of social network measurement instruments. *Social Networks*, 21(2):111–130, 1999.

28. L. C. Freeman and S. C. Freeman. A Semi-Visible College: Structural Effects on a Social Networks Groups. In M. M. Henderson and M. J. MacNaughton, editors, *Electronic Communication: Technology and Impacts*, number 52 in AAAS Selected Symposium Series, pages 77–85. Westview Press, New York, 1980.

29. L. A. Goodman. Statistical methods for the preliminary analysis of transaction flows. *Econometrica: Journal of the Econometric Society*, pages 197–208, 1963.

30. V. Hlebec and A. Ferligoj. Reliability of Social Network Measurement Instruments. *Field Methods*, 14(3):288–306, 2002.

31. B. B. Hughes. Transaction data and analysis: In search of concepts. *International Organization*, 26(4):659–680, 1972.

32. J. R. Lincoln and M. L. Gerlach. *Japan's Network Economy: Structure, Persistence, and Change*, volume 24. Cambridge University Press, 2004.

33. F. Lorrain and H. C. White. Structural equivalence of individuals in social networks. *Journal of Mathematical Sociology*, 1(1):49–80, 1971.

34. M. C. Mahutga. The Persistence of Structural Inequality? A Network Analysis of International Trade, 1965–2000. *Social Forces*, 84(4): 1863–1889, 2006.

35. L. L. McQuitty. Multiple clusters, types, and dimensions from iterative intercolumnar correlational analysis. *Multivariate Behavioral Research*, 3 (4):465–477, 1968.

36. G. W. Milligan and M. C. Cooper. An examination of procedures for determining the number of clusters in a data set. *Psychometrika*, 50(2): 159–179, 1985.

37. F. Mosteller. Association and estimation in contingency tables. *Journal of the American Statistical Association*, 63(321):1–28, 1968.
38. C. G. Nelson. European Integration: Trade Data and Measurement Problems. *International Organization*, 28(3):399–433, 1974.
39. R. J. Nemeth and D. A. Smith. International trade and world-system structure: A multiple network analysis. *Review (Fernand Braudel Center)*, 8(4):517–560, 1985.
40. C. Nordlund. Identifying regular blocks in valued networks: A heuristic applied to the St. Marks carbon flow data, and international trade in cereal products. *Social Networks*, 29(1):59–69, 2007.
41. C. Nordlund. A deviational approach to blockmodeling of valued networks. *Social Networks*, 44:160–178, 2016.
42. R. A. Olshen and B. Rajaratnam. Successive normalization of rectangular arrays. *Annals of Statistics*, 38(3):1638, 2010.
43. A. K. Romney and K. Faust. Predicting the structure of a communications network from recalled data. *Social Networks*, 4(4):285–304, 1982.
44. L. D. Sailer. Structural equivalence: Meaning and definition, computation and application. *Social Networks*, 1(1):73–90, 1978.
45. I. R. Savage and K. W. Deutsch. A Statistical Model of the Gross Analysis of Transaction Flows. *Econometrica*, 28(3):551–572, 1960.
46. J. E. Schwartz. An examination of CONCOR and related methods for blocking sociometric data. *Sociological Methodology*, 8:255–282, 1977.
47. J. Shetty and J. Adibi. The Enron email dataset database schema and brief statistical report. Information Sciences Institute Technical Report, University of Southern California, 4, 2004. URL http://citeseerx.ist.psu.edu/viewdoc/download?doi=10.1.1.296.9477\&rep=rep1\&type=pdf.
48. M. Škerlavaj, V. Dimovski, A. Mrvar, and M. Pahor. Intra-organizational learning networks within knowledge-intensive learning environments. *Interactive Learning Environments*, 18(1):39–63, 2010.
49. D. A. Smith and D. R. White. Structure and dynamics of the global economy: Network analysis of international trade 1965–1980. *Social Forces*, 70(4):857–893, 1992.
50. D. Snyder and E. L. Kick. Structural position in the world system and economic growth, 1955-1970: A multiple-network analysis of transnational interactions. *American Journal of Sociology*, 84(5):1096–1126, 1979.
51. S. Svensson and C. Nordlund. The building blocks of a Euroregion: Novel metrics to measure cross-border integration. *Journal of European Integration*, 37(3):371–389, 2015.
52. S. Wasserman and K. Faust. *Social Network Analysis: Methods and Applications*. Cambridge University Press, Nov. 1994.
53. D. R. White and K. P. Reitz. Graph and semigroup homomorphisms on networks of relations. *Social Networks*, 5(2):193–234, 1983.
54. E. P. Xing, W. Fu, and L. Song. A state-space mixed membership blockmodel for dynamic network tomography. *The Annals of Applied Statistics*, 4(2):535–566, 2010.
55. A. Žiberna. Generalized blockmodeling of valued networks. *Social Networks*, 29:105–126, 2007.
56. A. Žiberna. *Generalized Blockmodeling of Valued Networks (Posplošeno Bločno Modeliranje Omrežij z Vrednostmi Na Povezavah): Doktorska Disertacija*. PhD thesis, University of Ljubljana, Ljubljana, 2007.
57. A. Žiberna. Direct and indirect approaches to blockmodeling of valued networks in terms of regular equivalence. *Journal of Mathematical Sociology*, 32:57–84, 2008.
58. A. Žiberna. Evaluation of Direct and Indirect Blockmodeling of Regular Equivalence in Valued Networks by Simulations. *Metodološki Zvezki*, 6 (2):99–134, 2009.
59. A. Žiberna. Generalized blockmodeling of sparse networks. *Metodološki Zvezki*, 10(2):99–119, 2013.
60. A. Žiberna. Blockmodeling 0.3.1: An R package for generalized and classical blockmodeling of valued networks, 2018. URL https://r-forge.r-project.org/R/?group_id=203.
61. A. Žiberna, N. Kejžar, and P. Golob. A Comparison of Different Approaches to Hierarchical Clustering of Ordinal Data. *Metodološki Zvezki*, 1(1):57–73, 2004.

7

Treating Missing Network Data Before Partitioning

Anja Žnidaršič[1], Patrick Doreian[2,3], and Anuška Ferligoj[2,4]
[1]University of Maribor
[2]University of Ljubljana
[3]University of Pittsburgh
[4]NRU HSE Moscow

7.1 Introduction

The patterns of ties of a network are critical for revealing both macro and micro network structural features. This is especially true regarding the delineation of the macro structure of networks through network partitioning. Yet network data are prone to recording errors and/or missing data regardless of the substantive nature of the relationships measured. If specific real ties are not recorded and non-existent ties are recorded as if they are real, this creates major problems for analyzing network data. Also, correctly included ties can have incorrect values. Actors may refuse to respond regarding specific ties and can provide no information about ties to all other network members. The latter is known as actor non-response. It is crucial that the data used for clustering procedures are either error-free (a very rare event) or are treated appropriately when data were missing. Here, we continue an examination of the impact of actor non-response and treatments for it on the stability of partitions of actors obtained from different blockmodeling procedures. We use a set of real well-measured networks as the foundation for our analyses.

Missing network data take several forms (see Section 7.2 for more details). Previous studies tackled the problem of actor non-response and the consequences for network partitioning used to delineate the macro structure of networks. Žnidaršič *et al.* [33] examined binary networks and employed structural equivalence using direct blockmodeling [11]. Their study was extended through simulation of three widely-known macro-network structures (cohesive subgroups, core-periphery systems, and hierarchical networks) [34]. Indirect blockmodeling of

Advances in Network Clustering and Blockmodeling, First Edition.
Edited by Patrick Doreian, Vladimir Batagelj, and Anuška Ferligoj.

valued networks was employed. The set of possible treatments for binary networks was adjusted to deal with valued networks and extended by new ones (including imputations of the mean value of incoming ties and medians of the three nearest neighbors based on incoming ties). In binary networks, some attention has been given also to the impact of item non-response and its treatments on the stability of blockmodeling based on structural equivalence [32].

The chapter is organized as follows: Section 7.2 discusses types of missing network data, Section 7.3 focuses on treatments of missing data due to actor non-response, and Section 7.4 presents the simulation study design with key characteristics of blockmodeling and a set of real networks. Results are presented in Section 7.5, with Section 7.6 summarizing the results and presenting conclusions with an emphasis on recommendations for network researchers.

7.2   Types of Missing Network Data

Errors in social network research design can be divided into three broad categories: boundary specification problems, questionnaire design, and errors due to respondents [33]. The first two categories belong in the domain of researchers responsible for designing the best possible data collection instruments and being careful in the selection of respondents. Boundary specification problems concern rules of inclusion or exclusion of actors into studied networks (see, for example, [10, 21, 22] regarding the problems of getting network boundaries wrong). Sources of errors in measurement instruments include fixing the number of possible nominations [15, 21, 32]. There is also the choice between using free recall or rosters of actors, e.g. [6, 8, 13, 14], which affects the collected data. Finally, there is the choice of seeking directed or symmetric ties for relations [12, 30]. Care is needed in making these choices. Once made, mistakes made in these choices cannot be rectified: poorly constructed data collection instruments lead to poor quality data.

The third category consists of errors due solely to actors regardless of instrument design. There are three subcategories: complete actor non-response, item non-response (regarding specific ties), and reporting errors in the recorded ties. Here, we focus primarily on actor non-response in Sections 7.2.3 and 7.3. We consider briefly the other problems in Sections 7.2.1 and 7.2.2. Throughout these sections, we consider possible remedies.

7.2.1   Measurement Errors in Recorded (Or Reported) Ties

Measurement errors for binary networks, as defined by Holland and Leinhardt [15], occur when there are missing or extra ties. This can be extended to valued networks when the strength of the tie differs from the true value (including valued ties not being recorded). Figure 7.1 shows a small valued network with ten actors as a demonstration of this problem. Figure 7.1a has the matrix array on the left with a graphical representation of this network on the right. Tie values in the graph are represented by different arc widths and grey levels. All three types of actor-induced errors are illustrated using this network as described below.

Figure 7.2 presents the demonstration network but with ten item-response errors. In the matrix array, they are represented by using two triangles in the relevant squares. The upper triangle of each divided square has the true value. The lower triangle has the reported tie value. For example, the true value of the tie from A1 to A2 is 2 while the reported value is 5. The problem of incorrectly included ties involves A7 and A9. There is no tie from A7 to A9 but it was

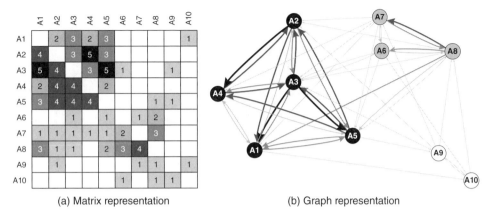

(a) Matrix representation (b) Graph representation

Figure 7.1 A demonstration network with ten actors.

	A1	A2	A3	A4	A5	A6	A7	A8	A9	A10
A1			3	2	3					1
A2	4		3	5	3					
A3	5	4		3	1	1		1		
A4	2	4	4			2	2			
A5	3		3	4				1	1	
A6		1			1		1	2		
A7	1	1	1	1	1	2		3	3	
A8	3	3	1		2	2	3	4		
A9		1					1	1		1
A10								1	1	

Figure 7.2 The matrix representation of the demonstration network with erroneously reported tie values.

reported as having the value 3. In contrast, there is a tie from A10 to A6 but it is absent from the reported ties.

These types of measurement errors are hard to detect, especially the incorrectly reported values of real ties. One option for doing this is to compare the responses to an original and a reversed question. Marouf and Doreian [23] asked employees in a company about who they went to for advice and also who came to them for advice. Ideally, the reported tie value by one respondent would be confirmed by other respondent. Confirming the presence of a tie is easier than confirming its value.

An and Schramski [1] emphasized that a significant number of reported exchange ties (e.g. information, goods, or services) tend to be disputed in the reports of senders and receivers. In such disagreements, these authors argue that neither eliminating contested reports nor symmetrizing them is appropriate. Instead, they propose measuring actors credibility based on their asymmetric connections and using deterministic or stochastic methods to estimate relations between pairs of actors.

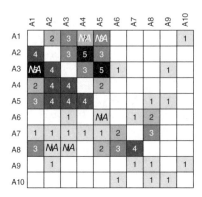

Figure 7.3 The matrix representation of the demonstration network with item non-response.

7.2.2 Item Non-Response

Item non-response occurs when one or more actors participate in a study, but provide no information on a subset of their outgoing ties. Figure 7.3 presents the demonstration network with six missing ties.

Studies of item non-response are limited. Borgatti *et al.* [7] stressed how absent ties "can lead to a radically different understanding of the network and misleading measurements of network indices such as centrality". Similarly, Huisman [17] emphasized the risks of severely biased results of network analyses if missing ties are ignored.

Huisman [17] studied the impact of three imputation methods (reconstruction, hot deck,[1] and preferential attachment) to estimate the mean degree, reciprocity, values of the clustering coefficient, assortativity, and inverse geodesic distances. He emphasized that, for directed networks, reconstruction was the best among the imputation methods he studied. In contrast, both preferential attachment and hot deck imputations created greater bias.[2]

The impacts of four treatments (reconstruction, imputations based on modal values, combination of the reconstruction and imputations based on modal values, and null tie imputations) for absent ties or item non-response on the results of blockmodeling based on structural equivalence were studied with four real networks by Žnidaršič *et al.* [32]. Their main conclusions were (i) the combination of reconstruction and imputations based on modal values is the best overall treatment method for absent ties, especially in networks with high reciprocity with symmetrical blockmodel structures, and (ii) for networks with low reciprocity, imputation based on modal values performed well. Null tie imputation was the worst treatment for absent ties.

7.2.3 Actor Non-Response

Actor non-response occurs where actors provide no information on *any* of the ties to *all* other members of the network. In a network with n actors where m actors provide no information

---

[1] Both categorical attributes and structural properties are used to find a completely observed "donor" actor by minimizing the absolute differences between an actor with incomplete data and the donor actor. The donor actor is used to replace the missing ties (or missing actors in case of actor non-response).

[2] Also, for item non-response, the hot deck method was unable to find donor actors to replace the missing ties if the proportion of missing ties was large.

	A1	A2	A3	A4	A5	A6	A7	A8	A9	A10
A1		2	3	2	3					1
A2	4		3	5	3					
A3	5	4		3	5	1			1	
A4	2	4	4		2					
A5	NA	NA	NA	NA	NA	NA	NA	NA	NA	NA
A6			1		1			1	2	
A7	NA	NA	NA	NA	NA	NA	NA	NA	NA	NA
A8	3	1	1		2	3	4			
A9	NA	NA	NA	NA	NA	NA	NA	NA	NA	NA
A10						1			1	1

Figure 7.4   The matrix representation of the demonstration network with three non-respondents.

on their outgoing tie values, the actor response rate, the "relational response rate" [19], is $1 - m/n$. With three non-respondents in the demonstration network, the relational response rate is equal to 70%. Žnidaršič et al. [33, 34] stress that observed information from respondents to non-respondents, their incoming ties, are still available and can be very useful.

Figure 7.4 presents the demonstration network with three non-respondents ($A5$, $A7$, and $A9$) creating three rows of missing ties (marked by NA).

The effects of actor non-response on direct blockmodeling based on structural equivalence have been examined in case of binary networks [33] but only for a subset of the treatments presented in Section 7.3. For valued networks, the impact of missing actors on indirect block-modeling together with employed treatments was assessed based on simulated networks [34]. Here, we focus on actor non-response to investigate the impact of seven actor non-response treatments (presented in Section 7.3) on the results of clustering both real binary and valued networks.

7.3   Treatments of Missing Data (Due to Actor Non-Response)

Stork and Richards [30] suggested that actor non-response in network data can be treated in three different ways: (i) using a complete-case analysis, (ii) using an available-case analysis, and (iii) imputing data values. In the complete-case analysis (known also as "listwise" deletion) all non-respondents are removed from the network. The result is a smaller network with com-promised network boundaries as noted by Stork and Richards. Here, we omit the complete-case approach from our simulations.[3] This strategy is worthless for examining both the macro struc-ture via blockmodeling and assessing the micro positions of *all* network members including non-respondents.

Seven actor non-response treatments (reconstruction, imputations of the mean values of incoming ties, imputations of the modal values of incoming ties, reconstruction and imputations based on modal values of incoming ties, imputations of the total mean, imputations of the median of the three nearest neighbors based on incoming ties, and null tie imputations) are presented in the following subsections for the small demonstration network of ten actors as

---

[3] Some results on how deletion of actors severely changes the network characteristics are provided in [7, 18, 26, 27].

presented in Figure 7.1.[4] In addition to comparisons of imputed or estimated tie values, the impact of treatments on more general network characteristics is discussed.

If $A5$, $A7$, and $A9$ refused to respond three rows of missing data will result. Table 7.1, first information line, presents some summary information about this demonstration network. First, the average outgoing tie values of these non-respondents ($A5$, $A7$, and $A9$) are 1.89, 1.11, and 0.44. Second, this network has 49 arcs of which 16 would be lost if the non-response was ignored. Third, the mean tie value is 2.245. Finally, its weighted density[5] and weighted reciprocity[6] are 1.222 and 0.691, respectively. All of these values provide a foundation for assessing the treatments of actor non-response discussed below.

Throughout this chapter we use the following terms: (i) a "whole network" is the known network, (ii) a "measured network" is obtained from the whole network by removing all outgoing ties for those actors providing no data about their ties, and (iii) a "treated network" is obtained by employing the actor non-response treatments applied to a measured network. The issue is how close the treated network is to the whole network. We turn now to consider the non-response treatments as applied to the demonstration network of Figure 7.1.

7.3.1   Reconstruction

Replacing the missing outgoing ties by their observed incoming ties could be used [17, 30]. Unavailable rows of data for non-respondents are replaced by the corresponding columns for those actors. Of course, the resulting ties involving non-respondents and respondents are symmetric. For undirected networks, this is an available-case approach where the relationship between two individuals is measured by using the one report of the tie that exists in the data (see [30]). For directed networks, this imputation procedure estimates the missing tie from the incoming ties (17).

However, for two non-respondents, the reconstruction of ties between them cannot be done, therefore some additional imputations are required to estimate such tie values. In the simplest case, zeros are imputed. This treatment is called reconstruction in the following sections.

---

[4] There exist other approaches to missing network data in the context of exponential random graph (p^*) models (e.g. [20, 24, 35]) which are not considered here.

[5] Density describes the general level of linkage among the actors [37] in a network. For a directed binary network with n actors, it is defined as the number of arcs (m) divided by the number of all possible arcs in a network ($n \cdot (n-1)$). Scott [25] noted a lack of agreement on how density should be measured in valued graphs. Here, we use a simple extension of binary density and use the average of tie values:

$$densW = \frac{\sum_{i \neq j} v_{ij}}{n(n-1)} , \tag{7.1}$$

where v_{ij} is the value of the arc from actor i to actor j and n is the number of actors in the network. This weighted density measures the average tie strength.

[6] For valued networks, computing reciprocity for dyads is complicated. Of course, the difference between tie values in a dyad must be considered. However, the distribution of strengths has to be taken into account as well. Weighted reciprocity ($recW$) is defined as [28]:

$$recW = \frac{\sum_i \sum_{j \neq i} \min(v_{ij}, v_{ji})}{\sum_i \sum_{j \neq i} v_{ij}} , \tag{7.2}$$

where v_{ij} is the value of the tie from actor i to j. The upper bound of $recW$ is 1 when all tie values are symmetrical in their values. In the case of binary networks the above definition corresponds to the definition provided in [17]: $2|\mathcal{E}|/(2|\mathcal{E}| + |\mathcal{A}|)$, where $|\mathcal{E}|$ is number of edges and $|\mathcal{A}|$ the number of arcs.

Table 7.1 Characteristics of the whole demonstration network and the seven treated networks

Network	Magnitude of changed ties in treated network according to the whole one without diagonal								Average (imputed) tie values of outgoing ties of non-respondents			Network characteristics				QAP corr.
	−4	−3	−2	−1	0	1	2	3	A5	A7	A9	Arcs	recW	densW	Mean tie value	
Demonstration network									1.89	1.11	0.44	49	0.691	1.222	2.245	
Treated networks: RE			1	11	73	5			1.78	0.56	0.22	42	0.843	1.133	2.429	0.953
MEAN			4	4	69	9	4		1.22	1.33	1.44	54	0.591	1.278	2.130	0.882
MO	2	2	1	10	71	1	2	1	0.33	0.56	0.56	37	0.522	1.150	2.486	0.811
REMO			1	10	73	6			1.78	0.67	0.33	44	0.827	1.156	2.364	0.952
TM		3	2	1	74	10			1	1	1	59	0.623	1.178	1.797	0.878
kNNMedian				11	74	3	2		1.56	0.67	0.78	45	0.660	1.178	2.356	0.949
NTI	3	2	1	11	73				0	0	0	32	0.506	0.878	2.469	0.817

RE, reconstruction; MEAN, imputations of the mean values of incoming ties; MO, imputations of the modal values of incoming ties; REMO, reconstruction and imputations based on modal values of incoming ties; TM, imputations of the total mean; kNNMedian, imputations of median of three nearest neighbors based on incoming ties; NTI, null tie imputations; recW, weighted reciprocity; densW, weighted density; mean tie value, mean of tie values (without zeros).

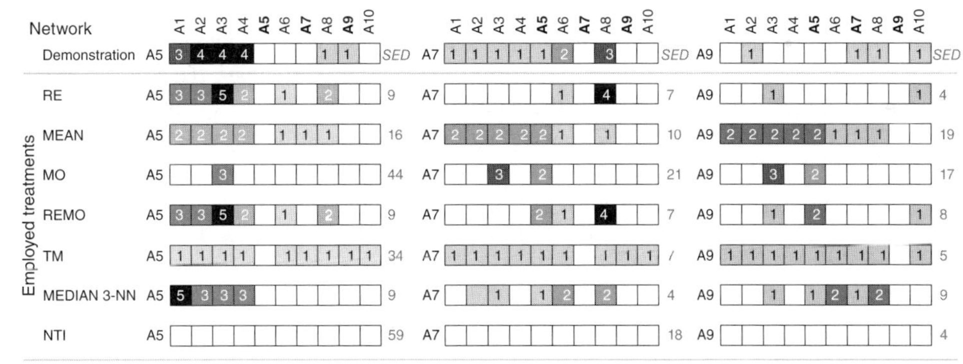

Figure 7.5 Results of seven actor non-response treatments for the demonstration network with three non-respondents. RE, reconstruction; MEAN, imputations of mean values of incoming ties; MO, imputations of modal values of incoming ties; REMO, reconstruction and modal values; TM, imputations of total mean; MEDIAN 3-NN, median of three nearest neighbors of incoming ties; NTI, null tie imputations; SED, squared Euclidean distances between the vectors of tie values of individual non-respondents and the corresponding vector of treated values.

The imputed tie values for *A*5, *A*7, and *A*9 in the demonstration network with the reconstruction procedure are presented in the second row in Figure 7.5. Comparison of tie values with those of the whole network (first row in the body of Figure 7.5) reveals that 17 tie values were changed. The majority (12) of missing tie values were decreased and five tie values were increased (first row in Table 7.1). The weighted reciprocity of this treated network is the highest (0.843) among all treatments because it decreased the number of ties by 22%. This is not surprising. As noted above, for the ties between non-respondents ($v_{5,7}$, $v_{5,9}$, $v_{7,5}$, $v_{7,9}$, $v_{9,5}$, $v_{9,7}$) zeros are imputed. For A5, the average tie value of imputed outgoing ties is 1.78, for A7 it is 0.56, and for A9 it is 0.22. The mean tie value (without zeros) is equal to 2.429, 8% higher compared to the whole network.

The right-most column of Table 7.1 shows the quadratic assignment procedure (QAP) correlation values of the treated networks with the whole network. While there is considerable variation in these values, all are significant, with p values less than 0.000. However, the three highest values are noteworthy, as is discussed below.

7.3.2   Imputations of the Mean Values of Incoming Ties

This treatment imputes the average value of incoming ties of an actor, known as the "item mean" (imputation) [17]. For each missing outgoing tie v_{ij} of the non-respondent i, the (rounded) mean value of all available incoming ties of actor j is imputed.

As emphasized by Žnidaršič et al. [33] for binary networks, this implies (due to rounding) imputing modal values of incoming ties which led them to introduce the term "imputations based on modal value of incoming ties" for binary networks. Although, the imputations of the mean values of incoming ties and the modal values of incoming ties are the same for binary networks, the differences between them in the case of valued networks can be substantial.

The imputed tie values based on imputations of the mean value of incoming ties for three non-respondents (A5, A7, and A9) are presented in the third row in Figure 7.5. For the three non-respondents, a total of 22 tie values (equal to 1 or higher) were imputed and two 1s were set to zero. Of these, 21 tie values differed from the known true values. This is the highest number of "changed" ties compared to the whole demonstration network, suggesting serious flaws with this imputation method. The weighted reciprocity for the treated network is 0.591 and the weighted density is 1.278 (see the second row in Table 7.1), the highest among all the treated networks. The average tie values of the imputed outgoing ties are 1.22, 1.33, and 1.44 for the non-respondents A5, A7, and A9, respectively. The values for A5 and A9 are poor.

7.3.3   Imputations of the Modal Values of Incoming Ties

Imputations based on the modal values of incoming ties take into account the available incoming ties of the non-respondent. For each missing outgoing tie value $v_{ij}(i \neq j)$ of actor j, the modal value of values on all available incoming ties to actor j is imputed.[7]

The imputed tie values using imputations of the modal values of incoming ties for three non-respondents (A5, A7, and A9) are presented in the fourth row in Figure 7.5. Multiple modal values for incoming ties to actor A3 exist: 1 and 3. The mean value of available incoming ties is 2, meaning that the difference of mean value to both modes is equal to 1. Therefore, one of the two modes is selected randomly. In our example in Figure 7.5 and Table 7.5, 3 was randomly selected for imputed ties from non-respondents to A3.

Only five tie values were imputed. Comparisons of the treated network and the demonstration network reveal that 18 ties were changed with the majority of tie values (15) being lowered, often

---

[7] If multiple modal values exist, the one closest to the average value of incoming ties is imputed.

considerably so. The weighted reciprocity of the treated network is the second lowest (0.522) among all treatments, decreasing by 24% compared to the corresponding value for the whole network. The average tie values of the imputed outgoing ties are 0.33, 0.56, and 0.56 (see the third row in Table 7.1) for the non-respondents $A5$, $A7$, and $A9$, respectively. All these values suggest major problems with this imputation method. Indeed, for $A5$ and $A7$, the average of imputed tie values is the lowest among all treatments except for null tie imputations.

7.3.4   Reconstruction and Imputations Based on Modal Values of Incoming Ties

As noted in Section 7.3.1, reconstructing ties between non-respondents cannot be done without making additional assumptions regarding ties between the non-respondents. We combined the reconstruction procedure with imputations based on the modal values of incoming ties, although any other imputation procedure (e.g. imputations of the total mean, imputations of the mean values of incoming ties) could be used.

Comparing the treated network (fifth row in Figure 7.5) to the whole network (first row in Figure 7.5) reveals that 17 tie values were changed to values differing from the known ties. Ten tie values were decreased by 1, one tie value was decreased by 2, and six tie values were increased by 1 (fourth row in Table 7.1). The average tie values of the imputed outgoing ties are 1.78, 0.67, and 0.33 for the non-respondent $A5$, $A7$, and $A9$, respectively. The weighted reciprocity is 0.827, the second highest among treatments considered here being 19.7% higher than for the whole network. The weighted density of the treated network compared to the whole network decreased by 5.4%, while the mean tie value (without zeros) increased by 5.3%.

7.3.5   Imputations of the Total Mean

For binary networks, the average number of ties in the network is used to impute values for the missing ties. If the threshold is set to 0.5, as suggested by Huisman [17], this implies imputing zeros for the missing ties in sparse networks and ones in denser networks.

The generalization for valued networks uses the (rounded) mean of all available tie values for imputing the missing ties. When there are k non-respondents in the network with n actors, the total mean is calculated using $\left\lfloor \sum_{i,j} v_{ij} / ((n-1) \cdot (n-k)) \right\rceil$. For the whole network in 7.1a the average available tie value $79/63 = 1.25$. So, the value of 1 was imputed instead of missing ties in the treated network in the sixth row in Figure 7.5.

As shown in Table 7.1, 16 tie values were changed by this imputation. The weighted density (1.178) and weighted reciprocity (0.623) both decreased compared to the whole network. The mean tie value (disregarding zeros) is 1.797 (see the fifth row in Table 7.1), the lowest among all treatments, being 20% lower than the corresponding value for the whole network.

7.3.6   Imputations of Median of the Three Nearest Neighbors based on Incoming Ties

The nearest-neighbor algorithm has been used widely in social surveys for estimating missing data. It was adjusted for use on network data by Žnidaršič *et al.* [34], who observed "the rationale for this imputation is that the non-respondents are treated individually and not as a group".

This treatment can be summarized in four steps. First, the Euclidean distances between actors are computed based on their incoming ties.[8] Second, for each non-respondent the three nearest neighbors (denoted by a, b, and c) are selected using the smallest calculated Euclidean distance. Third, for each missing outgoing tie v_{ij} of non-respondent i, the median of corresponding outgoing tie values of three nearest actors (labeled v_{aj}, v_{bj}, v_{cj}) is calculated. Finally, this value is imputed for the missing tie.

The imputed tie values for non-respondents $A5$, $A7$, and $A9$ for the demonstration network are shown in the seventh row in Figure 7.5. Compared to the tie values of the whole network, 16 tie values differed from the true values. Eleven tie values were decreased by 1, three were increased by 1, and two were increased by 2 (sixth row in Table 7.1). Compared to other treatments, this is the smallest number of tie values differing from the known values, a good feature. Furthermore, the network characteristics of this treated network are closest to those of the whole network, another indication of the utility of this imputation method. For $A5$, the average tie value on imputed outgoing ties is 1.56. For both $A7$ and $A9$, it is equal to 0.67 and 0.78, respectively (sixth row in Table 7.1). The weighted reciprocity and weighted density decreased by 4.5% and 3.6%, respectively, compared to the whole network. The mean tie value (without zeros) is equal to 2.356, also the closest to the mean tie value of the whole network.

7.3.7 Null Tie Imputations

This treatment simply imputes zeros for all missing tie values of each non-respondent. It is known as the worst treatment for obtaining blockmodel structures of both binary [33] and valued networks [34]. Although it is unlikely to be a good treatment, it is included for comparison with other treatments.

The imputed ties using null tie imputations are presented in the eighth row of Figure 7.5. A total of 17 tie values in this treated network differed for those in the whole network (seventh row of Table 7.1). Much worse, the treated network has only 32 arcs. The weighted reciprocity decreased by 26.8% with the weighted density decreasing by 28.2% compared to the whole network. Both these measures are the lowest among all the treated networks.

These simple descriptive results regarding the nature of the imputed values under the seven treatments suggest that blockmodels of the different treated networks will vary greatly. This is pursued in Section 7.3.8.

7.3.8 Blockmodel Results for the Whole and Treated Networks

The basic idea in evaluating the impact of actor non-response treatments on clustering is to compare the clustering results obtained for the whole and treated networks.

There are two distinct approaches, direct and indirect (described in Section 7.4.1), to blockmodeling network data. Both aim to partition actors into clusters based on a selected equivalence. The results reported below were obtained using the indirect approach for the whole and all seven treated demonstration networks.

---

[8] In our simulations, we assumed that loops are not allowed in the networks, therefore diagonal values were not taken into account when calculating distances. In addition, the standard Euclidean distance measure is "corrected" in a way that the sum is scaled up proportionally to the number of columns used (according to function *dist* in R).

Table 7.2   Cluster membership of the actors the whole demonstration network and the seven treated networks

Network		Actor's membership in clusters based on indirect blocmodeling			ARI
		Cluster 1	Cluster 2	Cluster 3	
Demonstration network		A1, A2, A3, A4, A5	A6, A7, A8	A9, A10	
Treated networks	RE	A1, A2, A3, A4, A5	A6, A7, A8	A9, A10	1
	MEAN	A2, A3, A4	A1, A5	A6, A7, A8, A9, A10	0.378
	MO	A1, A2, A4, A5	A3	A6, A7, A8, A9, A10	0.501
	REMO	A1, A2, A3, A4, A5	A6, A7, A8	A9, A10	1
	TM	A2, A3, A4	A1, A5	A6, A7, A8, A9, A10	0.378
	kNNMedian	A1, A2, A3, A4, A5	A6, A7, A8	A9, A10	1
	NTI	A1, A2, A3, A4	A5	A6, A7, A8, A9, A10	0.501

RE, reconstruction; MEAN, imputations of the mean values of incoming ties; MO, imputations of the modal values of incoming ties; REMO, reconstruction and imputations based on modal values of incoming ties; TM, imputations of the total mean; kNNMedian, imputations of median of three nearest neighbors based on incoming ties; NTI, null tie imputations; ARI, Adjusted Rand Index between the whole partition and the corresponding treated partition.

From the macro-structural perspective, the two partitions must be compared. The Adjusted Rand Index (ARI) is very useful for assessing the extent to which partitions coincide (or not). Its definition, based on the Rand Index [16], measures the concordance between two partitions and is corrected for chance [29], enabling comparisons of its values across different networks regarding their size and number of clusters in the underlying partition. The expected value of ARI is 0 and its maximal value is 1. General guidelines for interpreting the ARI values [29] are (i) ARI ≥ 0.9 indicates excellent agreement, (ii) $0.9 >$ ARI ≥ 0.8 suggests good agreement, (iii) $0.8 >$ ARI ≥ 0.65 indicates moderate agreement, and iv) ARI ≤ 0.65 indicates poor agreement. Based on these criteria, we claim agreement between two partitions is acceptable if ARI values are above 0.8.

Table 7.2 shows the known blockmodel partition of the whole network and the partitions obtained, using the same blockmodeling method, for the seven treated networks. Three imputation methods lead to blockmodels identical to the known partition: reconstruction, the combination of the reconstruction and imputations based on modal values of incoming ties, and imputations using the median of the three nearest neighbors based on incoming ties. The right-hand column of Table 7.2 reports the ARI values comparing the resulting blockmodels. It takes the maximum value of 1 for these treatments. These are the three treatments whose QAP correlations are close to 1 in Table 7.1. This suggests strongly that having a significant QAP correlation for a whole and treated network is *not* a strong enough criterion for an acceptable imputation method. The ARI values shown for the other four imputation treatments are totally unacceptable despite the high QAP correlations between these four treated networks and the whole network. In short, blockmodels from these four treated networks are worthless.

Figure 7.5 presents, in addition to the treated tie values, the squared Euclidean distances (SED) between the vectors of tie values of individual non-respondents and the corresponding vector of treated values. For non-respondents A5 and A7, their outgoing ties are the closest to the vector of outgoing ties treated by the median of the three nearest neighbors based on incoming ties, since the SED values are 6 and 3, respectively. For the non-respondent A9, the original tie values, according to these SED values, are closest to the vector of tie values treated by reconstruction and null tie imputations (SED = 4). This indicates that the median of three nearest neighbors based on incoming ties, most likely, is among the most successful treatments for estimation of missing values.

However, these results are for a network constructed solely for demonstrating some imputation treatments. We next examine some larger and real networks. We address which treatments are the more useful ones for these real networks and, perhaps more importantly, seek insights into why some treatments work better than others. Clearly, it is necessary to consider the combination of (i) the underlying structure of the networks partitioned, (ii) the nature and extent of the non-response, and (iii) the nature of the treatments for such missing data. This effort continues the research represented by the work of Žnidaršič et al. [32–34].

7.4   A Study Design Examining the Impact of Non-Response Treatments on Clustering Results

The impacts of the non-response treatments are based on clustering three distinct real networks after each has been subjected to actor non-response. We review briefly the basic distinction between indirect and direct blockmodeling in Section 7.4.1. Section 7.4.2 presents the basic design of our simulations, while in Section 7.4.3 the nature of the real networks used for our simulation study are presented.

7.4.1   Some Features of Indirect and Direct Blockmodeling

Two conceptually distinct approaches to blockmodeling are direct and indirect, as described by Batagelj et al. [4] and expanded by Doreian et al. [11]. The direct approach considers only the network data and searches for best-fitting partitions given a selected type of equivalence defined by using a set of permitted block types. A criterion function is used to evaluate the agreement between the "ideal" blocks, given a defined equivalence, and the empirically obtained blocks. For small networks, the direct approach is superior for identifying blockmodels. However, given that partitioning networks is an NP-hard problem, the direct approach is computationally burdensome, especially when networks are large.

The indirect approach (suitable for both valued and binary networks) involves two steps [11]. First, some measure of (dis)similarity between each pair of units is computed according to a selected equivalence. Second, a clustering algorithm is used to identify clusters of units. There are choices involving both the clustering algorithm used and the measures of (dis)similarity. Here, we considered only structural equivalence and used the compatible corrected Euclidean distance [4]. For clustering, we used Ward's agglomerative clustering algorithm [36] applied to these dissimilarities.

In Section 7.3.8, we described the ARI as one way of assessing the correspondence of two partitions of a network. This will be used for comparing the partitions of whole and treated real networks. We use also a second measure for binary networks: the proportion of incorrect block types in the blockmodel of a treated network compared to the blockmodel of corresponding whole network where all block types and their locations are specified. Consistent with Žnidaršič *et al.* [33], blockmodels of the treated networks are acceptable only if the measure, denoted by *mErrB*, is below 0.2.

7.4.2 Design of the Simulation Study

The simulations were conducted by using a combination of R along with the blockmodeling package [31] and the `Pajek` program [2, 3].

We start with a schematic outline of the simulation procedure for directed valued networks.

1. For each real network, do the following:
   1.1 Establish the partition of the whole network using indirect blockmodeling employing corrected Euclidean distance and Ward's clustering method.
   1.2 Construct the "observed" data for a wide range for the number of non-respondents to create the measured networks. This was done by randomly selecting actors to become non-respondents and deleting all of their outgoing ties.
   1.3 Employ each of the seven non-response data treatments (presented in Section 7.3) separately to impute values replacing missing data to create the treated networks.
   1.4 Establish a partition of each treated network using indirect blockmodeling with the corrected Euclidean distance and Ward's clustering method.
   1.5 Compare the partitions of the whole original and treated networks by the ARI.

Currently, the direct approach to partitioning networks is confined to binary networks. In an effort to examine the impact of actor non-response for the real networks, we modified the foregoing simulation study by establishing partitions by direct blockmodeling under structural equivalence (with the restriction of having each cluster include at least two vertices). The comparison of the partitions of the whole binary network and the treated networks was done by the ARI and the proportion of incorrect block types, *mErrB*.

7.4.3 The Real Networks Used in the Simulation Studies

We begin our analyses by using two real valued networks gathered by Marouf and Doreian [23]. These data on information and knowledge flows were collected in a Middle Eastern oil and gas company through a web survey in which all respondents were given a roster of the names of the other relevant organizational members. The questions were asked in terms of (i) the frequency with which an individual typically sought work-related information *from* others in the company, (ii) the frequency with which each individual gave work-related information *to* others, (iii) to whom did each individual typically turn for help in thinking through a new or challenging problem at work, and (iv) the frequency with which *others* turned to individuals for help in thinking through such new or challenging problems. The first and third relations were used to construct the *reported* networks. Most often, these would be the networks analyzed despite being unconfirmed. The second and fourth questions were used for constructing two confirmed networks.

In an ideal world, the relation of seeking advice and the transpose of being given advice on the same topic would correspond. In the Marouf and Doreian study [23], these relations were used to create "confirmed" networks for each of the two relations. There are two aspects to this. One concerns confirming the existence of a tie (regardless of the tie value). The second is obtaining confirmed values of the ties. The former leads to a binary network while the latter leads to a valued network. The confirmed tie regarding seeking work-related information is labeled as *SWRIc*. *PHPSc* is the label for the confirmed relation for providing help in problem solving.

Figure 7.6a shows the confirmed network of seeking work-related information (*SWRIc*) with a partition according to the departments (the central administrative group, a commercial affairs group, and three drilling teams labeled A, B, and C). The confirmed network of providing help in problem solving (*PHPSc*) partitioned according to these five departments is presented Figure 7.6b. The shading of the squares indicates the strengths of the ties: the darker the shading, the stronger the tie. The white squares denote null ties.

The third network used in our simulations for actor non-response is a student note-borrowing network (*netBorrowing*). Data were gathered among 15 undergraduate students attending lectures [5]. Males are represented by squares and females by circles. A fitted blockmodel using structural equivalence and indirect blockmodeling produced a partition with three clusters as presented in Figure 7.7.

Conti and Doreian [9] studied the evolution of social networks in a police academy located in a metropolitan American city. They collected network data regarding a variety of social relations at three stages of police training. One of their networks is the fourth network used in our simulations. Figure 7.8 presents a matrix representation of the "social knowledge of" relation at the second time point. The data are valued with the network labeled as *acad2vm*. Using indirect blockmodeling yielded a partition into four clusters, which is shown in Figure 7.8.

Table 7.3 lists some summary details for the two valued and three binary networks (two binarized and one originally binary network). The distribution of tie values is provided along with network sizes, reciprocity (*recW*), density (*densW*), and the average tie value.

7.5   Results

The impact of the seven non-response treatments for the real networks is presented for three types of analyses: (i) indirect blockmodeling of valued networks (in Section 7.5.1), (ii) indirect blockmodeling of binary networks (in Section 7.5.2), and (iii) direct blockmodeling of binary networks (in Section 7.5.3). In each subsection, the partition of actors in the whole networks is justified before the impacts of the actor non-response treatments are examined.

7.5.1   Indirect Blockmodeling of Real Valued Networks

According to the dendrograms presented in Figure 7.9, five clusters are appropriate when partitioning both the *SWRIc* and *PHPSc* networks using structural equivalence. Horizontal dashed lines represent cutting of the dendrogram branches. The cluster memberships of the actors are represented by grey rectangles.

Figure 7.10 presents matrices of *SWRIc* and *PHPSc* networks with reordered vertices according to partitions into five clusters.

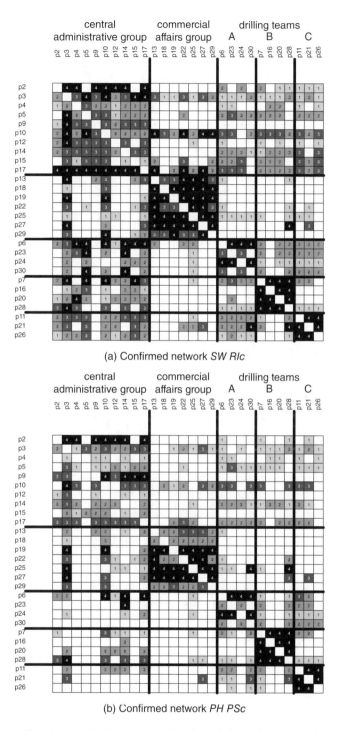

(a) Confirmed network *SW RIc*

(b) Confirmed network *PH PSc*

Figure 7.6 The confirmed networks for seeking work-related information and providing help in problem solving partitioned by work units.

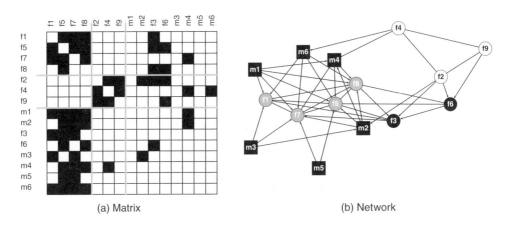

Figure 7.7   Matrix and graphical representations of the note-borrowing network with three clusters.

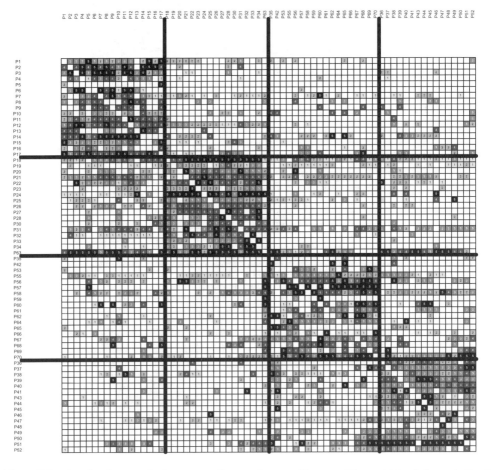

Figure 7.8   A valued network from a police academy with a partition obtained from indirect blockmodeling.

Table 7.3 Characteristics of three real networks and binarized version of valued ones

Network	Number of ties with values						Network characteristics			
	0	1	2	3	4	5	N	$recW$	$densW$	Average tie value
Valued networks										
SWRIc	312	88	146	98	112		28	0.822	1.484	2.527
PHPSc	433	96	103	56	68		28	0.709	0.981	2.297
acad2vm	2834	230	408	532	362	258	68	0.580	1.18	3.006
Binarized or binary networks										
SWRIc_bin	312	444					28	0.973	0.587	1.000
PHPSc_bin	433	323					28	0.786	0.427	1.000
netBorrowing	169	56					15	0.464	0.267	1.000

N, number of actors; *recW*, weighted reciprocity; *densW*, weighted density; mean tie value, mean of tie values (without zeros).

We now consider imposing various levels of actor non-response, treating them in seven ways and examining the consequences for the resulting blockmodels. The results of these simulations and treatments for the *SWRIc* relation are presented in Figure 11a. The generation of missing data was repeated 28 times for networks with one missing actor (each actor was assigned to be a non-respondent) and 100 times for each combination of two or more (3, 4, 5, 6, 8, 10, 12, and 14) missing actors. Together 928 measured networks were generated and, after employment of seven treatments, we obtained 6496 treated networks. Throughout, the figures show the plots of the percentage of missing actors (ranging from 0 to 50%) on the horizontal axis with the mean values of the ARI (*mARI*) on the vertical axis.

As expected, the trajectories for *mARI* decline as the non-response gets more severe. However, the performance of the treatments differ greatly. Overall, using the median of the three nearest neighbors based on incoming ties is the superior imputation treatment. The values of *mARI* are above 0.8 for 36% non-respondents or less, indicating good agreement between the original and treated partitions. Both reconstruction treatments perform well for non-response rates of 25% or less. They are practically interchangeable for 21% non-respondents or less. This is not surprising given the relatively high value (0.822) of weighted reciprocity (*recW*) reported in Table 7.3. For higher percentages of non-respondents, the combination of the reconstruction and imputations based on modal values performs slightly better. However, the performance of these treatments diminishes rapidly when actor non-response gets higher.

By far the worst treatment is null tie imputation. It is unacceptable for even three non-respondents (11%) with *mARI* values below 0.8. These values diminish quickly as the non-response problem gets worse. The next two worst treatments are imputations based on the modal values of incoming ties and imputations of the total mean, since their *mARI* values are below 0.8 for four (14%) and five (18%) non-respondents, respectively. For networks having this structure, these three imputation treatments are of no real value.

The corresponding results for the indirect blockmodeling of the valued network *PHPSc* are presented in Figure 7.11b. Again, the median of the three nearest neighbors based on incoming ties is the best treatment, since the values of *mARI* are above 0.8 for 36% of non-respondents.

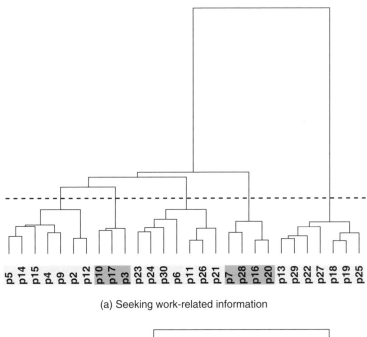

(a) Seeking work-related information

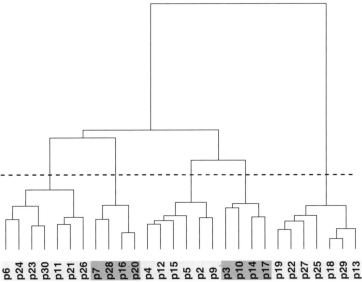

(b) Providing help in problem solving

Figure 7.9   Dendrograms for the indirect blockmodeling of the confirmed (valued) networks of seeking work-related information and of providing help in problem solving.

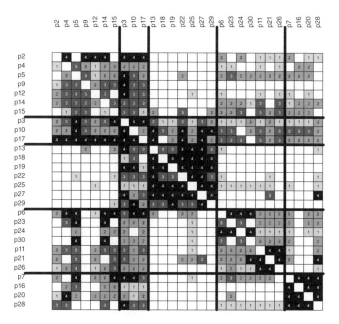

(a) Seeking work-related information

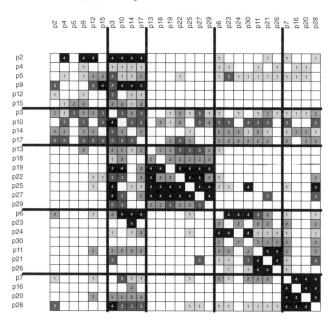

(b) Providing help in problem solving

Figure 7.10   The two valued networks with partitions obtained from indirect blockmodeling.

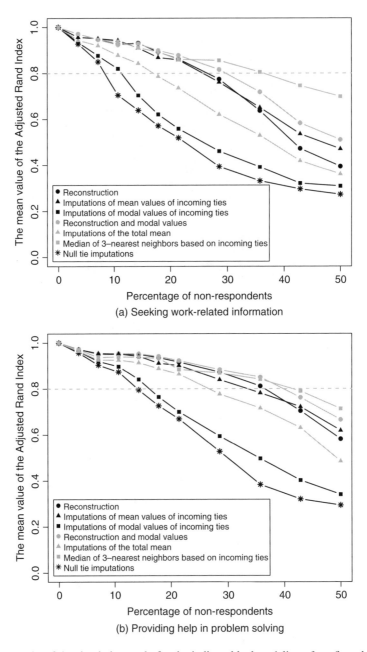

(a) Seeking work-related information

(b) Providing help in problem solving

Figure 7.11   Results of the simulation study for the indirect blockmodeling of confirmed (valued) networks of seeking work-related information and of providing help in problem solving.

Indeed, for 20% of non-respondents or less, the values of *mARI* are above 0.9, indicating excellent agreement between the two partitions. The reconstruction, the reconstruction in combination with imputations based on modal values, and imputations of the mean values of incoming ties are the next best treatments, performing satisfactorily for 36% of non-respondents or less. For higher percentages of non-respondents the combination of reconstruction and imputations based on modal values performs better than the other two treatments.

Again, the three worst treatments are the null tie imputations, imputations based on the modal values of incoming ties, and imputations of the total mean, since their *mARI* values are below 0.8 for five non-respondents and more, and the trajectories of their values decrease in the most extreme fashion.

The results for the two valued networks in Figure 7.6 have some intriguing similarities and differences for the analysis of the *SWRIc* and *PHPSc* networks. The performance trajectories for the ARI measures are more sharply differentiated for the lower panel of Figure 7.11 into two groups, a reminder that the actual structure of the network matters. Yet, only one imputation method, the median of the three nearest neighbors based on incoming ties, is clearly superior. This is fully consistent with the results reported by Žnidaršič *et al.* [34].

According to the dendrogram presented in Figure 7.12, four clusters (represented by grey rectangles according to cutting denoted by the dashed line) are appropriate when partitioning the police academy network using indirect blockmodeling.

Figure 7.8 showes a blockmodel partition with four clusters. The academy formed four squads used for para-military training, going to shooting ranges and driving training. Members of the squads identified strongly with their squads. Of some interest is that the ARI value comparing the partition in the squads and blockmodeling partition is 0.88. Conti and Doreian [9] determined, using QAP methods, that squad membership had a strong impact on relation formation at the time point for these data.

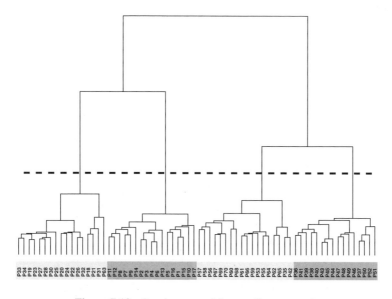

Figure 7.12 Dendrogram of the *acad2vm* network.

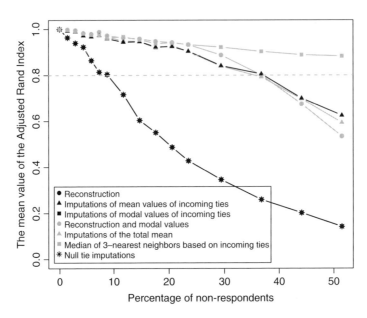

Figure 7.13   Results of the simulation study for the indirect blockmodeling of valued network from a police academy. The trajectories for reconstruction and for reconstruction and modal values are indistinguishable. The same holds for null tie imputation and imputing the mode of incoming ties.

Figure 7.13 shows the ARI trajectories for the police academy data. Yet again, the median of the three nearest neighbors outperforms all of the other imputation treatments. Indeed, for up to 30% of actor non-response, the ARI values are above 0.9 and, for higher levels of actor non-response, they are well above 0.8. The ARI trajectories for the two reconstruction imputation methods are virtually identical and cannot be distinguished in Figure 7.13. Up to slightly less than 30% of actor non-response, these two methods are the next best. Imputation of the total mean also performs well. Next comes imputations of the mean of incoming ties. After about 37% actor non-response, the only acceptable treatment is the median of the three nearest neighbors. Imputing the mode of incoming ties and null tie imputation are virtually identical and are unacceptable when the level of actor non-response exceeds slightly less than 10% actor non-response. They are also outperformed by all other imputation methods even for low levels of non-response.

7.5.2   Indirect Blockmodeling on Real Binary Networks

We next consider the impact of actor non-response treatment on indirect blockmodeling for three networks. The first is the student note-borrowing network (Figure 7.7). The other two networks are the binarized versions of networks *SWRIc* and *PHPSc*.

The dendrogram presented in Figure 7.14a reveals that actors belong to three clusters as presented in Figure 7.7, consistent with the original analysis [5]. Figure 7.14b presents the results of the actor non-response simulations with indirect blockmodeling into three clusters. The generation of missing data was repeated 15 times for networks with one missing actor (each actor was

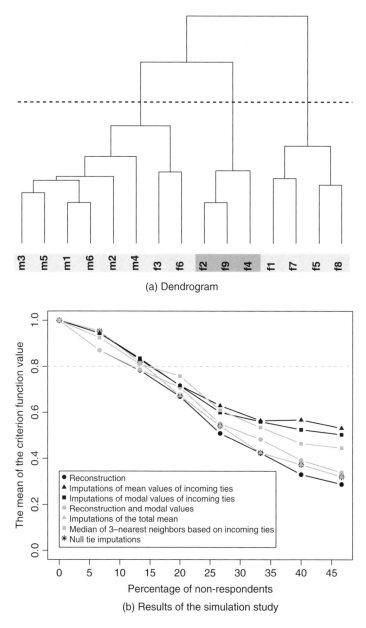

(a) Dendrogram

(b) Results of the simulation study

Figure 7.14 Dendrogram and results of the simulation study for the indirect blockmodeling of the note-borrowing network.

assigned to be a non-respondent), 50 times for a combinations of two non-respondents, and 100 times for combinations of three to seven missing actors. Together, 928 measured networks were generated. After employing the seven treatments, we obtained 6496 treated networks. Again, the trajectories of the mean ARI values are plotted against the percentage of missing respondents.

In general, the results here are less encouraging regarding the efficacy of treatments for actor non-response. Three best treatments, with interchanging rank of best performance over the whole range of non-respondents, are imputations of the mean values of incoming ties, imputations of modal values of incoming ties, and the imputation of the median of the three nearest neighbors based on incoming ties. Yet agreement between the partitions, for all treatments, is unacceptable for three (20%) non-respondents or more: in this range, all *mARI* values are below 0.8.

The values of *mARI* for the reconstruction procedures are below those for all other treatments. We suspect the reason for reconstruction performing so badly is due to the low reciprocity (0.464) of the network. Given its low density (0.267), imputations using the total mean amounts to using null tie imputations known to be poor in general. One message is clear for networks with this structure: avoid actor non-response, a claim relevant for all networks.

As noted above, the oil company networks *SWRIc* and *PHPSc* were binarized so that each tie value of 1 or more regardless of its magnitude was set to 1. The density of the binarized *SWRIc* network is 0.587, while its reciprocity is equal to 0.973. The corresponding numbers for the binarized *PHPSc* network are 0.427 and 0.786.

The dendrogram shown in Figure 7.15a suggests five clusters are reasonable for *SWRIc*. In contrast, a partition into four clusters is the most appropriate for *PHPSc*. This implies binarizating valued networks may be problematic, something depending on the underlying structures of networks subjected to this treatment. It seems that binarization destroys the blockmodeling structure by reducing the high variability of the original tie values. The weighted reciprocity values are higher in the binarized networks compared to the corresponding valued versions, especially for *SWRIc* (see Table 7.3). Even so, we explore the binarized networks further with regard to actor non-response.

Figure 7.16a presents the binarized *SWRIc* network with the five clusters from indirect blockmodeling. Comparing the partitions based on indirect blockmodeling of the valued network (Figure 7.9a) and the binarized version of the *SWRIc* network (Figure 7.15a) shows they are quite different. The value of ARI confirms this since its value is 0.601. In short, the two partitions do not correspond. The four-cluster partition of the binarized *PHPSc* is presented in Figure 7.16b. The ARI for partitions based on indirect blockmodeling of the valued (Figure 7.9b) and binarized (Figure 7.15b) *SWRIc* is 0.763, below the 0.8 threshold for being viewed as consistent partitions. Clearly, binarization of valued networks can severely change the blockmodel structure and the partition of the actors, especially if the variation in the valued tie values is large.

Figure 7.17a presents the simulation results concerning indirect blockmodeling of the *SWRIc* network into five clusters. The imputation treatments do not fare as well as for the valued version of this network. Using the median of the three nearest neighbors performs the best for 21% of non-respondents or more, although the ARI values are below the desired threshold of 0.8. This treatment is acceptable up to 11% of non-respondents. Acceptable treatments for up to 11% and 14% of non-respondents are reconstruction plus modal values and simple reconstruction. All other treatment methods perform even worse, with some being unacceptable even for only two non-respondents. There is little point in comparing how badly they perform relative to each other. One reason for their wretched performance is that the binarized version of this network

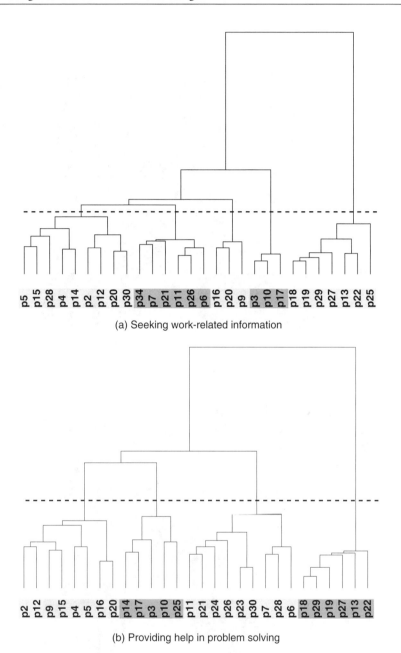

(a) Seeking work-related information

(b) Providing help in problem solving

Figure 7.15 Dendrograms for the indirect blockmodeling of the binarized confirmed networks of seeking work-related information and of providing help in problem solving.

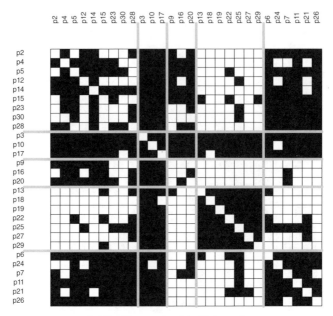

(a) Seeking work-related information

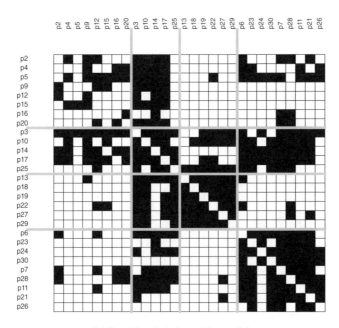

(b) Providing help in problem solving

Figure 7.16   Binarized networks with partitions from indirect blockmodeling.

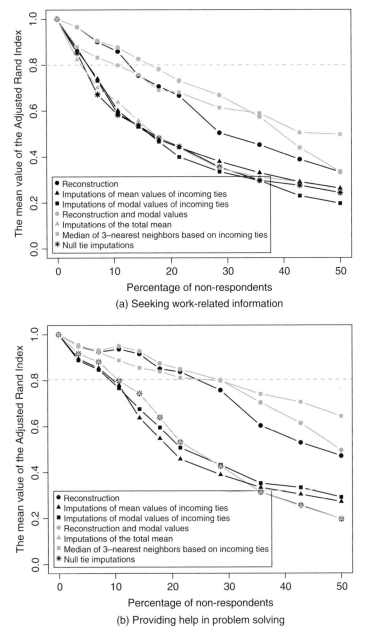

Figure 7.17   Results of the simulation study for the indirect blockmodeling of binarized confirmed networks of seeking work related information and of providing help in problem solving.

is almost completely symmetrical, with reciprocity equal to 0.972. If so, this suggests that only three imputation treatments have any value for highly symmetric networks.

The impacts of actor non-response treatments using indirect blockmodeling of the *PHPSc* network into four clusters are presented in Figure 7.17b. Again, the median of the three nearest neighbors performs the best for larger percentages of non-respondents (30% or more). Up to 30% of non-respondents, its *mARI* values are above the threshold of 0.8, indicating acceptable fit between partitions. Both reconstruction procedures perform in an acceptable fashion for up to slightly more than a 20% level of non-response. Of the two, combinations of reconstruction and imputations of the modal values perform slightly better. All other treatments perform badly for more than 10% of non-respondents, with *mARI* trajectories declining more sharply.

For both of the work-related networks, the median of the three nearest neighbors performs well. In these binarized networks, it performed better for the *PHPSc* network, as it was less affected by the binarization of the two networks. Both of the treatments involving reconstruction perform well, but for lower levels of non-response compared to the superior treatment. Overall, the results reported in this section suggest there are, at most, three acceptable imputation treatments.

We turn to consider direct blockmodeling of these three binary networks, again using structural equivalence.

7.5.3   Direct Blockmodeling of Binary Real Networks

Under direct blockmodeling, a set of permitted block types is specified for the selected type of equivalence. For structural equivalence, only null blocks and complete blocks can be fitted to data. The value of the criterion function is the number of inconsistencies in empirical blocks compared to the ideal blocks.

The first binary network we consider is the note-borrowing network. A narrower set of treatments has already been examined for that network [33]. There, it was established that a blockmodel with three clusters delineates the macro-structure of the network (with a criterion function having 28 inconsistencies).

Continuing our general strategy, Figure 7.18a plots the value of *mARI* against the percentage of actor non-response. Again, the median of the three nearest neighbors performs the best, although *mARI* values are below 0.8 for a quarter of non-respondents and more. The second-best treatment couples reconstruction with imputations of modal values of incoming ties, but only for up to about 12% of non-response levels. While all of the treatment methods perform in an adequate fashion at low levels of non-response, their trajectories drop rapidly thereafter. Null tie imputation performs the worst due to the low network density (0.266), making this treatment equivalent to using the total mean. For this binarized network, reconstruction is the second worst performer, a departure from earlier results. But this is not surprising for such a non-symmetrical network (reciprocity is only 0.464).

The second criterion for assessing the adequacy of identified blockmodels is the proportion of incorrect block types, *mEErB*. On this measure (Figure 7.18b), the differences among treatments are much smaller. Yet, the median of the three nearest neighbors remains the best treatment over a wider range of actor non-response. Up to 47% of non-response, only 10% of block types in the blockmodel are incorrectly identified. For lower levels of non-response the median of the three nearest neighbors does much better. The worst treatments are imputations of the total mean and

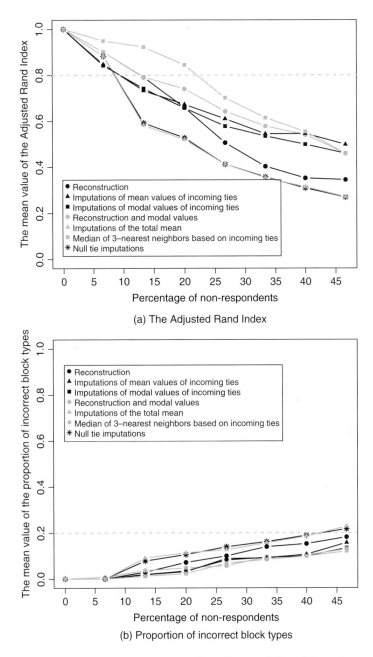

Figure 7.18 Results of the simulation study for the direct blockmodeling of the student note-borrowing network.

null tie imputations, where for 40% of non-respondents, on average, 20% of the block types are identified incorrectly.

Direct blockmodeling based on structural equivalence of the binarized *SWRIc* network established a partition with five clusters. The value of the criterion function is 98.[9] The binarized network of *SWRIc* with five clusters is presented in Figure 7.19a.

Figure 7.20a presents the simulation results for this network. Their most distinctive feature is the sharp separation of the trajectories of *mARI* plotted against the percentage of actor non-response. Four imputation treatments are spectacularly bad for revealing position memberships of actors. Imputations of the mean value of incoming ties, imputations of the modal value of incoming ties, imputations of the total mean, and null tie imputation are unacceptable for this network. A surprise comes with the acceptable treatments. The best treatment is reconstruction with modal values. It performs well for up to about 28% of non-respondents, although the trajectory drops sharply thereafter. The second-best treatment, up to about 18% non-response levels, is the median of the three nearest neighbors based on incoming ties. Reconstruction does well up to slightly less than 20% non-response and is better than using the median of the three nearest neighbors up to that level. This is due to the network being very symmetric.

Regarding the identification of the correct block types, Figure 7.20b shows the best treatments are the median of the three nearest neighbors based on incoming ties and the combination of reconstruction procedure and imputations based on modal value of incoming ties. They perform very well over the entire range of non-response. Not quite as good, but still acceptable for the whole range of non-respondents, are reconstruction and imputations based on modal and median values of incoming ties.

The binarized network of *PHPSc* with partition into four clusters is presented in Figure 7.19b. Based on this, direct blockmodeling, using structural equivalence, of the binarized *PHPSc* network was performed with four clusters as the dendrogram in Figure 7.15b suggests four clusters. The value of the criterion function for this partition is 143.

Figure 7.21b reveals, as for the binarized version of the *SWRIc*, that four treatments, imputations of the mean and modal values of incoming ties, imputations of the total mean, and null tie imputations, perform poorly in revealing position memberships. For 21% of non-respondents or less both reconstruction procedures perform well. Again, using the median of the three nearest neighbors based on incoming ties is better than both reconstruction procedures being acceptable up to 35% non-respondents.

In terms of *mErrB* values in Figure 7.21b, the best treatment is reconstruction combined with using modal values of incoming ties. Even for 50% of non-respondents values of *mErrB* are around 0.1. Using the median of the three nearest neighbors based on incoming ties is acceptable also across the whole range of non-respondents. Null tie imputations and imputations of the total mean were unable to identify the macro structure of the network for 18% of non-respondents or more.

---

[9] For the partition into four clusters, the criterion function had a value of 114. Worse, there were two equally well-fitting partitions. Using six clusters, the value of the criterion function dropped to 92 but, again, there were two equally well-fitting partitions. It is well known that the value of the criterion function for structural equivalence decreases monotonically with the number of clusters. Given the other partitions of this network, using more clusters seems unadvisable. Hence our choice to work with five clusters.

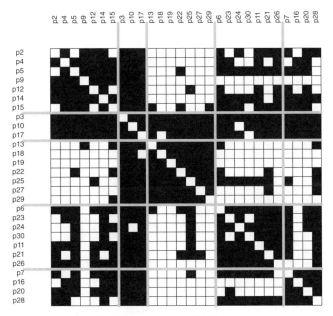

(a) Seeking work-related information

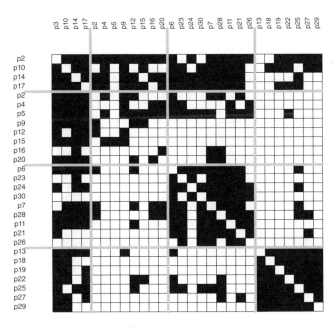

(b) Providing help in problem solving

Figure 7.19   Networks with partitions from direct blockmodeling based on structural equivalence.

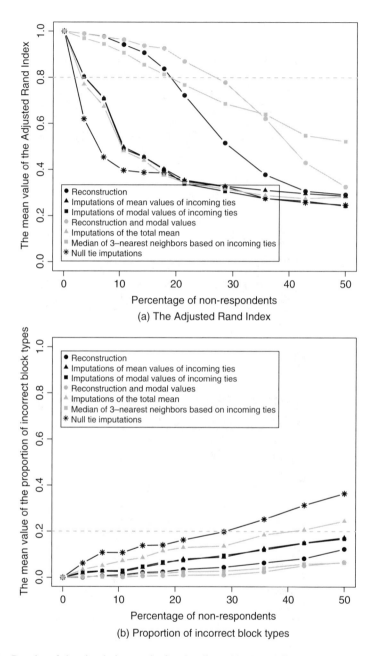

Figure 7.20 Results of the simulation study for the direct blockmodeling of the binarized confirmed network of seeking work-related information.

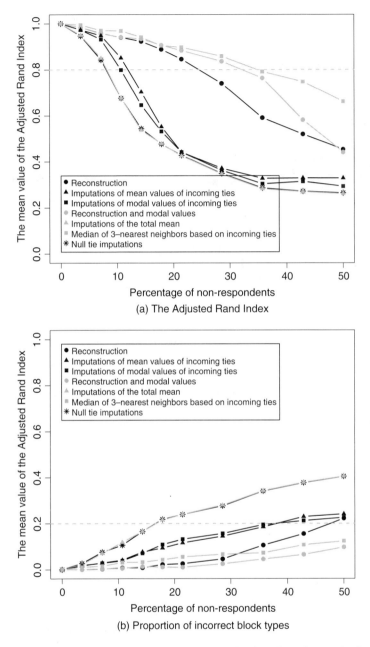

(a) The Adjusted Rand Index

(b) Proportion of incorrect block types

Figure 7.21 Results for the direct blockmodeling of the binarized confirmed network of providing help in problem solving.

Table 7.4   Summary of impact of actor non-response treatments on clustering

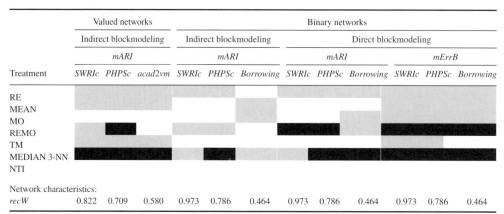

	Valued networks			Binary networks								
	Indirect blockmodeling			Indirect blockmodeling			Direct blockmodeling					
	mARI			mARI			mARI			mErrB		
Treatment	SWRIc	PHPSc	acad2vm	SWRIc	PHPSc	Borrowing	SWRIc	PHPSc	Borrowing	SWRIc	PHPSc	Borrowing
RE												
MEAN												
MO												
REMO												
TM												
MEDIAN 3-NN												
NTI												
Network characteristics:												
recW	0.822	0.709	0.580	0.973	0.786	0.464	0.973	0.786	0.464	0.973	0.786	0.464

RE, reconstruction; MEAN, imputations of mean values of incoming ties; MO, imputations of modal values of incoming ties; REMO, reconstruction and modal values; TM, imputations of total mean; MEDIAN 3-NN, median of three nearest neighbors based on incoming ties; NTI, null tie imputations.

7.6   Conclusions

The results from the simulations are complex. Table 7.4 provides a summary for all the studied networks and clustering procedures. For each network used in clustering, the performance of actor non-response treatments based on both the ARI (*mARI*) and the proportion of incorrect block types (*mErrB*) is presented by different shades. Bad performances are denoted by white, moderate performances by grey, and good performances by black. We note the row for the median of the three nearest neighbors, which performed well for all but one of the 12 networks examined. Below the table, weighted reciprocity is provided since the selection of the best treatment depends of the level of the symmetry and/or correspondence of tie values.

Based on the results and the summary given in Table 7.4, together with the results of extensive simulation studies on three network models (cohesive subgroups, core periphery, and hierarchical model) presented by Žnidaršič et al. [34], the following recommendations can be given:

- *Never* disregard the non-respondents of the network. Ignoring incoming ties is an absurd strategy despite it being a frequent response to the presence of actor non-response. Doing so borders on irresponsibilty.
- Use the best possible imputation method for missing data.
- Regardless of the hypothesized blockmodel structure, the clustering procedure employed, and the level of reciprocity the most preferable actor non-response treatment is using the median of the *k* nearest neighbors based on incoming ties.
- We recommend strongly the inclusion of this imputation method for treating actor non-response and plead for its inclusion in all social network analytic packages.
- The reconstruction procedure combined with imputations based on modal values performs well, especially if a network is highly symmetrical.

- For direct blockmodeling of binarized networks, the macro structure is more stable than micro-level memberships for non-respondents in clusters.
- However, the binarization of the valued network can *severely* change the underlying block-model structure, especially when the variability of tie values is high. We claim this data analytic strategy *must* be avoided when partitioning networks.

The main limitation of this study is the small set of relatively small real-world networks we examined, despite the police academy data being the largest network we considered. Even so, if these results are combined with similar conclusions drawn by Žnidaršič *et al.* [34], the above conclusions must be emphasized. Clearly, the structure of the empirical network has great relevance for assessing the consequences of having measurement errors [34]. Using *both* real and completely simulated networks is, and will continue to be, useful in this effort. Actor non-response is eminently treatable but other missing data problems remain. The impact of *item* non-response on the results of network partitioning may be more consequential, as it is harder to detect and treat. Even more crucial is an effort to determine the influence of measurement errors in the form of misreported tie strengths of valued networks for partitioning outcomes.

Acknowledgements

This work was supported in part by the Slovenian Research Agency (research program P5-0168) and by the Russian Academic Excellence Project '5-100'.

References

1. W. An and S. Schramski. Analysis of contested reports in exchange networks based on actors' credibility. *Social Networks*, 40:25–33, 2015.
2. V. Batagelj and A. Mrvar. *Pajek 5.01*. 1996-2017. URL http://mrvar.fdv.uni-lj.si/pajek/. (February 1, 2017).
3. V. Batagelj and A. Mrvar. *Pajek and Pajek-XXL, Program for Analysis and Visualization of Large Networks, Reference Manual, List of commands with short explanation, version 5.01*. 1996-2017. URL mrvar.fdv.uni-lj.si/pajek/pajekman.pdf.
4. V. Batagelj, A. Ferligoj, and P. Doreian. Direct and indirect methods for structural equivalence. *Social Networks*, 14(1-2):63–90, 1992.
5. V. Batagelj, A. Mrvar, A. Ferligoj, and P. Doreian. Generalized blockmodeling with Pajek. *Metodološki Zvezki*, 1:455–467, 2004.
6. D. C. Bell, B. Belli-McQueen, and A. Haider. Partner naming and forgetting: Recall of network members. *Social Networks*, 29(2):279–299, 2007.
7. S. P. Borgatti, K. M. Carley, and D. Krackhardt. On the robustness of centrality measures under conditions of imperfect data. *Social Networks*, 28(2):124–136, 2006.
8. D. D. Brewer and C. M. Webster. Forgetting of friends and its effects on measuring friendship networks. *Social Networks*, 21(4):361–373, 2000.
9. N. Conti and P. Doreian. Social engineering and race in a police academy. *Social Networks*, 32(1):30–43, 2010.
10. P. Doreian and K. L. Woodard. Defining and locating cores and boundaries of social networks. *Social Networks*, 16(4):267–293, 1994.
11. P. Doreian, V. Batagelj, and A. Ferligoj. *Generalized Blockmodeling*. Cambridge University Press, New York, 2005.
12. A. Ferligoj and V. Hlebec. Evaluation of social network measurement instruments. *Social Networks*, 21(2):111–130, 1999.

13. V. Hlebec. Recall versus recognition: Comparison of the two alternative procedures for collecting social network data. *Metodološki Zvezki*, 9: 121–128, 1993.

14. V. Hlebec. Meta-analiza zanesljivosti anketnega merjenja socialne opore v popolnih omrežjih. *Teorija in Praksa*, 38:63–76, 2001.

15. P. W. Holland and S. Leinhardt. The structural implications of measurement error in sociometry. *Journal of Mathematical Sociology*, 3 (1):85–11, 1973.

16. L. Hubert and P. Arabie. Comparing Partitions. *Journal of Classification*, 2:193–218, 1985.

17. M. Huisman. Effects of missing data in social networks. *Journal of Social Structure*, 10(1), 2009.

18. M. Huisman and C. Steglich. Treatment of non-response in longitudinal network studies. *Social Networks*, 30(4):297–308, 2008.

19. D. Knoke and S. Yang. *Social networks analysis*. Sage Publications, Los Angeles, 2nd edition, 2008.

20. J. H. Koskinen, G. L. Robins, P. Wang, and P. E. Pattison. Bayesian analysis for partially observed network data, missing ties, attributes and actors. *Social Networks*, 35(4):514–527, 2013.

21. G. Kossinets. Effects of missing data in social networks. *Social Networks*, 28(3):247–268, 2006.

22. E. Laumann, P. Marsden, and D. Prensky. The boundary specification problem in network analysis. In R. Burt and M. Minor, editors, *Applied Network Analysis: A Methodological Introduction*, pages 18–34. Sage Publications, London, 1983.

23. L. Marouf and P. Doreian. Understanding information and knowledge flows as network processes in an oil company. *Journal of Information and Knowledge Management*, 9(2):105–118, 2010.

24. G. Robins, P. Pattison, and J. Woolcock. Missing data in networks: exponential random graph (p*) models for networks with non-respondents. *Social Networks*, 26(3):257–283, 2004.

25. J. Scott. *Social Network Analysis*. SAGE, London, 3rd edition, 2013.

26. J. A. Smith and J. Moody. Structural effects of network sampling coverage I: Nodes missing at random. *Social Networks*, 35(4):652–668, 2013.

27. J. A. Smith, J. Moody, and J. H. Morgan. Network sampling coverage II: The effect of non-random missing data on network measurement. *Social Networks*, 48:78–99, 2017.

28. T. Squartini, F. Picciolo, F. Ruzzenenti, and D. Garlaschelli. Reciprocity of weighted networks. *Scientific Reports*, 3, 2729, 2013.

29. D. Steinley. Properties of the Hubert-Arabie Adjusted Rand Index. *Psychological Methods*, 9(3):386–396, 2004.

30. D. Stork and W. D. Richards. Nonrespondents in communication network studies: problems and possibilities. *Group and Organization Management*, 17:193–209, 1992.

31. A. Žiberna. Blockmodeling 0.1.8: An R package for generalized and classical blockmodeling of valued networks, 2010.

32. A. Žnidaršič, P. Doreian, and A. Ferligoj. Absent ties in social networks, their treatments, and blockmodeling outcomes. *Metodološki Zvezki*, 9: 119–138, 2012.

33. A. Žnidaršič, A. Ferligoj, and P. Doreian. Non-response in social networks: the impact of different non-response treatments on the stability of blockmodels. *Social Networks*, 34:438–450, 2012.

34. A. Žnidaršič, A. Ferligoj, and P. Doreian. Actor non-response in valued social networks: The impact of different non-response treatments on the stability of blockmodels. *Social Networks*, 34:46–56, 2017.

35. C. Wang, C. T. Butts, J. R. Hipp, R. Jose, and C. M. Lakon. Multiple imputation for missing edge data: A predictive evaluation method with application to add health. *Social Networks*, 45:89–98, 2016.

36. J. Ward, Joe. H. Hierarchical grouping to optimize an objective function. *Journal of the American Statistical Association*, 58:236–244, 1963.

37. S. Wasserman and K. Faust. *Social Network Analysis: Methods and Applications*. Cambridge University Press, Cambridge, 2nd edition, 1998.

8

Partitioning Signed Networks

Vincent Traag[1], Patrick Doreian[2,3], and Andrej Mrvar[2]

[1]CWTS, University of Leiden
[2]University of Ljubljana
[3]University of Pittsburgh

We are concerned with signed networks, where each link is associated with either a positive (+) or negative sign (−). More generally, weights w_{ij} could be used. Although weights are often assumed to be positive, we explicitly allow them also to be negative. For simplicity, we deal primarily with non-weighted networks, but most concepts used here can be adapted easily to the weighted case.

8.1 Notation

While we try to be as consistent as possible with the general notation used throughout this book, we require some additional notation because signed networks have signs for arcs and edges. We denote a directed signed network by $\mathscr{G} = (\mathscr{V}, \mathscr{A}^-, \mathscr{A}^+)$ where $\mathscr{A}^- \subseteq \mathscr{V} \times \mathscr{V}$ are the negative links and $\mathscr{A}^+ \subseteq \mathscr{V} \times \mathscr{V}$ the positive links. We assume that $\mathscr{A}^- \cap \mathscr{A}^+ = \emptyset$, so that no link is both positive and negative. We exclude loops on nodes. Many studied signed networks are directed. Some are not, including the network we study here. Similarly, an undirected signed network is denoted by $\mathscr{G} = (\mathscr{V}, \mathscr{E}^-, \mathscr{E}^+)$, where $\mathscr{E}^- \subseteq \mathscr{V} \times \mathscr{V}$ are the negative links and $\mathscr{E}^+ \subseteq \mathscr{V} \times \mathscr{V}$ the positive links. As for the directed case, $\mathscr{E}^- \cap \mathscr{E}^+ = \emptyset$.

We present our initial discussion in terms of directed signed networks. However, if we restrict ourselves to undirected graphs, then $(i : j) \in \mathscr{E}^\pm$ is identical to $(j : i) \in \mathscr{E}^\pm$. Also, we assume that there are no self-loops, i.e. no $(i : i)$ exists. For edges, the signs on them are symmetrical by definition.

We define the adjacency matrices A^+ and A^-. We set $A_{ij}^+ = 1$ whenever $(i : j) \in \mathscr{A}^+$ and $A_{ij}^+ = 0$ otherwise. Similarly, $A_{ij}^- = 1$ whenever $(i : j) \in \mathscr{A}^-$ and $A_{ij}^- = 0$ otherwise. We denote

Advances in Network Clustering and Blockmodeling, First Edition.
Edited by Patrick Doreian, Vladimir Batagelj, and Anuška Ferligoj.
© 2020 John Wiley & Sons Ltd. Published 2020 by John Wiley & Sons Ltd.

the *signed* adjacency matrix $A = A^+ - A^-$. This can be summarized as follows

$$A_{ij} = \begin{cases} -1 & \text{if } (i,j) \in \mathscr{A}^-, \\ 1 & \text{if } (i,j) \in \mathscr{A}^+, \\ 0 & \text{otherwise.} \end{cases} \tag{8.1}$$

Note that we exclusively work with the *signed* adjacency matrix in this chapter, and A should not be confused with the ordinary adjacency matrix. The signed adjacency matrix for undirected networks is defined in a similar fashion. For undirected networks the signed adjacency matrix is symmetric, and $A = A^\top$.

The neighbors of a node are those nodes to which it is connected. The positive neighbors are $\mathscr{N}_v^+ = \{u \mid (v : u) \in \mathscr{E}^+\}$ and the negative neighbors similarly $\mathscr{N}_v^- = \{u \mid (v : u) \in \mathscr{E}^-\}$, and all the neighbors are simply the union of both $\mathscr{N}(v) = \mathscr{N}^+(v) \cap \mathscr{N}^-(v)$. The number of edges connected to a node is its degree. We distinguish between the positive degree $d_v^+ = |\mathscr{N}_v^+|$, negative degree $d_v^- = |\mathscr{N}_v^-|$, and total degree $d_v = |N_v| = d_v^+ + d_v^-$. Similar formulations are possible for directed signed networks.

Blockmodeling, as a way of partitioning social networks, started with a clear *substantive* rationale expressed in terms of social roles [27]. However, the availability of algorithms for partitioning (unsigned) networks [4, 6], based on ideas of structural equivalence, led to a rather mechanical application to simply partition social networks with a subsequent *ad hoc* interpretation of what was identified. Such algorithms are indirect in the sense of having network transformed to (dis)similarity measures for which partitioning methods are used. In contrast, a direct approach was proposed [14] in which the network data are clustered directly. This allows the inclusion of substantive ideas within the rubric of pre-specification.

Consistent with this, the approach known as structural balance theory has a clear substantive foundation. We briefly review the basics of balance theory as it connects directly to partitioning signed social networks. We then review some methods for partitioning networks in practice, and examine how they connect to balance theory. Finally, we briefly explore how structural balance evolves through time in an empirical example of international alliances and conflict.

8.2   Structural Balance Theory

The basis of structural balance theory is founded on considerations of cognitive dissonance. Heider [20] focused on so-called p-o-x triplets, considering the relations between an actor (p), another actor (o), and some object(x), and claimed such triplets tend to be consistent in attitudes. For example, in this perspective, if someone (p) has a friend (o) who dislikes conservative philosophies (x), then p also tends to dislike conservative philosophies. This extends naturally to p-o-q triples for three actors denoted by p, o, and q. In the formulation involving three actors, well-known claims such as "a friend of a friend is a friend", "an enemy of a friend is an enemy", "a friend of an enemy is an enemy", and "an enemy of an enemy is a friend" are thought to hold. The notion of balance from Heider [20] was further formalized and extended to an arbitrary number of persons or objects by Cartwright and Harary [7]. They modeled relations between persons as a graph where nodes are persons and the relations between them links in the graph. The four possible triads for the undirected case are shown in Figure 8.1.

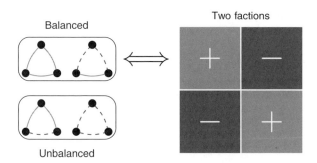

Figure 8.1 Structural balance. There are four possible configurations for having positive or negative links between three nodes (a triad). These are demonstrated on the left, where a solid line represents a positive link and a dashed line represents a negative link. The upper two triads are structurally balanced because the product of their signs is positive. Similarly, the lower two triads are not structurally balanced because the product of their signs is negative. If all triads (in a complete network) are structurally balanced, the network can be partitioned in two factions such that they are internally positively linked, with negative links between the two factions, as illustrated on the right.

For the remainder of the chapter, we restrict ourselves to undirected graphs. We first focus on complete graphs, where all links are present (excluding self-loops). Following [7], we provide the following definition.

Definition 8.1 *A triad i, j, k is called balanced whenever the product*

$$A_{ij}A_{jk}A_{ki} = 1. \tag{8.2}$$

A complete signed graph $\mathscr{G}$ is structurally balanced whenever all triads are balanced.

Of the four possible triads, two are balanced $(+++$ and $+--)$ and two are unbalanced $(++-$ and $---)$ according to this definition (see also Figure 8.1).

Harary [18] proved that if the graph $\mathscr{G}$ is structurally balanced, then it can be partitioned in two clusters such that there are only positive links within each cluster and negative links between them. Cartwright and Harary [7] called this observation the structure theorem, and Doreian and Mrvar [11] called it the *first* structure theorem.

Theorem 8.1 (Structure theorem, [18]) *Let $\mathscr{G} = (\mathscr{V}, \mathscr{E}^+, \mathscr{E}^-)$ be a complete signed graph. If and only if $\mathscr{G}$ is balanced can $\mathscr{V}$ be partitioned into two disjoint subsets $\mathscr{V}_1$ and $\mathscr{V}_2$ such that a positive edge $e \in \mathscr{E}^+$ either in $\mathscr{V}_1 \times \mathscr{V}_1$ or $\mathscr{V}_2 \times \mathscr{V}_2$ while a negative edge $e \in \mathscr{E}^-$ falls in $\mathscr{V}_1 \times \mathscr{V}_2$.*

Proof: Assume $\mathscr{G}$ is balanced. Consider some node $v \in \mathscr{V}$ and set $\mathscr{V}_1 = v \cup \mathscr{N}^+(v)$ as well as the set $\mathscr{V}_2 = \mathscr{V} \setminus \mathscr{V}_1$. Consider an edge $(u : w) \in \mathscr{V}_2 \times \mathscr{V}_2$. Then $(u : v) \in \mathscr{E}^-$ and $(w : v) \in \mathscr{E}^-$ by definition of $\mathscr{V}_2$ so that $(u : w) \in \mathscr{E}^+$ by structural balance. Hence all edges in $\mathscr{V}_2$ are positive. Similarly, any edge $(u : w) \in \mathscr{V}_1 \times \mathscr{V}_1$ is positive. Hence, we can partition $\mathscr{V}$ into the stated disjoint sets $\mathscr{V}_1$ and $\mathscr{V}_2$. In reverse, any triad is easily seen to be balanced if $\mathscr{V}$ is partitioned as stated in the theorem. ∎

While the above is limited to complete graphs, it can be generalized to incomplete graphs. For this we first need to introduce another definition for structural balance.

Definition 8.2 (Structural Balance) *Let $\mathcal{G} = (\mathcal{V}, \mathcal{E}^+, \mathcal{E}^-)$ be a signed graph and A the signed adjacency matrix. Let $C = v_1 v_2 \ldots v_k v_1$ be a cycle consisting of nodes v_i with $v_{k+1} = v_1$. Then the cycle C is called balanced whenever*

$$\text{sgn}(C) := \prod_{i=1}^{k} A_{v_i v_{i+1}} = 1. \tag{8.3}$$

A signed graph $\mathcal{G}$ is called balanced if all its cycles C are balanced.

Stated differently, sgn(C) is the sign of the cycle which is balanced if its sign is positive. If a cycle contains m^- negative edges, then $\text{sgn}(C) = (-1)^{m^-}$. In other words, a cycle is balanced if it contains an even number of negative links. Note that for a cycle of length three, this coincides exactly with the definition of a balanced triad.

The sign of a cycle can be decomposed in the sign of subcycles if the cycle has a *chord*: an edge between two nodes of the cycle (see Figure 8.2).

Theorem 8.2 *Let $C = v_1 v_2 \ldots v_k v_1$ be a cycle with a chord between nodes v_1 and v_r in C. Then let $C_1 = v_1 v_2 \ldots v_r v_1$ and $C_2 = v_1 v_k \ldots v_r v_1$ be the induced subcycles. Then sgn(C) = sgn(C_1)sgn(C_2).*

Proof: We denote by m_1^- the number of negative links of C_1 and similarly m_2^- for C_2 and m^- for C. Suppose that the link $(v_1 : v_r)$ is not a negative link. Then the number of negative links in C is $m^- = m_1^- + m_2^-$ so that $\text{sgn}(C) = (-1)^{m^-} = (-1)^{m_1^-}(-1)^{m_2^-} = \text{sgn}(C_1)\text{sgn}(C_2)$. Suppose that $(v_1 : v_r)$ is a negative link. Then $m^- = (m_1^- - 1) + (m_2^- - 1)$ so that $\text{sgn}(C) = (-1)^{m^-} = (-1)^{m_1^-}(-1)^{m_2^-}(-1)^{-2} = \text{sgn}(C_1)\text{sgn}(C_2)$. ■

In other words, it is not necessary to determine the structural balance of all cycles, and we can restrict ourselves to the balance of chordless cycles. In fact, this statement can be made stronger, and holds for any combination of cycles. With a combination of cycles, we mean the

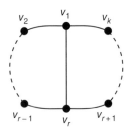

Figure 8.2 Chords and cycles. This illustrates a cycle $v_1 \ldots v_k$ with a chord between nodes v_1 and v_r. There are two subcycles: one following the left path and the other following the right path in the illustration. These two subcycles have a single common edge: $v_1 v_r$. The sign of the large cycle is then the product of the sign of the two subcycles.

symmetric difference of the edges of the two cycles. To define this properly it is more convenient to denote a cycle by the set of its edges (in no particular order). That is, we define a cycle $C = \{e_1, e_2, \ldots, e_k\}$ where the edges form a cycle, i.e. the subgraph of C is a cycle.

Definition 8.3 *Let* $C_1 = \{e_1, e_2, \ldots, e_k\}$ *and* $C_2 = \{f_1, f_2, \ldots, f_k\}$ *be two cycles. Then we define the* symmetric difference *as*

$$C_1 \triangle C_2 = (C_1 \cup C_2) \setminus (C_1 \cap C_2). \tag{8.4}$$

We also refer to this as the combination *of two cycles.*

Note that the combination of two cycles may actually be a set of multiple edge-disjoint cycles. Now we can prove the stronger statement on the combination of cycles.

Theorem 8.3 *Let* $C_1 = \{e_1, e_2, \ldots, e_k\}$ *and* $C_2 = \{f_1, f_2, \ldots, f_k\}$ *be two cycles. If* $C = C_1 \triangle C_2$ *then* $\text{sgn}(C) = \text{sgn}(C_1)\text{sgn}(C_2)$.

Proof: Let us denote the number of negative links in a set C by $m^-(C) = |C \cap E^-|$. Let $S = C_1 \cup C_2$ be the union of the two cycles, and $T = C_1 \cap C_2$ be the overlap of the two cycles. Then $m^-(S) + m^-(T) = m^-(C_1) + m^-(C_2)$, and $m^-(C) = m^-(S) - m^-(T) = m^-(C_1) + m^-(C_2) - 2m^-(T)$. Hence

$$\text{sgn}(C) = (-1)^{m^-(C)} \tag{8.5}$$

$$= (-1)^{m^-(C_1)+m^-(C_2)-2m^-}(T) \tag{8.6}$$

$$= (-1)^{m^-(C_1)}(-1)^{m^-(C_2)}(-1)^{2m^-}(T) \tag{8.7}$$

$$= (-1)^{m^-(C_1)}(-1)^{m^-(C_2)} \tag{8.8}$$

$$= \text{sgn}(C_1)\text{sgn}(C_2) \tag{8.9}$$

since $(-1)^{2m} = 1$ for any integer m. ∎

In other words, if we know the balance of some limited number of cycles, we can determine the balance of all cycles. These "limited number of cycles" are called the *fundamental cycles*. Any cycle can be obtained as a combination of two (or more) fundamental cycles. This implies that if all fundamental cycles are balanced, then the graph as a whole is balanced. We do not consider fundamental cycles in more detail, but this notion underlies the technique by Altafini [2] which we consider in Section 8.3.1.2.

Similar to the sign of cycles, we can define the sign of a path.

Definition 8.4 *Let* $P = v_1 v_2 \ldots v_k$ *be a path in a signed graph* $\mathcal{G}$ *with signed adjacency matrix* A. *The sign of the path* P *is then defined as*

$$\text{sgn}(P) := \prod_{i=1}^{k-1} A_{v_i v_{i+1}}. \tag{8.10}$$

Paths in signed networks are either positive or negative. A cycle can be decomposed in two paths so the sign of a cycle is the product of the sign of the two paths. Hence, a cycle is balanced if the two paths have the same sign.

As before, the graph $\mathcal{G}$ can be partitioned in two clusters with positive links within clusters and negative links between clusters.

Theorem 8.4 (Structure theorem, [18]) *Let $\mathcal{G} = (\mathcal{V}, \mathcal{E}^+, \mathcal{E}^-)$ be a connected signed graph and A the signed adjacency matrix. Then $\mathcal{G}$ is balanced if and only if $\mathcal{G}$ can be partitioned into two disjoint subsets $\mathcal{V}_1$ and $\mathcal{V}_2$ such that a positive edge $e \in \mathcal{E}^+$ falls either in $\mathcal{V}_1 \times \mathcal{V}_1$ or $\mathcal{V}_2 \times \mathcal{V}_2$ while a negative edge $e \in \mathcal{E}^-$ falls in $\mathcal{V}_1 \times \mathcal{V}_2$.*

Proof: First, assume $\mathcal{G}$ is balanced. Then select any $v \in \mathcal{V}$ and set $\mathcal{V}_1 = \{u | \mathrm{sgn}(u - v \text{ path}) = 1\}$, that is, all the nodes that can be reached through a positive path. Define $\mathcal{V}_2 = \mathcal{V} \setminus \mathcal{V}_1$. Let $e = (u : w) \in \mathcal{E}^-$. Suppose $e \in \mathcal{V}_1 \times \mathcal{V}_1$. By construction of $\mathcal{V}_1$, then both u and w have a positive path to v, so that the path $u - w$ through v is also positive. But if $(u : w)$ is negative, it would be contained in a negative cycle, contradicting balance. Hence $e \notin \mathcal{V}_1 \times \mathcal{V}_1$. Similarly, suppose that $e \in \mathcal{V}_2 \times \mathcal{V}_2$. Then both the $u - v$ path and the $w - v$ path are negative (otherwise u and w would be in $\mathcal{V}_1$). The $u - w$ path through v is then positive since the product of the two negative paths is positive. Again, since $(u : w) \in \mathcal{E}^-$ this contradicts balance. Hence, all negative edges lie between $\mathcal{V}_1$ and $\mathcal{V}_2$. Finally, let $e = (u : w) \in \mathcal{E}^+$ with $u \in \mathcal{V}_1$ and $w \in \mathcal{V}_2$. Then there is a positive $u - v$ path and a negative $w - v$ path, so that the $u - w$ path through v is negative, which combined with the positive edge $(u : w)$ leads to a negative cycle, contradicting balance. Hence, positive edges lie within $\mathcal{V}_1$ and $\mathcal{V}_2$. We conclude that if $\mathcal{G}$ is balanced, it can be partitioned as stated. Vice versa, suppose $\mathcal{G}$ can be partitioned into the two states subsets $\mathcal{V}_1$ and $\mathcal{V}_2$. Let C be a cycle. If C is contained within $\mathcal{V}_1$ or $\mathcal{V}_2$ it is completely positive, so that $\mathrm{sgn}(C) = 1$. Suppose C has some node $u \in \mathcal{V}_1$ and $v \in \mathcal{V}_2$. Then any $u - v$ path contains an odd number of negative links, and is hence negative, so that the cycle C is positive. Hence, all cycles are balanced, and so $\mathcal{G}$ is balanced. ∎

8.2.1 Weak Structural Balance

Classical structural balance theory predicts that a balanced network can be partitioned into two clusters. However, as suggested by Davis [9] and Cartwright and Harary [8], we can generalize this notion of structural balance by redefining the notion of an unbalanced triad or cycle. Consider, for example, the (unbalanced) triad with three negative links. The three nodes can be partitioned into three clusters: trivially, all links between clusters are negative and all positive links are within clusters. There is a simple characterization of networks that can be partitioned in such a way: no cycle can contain exactly one negative link. Davis [9] established this only for complete graphs, and Cartwright and Harary [8] extended it to sparse graphs. We call signed networks with this property *weakly* structurally balanced (or weakly balanced).

Definition 8.5 *A cycle $C = v_1 v_2 \ldots v_k v_1$ is termed weakly balanced if it does not contain exactly a single negative link. A signed graph $\mathcal{G}$ is called weakly balanced if all its cycles C are weakly balanced.*

Following this, we can call the previous definition *strong* structural balance. Any graph that is strongly structurally balanced is also weakly structurally balanced: a cycle with a positive sign must contain an even number of negative links. It cannot have exactly one. The reverse does not hold: a weakly structurally balanced cycle can have three negative links, which is not allowed in strong structural balance.

Lemma 8.5 *Let $C = v_1 v_2 \ldots v_k v_1$ be a cycle with a chord between nodes v_1 and v_r in C. Then let $C_1 = v_1 v_2 \ldots v_r v_1$ and $C_2 = v_1 v_k \ldots v_r v_1$ be the induced subcycles. Then C is weakly balanced if C_1 and C_2 are weakly balanced.*

Proof: We denote by $m_1^- \neq 1$ the number of negative links of C_1 and, similarly, $m_2^- \neq 1$ for C_2 and m^- for C. Suppose that the link $(v_1 : v_r)$ is not a negative link, then the number of negative links in C is $m^- = m_1^- + m_2^- \neq 1$, implying C is weakly balanced. Suppose that $(v_1 : v_r)$ is a negative link. Then both $m_1^- \geq 2$ and $m_2^- \geq 2$, and $m^- = (m_1^- - 1) + (m_2^- - 1) \geq 2$ so that C is weakly balanced. ∎

The inverse does not hold. This can readily be seen by considering an all-positive cycle with a single negative chord. The all-positive cycle, clearly, is weakly balanced, but the induced sub-cycles contain exactly one single negative link, and are therefore not weakly balanced. The theorem on chordless cycles for weak balance is hence a weaker statement than the corresponding theorem for strong structural balance. Nonetheless, we can still limit ourselves to considering chordless cycles for determining whether a graph is weakly structurally balanced.

Theorem 8.6 *Let $\mathcal{G}$ be a signed network. Then $\mathcal{G}$ is weakly structurally balanced if and only if all chordless cycles are weakly structurally balanced.*

Proof: If $\mathcal{G}$ is weakly balanced, all cycles are balanced, so that trivially all chordless cycles are balanced. Vice versa, assume all chordless cycles are weakly balanced. We use induction on $|C|$. All chordless cycles C are balanced by assumption, providing our inductive base for $|C| = 3$ (because triads are chordless by definition). Assume all cycles with $|C| < r$ are balanced, then consider cycle C of length r. If C contains a chord, we can separate C in cycles C_1 and C_2, which are balanced by our inductive assumption. Then, by Lemma 8.5 cycle C is balanced. Hence, all cycles are weakly balanced. ∎

To determine whether a graph is weakly structurally balanced, we need only consider the chordless cycles rather than all cycles. Computationally, this is important.

Similar to strong structural balance, we can partition a weakly structurally balanced graph, but now in possibly more than two clusters. This is called the *second* structure theorem by Doreian and Mrvar [11].

Theorem 8.7 (Clusterability theorem, [8]) *Let $\mathcal{G} = (\mathcal{V}, \mathcal{E}^+, \mathcal{E}^-)$ be a connected signed graph. Then $\mathcal{G}$ is weakly structurally balanced if and only if $\mathcal{G}$ can be partitioned into disjoint subsets $\mathcal{V}_1, \mathcal{V}_2, \ldots, \mathcal{V}_r$ such that a positive edge $e \in \mathcal{E}^+$ falls in $\mathcal{V}_c \times \mathcal{V}_c$ while a negative edge $e \in \mathcal{E}^-$ falls in $\mathcal{V}_c \times \mathcal{V}_d$ for $c \neq d$.*

Proof: Suppose $\mathcal{G}$ is weakly balanced. Let $\mathcal{G}^+ = (\mathcal{V}, \mathcal{E}^+)$ be the positive part of the signed graph, and let the clusters be defined by the connected components of $\mathcal{G}^+$. Any positive edge then clearly cannot fall between clusters because different connected components cannot be connected through a positive link. Consider then some negative link $(u : v) \in \mathscr{E}^-$. Suppose that u and v are both in some $\mathscr{V}_c$. Then there exists a positive $u - v$ path because they are in the same component, thus yielding a cycle with exactly a single negative link, contradicting weak balance. Hence, any negative link falls between clusters. Vice versa, suppose $\mathscr{G}$ is split into clusters as stated in the theorem. Any cycle completely contained within a cluster has only positive links. Consider a cycle through u and v where $u \in \mathscr{V}_c$ and $v \in \mathscr{V}_d, d \neq c$. Then any path between u and v must contain at least a single negative link, so that any cycle must contain at least two negative links.   ∎

It is easy to see when a complete signed graph is weakly structurally balanced: it must not contain the $+ + -$ triad.

In summary, signed networks which are strongly structurally balanced can be partitioned in two clusters. Signed networks which are weakly structurally balanced can be partitioned in multiple clusters. Clearly, all signed networks which are strongly structurally balanced are also weakly structurally balanced, but not vice versa. One obvious question is whether strong or weak structural balance is more realistic. This led to partitioning signed networks, which we examine in the next section.

8.3 Partitioning

The previous section introduced the general idea and structure theorems for structural balance. However, these conditions are rather strict: *no* cycle can exist with an odd number of negative links (strong balance) or a single negative link (weak balance). Empirically, this is rather unrealistic to achieve exactly, but we might come close. This was suggested by Cartwright and Harary [7], when introducing the notion of structural balance, who suggested counting the number of cycles that are balanced and measuring the proportion of balanced cycles, termed the *degree of balance*:

$$b(\mathscr{G}) = \frac{c^+(\mathscr{G})}{c(\mathscr{G})} \tag{8.11}$$

where $c^+(\mathscr{G})$ is the number of balanced cycles and $c(\mathscr{G})$ is the total number of cycles. This measure is used infrequently because it is computationally intensive to list all cycles [23]. The number of cycles in a graph increases exponentially with its size. Depending on the so-called cyclomatic number, $\mu = m - n + 1$, there are between μ and 2^μ cycles [36], which Harary [19] also uses to define bounds on the degree of balance. However, this number provides little insight into the structure of the network.

A more useful measure was suggested by Harary [19]: the smallest number of ties to be deleted in order to make the network (weakly) balanced. This is the same as the number of ties whose reversal of signs leads to a balanced network. This is known as the *line index of imbalance*. Computing the line index of imbalance is computationally intensive as it is an NP-hard problem. Initially the definition was restricted to strong structural balance. Doreian and Mrvar [11] were the first to introduce this in the context of clustering for weak structural balance.

8.3.1 Strong Structural Balance

Given a partition into two subsets, $\mathcal{V}_1$ and $\mathcal{V}_2$, we can measure the number of edges that are in conflict with structural balance. The number of negative edges within $\mathcal{V}_1$ are

$$C^-(\mathcal{V}_1) = \frac{1}{2} \sum_{i \in \mathcal{V}_1, j \in \mathcal{V}_1} A_{ij}^- \tag{8.12}$$

and similarly so for $\mathcal{V}_2$, while the positive edges between $\mathcal{V}_1$ and $\mathcal{V}_2$ are

$$C^+(\mathcal{V}_1, \mathcal{V}_2) = \sum_{i \in \mathcal{V}_1, j \in \mathcal{V}_2} A_{ij}^+ \tag{8.13}$$

so that the total number of edges inconsistent with structural balance for a partition into $\mathcal{V}_1$ and $\mathcal{V}_2$ is

$$C(\mathcal{V}_1, \mathcal{V}_2) = C^-(\mathcal{V}_1) + C^-(\mathcal{V}_2) + C^+(\mathcal{V}_1, \mathcal{V}_2). \tag{8.14}$$

This is the *line index of imbalance* mentioned earlier. A graph $\mathcal{G}$ is then structurally balanced if and only if the minimum line index of imbalance is zero.

8.3.1.1 Spectral Theory

Given a partition into $\mathcal{V}_1$ and $\mathcal{V}_2$, let $x_i = 1$ if $i \in \mathcal{V}_1$ and $x_i = -1$ if $i \in \mathcal{V}_2$. Then for an edge $(i : j)$, if $x_i = x_j$ then $x_i A_{ij} x_j = A_{ij}$, while for $x_i \neq x_j$ we have $x_i A_{ij} x_j = -A_{ij}$. Hence

$$x^\mathsf{T} A x = \sum_{x_i = x_j} (A_{ij}^+ - A_{ij}^-) + \sum_{x_i \neq x_j} (A_{ij}^- - A_{ij}^+) \tag{8.15}$$

$$= 2m - \sum_{x_i = x_j} (A_{ij}^+ + A_{ij}^-) - \sum_{x_i \neq x_j} (A_{ij}^+ + A_{ij}^-)$$
$$+ \sum_{x_i = x_j} (A_{ij}^+ - A_{ij}^-) + \sum_{x_i \neq x_j} (A_{ij}^- - A_{ij}^+) \tag{8.16}$$

$$= 2m - 2 \sum_{x_i = x_j} A_{ij}^- - 2 \sum_{x_j \neq x_j} A_{ij}^+ \tag{8.17}$$

So that $x^\mathsf{T} A x = 2(m - C(\mathcal{V}_1, \mathcal{V}_2))$ gives (twice) the number of edges that are consistent with balance, the inverse of the line index of imbalance. Note that this also implies that if x_i is the partition corresponding to structural balance, then $x_i A_{ij} x_j > 0$ for all i, j.

Theorem 8.8 *Let $\mathcal{G}$ be a connected signed graph and let u be the dominant eigenvector of the signed adjacency matrix A. Then $\mathcal{G}$ is balanced if and only if $\mathcal{V}_1 = \{i \in \mathcal{V} | u_i \geq 0\}$ and $\mathcal{V}_2 = \mathcal{V} \setminus \mathcal{V}_1$ defines the split into two clusters as in Theorem 8.4.*

Proof: If the split defines a correct partition, then obviously $\mathcal{G}$ is balanced (Theorem 8.4). In reverse, suppose $\mathcal{G}$ is balanced. Let u be the dominant eigenvector. Suppose that $u_i A_{ij} u_j < 0$ for some i,j. Then let x be another vector with $|x_i| = |u_i|$ for all i and $x_i A_{ij} x_j \geq 0$ for all i,j, which is possible by structural balance of $\mathcal{G}$. Then $\| x \| = \| u \|$ and

$$u^\top A u = \sum_{ij} u_i A_{ij} u_j \qquad (8.18)$$

$$< \sum_{ij} |u_i A_{ij} u_j| \qquad (8.19)$$

$$= \sum_{ij} |x_i A_{ij} x_j| \qquad (8.20)$$

$$= \sum_{ij} x_i A_{ij} x_j = x^\top A x, \qquad (8.21)$$

which contradicts the fact that u is the dominant eigenvector. Hence, $u_i A_{ij} u_j \geq 0$ for all i,j and it defines a correct partition.   ∎

The vector space constrained to $|x_i| = 1$ is rather difficult to optimize. Taking general vectors with $\| x \| = 1$, the dominant eigenvector x maximizes this and the largest eigenvalue of the adjacency matrix $\lambda_n(A)$ gives a lower bound of the line index of imbalance.

Kunegis *et al.* [26] suggest using the signed Laplacian [21] for measuring structural balance. It is defined as

$$\mathcal{L} = D - A \qquad (8.22)$$

where A is the signed adjacency matrix as defined earlier and $D = \mathrm{diag}(d_1, \ldots, d_n)$ the diagonal matrix of total degrees. The rows of $\mathcal{L}$ sum to twice the negative degrees $2(d_1^-, \ldots, d_n^-)$ because $\sum_j A_{ij} = d_i^+ - d_i^-$, so that $(d_i^+ + d_i^-) - (d_i^+ - d_i^-) = 2d_i^-$. Furthermore, the Laplacian is positive-semidefinite, i.e. $x^\top \mathcal{L} x \geq 0$ for all x. We can show this as follows. Writing this out, we obtain

$$x^\top \mathcal{L} x = \sum_{ij} x_i \mathcal{L}_{ij} x_j \qquad (8.23)$$

$$= \sum_{ij} x_i \delta_{ij} d_i x_j - \sum_{ij} x_i A_{ij} x_j. \qquad (8.24)$$

Since $d_i = \sum_{ij} |A_{ij}|$, we can write $\sum_{ij} x_i \delta_{ij} d_i x_j = \sum_{ij} |A_{ij}| x_i^2$ and obtain

$$= \sum_{ij} |A_{ij}| x_i^2 - \sum_{ij} x_i A_{ij} x_j. \qquad (8.25)$$

Clearly $\sum_{ij} |A_{ij}| x_i^2 = \sum_{ij} |A_{ij}| x_j^2$ so that we get

$$= \frac{1}{2} \left(\sum_{ij} |A_{ij}| x_i^2 + \sum_{ij} |A_{ij}| x_j^2 - 2 \sum_{ij} x_i A_{ij} x_j \right), \qquad (8.26)$$

which can be nicely expressed as a square (because $A_{ij}^2 = |A_{ij}|$ and $|A_{ij}|^2 = |A_{ij}|$)

$$= \frac{1}{2} \sum_{ij} |A_{ij}| (x_i - A_{ij} x_j)^2 \geq 0. \qquad (8.27)$$

Now suppose $\mathcal{G}$ is strongly balanced so that we can partition the nodes into $\mathcal{V}_1$ and $\mathcal{V}_2$ without violating balance. Let $x_i = 1$ if $i \in \mathcal{V}_1$ and $x_i = -1$ if $i \in \mathcal{V}_2$. Then for any edge $(i : j)$, if $x_i = x_j$ then by strong balance $A_{ij} = 1$, while if $x_i = -x_j$ we have $A_{ij} = -1$. Hence $|A_{ij}|(x_i - A_{ij}x_j)^2 = 0$ and the smallest eigenvalue of the Laplacian is 0. Vice versa, if the Laplacian is 0, $\mathcal{G}$ is balanced: the term $|A_{ij}|(x_i - A_{ij}x_j)^2 = 0$ can only be 0 for all ij if A is balanced.

More generally, given a partition into $\mathcal{V}_1$ and $\mathcal{V}_2$, let $x_i = 1$ if $i \in \mathcal{V}_1$ and $x_i = -1$ if $i \in \mathcal{V}_2$. Then for an edge $(i : j)$, if $x_i = x_j$ then $|A_{ij}|(x_i - A_{ij}x_j)^2 = 4A_{ij}^-$ while if $x_i = -x_j$ then $|A_{ij}|(x_i - A_{ij}x_j)^2 = 4A_{ij}^+$. Hence,

$$x^{\top}\mathcal{L}x = \frac{1}{2}\sum_{ij}|A_{ij}|(x_i - A_{ij}x_j)^2 \tag{8.28}$$

$$= \frac{1}{2}\left(\sum_{x_i=x_j}4A_{ij}^- + \sum_{x_i \neq x_j}4A_{ij}^+\right) \tag{8.29}$$

$$= 2C(\mathcal{V}_1, \mathcal{V}_2) \tag{8.30}$$

and the vector x gives (twice) the line index of imbalance.

The vector space constrained to $|x_i| = 1$ is rather difficult to optimize. Taking general vectors with $\| x \| = 1$, the minimal eigenvector $x = u$ minimizes $x^{\top}\mathcal{L}x$. Consequentially, the smallest eigenvalue of the Laplacian $\lambda_1(\mathcal{L})$ gives a lower bound of the line index of imbalance, as $x^{\top}\mathcal{L}x \geq u^{\top}\mathcal{L}u$ where u is the smallest eigenvector. The partition induced by u, however, taking $x = \text{sgn}(u)$, i.e. $x_i = \text{sgn}(u_i)$, gives an upper bound, as the minimum index of imbalance is at most the index of an actual partition. Hence, we obtain

$$\lambda_1(\mathcal{L}) \leq 2C(\mathcal{V}_1, \mathcal{V}_2) \leq \tilde{\lambda}_1(\mathcal{L}) \tag{8.31}$$

where $\tilde{\lambda}_1(\mathcal{L}) = \text{sgn}(u)^{\top}\mathcal{L}\text{sgn}(u)$.

We thus obtain the identity that $x^{\top}Ax = 2m - x^{\top}\mathcal{L}x$ and that maximizing $x^{\top}Ax$ is equivalent to minimizing $x^{\top}\mathcal{L}x$. However, the eigenvectors of the adjacency matrix and the Laplacian are, in general, not identical. In the case of balanced graphs though, the largest eigenvector of the adjacency matrix and the smallest eigenvector of the Laplacian provide identical information: the partition into $\mathcal{V}_1$ and $\mathcal{V}_2$.

8.3.1.2 Switching

One interesting observation in signed graph theory is that we can change the sign of some links without affecting balance. More precisely, we can switch the signs of edges across a cut without changing structural balance. Switching signs was introduced originally by Abelson and Rosenberg [1], who used it to calculate the line index of imbalance (although they called it the "complexity" of a signed graph). This was later used by Zaslavsky [37] in a formal graph-theoretical setting. More recently, Iacono et al. [24] use sign switches in an algorithm for calculating the line index of imbalance.

Definition 8.6 (Switching) *Let $\mathcal{G} = (\mathcal{V}, \mathcal{E}^+, \mathcal{E}^-)$ be a signed graph with signed adjacency matrix A and let $\mathcal{V}_1$ and $\mathcal{V}_2$ be a partition of $\mathcal{V}$. Then let $s_i = 1$ if $i \in \mathcal{V}_1$ and $s_i = -1$ if $i \in \mathcal{V}_2$,*

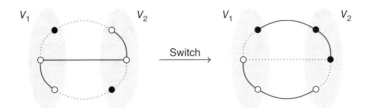

Figure 8.3 Switching. On the left, there are three edges crossing the partition into $\mathcal{V}_1$ and $\mathcal{V}_2$: two negative (the dashed lines) and one positive (the solid line). When we switch according to the partition $\mathcal{V}_1$ and $\mathcal{V}_2$, this implies that we switch the signs of the edges crossing the partition, but leave all the other signs unchanged. This is illustrated on the right. All cycles keep the same sign after the switching. In this case this reduces the number of negative links and simplifies finding the balanced partition. The balanced partition is indicated by black and white nodes in both cases. On the right, the black and white are reversed for $\mathcal{V}_2$, corresponding to the switching of the balanced partition by $\mathcal{V}_1$ and $\mathcal{V}_2$ as explained in Theorem 8.10.

with $S = \text{diag}(s)$ and define $\hat{A} = SAS$ so that $\hat{A}_{ij} = s_i A_{ij} s_j$. Then the graph $\hat{\mathcal{G}}$ defined by $\hat{A}$ is called a switching of $\mathcal{G}$ defined by the partition $\mathcal{V}_1$ and $\mathcal{V}_2$.

Hence, for a link $(i : j)$ with $i \in \mathcal{V}_1$ and $j \in \mathcal{V}_2$, then $\hat{A}_{ij} = -A_{ij}$, while if both $i, j \in \mathcal{V}_1$ (or $i, j \in \mathcal{V}_2$), $\hat{A}_{ij} = A_{ij}$. In other words, switching means we invert the signs of links across the cut by the partition $\mathcal{V}_1$ and $\mathcal{V}_2$, as illustrated in Figure 8.3. Most importantly, any switching preserves the balance of any cycle.

Theorem 8.9 Let $\mathcal{G} = (\mathcal{V}, \mathcal{E}^+, \mathcal{E}^-)$ be a signed graph and let $\hat{\mathcal{G}}$ be a switched signed graph. Denote by $\text{sgn}_{\mathcal{G}}(C)$ the sign of some cycle C with respect to $\mathcal{G}$. Then for any cycle C, $\text{sgn}_{\mathcal{G}}(C) = \text{sgn}_{\hat{\mathcal{G}}}(C)$.

Proof: Let C be a cycle and let $\mathcal{V}_1$ and $\mathcal{V}_2$ be a partition of $\mathcal{G}$. Let $m_{\text{cut}}^{\pm}$ be the number of positive/negative links across the cut between $\mathcal{V}_1$ and $\mathcal{V}_2$ in $\mathcal{G}$ and $m_{\text{within}}^{\pm}$ the number of positive/negative links within $\mathcal{V}_1$ or $\mathcal{V}_2$. Then $\text{sgn}_{\mathcal{G}}(C) = (-1)^{m_{\text{cut}}^- + m_{\text{within}}^-}$. Hence there are $\hat{m}_{\text{cut}}^- = m_{\text{cut}}^+$ and $\hat{m}_{\text{cut}}^+ = m_{\text{cut}}^-$ across the cut in $\hat{\mathcal{G}}$, while $\hat{m}_{\text{within}}^{\pm} = m_{\text{within}}^{\pm}$. By definition $m_{\text{cut}}^+ + m_{\text{cut}}^-$ is even since any cycle must cross $\mathcal{V}_1$ and $\mathcal{V}_2$ an even number of times. In other words, $(-1)^{m_{\text{cut}}^+ + m_{\text{cut}}^-} = 1$ and hence $(-1)^{m_{\text{cut}}^+} = (-1)^{m_{\text{cut}}^-}$. We thus obtain

$$\text{sgn}_{\hat{\mathcal{G}}}(C) = (-1)^{\hat{m}_{\text{cut}}^- + \hat{m}_{\text{within}}^-} \tag{8.32}$$

$$= (-1)^{m_{\text{cut}}^+} (-1)^{m_{\text{within}}^-} \tag{8.33}$$

$$= (-1)^{m_{\text{cut}}^-} (-1)^{m_{\text{within}}^-} \tag{8.34}$$

$$= \text{sgn}_{\mathcal{G}}(C) \tag{8.35}$$

∎

Recall that the line index of imbalance is the minimum number of signs that would need to be changed to make the graph structurally balanced. So, if the balance of the cycles does not change, then neither would the minimum number of sign changes required, and hence the line index of imbalance remains the same.

Theorem 8.10 *Let* $\sigma_i = \{-1,1\}$ *be a partition of* $\mathcal{G}$ *and let S be a switching of* $\mathcal{G}$. *Then the switched partition* $\hat{\sigma} = \sigma S$ *has the same imbalance for the switched graph* $\hat{\mathcal{G}}$.

Proof: The imbalance the partition σ on $\mathcal{G}$ is $\sigma A \sigma^\mathsf{T}$, and for the switched partition and graph we have

$$\hat{\sigma} \hat{A} \hat{\sigma}^\mathsf{T} = \sigma SSASS\sigma^\mathsf{T} = \sigma A \sigma^\mathsf{T} \qquad (8.36)$$

because $SS = I$. ∎

Even if the partition itself is not balanced, switching is defined for any partition. If $\mathcal{G}$ is balanced, we can take the balanced partition $\mathcal{V}_1$ and $\mathcal{V}_2$, in which case all the negative links become positive (because they fall between $\mathcal{V}_1$ and $\mathcal{V}_2$), so that we end up with a completely positive graph. In reverse, the same thing holds: if we can find a switching S such that SAS is completely positive, $\mathcal{G}$ is balanced, and the switching S defines the optimal partition (see also Hou *et al.* [21]). The same principle does not hold for weak structural balance. For example, a triad with three negative links contains a single one after switching so that the original was weakly balanced but the switched one is not.

When Abelson and Rosenberg [1] introduced the idea of switching, they considered a node with the maximal difference of the positive and negative degree: $d_i^+ - d_i^-$. Switching the signs of all its links would then decrease the total number of negative links, while the balance would remain unchanged. The final number of negative links then gives an upper bound on the number of negative links that would need to be removed (or switched) in order to yield structural balance. In other words, it provides an upper bound on the line index of imbalance.

More recently, a rather similar approach was used by Iacono *et al.* [24]. They follow the same procedure as Abelson and Rosenberg [1] for reducing the number of negative links to arrive at an upper bound for the line index of imbalance. The optimal solution may contain even fewer negative links. Iacono *et al.* [24] also provide a way to arrive at a lower bound. The key idea is to associate each negative link to an edge-independent unbalanced cycle, which is easier if the graph contains few negative links. This procedure relies on the fundamental cycles we briefly encountered earlier. Clearly at least one link must change for each edge-independent unbalanced cycle. Even though some cut set may reduce the number of negative links, no cut set can reduce it more than the number of unbalanced edge-independent cycles. Hence, this provides a lower bound on the line index of imbalance.

8.3.2 Weak Structural Balance

The previous subsection dealt only with a split in two factions. We can provide similar definitions for a split in multiple factions. In particular, the number of inconsistencies with structural balance for a given partition into $\mathcal{V}_1, \mathcal{V}_2, \ldots, \mathcal{V}_q$ is

$$C = \frac{1}{2} \sum_{\mathcal{V}_i \neq \mathcal{V}_j} C^+(\mathcal{V}_i, \mathcal{V}_j) + \sum_{\mathcal{V}_i} C^-(\mathcal{V}_i). \qquad (8.37)$$

Note that if a network contains only positive or only negative links, the minimum line index of imbalance is, by definition, 0. For a network of only positive links, the trivial partition consisting of a single cluster provides such a solution. Similarly, for a network of only negative links, the trivial partition consisting of each node in its own cluster, commonly called the singleton partition, achieves zero imbalance.

However, if all links are negative there is an interesting problem: finding the minimum number of factions required for obtaining an imbalance of 0. Having n factions (clusters), with each node in its own faction with an imbalance measure of 0, most often, has little value. It is reasonable to think that this measure could be achieved with fewer factions. For example, for a bipartite graph with all negative links, we have to use only two factions. This minimum number of factions necessary to obtain an imbalance of 0 is known also as the chromatic number: the minimum number of colors necessary to color each node such that two nodes that are connected have different colors. This is a much studied area of research in graph theory. It is an NP-complete problem. This connection was recognized by Cartwright and Harary [8]. The similar problem for positive links is oddly enough trivial: the maximum number of communities for which the imbalance is still 0 simply corresponds to the connected components.

8.3.3   Blockmodeling

The original blockmodel function proposed by Doreian and Mrvar [11] is exactly equivalent to the line index of imbalance. They also propose a more general form, however, weighting differently positive or negative violations of balance:

$$C = \alpha C^+ + (1 - \alpha)C^- \tag{8.38}$$

where $\alpha = 0.5$ returns (half) the original line index. However, this generality comes with costs. Without surprise, different values for α return different values of C. More consequentially, different partitions of the nodes can be returned. This implies there is no principled way for selecting a value of α and hence a partition. This issue was noted by Doreian and Mrvar [13]. It can be called "the alpha problem" which amounts to understanding the interplay of the number of positive and negative links in a signed network, the shape of the criterion function, and the role of α in determining partitions.

The blockmodeling approach partitions the nodes into positions and the links into blocks, which are the sets of links between nodes in the positions. There is only one type of blockmodel in accordance with structural balance: positive blocks on the main diagonal and negative blocks off the diagonal. Of course, for most empirical situations, the links contributing to the line index for imbalance are distributed across blocks. To address this, Doreian and Mrvar [12] examined other possible blockmodels. They considered two mutually antagonistic camps being mediated by a third group (either internally negative or not). So, rather than seeking a blockmodel consisting of diagonal positive blocks and off-diagonal negative blocks, they proposed blockmodels with positive and negative blocks appearing anywhere. For the empirical networks they studied, the results were better fits to the data, according to the line index, and more useful partitions. Unfortunately this comes at a price: if the number of clusters is left unspecified *a priori*, the best partition is the singleton partition (i.e. each node in its own cluster). This line of research is further studied in [5, 15].

Stochastic block models can also deal with negative links [25], but we do not discuss them further here.

8.3.4   Community Detection

Assuming structural balance holds for a network, the resulting partition is a set of clusters with primarily positive ties within them. Structural balance models would not be informative for networks without negative ties. Even so, the positively connected clusters may contain some further substructure. Most networks that contain only positive links can show a clear group structure, commonly called the community structure or modular structure, covered in Chapter 4. One of the most popular methods for community detection in networks with only positive links is known as modularity. It is defined as

$$Q = \sum_{ij} \left(A_{ij} - \frac{d_i d_j}{2m} \right) \delta(\sigma_i, \sigma_j) \tag{8.39}$$

where it is assumed that A_{ij} only contains positive entries and σ_i denotes the community of node i (i.e. if $\sigma_i = c$ it means that node i is in community c) and where $\delta(\sigma_i, \sigma_j) = 1$ if $\sigma_i = \sigma_j$ and otherwise $\delta(\sigma_i, \sigma_j) = 0$. Although this method suffers from a number of problems, most prominently the resolution limit [16], it seems to return sensible partitions for graphs with only positive links.

However, modularity suffers from a problem when some of the links are negative [17, 32]. In particular, imagine there are two fully connected subgraphs, the first with $n_1 = 5$ nodes and the second with only $n_2 = 2$ nodes while there are $n_1 n_2 = 10$ negative links between these two subgraphs. Using the ordinary definitions, the weighted degree for the first subgraph would be $d_i = 4 - 2 = 2$ because each node has four links to the other in the subgraph, and two negative links to the other subgraph. Similarly, the weighted degree for the second subgraph is $d_i = 1 - 5 = -4$ and the total weight is $m = \binom{5}{2} + \binom{2}{2} - 5 \cdot 2 = 1$. Hence, for any link within the first subgraph,

$$A_{ij} - \frac{d_i d_j}{2m} = 1 - \frac{2 \cdot 2}{2} = -1 \tag{8.40}$$

and for the second subgraph

$$A_{ij} - \frac{d_i d_j}{2m} = 1 - \frac{(-4)(-4)}{2} = -7 \tag{8.41}$$

and for any link in between

$$A_{ij} - \frac{d_i d_j}{2m} = 1 - \frac{(2)(-4)}{2} = 3. \tag{8.42}$$

This is rather surprising, as it says that two nodes that are positively connected should be split apart (their contribution is negative), while two negatively connected nodes should be kept together (their contribution is positive). Of course the correct partition here should be a partition into two communities: all nodes of the first subgraph form one community and all nodes of the second subgraph form the other. However, summing up the contributions in Equations (8.40) and (8.41) the quality of such a partition would be

$$\frac{5 \cdot 4}{2} \cdot (-1) + \frac{2 \cdot 1}{2} \cdot (-7) = -17 \tag{8.43}$$

while if there is only one single large partition, adding the contribution from Equation (8.42), we obtain

$$-17 + 5 \cdot 2 \cdot 3 = 13. \tag{8.44}$$

In short, modularity cannot be simply applied to signed networks, as the results are inconsistent with the correct partition (cf. [13]). Hence, modularity needs to be corrected in some way to account for the presence of negative links for it to be useful for signed networks.

Consistent with structural balance, we would expect negative links to be between communities, while positive links are within communities. Hence, if we define the quality of the partition on the positive part as

$$Q^+ = \sum_{ij} \left(A_{ij}^+ - \frac{d_i^+ d_j^+}{2m} \right) \delta(\sigma_i, \sigma_j) \tag{8.45}$$

and on the negative part as

$$Q^- = \sum_{ij} \left(A_{ij}^- - \frac{d_i^- d_j^-}{2m} \right) \delta(\sigma_i, \sigma_j) \tag{8.46}$$

then we should like to maximize Q^+ and minimize Q^-. We can do this by combining $Q = Q^+ - Q^-$, which then becomes

$$Q = \sum_{ij} \left[A_{ij} - \left(\frac{d_i^+ d_j^+}{2m^+} - \frac{d_i^- d_j^-}{2m^-} \right) \right] \delta(\sigma_i, \sigma_j) \tag{8.47}$$

where $A_{ij} = A_{ij}^+ - A_{ij}^-$ as throughout this chapter. In essence, this comes down to using a null model that is adapted to signed networks. More details can be found in Traag [31, Chapter 5].

More generally speaking, one could always define Q^+ for a partition on the positive subnetwork and Q^- on the negative subnetwork and then define a new quality function as $Q = Q^+ - Q^-$. For some methods this turns out to give quite nice results, for example for the Constant Potts Model (CPM) [33]. This method was introduced to circumvent any particular form of the resolution limit. The formulation (again assuming A_{ij} is only positive) is simple:

$$Q = \sum_{ij} [A_{ij} - \gamma] \delta(\sigma_i, \sigma_j). \tag{8.48}$$

Here γ plays the role of a resolution parameter, which needs to be chosen in some way. This parameter has a nice interpretation though, which could motivate a particular parameter setting, and functions as a sort of threshold. In any optimal partition, the density between any two communities is no higher than γ. Put differently, $e_{cd} \leq \gamma n_c n_d$, where e_{cd} is the number of edges between c and d, and n_c and n_d are the number of nodes in that community. Similarly, any community has a density of at least γ, i.e. $e_{cc} \geq \gamma \binom{n_c}{2}$. Even stronger, in fact, any subset of a community is connected to the rest of its community with a density of at least γ in an optimal partition.

If we extend our previous suggestion of combining the positive and negative parts we arrive at the following:

$$Q = Q^+ - Q^- \tag{8.49}$$

$$= \sum_{ij} [A_{ij}^+ - \gamma^+] \delta(\sigma_i, \sigma_j) - \sum_{ij} [A_{ij}^- - \gamma^-] \delta(\sigma_i, \sigma_j) \tag{8.50}$$

$$= \sum_{ij} [(A_{ij}^+ - A_{ij}^-) - (\gamma^+ - \gamma^-)] \delta(\sigma_i, \sigma_j) \tag{8.51}$$

which, by setting $\gamma = \gamma^+ - \gamma^-$, leads to

$$Q = \sum_{ij} [A_{ij} - \gamma] \delta(\sigma_i, \sigma_j). \tag{8.52}$$

In other words, for CPM, there is no need to treat negative links separately and we can immediately apply the same method.

Finally, for $\gamma = 0$, CPM is equivalent to optimizing the line index of imbalance. Indeed, note that we can write the line index of imbalance as

$$C = \frac{1}{2} \sum_{ij} [A_{ij}^- \delta(\sigma_i, \sigma_j) + A_{ij}^+ (1 - \delta(\sigma_i, \sigma_j))] \tag{8.53}$$

We can rewrite this as

$$C = \frac{1}{2} \sum_{ij} [A_{ij}^- \delta(\sigma_i, \sigma_j) + A_{ij}^+ (1 - \delta(\sigma_i, \sigma_j))] \tag{8.54}$$

$$= \frac{1}{2} \sum_{ij} [(A_{ij}^- - A_{ij}^+) \delta(\sigma_i, \sigma_j) + A_{ij}^+] \tag{8.55}$$

$$= m^+ - \frac{1}{2} \sum_{ij} A_{ij} \delta(\sigma_i, \sigma_j). \tag{8.56}$$

so that $C = m^+ - -\frac{1}{2}Q$ for the CPM definition of Q.

Given any particular quality function, the problem is always how to find a particular partition that maximizes this quality function. In general, this problem cannot be solved efficiently (it is NP-hard) and so we have to employ heuristics. One of the best performing algorithms for optimizing modularity is the so-called Louvain algorithm [3]. It can be adapted for taking into account negative links. In addition, it can also be adapted for CPM (and other quality functions). See https://pypi.python.org/pypi/louvain/ for a `Python` implementation designed for handling negative links and working with these various methods.

We do not discuss in detail how the algorithm works, but do discuss one particular element that needs to be changed for dealing with negative links. The basic ingredient of the algorithm is that it moves nodes to the best possible community. Ordinarily, in community detection, all communities are connected, and hence the algorithm only needs to consider moving nodes to neighboring communities. However, this property no longer holds when negative links are present. A trivial example is a fully connected bipartite graph with all negative links. In that case, none of the nodes in any community are connected at all. When only considering neighboring communities, the algorithm never considers moving a node to a community to which it is not connected. In the end, if the algorithm starts from a singleton partition (i.e. each node in its own community), it will remain there. So, we need to calculate the change in Q for all communities, even if it is not connected to that community. Unfortunately this increases the computational time required for running the algorithm. Nonetheless, the algorithm is quite fast. Of course, it only provides a lower bound on the optimal quality value. Hence, for minimizing the line index of imbalance it only provides an upper bound.

8.3.4.1 Temporal Community Detection

One concern when studying the evolution of balance is that we also would like to track the partition over time. For example, if we have two network snapshots and we try to detect the partition minimizing the imbalance, there is an arbitrary assignment to the clusters -1 and 1 (or 0 and 1) in the sense that simply relabeling the partition by exchanging the -1 and 1 yields exactly the same imbalance. For two communities this is still reasonably limited, but for more communities the problem may become more difficult, especially when dealing with many snapshots throughout time.

We rely on a method introduced by Mucha *et al.* [29] to do temporal community detection, while still accounting for negative links. Because this is not the core issue in this chapter, we discuss it only briefly. The idea is to create one large network, which contains all the snapshots of the same network. Then, each node represents a temporal node: a combination of a time snapshot and the original node. Without any links between the different snapshots, the large network would thus consist of as many connected components as there are snapshots (assuming each snapshot is connected). Each snapshot is commonly called a slice, and each link within a slice is called an intraslice link. We introduce additional interslice links, which connects two identical nodes in two consecutive time slices (i.e. they represent the same underlying node, but at a different time) with a certain strength, called the interslice coupling strength. This requires also some additional changes on the Louvain algorithm.

8.4 Empirical Analysis

Empirical research has shown that while few empirical networks are close to balance, at least they are much closer than can be expected at random. Hence, there is considerable evidence that structural balance holds to some extent, at least for weak balance. The evidence for strong structural balance is far more modest, with many exceptions present in the literature. In particular, the all negative triad was found relatively frequently by Szell *et al.* [30] contradicting strong structural balance. They found triads having a single negative link (which is the only triad that is weakly unbalanced) much more rarely. Overall, their evidence favors weak structural balance over strong structural balance. Contrary to dynamical models of sign change, they find that links almost never change sign. However, there is relatively little research into the dynamics of structural balance. Examples where this has been done include Hummon and Doreian [22], Doreian and Krackhardt [10], Marvel *et al.* [28], and Traag *et al.* [34].

We here briefly investigate the dynamics of the network of international relations, where structural balance is argued to play a role by Doreian and Mrvar [13]. We gathered data from the Correlates of War[1] (CoW) dataset, which collects a variety of information about international relations. We create a signed network based on their latest data on alliances (v4.1), representing the positive links, and the militarized interstate disputes (MID, v4.1), representing the negative links. To arrive at a single weight for each link, we sum the different weights on alliances and MIDs for each dyad (a dyad can be involved in multiple alliances and multiple MIDs at the same time). Each MID generates an undirected (negative) link for all states that are involved on different sides. For example, if the USA and the UK were in conflict with Egypt and the USSR, then this would generate four negative links: US–Egypt, UK–Egypt, US–USSR and UK–USSR. The MID weight is set to $\frac{HighestAct}{21}$ so that the weight is in [0, 1] (see CoW documentation for more details). Each alliance generates a link for all dyads involved in the alliance. The weighting is more complicated, since no *a priori* weights are assigned. We chose to weigh a defense pact with a weight of $\frac{10}{14}$, non-aggression by a weight of $\frac{2}{14}$, and both neutrality and an entente by a weight of $\frac{1}{14}$. The single weight is then the sum of the alliance weight minus the sum of the MID weight. Note that a dyad may be involved in multiple MIDs and/or multiple alliances at the same time, so that the individual weight of a link is not necessarily restricted to $[-1, 1]$.

We find that structural balance does not follow any singular trend, and certainly does not converge to structural balance, and remains stable. The same was found by Doreian and Mrvar [13] and Vinogradova and Galam [35], where an earlier version of the CoW data was used. We detect communities using CPM with $\gamma = 0$ and use the approach by Iacono *et al.* [24], which we abbreviate as IRSA (after the authors). We ran the Louvain algorithm for CPM both with an unlimited number of communities (corresponding to weak structural balance) and also with the number of communities restricted to two (corresponding to strong structural balance).

IRSA provides less stable results compared to the CPM estimates (see Figure 8.4). Perhaps with more computation time, more accurate results could be achieved. Even so, regardless of the method, no clear stability emerges. There are some large peaks of imbalance around World War II (WWII), which we discuss. But during the Cold War, and even after the Cold War, no particular convergence towards 0 imbalance is observed. This is not unreasonable, as the international system is subject to new shocks when new conflicts, some of which are major conflicts, emerge. Rather than settling at some level of balance, some unbalance remains in the system which never completely dissipates. Most often, the difference between strong and weak structural balance usually is not so large. This implies that a partition of nations into just two factions already explains much of the structure in international relations. At face value, this suggests that strong balance is at least a reasonable first approximation, and provides some evidence that strong balance is operating in the international system. It is likely that weak balance operates also, perhaps at different timescales. Nonetheless, there are some clear deviations in the patterns of imbalance.

In particular, both IRSA and CPM find that 1944 shows a large peak with an imbalance of 43.9 (CPM) or 47.4 (IRSA), whereas weak structural balance only has an imbalance of 3.14. For this time point, using weak balance may be more useful. This result is due to the large number of conflicts among various parties, which weak balance can accommodate, but which presents problems for strong balance (see Figure 8.5). Indeed, of the 1785 triads in this network, there are 411 strongly unbalanced triads, of which 406 are all-negative triads. The all-negative triad is considered unbalanced under strong balance, but balanced under weak balance. This leaves

---

[1] http://www.correlatesofwar.org/

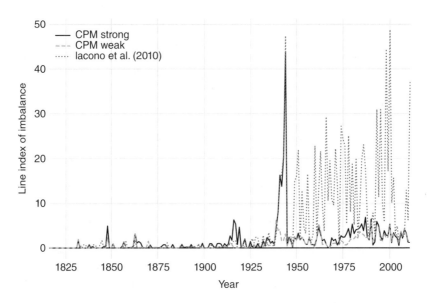

Figure 8.4 Balance timeline. The line index of imbalance using two different methods. The approach by Iacono *et al.* [24] only works for strong balance. The CPM approach can be applied to both the strong and weak balance. CPM seems to provide more stable results than the approach by Iacono *et al.* [24].

only five unbalanced triads under weak balance (although this does not preclude the existence of longer unbalanced cycles).

Many of the all-negative triads are attributable to conflict among nine different countries who were all in conflict with each other: France, Germany, Italy, Hungary, Bulgaria, Romania, Russia, Finland, and New Zealand. Many other countries were opposed to at least two others of this large conflict: Japan, for example, was in conflict with both Russia and New Zealand. These conflicts may be unrelated, but they serve to create an additional unbalanced triad (in the strong sense). The weakly unbalanced triads involve the UK and Turkey. The UK was allied with Portugal and Turkey, but Portugal was also allied with Spain (through the alliance between the dictators controlling both countries), which was in conflict with the UK. Turkey was allied with Germany, Hungary, and Iraq in addition to the UK while the UK was in conflict with both Germany and Hungary. At the same time, Germany was also in conflict with Hungary and Iraq, complicating things further. Clearly, WWII featured many dyadic conflicts, each with their own dynamics.

There is another interesting observation: weak structural balance is higher than strong structural balance for 1939. This should not be the case ordinarily, as the minimal imbalance in weak structural balance should always be lower than strong structural balance. This then seems due to the shift of alliances during WWII. Since the clustering also favors a certain continuity over time, it may be better to cluster countries in a more stable way, without accounting for short-term deviations. This is what seems to happen in 1939 (see Figure 8.6). In particular, Russia is still allied with Germany and Italy, while Russia is in conflict with the UK, France, and Belgium at that time. Similarly, Hungary is allied with Turkey, and Spain with Portugal. Surprisingly, the UK also had some conflict with the USA at that time according to the CoW data.

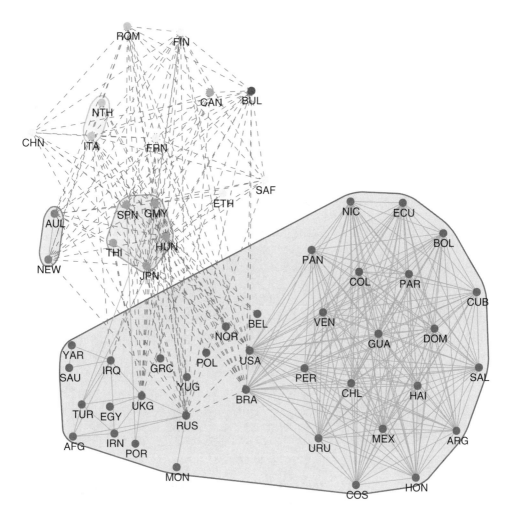

Figure 8.5 International relations 1944. The solid lines represent positive links and the dashed lines represent negative links. The countries that are clustered together are encircled.

At the height of the Cold War, we see the familiar division (see Figure 8.7). We also see the non-aligned states clustered outside of the familiar division. Yet some countries are clustered differently than what one would expect. For example, much of the Arab world is clustered with the West because of the alliance of Morocco and Libya with France, but note that Algeria is not as it was fighting a war of independence with France. Also, Yugoslavia is commonly seen as non-aligned in the Cold War, but here it is clustered with the West through its alliances with Greece and Turkey.

Finally, in more recent times the weak balance clustering seems increasingly unrealistic. This is due to the fact that even if some countries are only weakly positively connected, they are immediately considered as a single cluster. In 2010, for example, most of the world is grouped together in a single cluster, except Africa and some exceptions. Nonetheless, some

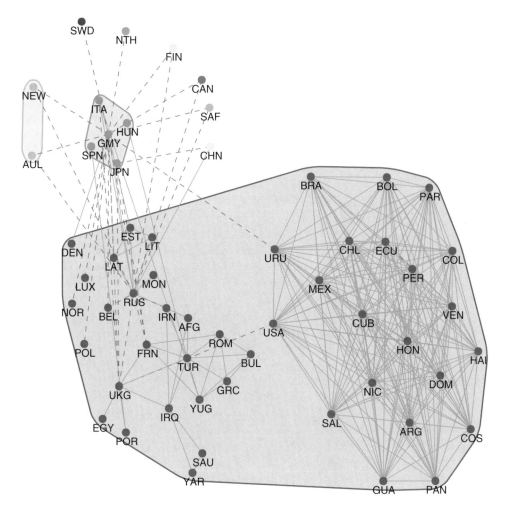

Figure 8.6 International relations 1939. The solid lines represent positive links and the dashed lines represent negative links. The countries that are clustered together are encircled.

clearly separate clusters exist. We therefore also detected clusters using CPM with $\gamma = 0.1$ for 2010. The results are shown in Figure 8.8. There are clearly different clusters in Africa, something missed completely when partitioning with weak structural balance. Africa is divided into a Central African bloc, a Western African bloc, and a Northern African bloc clustered with Arab nations in the Middle East, with the remainder of Africa scattered across other communities. The former USSR remains a separate community. The so-called West breaks into two communities: North and South America constitute a community whereas Europe becomes a separate community.

This is also interesting from another perspective. Structural balance emphasizes both that negative links ought not exist within clusters and positive links ought not exist between clusters. This seems too restrictive by ignoring the presence of some conflict within clusters along

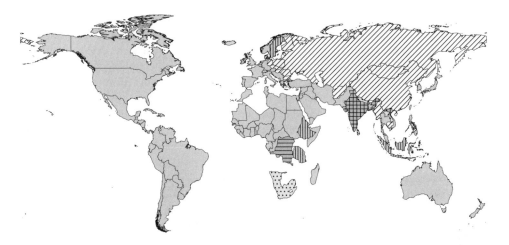

Figure 8.7   A map of the weak balance partition in 1962.

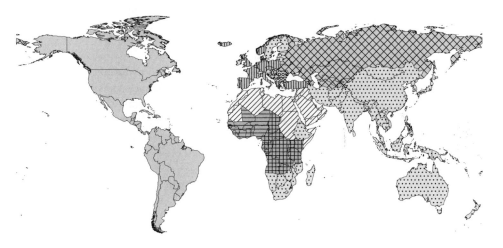

Figure 8.8   Map of CPM partition with $\gamma = 0.1$ in 2010.

with positive ties between clusters. Arguably, it makes more sense to allow for a few positive links between clusters without requiring them to be considered immediately a single cluster. Indeed, when using CPM with $\gamma = 0.1$ relatively less conflict happens within clusters, and most conflict takes place between clusters. Nonetheless, strong balance remains a reasonable first approximation.

8.5   Summary and Future Work

Partitioning signed networks raises methodological issues that differ from those involved in partitioning unsigned networks. Various approaches have been developed. We started our discussion with a consideration of structural balance as it provides a substantively driven framework for considering signed networks. Formulated in terms of exact balance, the initial results in the literature take the form of existence theorems, which we discussed in some detail. We distinguished strong structural balance and weak structural balance. Empirically, most signed networks are not exactly balanced. One of the underlying assumptions of classical structural balance theory is that signed networks tends towards balance. To assess such a claim, it is necessary to have a measure of the extent to which a network is balanced or imbalanced. We discussed some of the measures in the literature but focused primarily on the line index of imbalance. Obtaining this measure is an NP-hard problem. We provided theorems regarding obtaining this measure and its upper and lower bounds.

In discussing strong structural balance, we considered spectral theory and presented some results showing how this is another useful approach for obtaining measures of imbalance. In doing so, we revisited the concept of switching. For partitioning signed networks, we considered signed blockmodeling as a method, pointing out its value and serious limitations. We considered community detection and outlined ways in which is can be adapted usefully to partition signed networks. In discussing this we considered also the Constant Potts Model and how it can be used to partition signed networks. We discussed briefly the notion of temporal community detection.

With the formal results in place, we turned to an empirical example using data from the Correlates of War (CoW) data. We applied two methods to obtain partitions for different points in time. We made no attempt to assess which is the "best" partitioning method, for they all have strengths and weaknesses. However, we did initiate a discussion regarding the conditions under which one method may perform better than others, without being universally the "best" under all conditions. This included a discussion of the utility of weak balance and strong balance, the number of clusters, and the temporal dynamics of the empirical network we studied.

Our results, consistent with other results for the CoW networks produced by others, is that, temporally, signed networks can move towards balance at some time points and away from balance at others. The assumption that signed networks tend towards balance had unfortunate consequences. The more important question, substantively, is simple to state: What are the conditions under which these changes take place? To some extent, this mirrors the issue of when some methods work better than others. The two are related. Together, these issues will form a focus for our future work both analytically and substantively.

References

1. R. P. Abelson and M. J. Rosenberg. Symbolic psycho-logic: A model of attitudinal cognition. *Behavioral Science*, 3(1):1–13, 1958.
2. C. Altafini. Dynamics of opinion forming in structurally balanced social networks. *PLoS One*, 7(6):e38135, 2012.
3. V. D. Blondel, J.-L. Guillaume, R. Lambiotte, and E. Lefebvre. Fast unfolding of communities in large networks. *Journal of Statistical Mechanics*, 10008(10):6, 2008.
4. R. L. Breiger, S. A. Boorman, and P. Arabie. An algorithm for clustering relational data with applications to social network analysis and a comparison with multidimensional scaling. *Journal of Mathematical Psychology*, 12:328–383, 1975.
5. M. Brusco, P. Doreian, A. Mrvar, and D. Steinley. Two algorithms for relaxed structural balance partitioning: linking theory, models, and data to understand social network phenomena. *Sociological Methods and Research*, 40(1):57–87, 2011.
6. R. S. Burt. Positions in social networks. *Social Forces*, 55:93–122, 1976.
7. D. Cartwright and F. Harary. Structural balance: a generalization of Heider's theory. *Psychological Review*. 63(5):277–293, 1956.

8. D. Cartwright and F. Harary. On the Coloring of Signed Graphs. *Elemente der Mathematik*, 23(4):85–89, 1968.

9. J. A. Davis. Clustering and structural balance in graphs. *Human Relations*, 20(2):181–187, 1967.

10. P. Doreian and D. Krackhardt. Pre-transitive balance mechanisms for signed networks. *Journal of Mathematical Sociology*, 25(1):43–67, 2001.

11. P. Doreian and A. Mrvar. A partitioning approach to structural balance. *Social Networks*, 18(2):149–168, 1996.

12. P. Doreian and A. Mrvar. Partitioning signed social networks. *Social Networks*, 31(1):1–11, 2009.

13. P. Doreian and A. Mrvar. Structural balance and signed international relations. *Journal of Social Structure*, 16(2):149–168, 2015.

14. P. Doreian, V. Batagelj, and A. Ferligoj. *Generalized Blockmodeling*, volume 23. Cambridge University Press, Cambridge, 2004.

15. R. Figueiredo and G. Moura. Mixed integer programming formulations for clustering problems related to structural balance. *Social Networks*, 35(4): 639–651, 2013.

16. S. Fortunato and M. Barthélemy. Resolution Limit in Community Detection. *Proceedings of the National Academy of Sciences of the USA*, 104(1):36, 2007.

17. S. Gómez, P. Jensen, and A. Arenas. Analysis of community structure in networks of correlated data. *Physical Review E*, 80(1):016114, 2009.

18. F. Harary. On the notion of balance of a signed graph. *Michigan Mathematics Journal*, 2(2):143–146, 1953.

19. F. Harary. On the measurement of structural balance. *Behavioral Science*, 4(4):316–323, 1959.

20. F. Heider. Attitudes and Cognitive Organization. *Journal of Psychology*, 21(1):107–112, 1946.

21. Y. Hou, J. Li, and Y. Pan. On the Laplacian eigenvalues of signed graphs. *Linear Multilinear Algebra*, 51(1):21–30, 2003.

22. N. P. Hummon and P. Doreian. Some dynamics of social balance processes: bringing Heider back into balance theory. *Social Networks*, 25(1):17–49, 2003.

23. N. P. Hummon and T. J. Fararo. Assessing hierarchy and balance in dynamic network models. *Journal of Mathematical Sociology*, 20(2-3): 145–159, 1995.

24. G. Iacono, F. Ramezani, N. Soranzo, and C. Altafini. Determining the distance to monotonicity of a biological network: a graph-theoretical approach. *IET Systems Biology*, 4(3):223–235, 2010.

25. J. Q. Jiang. Stochastic block model and exploratory analysis in signed networks. *Physical Review E*, 91(6):062805, 2015.

26. J. Kunegis, S. Schmidt, and A. Lommatzsch. Spectral analysis of signed graphs for clustering, prediction and visualization. Proceedings of the SIAM International Conference on Data Mining, SDM 2010, April 29 - May 1, 2010, Columbus, Ohio, USA, 2010.

27. F. Lorrain and H. C. White. Structural equivalence of individuals in social networks. *Journal of Mathematical Sociology*, 1(1):49–80, 1971.

28. S. A. Marvel, J. Kleinberg, R. D. Kleinberg, and S. H. Strogatz. Continuous-time model of structural balance. *Proceedings of the National Academy of Sciences of the USA*, 108(5):1771–6, 2011.

29. P. J. Mucha, T. Richardson, K. Macon, M. A. Porter, and J.-P. Onnela. Community structure in time-dependent, multiscale, and multiplex networks. *Science*, 14;328(5980):876–878, 2010.

30. M. Szell, R. Lambiotte, and S. Thurner. Multirelational organization of large-scale social networks in an online world. *Proceedings of the National Academy of Sciences of the USA*, 107(31):13636–13641, 2010.

31. V. A. Traag. *Algorithms and dynamical models for communities and reputation in social networks*. PhD thesis, Heidelberg University, 2013.

32. V. A. Traag and J. Bruggeman. Community detection in networks with positive and negative links. *Physical Review E*, 80(3):036115, 2009.

33. V. A. Traag, P. Van Dooren, and Y. Nesterov. Narrow scope for resolution-limit-free community detection. *Physical Review E*, 84(1): 016114, 2011.

34. V. A. Traag, P. Van Dooren, and P. De Leenheer. Dynamical models explaining social balance and evolution of cooperation. *PLoS One*, 8(4): e60063, 2013.

35. G. Vinogradova and S. Galam. Global alliances effect in coalition forming. *European Physical Journal B*, 25(87:266):43–67, 2014.

36. L. Volkmann. Estimations for the number of cycles in a graph. *Period. Math. Hungarica*, 33: 153, 1996.

37. T. Zaslavsky. Signed graph coloring. *Discrete Mathematics*, 39(2): 215–228, 1982.

9

Partitioning Multimode Networks

Martin G Everett[1] and Stephen P Borgatti[2]

[1]University of Manchester
[2]University of Kentucky

9.1 Introduction

Most networks examined so far involve connections between nodes all of the same type, known as one-mode networks. There are circumstances in which the nodes are of different types and the connections are only between different types of nodes, and not between nodes of the same type. We refer to these as multimode networks, some authors call these multiway networks. A simple example consists of nodes made up of authors and journals. An author is connected to a journal if they have published a paper in that journal. Since we have two types of nodes, authors and journals, this results in a two-mode network. There are many examples of two-mode networks, such as people attending events [16], legislators being members of committees [38], directors serving on boards [13], companies collaborating on projects [39] etc. In principle there is no reason to limit the node types to two, we could have three or more. An example of a three mode dataset would be criminal by crime by victim. Such datasets are less common and we shall concentrate at first on two mode datasets and discuss general multimode approaches later in the chapter.

For clarity of exposition, when considering two-mode data we shall refer to one mode as actors and the other mode as events. The resultant network will form a bipartite network, that is a network in which the nodes can be divided into two groups with edges only occurring between the groups and not within the groups (note that a one-mode network can also be bipartite). If we have n actors and m events we can represent the data as an $n \times m$ affiliation matrix $\mathbf{A}$, where $a(i,j) = 1$ if actor i attended event j, and 0 otherwise. If the data are valued, reflecting, for example, the time actor i was at event j, then we can replace the binary entry with the value. It is normal to ignore direction in two-mode data since in most cases the direction is from one mode to the other, for example actors choose events and not the other way round. One can envisage examples in which this is not the case, for example heterosexual actors selecting members of

Advances in Network Clustering and Blockmodeling, First Edition.
Edited by Patrick Doreian, Vladimir Batagelj, and Anuška Ferligoj.
© 2020 John Wiley & Sons Ltd. Published 2020 by John Wiley & Sons Ltd.

the opposite sex indicating whom they would be interested in dating. However, there are few techniques or datasets of this type and so we will not discuss the issue further, but just mention it is probably worthy of additional research work.

There have been two distinct approaches to dealing with two-mode data. The oldest method is to convert the data to one-mode and this is often referred to as projection. There are two possible projections for a two-mode affiliation matrix, one resulting in an actor by actor matrix the other in an event by event matrix. In these cases the relations are attended an event together and had an actor in common, respectively. We can capture more information in our projections if we record the number of events each pair of actors attended and the number of actors each pair of events had in common. These are given by $\mathbf{AA}^T$ and $\mathbf{A}^T\mathbf{A}$, respectively, where $\mathbf{A}^T$ represents the transpose of $\mathbf{A}$. As Breiger [11] pointed out in his 1974 paper these should not be seen as two independent data matrices but as dual representations of the data. However, it became common practice to always dichotomize the data and in many applications only one projection was considered. Clearly reducing valued data to binary in any situation results in loss of information and this is compounded here by ignoring one of the projections. A consequence of this approach is that there is significant data loss, resulting in an inferior analysis.

An alternative approach is to develop methods for analyzing the bipartite graph directly. The first systematic example of this approach was due to Borgatti [7] in 1989 and it was further developed by Borgatti and Everett [8]. They showed how to extend structural and regular equivalence to multimode data directly, and this was later extended to generalized blockmodelling by Doreian *et al.* [19]. In essence this was a simple matter of extending the known block structures for one mode data to block structures for multimode data. In practice these structures require very little modification and progress in the area of blockmodeling has until recently been entirely computational. In this chapter we shall apply both approaches to the same data set in order to gain some insight into how the techniques perform.

9.2 Two-Mode Partitioning

At the heart of the blockmodeling approach described in the previous section is a need to optimize a cost function which captures the extent to which a given partition of the rows and columns of the data matrix corresponds to a blockmodel. The resulting combinatorial optimization problem is unlikely to have a polynomial time solution and hence more heuristic methods are required. There is no consensus on what methods will perform best but van Rosmalen et al. [42] examined five different techniques on simulated data in which they optimize using a Euclidean metric. Their test datasets were relatively small with a maximum value of n and m set at 120 and up to seven clusters in the rows and columns. Their simulations indicated a two-mode version of the k-means method (see the paper for details) had the best overall performance and they then validated this finding on some empirical data.

The study by van Rosmalen *et al.* did not capitalize on the binary nature of any affiliation matrix (provided we have non-valued data). Brusco and Steinley have proposed an extension of variable neighbourhood search which did [12]. Variable neighbourhood searches are meta-heuristic methods in which increasingly large neighbourhoods of the current best solution are explored. Overall the performance of their algorithm was very similar to the two-mode k-means except in the situation that the block positions were known. In this case their algorithm was an improvement.

These are quite sophisticated and the articles contain pseudo-code which provides more details. Many of the applications apply fairly simple greedy algorithms that prioritize efficiency over the ability to avoid local minimum, but they often find acceptable solutions by using many different starting positions. The techniques we have discussed so far are very general and can be applied to many different types of data. We now look at methods specifically designed for social network type data.

9.3 Community Detection

In a vain effort to bring some consistency to terminology we suggest that the term community detection is used for the partitioning of a network into groups such that actors within a group are more closely connected to each other than those in other groups. We shall refer to these groups as communities. If we allow actors to be in more than one group, so that groups overlap and do not insist that all actors are assigned to any group but still have highly connected groups, we shall call these groups cohesive subgroups. A consequence of these definitions is that community detection is a special case of blockmodeling. Blockmodeling does partition the actors but allows for more general forms of blocks which do not have to reflect closely connected sets of actors.

In considering two-mode networks we shall consider the problem of partitioning both modes so that we find sets of actors and events. We will require the density of the submatrix containing these actors and events to be denser than the other submatrices containing either the actors or the events. It should be noted that some authors consider the event communities and the actor communities separately (see [24], for example). For single-mode networks the most commonly used and accepted technique (although it has some well-known short-comings [22]) is Newman's community detection, which optimizes modularity [36]. Barber [2] extended modularity to two-mode data and developed an algorithm specifically for this type of data. We outline these ideas below.

First we give the formula for modularity for a single-mode network in matrix form. Suppose a network with n nodes and m edges has adjacency matrix $\mathbf{A}$. Let $\mathbf{P}$ be a matrix of probabilities in which the i, j entry is the probability that actor i has an edge to actor j given that the edges are distributed at random (but with the expected degrees made to match those in $\mathbf{A}$). Given a partition of the nodes into c groups let $\mathbf{S}$ be the $n \times c$ indicator matrix in which the i, j entry is a 1 if actor i is a member of group j and 0 otherwise. Let $\mathbf{B} = \mathbf{A} - \mathbf{P}$ then the modularity Q is given by

$$Q = \frac{1}{2m} \, \mathrm{Tr}(\mathbf{S}^T \mathbf{B} \mathbf{S}) \tag{9.1}$$

where $\mathrm{Tr}(\mathbf{X})$ is the trace of matrix $\mathbf{X}$. In the two-mode version the adjacency matrix is replaced by an affiliation matrix $\mathbf{\check{A}}$ and the $\mathbf{P}$ matrix adapted to take account of the bipartite structure to form $\mathbf{\check{P}}$, so that $\mathbf{B}$ is replaced by $\mathbf{\check{B}}$. In addition, instead of a single $\mathbf{S}$ indicator matrix we need to have one matrix for each mode, which we call $\mathbf{R}$ and $\mathbf{T}$.

The resultant formula has the form

$$Q = \frac{1}{m} \, \mathrm{Tr}(\mathbf{R}^T \mathbf{\check{B}} \mathbf{T}) \tag{9.2}$$

If we have c communities and our bipartite network has p actors and q events then $\mathbf{R}$ is $p \times c$, $\mathbf{\check{B}}$ is $p \times q$, and $\mathbf{T}$ is $q \times c$. Barber also proposes an algorithm, BRIM (Bipartite, Recursive Induced Modules), which uses the singular vectors of $\mathbf{B}$ to recursively partition both actors and events

```
                                          1   1 1 1 1
                      1 2 3 4 5 6   7 8   9 1   0 2 3 4
                      -----------------------------------
 1   EVELYN     |  1 1 1 1 1 1 |   1 | 1 |         |
 2   LAURA      |  1 1 1   1 1 | 1 1 |   |         |
 3   THERESA    |    1 1 1 1 1 | 1 1 | 1 |         |
 4   BRENDA     |  1   1 1 1 1 | 1 1 |   |         |
 5   CHARLOTTE  |      1 1 1   |   1 |   |         |
 6   FRANCES    |      1   1 1 |   1 |   |         |
                      -----------------------------------
 7   ELEANOR    |            1 | 1 1 | 1 |         |
10   VERNE      |              | 1 1 | 1 1 |   1   |
 9   RUTH       |          1   | 1 1 | 1 |         |
                      -----------------------------------
 8   PEARL      |              |   1 | 1 |   1     |
17   OLIVIA     |              |     | 1 | 1 1     |
16   DOROTHY    |              |   1 | 1 |         |
18   FLORA      |              |     | 1 | 1 1     |
                      -----------------------------------
11   MYRNA      |              |   1 | 1 | 1 1     |
15   HELEN      |              | 1 1 |   1 | 1 1    |
12   KATHERINE  |              |   1 | 1 | 1 1 1 1  |
13   SYLVIA     |              | 1 1 | 1 | 1 1 1 1  |
14   NORA       |            1 |   1 |   1 1 | 1 1 1 1 |
                      -----------------------------------
```

Figure 9.1   Group assignment maximizing modularity.

into groups. The algorithm does not, however, provide a method to find the maximum value for c, the number of groups. To overcome this he suggests starting with $c = 1$, calculating Q, and then keep doubling c until Q decreases. At this stage use bisection to find the value of c which maximizes Q.

As an example we apply the technique to the Southern Women Data [16] and obtain four groups as given in the blocked affiliation matrix in Figure 9.1. In this data the rows correspond to 18 women and the columns to 14 social events attended by the women.

From the blocked affiliation matrix we can see that the top group of women were the main attendees of the first six events and the bottom group of women attended events 10, 12, 13, and 14. We see that most women attended the middle pair of events with Eleanor, Verne, and Ruth all attending events 7 and 8 whereas Pearl, Olivia, Dorothy, and Flora all attended event 9, with two of them attending event 11. These groupings are similar (but not exactly the same) to others found in this data [23].

Other authors have suggested alternative extensions for modularity (see Guimera *et al.* [24] and Murata [35] as examples) as well as other measures [43].

9.4   Dual Projection

A common criticism of projection methods are that information is lost in the process. This is definitely true if the projection is dichotomized or if only one projection is used. However, some authors have claimed (without proof) that there is always data loss even if both projections are used in their undichotomized form [31]. Everett and Borgatti [21] challenged this assumption and provided evidence that it is not true. They argued that in the vast majority of cases given two projections it is possible to recover the original matrix and hence no information is lost. This issue has been further explored by Kirkland [26] where he shows as the size of the matrices increases then cases of data loss decrease. He also gives examples of when data loss does occur but these matrices are highly structured and are unlikely to occur in real data. Everett and Borgatti therefore suggested constructing both projections then using methods which are applicable to proximity matrices on both projections and finally combining these, preferably using the original data. They call this approach the dual projection approach and show how it can be used for blockmodeling, in particular core-periphery models and briefly centrality. In a later paper [32] Melamed explored the method as applied to community detection.

The dual projection works best when there are robust methods for analyzing the projected matrices, that is, techniques that work well on proximity-type data. We shall provide two examples. Our first is a core-periphery partition, this is a blockmodel but as we shall see the periphery is not well connected and so it is not community detection. Borgatti and Everett [10] suggested a method for dividing a proximity matrix $\mathbf{P}$ into a core and periphery. This is a two-stage process. The first process is to find a vector $\mathbf{C}$ such that $||\mathbf{P} - \mathbf{CC}^T||_2$, that is, the Euclidean 2-norm, is minimized. This $\mathbf{C}$ then gives a core-periphery score for each object and we sort $\mathbf{C}$ to form $\mathbf{C}'$ so that its elements are in descending order. Let $\mathbf{I}_k$ be a vector in which the first k elements are 1 and the rest are 0. We next find the value of k for which the correlation of $\mathbf{I}_k$ and $\mathbf{C}'$ is the highest. We now assign the first k objects in $\mathbf{C}'$ to the core and the remainder to the periphery. In our dual projection method we use this on both projections and then map these back onto the affiliation matrix $\mathbf{A}$. Again using the Southern Women data we obtain the two-mode core periphery structure shown in Figure 9.2.

Looking at the partition in Figure 9.2 we see that the core events were the most popular, all with eight or more attendees, whereas the peripheral events all had six attendees or fewer. We see that peripheral actors attend core events and core actors attend peripheral events but not as much as core actors attending core events. The least dense area is peripheral actors attending peripheral events, which all gives some validation to the core-periphery structure.

```
                                    1 1 1 1 1
                        8 9 6 7 5   3 4 1 2 0 1 2 3 4
                      ---------------------------------
      1     EVELYN  | 1 1 1   1 | 1 1 1 1           |
      2      LAURA  | 1   1 1 1 | 1   1 1           |
      3    THERESA  | 1 1 1 1   | 1 1   1           |
      4     BRENDA  | 1   1 1 1 | 1 1 1             |
     14       NORA  |   1 1 1   |           1 1 1 1 1 |
      7    ELEANOR  | 1   1 1 1 |                   |
      9       RUTH  | 1 1   1 1 |                   |
     13     SYLVIA  | 1 1   1   |         1   1 1 1 |
                      ---------------------------------
      6    FRANCES  | 1   1   1 | 1                 |
      8      PEARL  | 1 1 1     |                   |
     10      VERNE  | 1 1   1   |             1     |
     12  KATHERINE  | 1 1       |         1   1 1 1 |
     11      MYRNA  | 1 1       |         1   1     |
      5  CHARLOTTE  |     1 1   | 1 1               |
     15      HELEN  | 1     1   |         1 1 1     |
     16    DOROTHY  | 1 1       |                   |
     17     OLIVIA  | 1         |             1     |
     18      FLORA  | 1         |             1     |
                      ---------------------------------
```

Figure 9.2   Dual projection core-periphery of Southern Women.

```
                                                  1  1  1  1  1
                             1  2  3  4  5  6  7  8  9  0  1  2  3  4
                           -----------------------------------------
     1      EVELYN |  1  1  1  1  1  1        1  1 |                 |
     2       LAURA |  1  1  1     1  1  1  1      |                 |
     3     THERESA |     1  1  1  1  1  1  1  1  |                 |
     4      BRENDA |  1     1  1  1  1  1  1      |                 |
     5   CHARLOTTE |        1  1  1        1      |                 |
     6     FRANCES |        1     1  1        1   |                 |
     7     ELEANOR |              1  1  1  1      |                 |
     8       PEARL |                 1        1  1 |                 |
     9        RUTH |              1        1  1  1 |                 |
                           -----------------------------------------
    10       VERNE |                    1  1  1 |        1           |
    11       MYRNA |                    1  1 |  1     1              |
    12   KATHERINE |                    1  1 |  1        1  1  1     |
    13      SYLVIA |                 1  1  1 |  1        1  1  1     |
    14        NORA |              1  1     1 |  1  1  1  1  1        |
    15       HELEN |                 1  1 |  1  1  1                 |
    16     DOROTHY |                    1  1 |                       |
    17      OLIVIA |                       1 |     1                 |
    18       FLORA |                       1 |     1                 |
                           -----------------------------------------
```

Figure 9.3 Dual projection community detection for the Davis data.

We can also use dual projection to do community detection. In order to not lose any structural information we need to partition the proximity matrices (note this is not the approach taken by Melamed [32] as he uses a dichotomized projection). Guimera *et al.* [24] do take this approach and use a simple extension to modularity to deal with the valued data. However, they only find clusters of women and attach the relevant events to the clustered women data. They also use the Davis data and report a straight split of the women into two groups, namely {Evelyn, Laura, Theresa, Brenda, Charlotte, Frances, Eleanor, Ruth}, and attach events 1 through 8 to this group using the weighted method. Clearly all the other women and events belong to the second group.

We partitioned both projections into two groups in order to obtain a comparison. Rather than use the modularity we used a fit function which used correlation between an ideal structure matrix of 1 for within group interaction and 0 for between group interaction with a Tabu search. The results are given in Figure 9.3.

The results in Figure 9.3 are in close agreement with the results obtained by Guimera *et al.*, the only difference is we have placed Pearl in the first group together with event 9. We also applied the Louvain method [6] to both projections; while the method is local it does have the advantage of finding the optimum number of clusters. For these Women data, this method reproduced the partition found by Guimera *et al.* into two groups. It partitioned the events into four groups with events 1 to 6 in the first group, 9 to 14 in a second group, and events 7 and 8 both in singleton clusters.

It should be noted that we cannot directly compare these methods with the two-mode modularity of Barber. First, we can decide to have a different number of partitions in each mode. Second, even if these are the same we do not necessarily have the groups defined by the diagonal blocks, i.e. we do not have communities made up of both women and events but we partition these separately. To see this we repeat the analysis above but with four groups in each partition to obtain the results shown in Figure 9.4.

```
                            1 1 1 1           1
                8 9 3 4 5 6 7   4 2 0 3   1 2   1
               ------------------------------------
 1     EVELYN | 1 1 1 1 1 1   |         | 1 1 |     |
 2      LAURA | 1   1   1 1 1 |         | 1 1 |     |
 3    THERESA | 1 1 1 1 1 1 1 |         |   1 |     |
 4     BRENDA | 1     1 1 1 1 |         | 1   |     |
 5  CHARLOTTE |     1 1 1   1 |         |     |     |
 6    FRANCES | 1   1   1 1   |         |     |     |
 7    ELEANOR | 1       1 1 1 |         |     |     |
 9       RUTH | 1 1     1   1 |         |     |     |
               ------------------------------------
17     OLIVIA |   1           |         |     | 1   |
18      FLORA |   1           |         |     | 1   |
               ------------------------------------
 8      PEARL | 1 1       1   |         |     |     |
16    DOROTHY | 1 1           |         |     |     |
               ------------------------------------
11      MYRNA | 1 1           |   1 1   |     |     |
10      VERNE | 1 1         1 |   1     |     |     |
15      HELEN | 1           1 |   1 1   |     | 1   |
12  KATHERINE | 1 1           | 1 1 1 1 |     |     |
13     SYLVIA | 1 1         1 | 1 1 1 1 |     |     |
14       NORA |   1       1 1 | 1 1 1 1 |     | 1   |
               ------------------------------------
```

Figure 9.4 Dual projection community detection for four groups.

Examining Figure 9.4 we see that two of the diagonal blocks are zero and so these do not correspond to mixed mode communities. We need to examine the partitions and not the blocks, although having both to examine helps us understand the data. For example, we can see that Olivia and Flora have been put together as they attended events 9 and 11 together and no other events.

9.5   Signed Two-Mode Networks

Heider's balance theory [25] has a long tradition in social networks and was formulated in network terms by Cartwright and Harary [14]. In the one-mode formulation the edges of the network are assigned either a positive or a negative sign reflecting positive or negative sentiment. In Heider's original formulation the actors showed positive or negative preferences to objects and so it is more analogous to two-mode data. In our formulation the actors attending events would see them as either positive or negative. A good example is a set of politicians or other actors voting on propositions or resolutions that they can vote for, against, or abstain. Data of this type was considered by Mrvar and Doreian [34], where the actors were supreme court judges, and Doreian *et al.* [20], where the actors were nation states voting in the UN. It follows that a two-mode signed network is a two-mode network of actors and objects (we use objects rather than events as this is more suggestive of the type of data that has been used) in which each edge has a positive or negative sign. In classic balance theory a balanced (one-mode) network can be partitioned into two sets with negative ties between the sets and positive ties within. Extended balance (Davis balance or clusterability) allows more than two clusters but still positive ties within clusters and negative ties between. Relaxed balance [18] again allows for any number of clusters but we now just require all the connections within a cluster to be of the same sign (positive or negative) and all the edges between pairs of clusters must also be of the same sign. In the two-mode case Mrvar and Doreian retain the idea of relaxed balance, but of course given the structure of the data no longer have within-cluster links. We now formalize these ideas but more details can be found in [20] and [18].

Let $\mathbf{A}$ be a signed affiliation matrix and suppose the rows are partitioned into k_1 clusters and the columns into k_2 clusters so that $\mathbf{A}$ is partitioned into $k_1 k_2$ blocks. Then we say the clustering is ideal if none of the blocks contain both positive and negative ties. Given a partition of $\mathbf{A}$ we can measure how close it is to the ideal by simply counting the number of positive and negative violations there are in the blocks, call these P and N. These are used to produce a measure of inconsistency given by $\alpha N + (1 - \alpha)P$. The α parameter allows us to weight either positive or negative violations more highly, with a value of 0.5 weighting them equally. Unfortunately this function can always be made zero by placing each node in its own unique cluster. In fact Mrvar and Doreian proved a stronger result showing that this function is monotonically decreasing with k_1 and k_2. It is therefore necessary to find strategies in which the blocks remain sufficiently large. For fixed values of k_1 and k_2 Mrvar and Doreian proposed a relocation algorithm which helps finding the minimum, but unfortunately it can easily get caught in local minima and needs many starts to reliably find a global minimum. As a consequence it becomes quite a challenge to partition such data, particularly given the fact we have two parameters k_1 and k_2 to contend with and the computational complexity of the task. It is possible that other factors can help determine in more detail the structure of the various blocks. This is explored in [20] but is really only feasible because of the small size of one of the modes in the datasets, consisting of nine supreme court judges. Some further guidance on this issue and some ideas are further examined in [18], which has far larger mode sizes, but no definitive approach is suggested. We conclude that while some early promising work has been done in this area there are many open issues worthy of further consideration.

9.6   Spectral Methods

One technique that has been developed for networks but has been largely ignored in the social network community is bipartite spectral co-partitioning [17]. This technique has the added advantage that it can handle valued two-mode networks. In this case we have an incidence matrix $\mathbf{A}$ in which $\mathbf{A}(i, j) = w$ indicates that i attended event j with weight w, where higher values represent a stronger association. An example would be that the actors are words, the events documents, and the entries in $\mathbf{A}(i, j)$ give the number of occasions word i was in document j. It was in this context that Dhillon proposed this method. We briefly outline the process but full details are given in his paper.

Given a weighted $n \times m$ affiliation matrix $\mathbf{A}$ then form $\mathbf{D}_1$, an $n \times n$ diagonal matrix with the row sums of $\mathbf{A}$ on the diagonal, and $\mathbf{D}_2$, an $m \times m$ diagonal matrix with the column sums on the diagonal. The algorithm proceeds as follows.

1. Form $\mathbf{A}_n = \mathbf{D}_1^{-\frac{1}{2}} \mathbf{A} \mathbf{D}_2^{-\frac{1}{2}}$.
2. Compute the second singular vectors of $\mathbf{A}_n$, $\mathbf{u}_2$, and $\mathbf{v}_2$. That is the eigenvectors of $\mathbf{A}_n \mathbf{A}_n^T$ and $\mathbf{A}_n^T \mathbf{A}_n$ corresponding to the second largest eigenvalue.

3. Form $\mathbf{z}_2 = \begin{bmatrix} \mathbf{D}_1^{-\frac{1}{2}}\mathbf{u}_2 \\ \mathbf{D}_2^{-\frac{1}{2}}\mathbf{v}_2 \end{bmatrix}$.

4. Run k-means on $\mathbf{z}_2$ to bipartition the data.

This clearly partitions the rows and columns into two and we can recursively apply the procedure to obtain a finer partition. Alternatively, Dhillon gives an extended version that allows us to find k groups by using additional singular vectors. Let $p = \lceil \log_2 k \rceil$ and instead of a vector $\mathbf{z}_2$ in step 3 create a matrix $\mathbf{Z}$, i.e.

$$\text{form } \mathbf{Z} = \begin{bmatrix} \mathbf{D}_1^{-\frac{1}{2}}\mathbf{U} \\ \mathbf{D}_2^{-\frac{1}{2}}\mathbf{V} \end{bmatrix}$$

where $\mathbf{U} = [\mathbf{u}_2 : \mathbf{u}_3 : \ldots : \mathbf{u}_{p+1}]$ and $\mathbf{V}$ is defined similarly. Again use k-means to cluster the rows of $\mathbf{Z}$ with the first n rows giving the partition of the rows and the last m rows giving the partition of the columns.

We note that this method produces groups made up of nodes from both modes, as in the Barber community detection discussed in Section 10.3. As an example we do a two-cluster split on the Davis data and the result is shown in Figure 9.5. As can be seen, this is very similar to the dual projection community detection shown in Figure 9.3 with only event 9 moved to a different group and thus this partition agrees with that found by Guimera et al. [24] and by the Louvain method in the dual projection as discussed above.

```
                              1 1 1 1 1
                  1 2 3 4 5 6 7 8   9 0 1 2 3 4
                  ---------------------------------
  1     EVELYN |  1 1 1 1 1 1   1 | 1              |
  2      LAURA |  1 1 1   1 1 1 1 |                |
  3    THERESA |    1 1 1 1 1 1 1 | 1              |
  4     BRENDA |  1   1 1 1 1 1 1 |                |
  5  CHARLOTTE |      1 1 1   1   |                |
  6    FRANCES |      1   1 1   1 |                |
  7    ELEANOR |        1 1 1 1   |                |
  8      PEARL |          1   1   | 1              |
  9       RUTH |          1   1 1 | 1              |
                  ---------------------------------
 10      VERNE |            1 1   | 1       1      |
 11      MYRNA |              1   | 1 1     1      |
 12  KATHERINE |              1   | 1 1     1 1 1  |
 13     SYLVIA |            1 1   | 1 1     1 1 1  |
 14       NORA |          1 1     | 1 1 1 1 1 1    |
 15      HELEN |            1 1   |   1 1 1        |
 16    DOROTHY |              1   | 1              |
 17     OLIVIA |                  | 1   1          |
 18      FLORA |                  | 1   1          |
                  ---------------------------------
```

Figure 9.5 Spectral bisection of the Davis data into two groups.

```
                                              1 1 1 1    1
                            1 2 3 4 5 6   7 8 9  0 4 2 3    1
                           ----------------------------------
     1     EVELYN       | 1 1 1 1 1 1 |   1 1 |          |     |
     2      LAURA       | 1 1 1   1 1 | 1 1   |          |     |
     3    THERESA       |   1 1 1 1 1 | 1 1 1 |          |     |
     4     BRENDA       | 1   1 1 1 1 | 1 1   |          |     |
     5  CHARLOTTE       |     1 1 1   | 1     |          |     |
     6    FRANCES       |     1   1 1 |   1   |          |     |
     7    ELEANOR       |         1 1 | 1 1   |          |     |
                           ----------------------------------
     8      PEARL       |         1   |   1 1 |          |     |
     9       RUTH       |         1   | 1 1 1 |          |     |
    10      VERNE       |             | 1 1 1 |     1    |     |
    15      HELEN       |             | 1 1   | 1   1    | 1   |
    16    DOROTHY       |             | 1 1   |          |     |
    14       NORA       |         1   | 1   1 | 1 1 1 1  | 1   |
                           ----------------------------------
    13     SYLVIA       |             | 1 1 1 | 1 1 1 1  |     |
    11      MYRNA       |             | 1 1   | 1   1    |     |
    12  KATHERINE       |             | 1 1   | 1 1 1 1  |     |
                           ----------------------------------
    17     OLIVIA       |             |   1   |          | 1   |
    18      FLORA       |             |   1   |          | 1   |
                           ----------------------------------
```

Figure 9.6   Spectral bisection of the Davis data into four groups.

We also obtained a split into four groups which involves using two of the singular vectors to obtain the partition shown in Figure 9.6.

As we found groups of women and events we need to compare this result with the modularity result shown in Figure 9.1. In this case the event split is nearly the same with just event 9 moved from being with event 11 to the community containing events 7 and 8. This leaves event 11 in a single cluster as found by the four split in the dual projection method. The first group of women is also similar but the effect of moving event 11 to a singleton cluster separates out Olivia and Flora into a community pair. This has a knock-on effect on the way the rest of the women are partitioned, and although there are some similarities there are also differences and it is difficult without additional information to decide which partition is the best.

It should be noted that the method can be extended to obtain different partitions of the rows and columns, see Kluger *et al.* [28]. This requires different normalization and three normalization schemes are proposed as well as a more sophisticated technique for partitioning the rows and the columns separately. The first suggested normalization is to make all the rows have the same mean and all the columns have the same mean (but not necessarily the same as the row mean). The second method suggests making all the row means the same as all the column means. The third method involves taking the log of the matrix $\mathbf{L}$ and then for each entry of $\mathbf{L}$, $\mathbf{L}(i,j) = \log \mathbf{A}(i,j)$, subtract off its row mean, its column mean, and the overall mean of $\mathbf{L}$. Finally in step 3 do not form $\mathbf{Z}$ but run k-means on $\mathbf{AV}$ and $\mathbf{A}^T\mathbf{U}$ separately where $\mathbf{U}$ and $\mathbf{V}$ are constructed from the singular vectors by selecting subsets that are the best projections. We do not give the details here and so the interested reader should consult their paper for details.

One further approach needs to be mentioned and that is the two-mode stochastic block models proposed by Larremore *et al.* [29]. In brief they assume a Poisson distribution and then search the likelihood space using a modified Kernigan–Lin algorithm. In its simple form this tends to sort out actors purely by degree but they use a degree correction procedure to counteract this. The degree correction explicitly takes into account the degree distribution of the data which allows for a wide variety of empirical degree distributions. When they use this technique on the Southern Women data they get exactly the same partition as the dual projection split shown in Figure 9.3.

9.7 Clustering

So far we have examined partitioning, i.e. we have insisted that every node is placed uniquely in one group. We now relax this condition and allow nodes to be placed in more than one group and do not insist that actors are assigned to any group. We shall only consider clustering in which we are trying to find subsets of nodes which are well connected to each other. As mentioned before we shall refer to these as cohesive subgroups. The standard one-mode definition of a clique as a maximal complete subgraph clearly can be generalized to a biclique as a maximally complete bipartite subgraph. Such structures have been considered in mathematics for over 400 years, although not usually as subgraphs. The use of bicliques as cohesive subgraphs in social networks was probably due to Borgatti and Everett [9] in 1997. It is a very simple matter to extend standard clique algorithms to find bicliques and the same techniques used to deal with overlap can be applied. It is also a simple matter to extend concepts such as k-plexes, n-cliques, n-clubs, n-clans etc. to the two-mode case. We also note that since the projections result in proximity data then the vast number of clustering algorithms that have been developed for this type of data can be used. We should also comment that a common measure in one-mode networks is the clustering coefficient (or transitivity index) that attempts to capture the extent to which a network is clustered. In this chapter we are concerned with uncovering structural patterns in terms of finding sets of nodes rather than providing graph invariants that try and capture the extent to which a network has a particular property. There have been a number of suggestions for extending the clustering coefficient (and other invariants) to two-mode data and the interested reader should consult [30] and [37].

One area in which there have been some developments is that of k-cores and their extension to two-mode. A k-core is an induced subgraph in which every node has degree k or more and was first proposed by Seidman [40] and extended to two-mode in [1]. A k-core is not a cohesive subgroup but any cohesive subgroups must be wholly contained in a k-core. The concept was extended to generalized k-cores in [3] and then to two-mode generalized cores by Cerinšek and Batagelj [15]. The idea of a generalized core is to extend the concept of degree to a property function defined on the nodes. For a network $N = (V, L, w)$ with node set V, edge set L, and a weight function $w : L \to \mathbb{R}^+$ a property function $f(v, C) \in \mathbb{R}_0^+$ is defined for all $v \in V$ and $C \subseteq V$. A subset C induces the subnetwork to which the evaluation of the property function is restricted. We can now give a formal definition of a generalized two-mode (p, q) core. Let $N = ((V_1, V_2), L, (f, g), w)$, $V = V_1 \cup V_2$ be a finite two-mode network – the sets V and L are finite. Let $P(V)$ be the power set of the set V. Let functions f and g be defined on the network $N: f, g : V \times P(V) \to \mathbb{R}_0^+$. A subset of nodes $C \subseteq V$ in a two-mode network N is a generalized two-mode core $C = \mathrm{Core}(p, q; f, g)$, $p, q \in \mathbb{R}_0^+$ if and only if in the subnetwork $K = ((C_1, C_2), L|C, w)$, $C_1 = C \cap V_1$, $C_2 = C \cap V_2$ induced by C it holds that for all $v \in C_1 : f(v, C) \geq p$ and for all $v \in C_2 : g(v, C) \geq q$, and C is the maximal such subset in V. Algorithms for finding generalized two-mode cores are relatively straightforward and efficient, and are based on the simple idea of deleting nodes that do not satisfy the criteria. This cited paper provides some examples drawn from web of science data.

9.8 More Complex Data

So far we have considered two-mode static data. If there are more than two modes then in some circumstances it is a straightforward matter to extend the two-mode case to more modes. As already mentioned, Borgatti and Everett [8] showed how to extend regular and structural equivalence to multimode data but they did not address the computational issues. Batagelj *et al.* [4] suggested a dissimilarity measure for three-mode structural equivalence and applied Wards algorithm to partition the data. They demonstrated the effectiveness of their approach on a three-way cognitive social structure.

If we have k-modes then we examine the $k(k-1)/2$ collection of two-mode networks between every pair of modes. Let $\mathbf{A}_{(r,s)}$ denote the two-mode affiliation matrix between mode r and mode s where $r < s$ and s runs from 1 to k. If we wanted to extend community detection then it is a simple matter to construct $\check{\mathbf{B}}_{(r,s)}$ and then define Q as

$$Q = \frac{1}{m} \sum_{i<j} \mathrm{Tr}(\mathbf{S}_i^T \, \check{\mathbf{B}}_{(i,j)} \, \mathbf{S}_j) \tag{9.3}$$

where $\mathbf{S}_i$ is the ith mode indicator matrix. One issue is that it is now not possible to use the spectral methods suggested described in Section 6.3 in order to find a maximum for Q. Clearly we can use other optimization methods but these will probably not be as efficient. One solution to this problem is to simply construct an adjacency matrix $\mathbf{Z}$ from $\check{\mathbf{B}}_{(i,j)}$. We form $\mathbf{Z}$ (given in blocked form) as follows:

$$\mathbf{Z}(i,j) = \begin{cases} \check{\mathbf{B}}_{(i,j)} & \text{if } i < j \\ \check{\mathbf{B}}_{(i,j)}^T & \text{if } j < i \\ \mathbf{0} & \text{if } i = j \end{cases}$$

So that $\mathbf{Z}$ is a square adjacency matrix in which all the modes have been included. The fact there are not connections within the modes is captured by the zero blocks on the diagonal. In this case we have

$$Q = \frac{1}{2m} \, \mathrm{Tr}(\mathbf{S}^T \mathbf{Z} \mathbf{S}) \tag{9.4}$$

where m is the number of edges and $\mathbf{S}$ is an indicator matrix over all the modes. We can now use spectral partitioning to maximize Q. An example of this approach for a three-mode network is given in [33].

Looking at the other methods discussed above there does not seem any reason why spectral bipartitioning cannot be extended in the same way but this approach does not seem to have been explored. In this case the constructed adajacency matrix would not have $\check{\mathbf{B}}_{(i,j)}$ as the blocks but would use $\mathbf{A}_{(i,j)}$. The one exception for extending to more than two modes in this way is dual projection. In this instance we would not construct a large adjacency matrix but would project all pairs of $\mathbf{A}_{(i,j)}$. In this case when we have more than two modes then the same mode appears in a number of different projection matrices. As a consequence each mode would have a number of different partitions and this would not generally be of use unless the goal is to find different partitions for different pairs of modes.

We mention one further complication that is multimode data that involves a time element. Both data and techniques for dealing with such data are not common. However, Tang *et al.* [41] examined how communities evolve over time in a dynamic multimode framework. They considered time-stamped data and they tried to make a smooth transition in terms of community detection from one time stamp to the next. They present an algorithm and an example that uses the Enron email corpus [27].

9.9 Conclusion

In this chapter we have examined partitioning and clustering in multimode network data. We briefly mentioned that a number of techniques have been developed for dealing with non-binary data or more precisely non-network type data. We did not explore these methods in much detail as they are generally well described elsewhere. We did not discussed software but most of the articles referenced that develop methods discuss implementations and point to available tools. In addition, we mainly discussed methodological issues and did not discussed applications. There are an increasing number of application areas that are using these methods, ranging from biology and information science through to sociology and political science. A good flavor of how to interpret some of these techniques can be gleaned from the examples in the book by Batagelj *et al.* [5]. It should be noted that this is a very active area for research and new methods and ideas are constantly being explored, particularly as new types of data emerge. The complexities of this type of data in terms of collecting, analyzing, and interpreting remain both challenging and deeply fascinating.

References

1. A. Ahmed, V. Batagelj, X. Fu, S. Hong, D. Merrick, and A. Mrvar. Visualisation and analysis of the internet movie database. In *APVIS 2007, 6th International Asia-Pacific Symposium on Visualization 2007, Sydney, Australia, 5–7 February 2007*, pages 17–24, 2007.
2. M. J. Barber. Modularity and community detection in bipartite networks. *Physical Review E*, 76(6):066102, 2007.
3. V. Batagelj and M. Zaveršnik. Fast algorithms for determining (generalized) core groups in social networks. *Advances in Data Analysis and Classification*, 5(2):129–145, 2011.
4. V. Batagelj, A. Ferligoj, and P. Doreian. *Indirect Blockmodeling of 3-Way Networks*, pages 151–159. Springer, Berlin, Heidelberg, 2007.
5. V. Batagelj, P. Doreian, A. Ferligoj, and N. Kejžar. *Understanding Large Temporal Networks and Spatial Networks: Exploration, Pattern Searching, Visualization and Network Evolution*. Wiley Series in Computational and Quantitative Social Science Series. John Wiley & Sons, New York, 2014.
6. V. D. Blondel, J.-L. Guillaume, R. Lambiotte, and E. Lefebvre. Fast unfolding of communities in large networks. *Journal of Statistical Mechanics: Theory and Experiment*, P10008, 2008.
7. S. Borgatti. Regular equivalence in graphs, hypergraphs, and matrices, 1989. Ph.D. dissertation.
8. S. P. Borgatti and M. G. Everett. Regular blockmodels of multiway, multimode matrices. *Social Networks*, 14(1-2):91–120, 1992.
9. S. P. Borgatti and M. G. Everett. Network analysis of 2-mode data. *Social Networks*, 19(3):243–269, 1997.
10. S. P. Borgatti and M. G. Everett. Models of core/periphery structures. *Social Networks*, 21(4):375–395, 1999.
11. R. L. Breiger. Duality of persons and groups. *Social Forces*, 53(2):181–190, 1974.
12. M. Brusco and D. Steinley. A variable neighborhood search method for generalized blockmodeling of two-mode binary matrices. *Journal of Mathematical Psychology*, 51(5):325–338, 2007.
13. R. S. Burt. Cooptive corporate actor networks – a reconsideration of interlocking directorates involving american manufacturing. *Administrative Science Quarterly*, 25 (4):557–582, 1980.

14. D. Cartwright and F. Harary. Structural balance – a generalization of Heider theory. *Psychological Review*, 63(5):277–293, 1956.

15. M. Cerinšek and V. Batagelj. Generalized two-mode cores. *Social Networks*, 42:80–87, 2015.

16. A. Davis, B. Gardner, and M. Gardner. *Deep South: A Sociological Anthropological Study of Caste and Class.* University of Chicago Press, Chicago, 1941.

17. I. S. Dhillon. Co-clustering documents and words using bipartite spectral graph partitioning. In *Proceedings of the Seventh ACM SIGKDD International Conference on Knowledge Discovery and Data Mining, San Francisco, CA, USA, August 26–29, 2001*, pages 269–274, 2001.

18. P. Doreian and A. Mrvar. Partitioning signed social networks. *Social Networks*, 31(1):1–11, 2009.

19. P. Doreian, V. Batagelj, and A. Ferligoj. Generalized blockmodeling of two-mode network data. *Social Networks*, 26(1):29–53, 2004.

20. P. Doreian, P. Lloyd, and A. Mrvar. Partitioning large signed two-mode networks: Problems and prospects. *Social Networks*, 35(2):178–203, 2013.

21. M. G. Everett and S. P. Borgatti. The dual-projection approach for two-mode networks. *Social Networks*, 35(2):204–210, 2013.

22. S. Fortunato and M. Barthelemy. Resolution limit in community detection. *Proceedings of the National Academy of Sciences of the USA.*, 104(1):36–41, 2007.

23. L. C. Freeman. Finding social groups: A meta-analysis of the southern women data. In R. Breiger, K. Carley, and P. Pattison, editors, *Dynamic Social Network Modeling and Analysis*, pages 39–97. The National Academies Press, Washington, DC, 2003.

24. R. Guimera, M. Sales-Pardo, and L. A. N. Amaral. Module identification in bipartite and directed networks. *Physical Review E*, 76(3):036102, 2007.

25. F. Heider. *The psychology of interpersonal relations.* John Wiley & Sons, New York, 1958.

26. S. Kirkland. Two-mode networks exhibiting data loss. *Journal of Complex Networks*, 6(2):297–316, 2018.

27. B. Klimt and Y. Yang. Introducing the enron corpus. In *CEAS 2004 – First Conference on Email and Anti-Spam, July 30–31, 2004, Mountain View, California, USA*, 2004.

28. Y. Kluger, R. Basri, J. T. Chang, and M. Gerstein. Spectral biclustering of microarray data: Coclustering genes and conditions. *Genome Research*, 13 (4):703–716, 2003.

29. D. B. Larremore, A. Clauset, and A. Z. Jacobs. Efficiently inferring community structure in bipartite networks. *Physical Review E*, 90(1):012805, 2014.

30. M. Latapy, C. Magnien, and N. Del Vecchio. Basic notions for the analysis of large two-mode networks. *Social Networks*, 30(1):31–48, 2008.

31. S. Lehmann, M. Schwartz, and L. K. Hansen. Biclique communities. *Physical Review E*, 78(1):016108, 2008.

32. D. Melamed. Community structures in bipartite networks: A dual–projection approach. *Plos One*, 9(5):e97823, 2014.

33. D. Melamed, R. L. Breiger, and A. J. West. Community structure in multi-mode networks: Applying an eigenspectrum approach. *Connections*, 33(1):18–23, 2013.

34. A. Mrvar and P. Doreian. Partitioning signed two-mode networks. *Journal of Mathematical Sociology*, 33(3):196–221, 2009.

35. T. Murata. Detecting communities from bipartite networks based on bipartite modularities. In *Proceedings of the 12th IEEE International Conference on Computational Science and Engineering, CSE 2009, Vancouver, BC, Canada, August 29–31, 2009*, pages 50–57, 2009.

36. M. E. J. Newman. Modularity and community structure in networks. *Proceedings of the National Academy of Sciences of the USA.*, 103(23):8577–8582, 2006.

37. T. Opsahl. Triadic closure in two-mode networks: Redefining the global and local clustering coefficients. *Social Networks*, 35(2):159–167, 2013.

38. M. A. Porter, P. J. Mucha, M. E. J. Newman, and C. M. Warmbrand. A network analysis of committees in the US House of representatives. *Proceedings of the National Academy of Sciences of the USA*, 102(20):7057–7062, 2005.

39. T. Roediger-Schluga and M. J. Barber. R&d collaboration networks in the european framework programmes: Data processing, network construction and selected results. *International Journal of Foresight and Innovation Policy*, 4(3-4):321–347, 2008.

40. S. B. Seidman. Network structure and minimum degree. *Social Networks*, 5(3):269–287, 1983.

41. L. Tang, H. Liu, J. Zhang, and Z. Nazeri. Community evolution in dynamic multi-mode networks. In *Proceedings of the 14th ACM SIGKDD International Conference on Knowledge Discovery and Data Mining, Las Vegas, Nevada, USA, August 24–27, 2008*, pages 677–685, 2008..

42. J. van Rosmalen, P. J. F. Groenen, J. Trejos, and W. Castillo. Optimization strategies for two-mode partitioning. *Journal of Classification*, 26(2):155–181, 2009.

43. Y. Xu, L. Chen, B. Li, and W. Liu. Density-based modularity for evaluating community structure in bipartite networks. *Information Sciences*, 317: 278–294, 2015.

10

Blockmodeling Linked Networks

Aleš Žiberna
Faculty of Social Sciences, University of Ljubljana

10.1 Introduction

This chapter introduces the blockmodeling of linked networks. Essentially, it is shown that the multilevel blockmodeling approach [36, 38] can also be used for other types of linked networks. The term "linked networks" describes a collection of one-mode networks, where the nodes from different one-mode networks are connected through two-mode networks. While, in principle, all one-mode and two-mode networks can be multi-relational, only single-relational networks are considered in examples. We could also say that linked networks are collections of networks defined on different sets of nodes, where all sets of nodes must somehow be connected. Some examples of linked networks are:

Multilevel networks Multilevel networks [28] are composed of one-mode networks representing ties among nodes at a given level and two-mode networks that tie nodes from different levels. In a multilevel context, the typical situation has persons as first-level nodes, organizations as second-level nodes, and the two-mode network represents the membership of persons in organizations. Lazega *et al.* [24] analyzed a multilevel network where the first-level one-mode network had collaboration relations among cancer researchers, the second-level one-mode network had collaboration ties among their research labs, and the two-mode network featured membership of researchers in the labs. The multilevel blockmodeling analysis of this network is presented in [36, 38]. Other examples of two-level networks can be found in [3–5].

Networks measured at several time points In this case, one-mode networks represent networks at specific points in time and two-mode networks join the same units at different points in time. Such networks are analyzed in [10, 20, 21].

Multilevel networks measured at several time points These networks are a combination of the first two network types. We could say these networks are essentially networks measured

---

Advances in Network Clustering and Blockmodeling, First Edition.
Edited by Patrick Doreian, Vladimir Batagelj, and Anuška Ferligoj.
© 2020 John Wiley & Sons Ltd. Published 2020 by John Wiley & Sons Ltd.

at several time points, where the networks at each time point constitute a multilevel network. Such a network is analyzed in [23].

Meta-networks based on the Precedence, Commitment, Assignment, Network, and Skill (PCANS) model Meta-matrices or meta-networks are collections of networks on several sets of nodes, e.g., in [19] these sets are individuals, tasks, and resources, but in some cases are extended by locations, events, etc. It is not necessary for all possible one-mode and two-mode networks to be present. Some more examples can be found in [8, 9].

The main goal of blockmodeling linked networks is to cluster nodes from all sets while taking all available information into account. By blockmodeling linked networks, we wish to blockmodel all one-mode and two-mode networks simultaneously so that we only obtain one clustering for nodes from each set (although they occur in one one-mode network and at least one two-mode network). We might also say that we wish to cluster nodes from all sets in such a way that each cluster contains only nodes from one set and that links among clusters from different sets (through two-mode networks) can be modeled.

Approaches to blockmodeling some types of linked networks have been developed. For networks measured at several time points, approaches within stochastic blockmodeling have been developed [1, 26, 32, 33]. However, these are not general enough to apply to other types of linked networks.

Žiberna [36] presented the multilevel blockmodeling approach, which is much more general and can be used with any linked network. While the approach is currently only implemented within the generalized blockmodeling framework [12, 34], in the blockmodeling R package [37], the approach itself is general and can apply to other blockmodeling approaches (e.g. stochastic blockmodeling [2, 11, 17, 18, 29, 30]), provided the issues discussed in the next section are addressed.

The structure of this chapter is as follows. In the next section, the approach to blockmodeling linked networks is developed based on the multilevel blockmodeling approach [36, 38]. Then examples of blockmodeling for two types of linked networks (a multilevel network and a network measured at several time points) are presented. The chapter finishes with conclusions.

10.2 Blockmodeling Linked Networks

The approach to blockmodeling linked networks is based on multilevel blockmodeling. Žiberna [36] presented three ways of blockmodeling multilevel networks:

Separate analysis This is not a true multilevel approach as it simply means blockmodeling each level separately. It should be followed by checking blockmodel(s) that the partitions obtained at different levels induce on the two-mode network(s) and checking the agreement of these partitions (for details, see [36, pp. 48–49]).

Conversion approach The idea of this approach is to convert all one-mode networks to the same level by using the two-mode networks joining them. In a simple case, the transformation is a simple task of pre- and post-multiplying the one-mode network with a suitable two-mode network (transposed as required), while sometimes some "averaging" rules must also be employed. See [36, pp. 49–50]) for more details.

A true multilevel approach Using this approach all one-mode and two-mode networks are blockmodeled simultaneously so that we only obtain one clustering for nodes from each level (even though they occur in one one-mode network and at least one two-mode network).

Not all of the above approaches generally apply to all linked networks. Substantial and technical reasons may make the *conversion approach* problematic for linked networks. Substantially, one should always think about the meaning of any derived networks. Technically, problems might occur if there are isolates in the two-mode networks which seem more likely in certain types of linked networks.[1]

10.2.1 Separate Analysis

Separate analysis is not in fact a special approach designed for a special kind of network(s), and is therefore directly applicable. As in the multilevel case, it is also practically a necessary first step in the analysis of linked networks. While, in the multilevel context, the two-mode networks were only used as a tie between the two levels and not directly blockmodeled [36, 38], this might be relevant for some linked networks, especially those from the metamatrix approach or generally where one-mode networks for some sets of nodes are missing [27].

10.2.2 A True Linked Blockmodeling Approach

In this chapter, most attention is given to *the true linked blockmodeling approach*, which is the same as the true multilevel approach suggested by Žiberna [36]. For this approach all one-mode networks of within node-set ties and two-mode networks of between node-sets must first be combined to form a single linked network. This linked network therefore contains all ties between the nodes from all sets. These ties most often represent several relations. The main idea is to use blockmodeling approaches on this linked (and possibly multirelational) network under the restriction that nodes from different sets cannot be placed in the same cluster.

As may be seen from this description, the approach is quite general and not specific to any particular blockmodeling approach. Nevertheless, it is currently only implemented within generalized blockmodeling [37]. While the idea is quite simple, several issues have to be addressed before the approach can be used. The most important ones are:

Multirelational blockmodeling From a substantial point of view, in most cases we cannot assume that all one-mode and two-mode networks contain only ties from one relation. In addition, in some cases even the one-mode networks themselves can be seen as multi-relational. From a computational point of view, the use of multiple relations at least for the current implementation, within generalized blockmodeling [37], allows and simplifies the use of different generalized blockmodeling approaches (e.g. binary, valued, homogeneity (see [34] and Chapter 6 on blockmodeling valued networks for details) and the weighting of different parts of the linked network.

---

[1] Isolates might also occur in multilevel networks.

Different sets of nodes To prevent nodes from different levels being clustered together, they must be treated as different sets of nodes. Then each set can be partitioned into its own clusters to ensure that clusters contain only nodes from one set/level.

Multicriteria optimization The problem is how to combine different criteria for blockmodeling. For generalized blockmodeling, different criteria represent different criteria functions for various parts of the linked network (one-mode and two-mode networks). A naive approach would be to blockmodel the whole linked network as a single network (with the same blockmodeling approach and with the constraint from the previous paragraph). Such an approach would usually cause the partition of node-sets to be predominately determined by parts of the network with relatively large criterion functions. These are usually larger parts with more ties. A true multiobjective blockmodeling method, as suggested by Brusco *et al.* [6], seems the most appropriate. However, a simpler approach is used here (as in [36, 38]). The multi-criteria clustering problem is transformed into a single-criteria one by a weighted sum approach [13, 14]. Yet, the problem of how to choose suitable weights remains.

10.2.3   Weighting of Different Parts of a Linked Network

Of these three approached listed above, only the third one is problematic, especially within generalized blockmodeling (within generalized blockmodeling the first two were addressed in [36]). In generalized blockmodeling, the value of the criterion function depends greatly at least on the size of the network (number of possible ties), the specific generalized blockmodeling approach [34] used (e.g. binary, valued, homogeneity) and the allowed block types. In the case of blockmodeling valued networks, the values of the ties (especially their range and variability) also strongly determine the value of the criterion function. The current implementation of blockmodeling R package allows the weights to be set per relation or per block. Therefore, the easiest way is to specify each (one-mode or two-mode) network or its relation as a separate relation and then specify weights by relations.

The current recommendation (see [38]) is to select weights so that they are inversely proportional to the average inconsistency of random partitions for a selected part of the linked network using the same blockmodeling approach, allowed block types (per block if desired), and number of clusters as are to be used in the linked blockmodeling.

10.3   Examples

Two examples are given in this section. The first one applies the linked blockmodeling approach to a co-authorship network measured in two time points, while the second applies it to a multi-level network of advice seeking among agents and contracts among their companies.

10.3.1   Co-authorship Network at Two Time-points

This example shows how the linked blockmodeling approach can be used for networks measured at several time points, specifically the co-authorship networks of Slovenian researchers registered with the Slovenian Research Agency (ARRS) in the discipline "Process engineering" measured in two 10-year periods (1991–2000 and 2001–2010) which were also used in [10]. When needed, the shorthand notation for the periods, t1 and t2, are used. Data were obtained from the Co-operative Online Bibliographic System and Services (COBISS) and the Slovenian Current Research Information System (SICRIS), maintained by the Institute of Information Science (IZUM) and ARRS. A tie between two researchers in this network is defined as co-authorship of at least one relevant bibliographic unit according to ARRS: original, short, or review articles, published scientific conference contributions, monographs or parts of monographs, scientific or documentary films, sound or video recordings, complete scientific databases, corpus and patents [10]. Similar data were analyzed in several works by Kronegger and Ferligoj and colleagues [15, 16, 20–22, 25]. The specific discipline was mainly chosen based on the size of the network among the data used in [10]. Where a network is too small blockmodeling approaches usually do not provide much added value compared to looking at the original data, while the use of large networks is problematic due to the time complexity of the generalized blockmodeling algorithm (using a relocation algorithm) as implemented in [37]. In order to reduce the size of the data, only nodes with a degree of at least 2, at a minimum of one time point were retained. This left us with 71 units in the period 1991–2000 (period 1) and 68 units in the period 2001–2010 (period 2). In addition to the co-authorship networks in both periods, a two-mode network connecting the same authors in different time periods was created. The linked network representation of this network is presented in Figure 10.1. The thick lines separate sets of authors at different time points. The upper/left nodes represent authors in period 1 and the lower/right those in period 2.

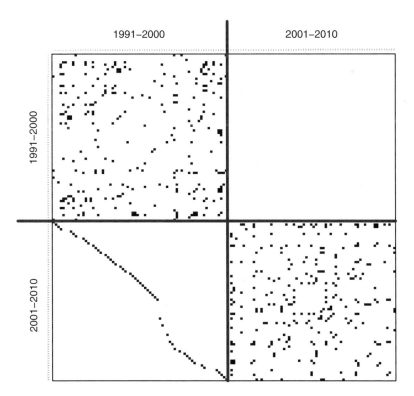

Figure 10.1   The linked network representation of the co-authorship network in two time points.

The co-authorship networks are shown as the two diagonal blocks. The lower left part of the network ties the same authors from different time points. Therefore, there is at most one tie in each row and each column in this two-mode network. An empty column here means the column author was only observed in the first time period, but not in the second, while the empty row means the opposite → the row unit was not observed in the first period, but was present in the second.

This linked network is blockmodeled simultaneously in the linked blockmodeling approach. However, as a starting point and to gain an impression about a suitable number of clusters, a separate analysis is first performed. For all blockmodeling tasks, binary blockmodeling according to structural equivalence with differential weights as suggested in [35] for sparse networks is used. In accordance with [35], the weight of the null blocks is set to 1 and that of the complete blocks to $d/(1 - d)$, here d is the density of the analyzed network. The reason for choosing this approach is that finding a smaller number of clusters that have a similar pattern of non-ties is preferred to finding many clusters of almost totally structurally equivalent nodes (relatively "clean" null blocks are preferred to relatively "clean" complete blocks). Of course, any (generalized) blockmodeling approach could be selected, if desired (and appropriate).

10.3.1.1 Separate Analysis

The results for blockmodeling the period 1990–2000 are presented in Figure 10.2. The first plot represents the inconsistencies by number of clusters via a scree plot, while the others show partitioned matrices for different numbers of clusters. Based on this figure, a five-cluster partition was chosen. Although a different algorithm was used, the basic multi-core-semiperiphery-periphery

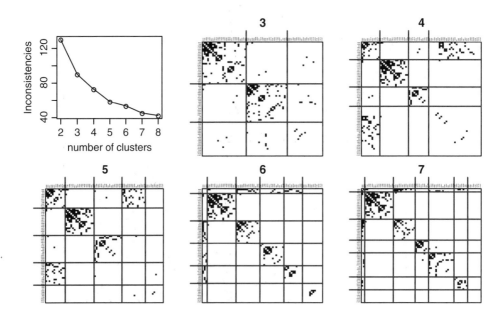

Figure 10.2 Blockmodeling of the first period (1991–2000) only.

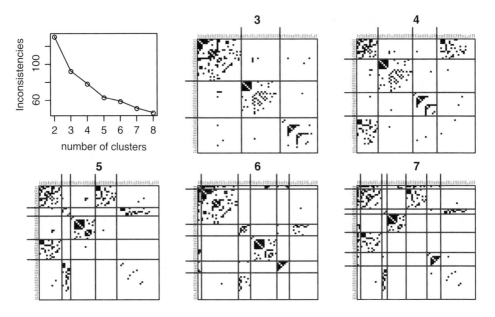

Figure 10.3   Blockmodeling of the second period (2001–2010) only.

structure as described and found in [10, 15, 20] can be observed at most resolutions (number of clusters). Based on the scree plot, the five-cluster partition seems the most appropriate. The five clusters are as follows

- A core group, a relatively densely group also connected to a peripheral group.
- Two relatively cohesive groups that are unconnected to other groups.
- A peripheral group, internally disconnected, connected only to the core group.
- A disconnected group, both internally and externally disconnected.

The same procedure is then applied to the second period (2001–2010). The results are presented in Figure 10.3. Based on the scree plot, five clusters can again be chosen. The clusters are, however, now different:

- A core group, a relatively densely group also connected to a peripheral group.
- A weakly connected group (two unconnected stars) that has ties to its own periphery group.
- A relatively cohesive group that is unconnected to other groups.
- Two peripheral groups, internally disconnected, connected only to either the first group (main core) or the second group.

Based on these two results (for period 1991–2000 and for period 2001–2010), we can also present the "linked" view of these two partitions, that is by partitioning the linked network presented in Figure 10.1 by these two partitions. The result is shown in Figure 10.4. The additional information compared to the results seen in Figures 10.2 and 10.3 is mainly the partitioned two-mode network. No clear pattern is present, although the members of the first two clusters in the second period are members of only specific clusters from the first period.

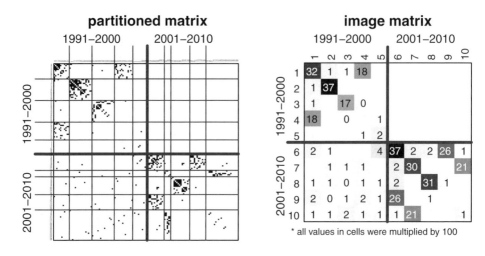

Figure 10.4 Linked view of the separately obtained partitions.

10.3.1.2 A True Linked Blockmodeling Approach

As a true linked approach is much more time-consuming than the separate analysis due to its use of twice the number of both nodes and clusters[2] and the possibility of different combinations of cluster numbers at both time points, not all numbers of clusters that are possible are explored. Rather, the number of clusters as found appropriate in the separate analysis is used and then updated if required. Therefore, five clusters are used in each period. The weights of different parts of the network (both one-mode networks for each time period and the two-mode network joining them) were computed as inversely proportional to the average inconsistency of 1000 random partitions (selected from a multinomial distribution) based on the same approach as used in true linked blockmodeling, in this case that is binary blockmodeling according to structural equivalence with different weights for null and complete blocks (the same approach as for the separate analysis and thus described there). This weighting is called "original" in this chapter. The actual weights are presented in Table 10.1.

The results are presented in Figure 10.5. However, the blocks in the one-mode network in this result are relatively "messy", especially in the second period, while the two-mode network has practically minimal inconsistency (the minimum number of complete blocks and no

Table 10.1 The weights used in a true linked approach

	t1 × t1	t2 × t2	t2 × t1
Original	1.00	0.98	5.67
Two-mode halved	1.00	0.98	2.84

---

[2] Given the restriction that each time period has its own clusters, the complexity is lower than it would be if we had blockmodeled an ordinary one-mode network of that size.

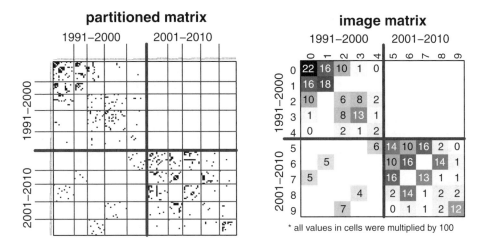

Figure 10.5 Results of linked blockmodeling with five clusters in each period and the "original" weighting.

inconsistencies in the null blocks). This indicates the weight of the two-mode network was set too high, which might also seriously hamper the search procedure for a good partition [36].

To rectify this, the weight for the two-mode network was halved, as shown in Table 10.1. The results of using these "two-mode halved" weights are given in Figure 10.6. These results are much better because the one-mode networks are much clearer while the two-mode network is not much worse. In fact, the criterion function value for this partition using the "original" weights is significantly lower than the value or the criterion function obtained when searching

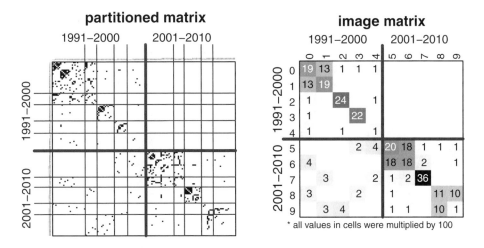

Figure 10.6 Results of linked blockmodeling with five clusters in each period and the "two-mode halved" weighting.

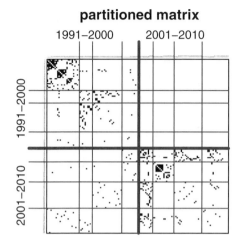

partitioned matrix

image matrix

* all values in cells were multiplied by 100

1991–2000 / 2001–2010 (image matrix)

	0	1	2	3	4	5	6	7
0	25	1	1	1				
1	1	11	14	1				
2	1	14	2	0				
3	1	1	0	1				
4	0			5	33	3	16	11
5	0	1	0	3	3	24		2
6			3		16		3	2
7	3	4			11	2	2	2

Figure 10.7 Results of linked blockmodeling with four clusters in each period and the "two-mode halved" weighting.

for the partition using the "original" weights. This just further demonstrates the claim made by Žiberna [36] that excessively high weights for the two-mode networks can deteriorate the results of optimization via the relocation algorithm.

The results of the linked approach using the "two-mode halved" weights (presented in Figure 10.6) show that at both time points the two groups have a very similar pattern of ties in the one-mode networks. Because of this, the number of groups was reduced to four at both time points. Results using the "two-mode halved" weights are shown in Figure 10.7. In this example, this is used as the final solution, although a higher number of clusters and different approaches to generalized blockmodeling could also be explored. The groups in the period 1991–2000 are:

- A relatively large cohesive group that is almost unconnected to other groups. A closer view reveals this group could be further partitioned into at least three more densely connected cohesive groups, which would, however, also be more loosely connected to each other and almost unconnected to others, which is why they are here grouped into a single group in addition to mainly being in the same groups at the second time point.
- A loosely connected group with relatively strong (for inter-group) ties to a peripheral group.
- A peripheral group, internally disconnected, connected only to the previously mentioned core group.
- A disconnected group, both internally and externally disconnected.

In the period 2001–2010, the groups are:

- A core group that is internally connected and also has the strongest ties (among all inter-group groups) to all other groups (although the tie to the other cohesive group is very low).
- An internally relatively densely connected cohesive group with non-existent or low ties to other groups.

- Two peripheral groups, mainly internally disconnected, connected only to the core group. They mainly differ in their ties to the groups in the previous period.

The transitions of groups between the periods are as follows:

- Most units from the first group (cohesive group) from t1 move to the second periphery group in t2.
- The core group of t1 (second group) also mainly moves (together with the previously mentioned group) into the second periphery group of t2.
- The peripheral group of t1 (third group) moves into the first peripheral group in t2.
- The disconnected group of t1 splits into the core group and the cohesive group of t2.
- The core and cohesive groups of t2 also obtain several new authors not present in t1.

Perhaps the most noteworthy result of this analysis is that the most cohesive groups are almost totally different in the two periods. Members of the cohesive or core groups of the first period move to the periphery or are no longer present in the second period. However, a few units also move to the cohesive group in the second period.

In more general terms, this example shows how the linked blockmodeling approach can be used to blockmodel networks measured at several time points. The most beneficial result of this approach compared to separate analysis is that the transitions of the groups in time are made much clearer.

10.3.2   A Multilevel Network of Participants at a Trade Fair for TV Programs

The data used in this example come from a trade fair for TV programs held in Eastern Europe in 2012 as gathered and analyzed by Brailly and colleagues [4, 5]. There are two sets of nodes in this dataset, persons (agents of the companies/organizations,[3] buyers and sellers of TV programs) and the organizations in the business of TV programs, namely the broadcasters, distributors, independent buyers, and media groups (in principle also producers, but they were not observed in this time period).

Among the persons an advice relation is recorded, while among the organizations a contract relation (signing a deal) is recorded. A contract relation is recorded as an asymmetric relation with the direction based on who reported it. However, since it is intrinsically a symmetric relation, it was symmetrized prior to the analysis. Only units that have reported at least one tie were retained, that is 97 persons and 105 companies. The unordered linked/multilevel network is presented in Figure 10.8.

In addition to these relational data, two attribute variables are used. One is the task of a person (seller or buyer), while the other is the type of company (broadcaster, distributor, independent buyer or media group). While both variables were collected from the persons, the companies are assigned aggregates of their employees.[4] See [4] and especially [5] for more details about the data and its collection.

---

[3] These two terms are used interchangeably.
[4] The type of company is the same for all its employees, while the task of employees can for a company also be "both" in addition to the original categories.

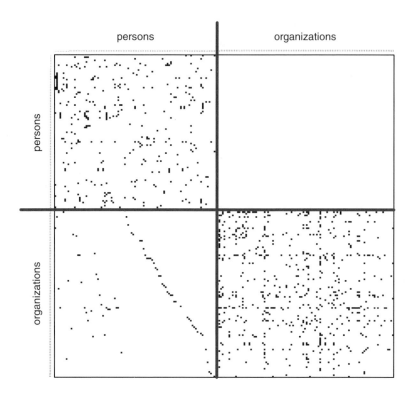

Figure 10.8 Original (unpartitioned) multilevel network of advice (persons × persons), contracts (organization × organization), and affiliation (organizations × persons) at a TV show trade fair.

For this network, the homogeneity sum of squares blockmodeling introduced in [34] and described in Chapter 5 according to structural equivalence was selected. The reason for choosing this algorithm is that it produces denser complete blocks than binary blockmodeling with differential weights according to the average density.[5] Based on a comparison with the attribute data, it seems that this better captures the roles the units play in the network.

10.3.2.1 Separate Analysis

First, separate analysis was performed. The results are presented in Figure 10.9.

Based on the scree plot, we cannot easily identify the appropriate number of clusters, although three and five seem suitable. Five clusters were selected, mainly due to them having fewer nodes in the largest, disconnected cluster, which is later demonstrated to be "mixed".

For this partition, we also explored the association with selected attribute variables, as presented in Table 10.2. The description of clusters based on both the partitioned matrix (Figure 10.9) and attribute data is as follows:

---

[5] Using binary blockmodeling according to structural equivalence with an equal weight of all blocks resulted in only singleton clusters and one giant cluster for up to eight clusters.

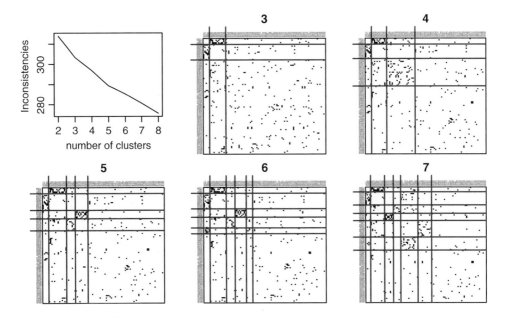

Figure 10.9   Blockmodeling of the advice relation among persons.

Table 10.2   Association of partition of the advice network and attribute variables: the number of persons from each cluster with each task and organization type (based on the organization type of each person's organization)

Cluster	Task		Organization type			
	Buyer	Seller	Broadcaster	Distributor	Independent buyer	Media group
1	5	0	4	0	0	1
2	1	13	1	10	0	3
3	5	2	1	2	3	1
4	1	9	1	6	0	3
5	30	31	24	23	6	8

- Cluster 1, first cluster of buyers, connected in both ways only to Cluster 2 (cluster of sellers).
- Cluster 2, first cluster of sellers, connected in both ways only to the first buyers' cluster.
- Cluster 3, more mixed clusters with five buyers and two sellers, connected only to cluster 4, with more outgoing (advice-seeking) ties than ingoing.
- Cluster 4, the second sellers' cluster, mainly connected to cluster 3, as described below.
- Cluster 5, the cluster of remaining units with about equal representation of both buyers and sellers, a disconnected cluster.

All clusters are internally disconnected and all but the last one (with the third also being more mixed) are composed of either mainly buyers or mainly sellers. Also, it can be observed

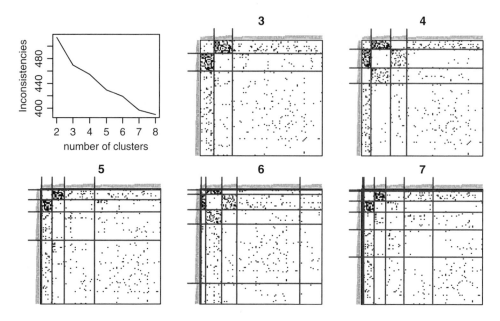

Figure 10.10   Results of blockmodeling of the contract network among organizations.

that diagonal blocks (except the largest one) are sparser than other null-like blocks.[6] Moreover, the buyers' clusters are not connected among themselves, neither are sellers' clusters (null-like blocks are sparser), indicating that buyers and sellers only seek advice from the other group. In addition, buyers seek more advice than sellers, which makes sense since they need to get to know about the programs they might purchase.

Now we move on the second level: organizations. The results of blockmodeling the contract network of organizations are shown in Figure 10.10. The scree plot again does not provide a definitive answer, although numbers of clusters 3, 5, and 7 seem more appropriate. The five-cluster partition was selected for further use. For this partition, we also explored the association with selected attribute variables, as presented in Table 10.3. The description of clusters based on the relational and attribute data is as follows:

- Cluster 1, a very active broadcaster/buyer, which had made deals with all companies from cluster 4 and the majority of cluster 3 (both mainly sellers, non-broadcasters).
- Cluster 2, the only cluster of buyers made up of nine distributors and one media group. They are linked to cluster 3 (non-broadcasters).
- Cluster 3, the first cluster of sellers (10/11), non-broadcasters, linked to clusters 1 (one very active broadcaster) and 2 (buyers, mainly broadcasters).
- Cluster 4, mainly sellers, non-broadcasters, all members have made deals with one broadcaster (cluster 1).

---

[6] When using homogeneity blockmodeling without modifications we do not classify blocks into null blocks.

Table 10.3 Association of partition of the contract network and attribute variables. The first part of the table (Task) shows the number of organizations from each cluster that have only buyers, only sellers or both types of agents. The second part (Organization type) shows the number of organizations of each organization type for each cluster

Cluster	Task			Organization type			
	Buyer	Both	Seller	Broadcaster	Distributor	Independent buyer	Media group
1	1	0	0	1	0	0	0
2	9	1	0	9	0	0	1
3	1	1	9	0	4	1	6
4	2	3	21	1	17	3	5
5	31	2	24	17	19	14	7

- Cluster 5, the cluster of remaining units with about equal representation of both buyers and sellers, of all organizational types (although the proportion of independent buyers is much higher than for overall), a disconnected cluster.

Like with the network of persons, most contracts are (as expected) between buyers/broadcasters and sellers/non-broadcasters.

The partitions of persons and organizations are also presented together on the multilevel network to provide a multilevel/linked view of these two solutions in Figure 10.11. Even here, we notice quite some correspondence among the clusters of persons and the clusters of organizations. If we disregard the biggest, mixed (in some sense undetermined) cluster in each level, we can observe than for each cluster of organizations, its members usually (in some cases also solely) come from only two clusters of persons. Something similar can be said if we exchange

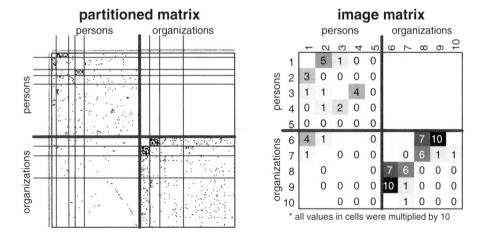

Figure 10.11 Results of the separate analysis presented on a multilevel network of a trade fair.

the roles of clusters of organizations and clusters of persons. This was, of course, expected, especially since at both levels the clusters differ substantially in terms of the task the persons play and the type of organization.

10.3.2.2   A True Linked Blockmodeling Approach

Based on these preliminary results, we conducted a true multilevel analysis by simultaneously partitioning the multilevel network into five clusters on both levels. The results are presented in Figure 10.12. However, here note that the last two clusters of persons have a very similar pattern of advice ties. Therefore, a solution with only four clusters of persons and five clusters of organizations was obtained. This solution is shown in Figure 10.13. The association with the attribute data is presented in Table 10.4. This result is discussed in greater detail bellow. The descriptions of the clusters of persons are as follows:

- Cluster 0, the most active buyers-only cluster, coming from broadcasters and media groups. It is mainly linked in both directions to the sole sellers-only cluster, cluster 2. Only a few internal ties exist.
- Cluster 1, a buyers/broadcasters-only cluster, which also has ties in both directions to seller cluster 2. Two of its members also seek advice from all other members. While this seems very strange based on the previous analysis, a glimpse at the multilevel view helps explain this situation. All members of this cluster are agents of the same organization and therefore not competitors.
- Cluster 2, a relatively large (20) sellers-only cluster (10/11) of non-broadcasters, linked in both directions to both clusters of buyers.
- Cluster 3, a mixed cluster with mainly seemingly sporadic ties.

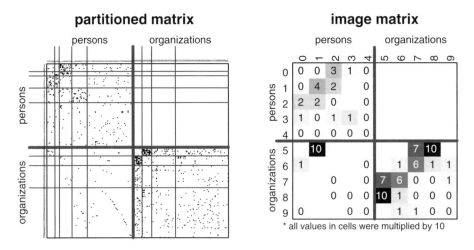

Figure 10.12   Results of a true multilevel/linked approach with five clusters on each level for the trade fair network.

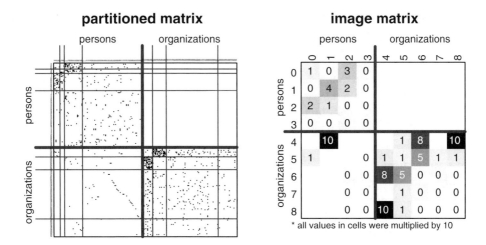

Figure 10.13 Results of a true multilevel/linked approach with four clusters of persons and five clusters of organizations for the trade fair network.

Table 10.4 Association of the multilevel partition and the attribute variables. The upper part (Persons) shows the frequencies of the types of agents by tasks and the types of their organizations for each cluster of persons (see caption of Table 10.2 for details). The lower part (Organizations) shows similar information aggregated to organization for each cluster of organizations (see caption of Table 10.3 for details)

	Cluster	Task			Organization type			
		Buyer	Both	Seller	Broadcaster	Distributor	Independent buyer	Media group
Persons	0	7	0	0	4	0	0	3
	1	5	0	0	5	0	0	0
	2	0	0	20	0	14	1	5
	3	30	0	35	22	27	8	8
Organizations	4	1	0	0	1	0	0	0
	5	9	2	0	10	0	0	1
	6	1	3	11	0	5	2	8
	7	31	2	24	17	19	14	7
	8	2	0	19	0	16	2	3

Regarding the partitioned matrix, it should be noted that the first organization cluster only contains one unit and is "hidden" within the line separating persons and organizations. The descriptions of the clusters of organizations are as follows:

- Cluster 4, a single buyer/broadcaster, has contracts with most organizations from clusters 6 and 8 (mainly sellers, non-broadcasters).

- Cluster 5, another cluster of buyers and broadcasters (and one media group) with many links to cluster 6 (mainly sellers, non-broadcasters).
- Cluster 6, mainly sellers and non-broadcasters, as described above, in contrast to clusters 4 and 5 (buyers/broadcasters).
- Cluster 7, a mixed cluster with mainly seemingly sporadic ties, with a few more ties (higher block density) to cluster 5 (buyers/broadcasters).
- Cluster 8, mainly sellers/non-broadcasters, in relational terms similar to the previous cluster, except that all units have contracts with cluster 4 (a single buyer/distributor).

Some connections among the clusters of persons and organizations are also very clear, especially with buyer (person) clusters. The first person cluster (buyers) is only composed of agents of organizations from the second organization cluster (broadcasters and one media group), while these organization cluster members also have agents in the mixed cluster of persons. The second cluster of persons (cluster 1) is only composed of the agents of a single broadcaster and vice versa, also explaining the unexpected advice ties within this cluster. The seller/person cluster (cluster 2) is mainly[7] composed of agents from both clusters of organizational sellers/non-broadcasters. The two clusters of non-broadcasters (organizations) also largely comprise agents from the person cluster made up of sellers (cluster 2). The mixed cluster of persons is composed of the agents of organizations from most organization clusters, while organizations from the mixed cluster of organizations have agents who are almost exclusively members of the mixed cluster of persons.

The main conclusion arising from this analysis is that both agents and organizations are divided mainly on the dimension buyers/broadcasters–sellers/non-broadcasters. In both one-mode networks, most ties are between clusters from the opposite side of this divide (the exception being the advice ties among agents of a single broadcaster). While the advice network is constructed as asymmetric, at least on the cluster level, most ties are quite reciprocal, although buyers also tend to be somewhat more active than sellers. All clusters of persons (except mixed) are quite clearly mainly tied to only one or two clusters of organizations, while the ties are not so clear from the view of the non-broadcasters' organizations' clusters.[8]

10.4 Conclusion

The blockmodeling of linked networks was introduced in this chapter. A linked network is a collection of two or more one-mode networks tied by two-mode networks. We might also say that linked networks are networks on several sets of nodes where ties both within and between node sets are possible. The aim of blockmodeling linked networks is to blockmodel the whole linked network in such a way that each node set is partitioned separately so that each cluster contains nodes from only one node-set. Special cases of linked networks include multilevel networks [28], networks measured at several time points, and meta-networks/matrices [19].

---

[7] In addition, one agent is from the mixed cluster of organizations and four agents are not tied to any organization from the contract network.
[8] This simply shows that non-mixed clusters of persons are much less tied to the cluster of mixed organizations than the non-mixed clusters of organizations are to the cluster of mixed persons.

As multilevel networks are a special case of linked networks, it was argued that essentially the same approach used for blockmodeling multilevel networks [36, 38] can be also used for linked networks. Separate analysis was suggested as the initial step in analysis for linked networks, followed by a true linked blockmodeling approach where the whole linked network is partitioned simultaneously. For both approaches, any blockmodeling approach could in principle be used, although generalized blockmodeling [12, 34, 35] was used in this paper.

The most problematic issue in a true linked blockmodeling approach is "negotiating" the influence of different parts of the whole linked network on the final solution. In generalized blockmodeling, this is translated in the how criterion values of different parts of the network are taken into account when searching for the combined partition. In this chapter, the weighted sum approach to multi-criteria optimization [13, 14] is suggested, with the weights selected so that they are inversely proportional to the average inconsistency of random partitions for a selected part of the linked network using the same settings as in [38].

The main benefit of using the linked blockmodeling approach over separate analysis is that, in addition to modeling ties among clusters of the same node set, ties between clusters of different node sets can also be modeled. This usually makes the ties between clusters from different node sets much clearer, as shown in the co-authorship temporal network example and multilevel network of ties among organizations and their agents at a trade fair for TV programs.

Of course, several problems remain unsettled. One concerns how to combine multiple criteria for blockmodeling linked networks (one for (each relation of) each part of a network). Here, the weighted sum approach was used and even within this approach more research is needed on how to set the appropriate weights. However, a true multi-criteria approach like that used in [6] might be better. In addition, one limitation when applying this approach within generalized blockmodeling is its computational complexity, making its use on larger networks (a few hundred nodes) impractical, if not impossible. A possible solution would be to implement the linked approach within some fast version of stochastic blockmodeling [e.g. [2, 11, 17]] or within an approach based on a two-mode k-means algorithm [7, 31].

Acknowledgements

The author acknowledges financial support from the Slovenian Research Agency (research core funding No. P5-0168 and project "Blockmodeling multilevel and temporal networks" No. J7-8279).

References

1. E. M. Airoldi, D. M. Blei, S. E. Fienberg, and E. P. Xing. Combining stochastic block models and mixed membership for statistical network analysis. In E. M. Airoldi, D. M. Blei, S. E. Fienberg, A. Goldenberg, E. P. Xing, and A. X. Zheng, editors, *Statistical Network Analysis: Models, Issues, and New Directions*, number 4503 in Lecture Notes in Computer Science, pages 57–74. Springer, Berlin, Heidelberg, 2007.
2. E. M. Airoldi, D. M. Blei, S. E. Fienberg, and E. P. Xing. Mixed membership stochastic blockmodels. *Journal of Machine Learning Research*, 9: 1981–2014, 2008.
3. E. Bellotti. Getting funded. Multi-level network of physicists in Italy. *Social Networks*, 34(2): 215–229, 2012.
4. J. Brailly. Dynamics of networks in trade fairs–A multilevel relational approach to the cooperation among competitors. *Journal of Economic Geography*, 16(6): 1279–1301, 2016.
5. J. Brailly, G. Favre, J. Chatellet, and E. Lazega. Embeddedness as a multilevel problem: A case study in economic sociology. *Social Networks*, 44: 319–333, 2016.

6. M. Brusco, P. Doreian, D. Steinley, and C. B. Satornino. Multiobjective blockmodeling for social network analysis. *Psychometrika*, 78(3): 498–525, 2013.

7. M. J. Brusco and P. Doreian. A real-coded genetic algorithm for two-mode KL-means partitioning with application to homogeneity blockmodeling. *Social Networks*, 41: 26–35, May 2015.

8. K. M. Carley and V. Hill. Structural change and learning within organizations. In A. Lomi, editor, *Dynamics of Organizational Societies: Models, Theories and Methods*. MIT Press/AAAI Press/Live Oak, Live Oak, 2001.

9. K. M. Carley, Y. Ren, and D. Krackhardt. Measuring and modeling change in C3I architectures. In *Proceedings of the 2000 Command and Control Research and Technology Symposium*, Monterey, 2000.

10. M. Cugmas, A. Ferligoj, and L. Kronegger. The stability of co-authorship structures. *Scientometrics*, 106(1): 163–186, 2016.

11. J. J. Daudin, F. Picard, and S. Robin. A mixture model for random graphs. *Statistics and Computing*, 18(2): 173–183, 2008.

12. P. Doreian, V. Batagelj, and A. Ferligoj. *Generalized Blockmodeling*. Cambridge University Press, New York, 2005.

13. M. Ehrgott and M. M. Wiecek. Multiobjective Programming. In J. Figueira, S. Greco, and M. Ehrgott, editors, *Multiple Criteria Decision Analysis: State of the Art Surveys*, pages 667–722. Springer Verlag, Boston, Dordrecht, London, 2005.

14. A. Ferligoj and V. Batagelj. Direct multicriteria clustering algorithms. *Journal of Classification*, 9(1): 43–61, 1992.

15. A. Ferligoj and L. Kronegger. Clustering of attribute and/or relational data. *Metodološki Zvezki*, 6(2): 135–153, 2009.

16. A. Ferligoj, L. Kronegger, F. Mali, T. A. B. Snijders, and P. Doreian. Scientific collaboration dynamics in a national scientific system. *Scientometrics*, 104(3): 985–1012, 2015.

17. G. Govaert and M. Nadif. Block clustering with Bernoulli mixture models: Comparison of different approaches. *Computational Statistics & Data Analysis*, 52(6): 3233–3245, 2008.

18. P. W. Holland, K. B. Laskey, and S. Leinhardt. Stochastic blockmodels: First steps. *Social Networks*, 5(2): 109–137, 1983.

19. D. Krackhardt and K. M. Carley. A PCANS model of structure in organizations. In *Proceedings of the 1998 International Symposium on Command and Control Research and Technology*. Evidence Based Research, Vienna, VA, 1998.

20. L. Kronegger, A. Ferligoj, and P. Doreian. On the dynamics of national scientific systems. *Quality & Quantity*, 45(5): 989–1015, 2011.

21. L. Kronegger, F. Mali, A. Ferligoj, and P. Doreian. Collaboration structures in Slovenian scientific communities. *Scientometrics*, 90(2): 631–647, 2012.

22. L. Kronegger, F. Mali, A. Ferligoj, and P. Doreian. Classifying scientific disciplines in Slovenia: A study of the evolution of collaboration structures. *Journal of the Association for Information Science and Technology*, 66(2): 321–339, 2014.

23. E. Lazega. Synchronization costs in the organizational society: Intermediary relational infrastructures in the dynamics of multilevel networks. In E. Lazega and T. A. Snijders, editors, *Multilevel Network Analysis for the Social Sciences*, pages 47–77. Springer International Publishing, Cham, 2016.

24. E. Lazega, M.-T. Jourda, L. Mounier, and R. Stofer. Catching up with big fish in the big pond? Multi-level network analysis through linked design. *Social Networks*, 30(2): 159–176, 2008.

25. F. Mali, L. Kronegger, P. Doreian, and A. Ferligoj. Dynamic scientific co-authorship networks. In A. Scharnhorst, K. Börner, and P. van den Besselaar, editors, *Models of Science Dynamics*, Understanding Complex Systems, pages 195–232. Springer, Berlin, Heidelberg, Jan. 2012.

26. C. Matias and V. Miele. Statistical clustering of temporal networks through a dynamic stochastic block model. *Journal of the Royal Statistical Society: Series B (Statistical Methodology)*, 2016.

27. I.-C. Moon, E. J. Kim, and K. M. Carley. Automated Influence Network Generation and the Node Parameter Sensitivity Analysis. Technical report, DTIC Document, 2008.

28. Multilevel Network Modeling Group. What Are Multilevel Networks. Technical report, University of Manchester, May 2012.

29. K. Nowicki and T. A. B. Snijders. Estimation and prediction for stochastic blockstructures. *Journal of the American Statistical Association*, 96: 1077–1087, 2001.

30. T. A. B. Snijders and K. Nowicki. Estimation and prediction for stochastic blockmodels for graphs with latent block structure. *Journal of Classification*, 14: 75–100, 1997.

31. J. van Rosmalen, P. J. F. Groenen, J. Trejos, and W. Castillo. Optimization strategies for two-mode partitioning. *Journal of Classification*, 26: 155–181, 2009.

32. E. P. Xing, W. Fu, and L. Song. A state-space mixed membership blockmodel for dynamic network tomography. *Annals of Applied Statistics*, 4(2): 535–566, 2010.

33. K. S. Xu and A. O. Hero. Dynamic stochastic blockmodels: Statistical models for time-evolving networks. In D. Hutchison, T. Kanade, J. Kittler, J. M. Kleinberg, F. Mattern, J. C. Mitchell, M. Naor, O. Nierstrasz, C. Pandu Rangan, B. Steffen, M. Sudan, D. Terzopoulos, D. Tygar, M. Y. Vardi, G. Weikum, A. M. Greenberg, W. G. Kennedy, and N. D. Bos, editors, *Social Computing, Behavioral-Cultural Modeling and Prediction*, volume 7812, pages 201–210. Springer, Berlin, Heidelberg, 2013.

34. A. Žiberna. Generalized blockmodeling of valued networks. *Social Networks*, 29: 105–126, 2007.

35. A. Žiberna. Generalized blockmodeling of sparse networks. *Metodološki Zvezki*, 10(2): 99–119, 2013.

36. A. Žiberna. Blockmodeling of multilevel networks. *Social Networks*, 39: 46–61, 2014.

37. A. Žiberna. Blockmodeling 0.3.1: An R package for generalized and classical blockmodeling of valued networks, 2018.

38. A. Žiberna and E. Lazega. Role sets and division of work at two levels of collective agency: The case of blockmodeling a multilevel (inter-individual and inter-organizational) network. In E. Lazega and T. A. B. Snijders, editors, *Multilevel Network Analysis for the Social Sciences*, number 12 in Methodos Series, pages 173–209. Springer International Publishing, 2016.

11

Bayesian Stochastic Blockmodeling

Tiago P. Peixoto
Department of Mathematical Sciences and Centre for Networks and Collective Behaviour,
University of Bath, United Kingdom, and ISI Foundation, Turin, Italy

This chapter provides a self-contained introduction to the use of Bayesian inference to extract large-scale modular structures from network data, based on the stochastic blockmodel (SBM), as well as its degree-corrected and overlapping generalizations. We focus on nonparametric formulations that allow their inference in a manner that prevents overfitting and enables model selection. We discuss aspects of the choice of priors, in particular how to avoid underfitting via increased Bayesian hierarchies, and we contrast the task of sampling network partitions from the posterior distribution with finding the single point estimate that maximizes it, while describing efficient algorithms to perform either one. We also show how inferring the SBM can be used to predict missing and spurious links, and shed light on the fundamental limitations of the detectability of modular structures in networks.

11.1   Introduction

Over the past decade and a half there has been an ever-increasing demand to analyze network data, in particular those stemming from social, biological, and technological systems. Often these systems are very large, comprising millions or even billions of nodes and edges, such as the World Wide Web, and the global-level social interactions among humans. A particular challenge that arises is how to describe the large-scale structures of these systems in a way that abstracts away from low-level details, allowing us to focus instead on "the big picture." Differently from systems that are naturally embedded in some low-dimensional space – such as the population density of cities or the physiology of organisms – we are unable just to "look" at a network and readily extract its most salient features. This has prompted much of activity in

Advances in Network Clustering and Blockmodeling, First Edition.
Edited by Patrick Doreian, Vladimir Batagelj, and Anuška Ferligoj.
© 2020 John Wiley & Sons Ltd. Published 2020 by John Wiley & Sons Ltd.

developing algorithmic approaches to extract such global information in a well-defined manner, many of which are described in the remaining chapters of this book. Most of them operate on a rather simple ansatz, where we try to divide the network into "building blocks," which then can be described at an aggregate level in a simplified manner. The majority of such methods go under the name "community detection," "network clustering" or "blockmodeling." In this chapter we consider the situation where the ultimate objective when analyzing network data in this way is to *model* it, i.e. we want to make statements about possible generative mechanisms that are responsible for the network formation. This overall aim sets us in a well-defined path, where we get to formulate probabilistic models for network structure, and use principled and robust methods of statistical inference to fit our models to data. Central to this approach is the ability to distinguish structure from randomness, so that we do not fool ourselves into believing that there are elaborate structures in our data which are in fact just the outcome of stochastic fluctuations, which tends to be the Achilles' heel of alternative nonstatistical approaches. In addition to providing a description of the data, the models we infer can also be used to generalize from observations, and make statements about what has *not* yet been observed, yielding something more tangible than mere interpretations. In what follows we will give an introduction to this inference approach, which includes recent developments that allow us to perform it in a consistent, versatile and efficient manner.

11.2   Structure Versus Randomness in Networks

If we observe a random string of characters we will eventually encounter every possible substring, provided the string is long enough. This leads to the famous thought experiment of a large number of monkeys with typewriters: assuming that they type randomly, for a sufficiently large number of monkeys any output can be observed, including, for example, the very text you are reading. Therefore, if we are ever faced with this situation, we should not be surprised if a such a text is in fact produced and, most importantly, we should not offer its simian author a place in a university department, as this occurrence is unlikely to be repeated. However, this example is of little practical relevance, as the number of monkeys necessary to type the text "blockmodeling" by chance is already of the order of 10^{18}, and there are simply not that many monkeys.

Networks, however, are different from random strings. The network analogue of a random string is an Erdős-Rényi random graph [22] where each possible edge can occur with the same probability. But differently from a random string, a random graph can contain a wealth of structure before it becomes astronomically large, particularly if we *search* for it. An example of this is shown in Figure 11.1 for a modest network of 5000 nodes, where its adjacency matrix is visualized using three different node orderings. Two of the orderings seem to reveal patterns of large-scale connections that are tantalizingly clear, and indeed would be eagerly captured by many network clustering methods [39]. In particular, they seem to show groupings of nodes that have distinct probabilities of connections to each other, in direct contradiction to the actual process that generated the network, where all connections had the same probability of occurring. What makes matters even worse is that Figure 11.1 shows only a very small subset of all orderings that have similar patterns, but are otherwise very distinct from each other. Naturally, in the same way we should not confuse a monkey with a proper scientist in our previous example, we should not use any of these node groupings to explain why the network has its structure. Doing so should be considering *overfitting* it, i.e. mistaking random fluctuations for generative structure, yielding an overly complicated and ultimately wrong explanation for the data.

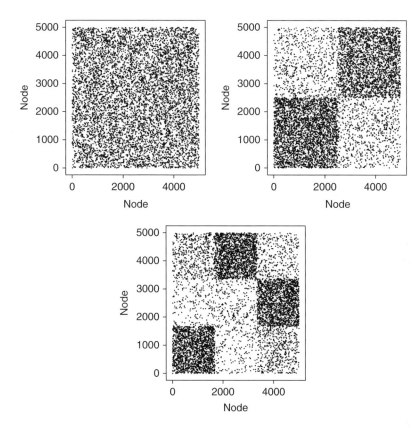

Figure 11.1 The three panels show the same adjacency matrix, with the only difference between them being the ordering of the nodes. The different orderings show seemingly clear, albeit very distinct, patterns of modular structure. However, the adjacency matrix in question corresponds to an instance of a fully random Erdős–Rényi model, where each edge has the same probability $p = \langle k \rangle / (N - 1)$ of occurring, with $\langle k \rangle = 3$. Although the patterns seen in the second and third panels are not mere fabrications – as they are really there in the network – they are also not meaningful descriptions of this network, since they arise purely out of random fluctuations. Therefore, the node groups that are identified via these patterns bear no relation to the generative process that produced the data. In other words, the second and third panels correspond each to an *overfit* of the data, where stochastic fluctuations are misrepresented as underlying structure. This pitfall can lead to misleading interpretations of results from clustering methods that do not account for statistical significance.

The remedy to this problem is to think probabilistically. We need to ascribe to each possible explanation of the data a probability that it is correct, which takes into account modeling assumptions, the statistical evidence available in the data, as well any source of prior information we may have. Imbued in the whole procedure must be the principle of parsimony – or Occam's razor – where a simpler model is preferred if the evidence is not sufficient to justify a more complicated one.

In order to follow this path, before we look at any network data, we must first look in the "forward" direction, and decide which mechanisms generate networks in the first place. Based on this, we will finally be able to look "backwards," and tell which particular mechanism generated a given observed network.

11.3 The Stochastic Blockmodel

As mentioned in the introduction, we wish to decompose networks into "building blocks" by grouping together nodes that have a similar role in the network. From a generative point of view, we wish to work with models that are based on a partition of N nodes into B such building blocks, given by the vector $\boldsymbol{b}$ with entries

$$b_i \in \{1, \dots, B\},$$

specifying the group membership of node i. We wish to construct a generative model that takes this division of the nodes as parameters and generates networks with a probability

$$P(A|\boldsymbol{b}),$$

where $A = \{A_{ij}\}$ is the adjacency matrix. But what shape should $P(A|\boldsymbol{b})$ have? If we wish to impose that nodes that belong to the same group are statistically indistinguishable, our ensemble of networks should be fully characterized by the number of edges that connects nodes of two groups r and s,

$$e_{rs} = \sum_{ij} A_{ij} \delta_{b_i,r} \delta_{b_j,s}, \tag{11.1}$$

or twice that number if $r = s$. If we take these as conserved quantities, the ensemble that reflects our maximal indifference towards any other aspect is the one that maximizes the entropy [48]

$$S = -\sum_A P(A|\boldsymbol{b}) \ln P(A|\boldsymbol{b}) \tag{11.2}$$

subject to the constraint of Equation (11.1). If we relax somewhat our requirements, such that Equation (11.1) is obeyed only for expectations then we can obtain our model using the method of Lagrange multipliers, using the Lagrangian function

$$F = S - \sum_{r \leq s} \mu_{rs} \left(\sum_A P(A|\boldsymbol{b}) \sum_{i<j} A_{ij} \delta_{b_i,r} \delta_{b_j,s} - \langle e_{rs} \rangle \right) - \lambda \left(\sum_A P(A|\boldsymbol{b}) - 1 \right) \tag{11.3}$$

where $\langle e_{rs} \rangle$ are constants independent of $P(A|\boldsymbol{b})$, and μ and λ are multipliers that enforce our desired constraints and normalization, respectively. Obtaining the saddle point $\partial F / \partial P(A|\boldsymbol{b}) = 0$, $\partial F / \partial \mu_{rs} = 0$ and $\partial F / \partial \lambda = 0$ gives us the maximum entropy ensemble with the desired properties. If we constrain ourselves to simple graphs, i.e. $A_{ij} \in \{0, 1\}$, without self-loops, we have as our maximum entropy model

$$P(A|\boldsymbol{p}, \boldsymbol{b}) = \prod_{i<j} p_{b_i, b_j}^{A_{ij}} (1 - p_{b_i, b_j})^{1 - A_{ij}}, \tag{11.4}$$

with $p_{rs} = e^{-\mu_{rs}} / (1 + e^{-\mu_{rs}})$ being the probability of an edge existing between any two nodes belonging to groups r and s. This model is called the *stochastic blockmodel* (SBM), and has its roots in the social sciences and statistics [44, 72, 100, 105], but has appeared repeatedly in the

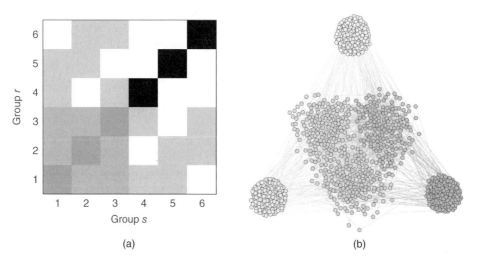

Figure 11.2 SBM: (a) the matrix of probabilities between groups p_{rs} defines the large-scale structure of generated networks and (b) a sampled network corresponding to (a), where the node colors indicate the group membership.

literature under a variety of different names [8–10, 12, 17, 102]. By selecting the probabilities $p = \{p_{rs}\}$ appropriately, we can achieve arbitrary mixing patterns between the groups of nodes, as illustrated in Figure 11.2. We stress that while the SBM can perfectly accommodate the usual "community structure" pattern [25], i.e. when the diagonal entries of p are dominant, it can equally well describe a large variety of other patterns, such as bipartiteness, core-periphery, and many others.

Instead of simple graphs, we may consider *multigraphs* by allowing multiple edges between nodes, i.e. $A_{ij} \in \mathbb{N}$. Repeating the same procedure, we obtain in this case

$$P(A|\lambda, b) = \prod_{i<j} \frac{\lambda_{b_i,b_j}^{A_{ij}}}{(\lambda_{b_i,b_j} + 1)^{A_{ij}+1}}, \tag{11.5}$$

with $\lambda_{rs} = e^{-\mu_{rs}}/(1 - e^{-\mu_{rs}})$ being the *average number* of edges existing between any two nodes belonging to group r and s. Whereas the placement of edges in Equation (11.4) is given by a Bernoulli distribution, in Equation (11.5) it is given by a geometric distribution, reflecting the different nature of both kinds of networks. Although these models are not the same, there is in fact little difference between the networks they generate in the *sparse limit* given by $p_{rs} = \lambda_{rs} = O(1/N)$ with $N \gg 1$. We see this by noticing how their log-probabilities become asymptotically identical in this limit, i.e.

$$\ln P(A|p, b) \approx \frac{1}{2} \sum_{rs} e_{rs} \ln p_{rs} - n_r n_s p_{rs} + O(1), \tag{11.6}$$

$$\ln P(A|\lambda, b) \approx \frac{1}{2} \sum_{rs} e_{rs} \ln \lambda_{rs} - n_r n_s \lambda_{rs} + O(1). \tag{11.7}$$

Therefore, since most networks that we are likely to encounter are sparse [66], it does not matter which model we use, and we may prefer whatever is more convenient for our calculations. With this in mind, we may consider yet another variant, which uses instead a Poisson distribution to sample edges [50],

$$P(A|\lambda, b) = \prod_{i<j} \frac{e^{-\lambda_{b_i,b_j}} \lambda_{b_i,b_j}^{A_{ij}}}{A_{ij}!} \times \prod_i \frac{e^{-\lambda_{b_i,b_i}/2} (\lambda_{b_i,b_i}/2)^{A_{ii}/2}}{(A_{ii}/2)!}, \tag{11.8}$$

where now we also allow for self-loops. Like the geometric model, the Poisson model generates multigraphs, and it is easy to verify that it also leads to Equation (11.7) in the sparse limit. This model is easier to use in some of the calculations that we are going to make, in particular when we consider important extensions of the SBM, therefore we will focus on it.[1]

The model above generates undirected networks. It can be very easily modified to generate directed networks instead, by making λ_{rs} an asymmetric matrix, and adjusting the model likelihood accordingly. The same is true for all model variations that are going to be used in the following sections. However, for the sake of conciseness we will focus only on the undirected case. We point out that the corresponding expressions for the directed case are readily available in the literature (e.g. [78, 84, 85]).

Now that we have defined how networks with prescribed modular structure are generated, we need to develop the reverse procedure, i.e. how to infer the modular structure from data.

11.4   Bayesian Inference: The Posterior Probability of Partitions

Instead of generating networks, our nominal task is to determine which partition b generated an observed network A, assuming this was done via the SBM. In other words, we want to obtain the probability $P(b|A)$ that a node partition b was responsible for a network A. By evoking elementary properties of conditional probabilities, we can write this probability as

$$P(b|A) = \frac{P(A|b)P(b)}{P(A)} \tag{11.9}$$

with

$$P(A|b) = \int P(A|\lambda, b)P(\lambda|b) \, d\lambda \tag{11.10}$$

being the *marginal likelihood* integrated over the remaining model parameters, and

$$P(A) = \sum_b P(A|b)P(b), \tag{11.11}$$

---

[1] Although the Poisson model is not strictly a maximum entropy ensemble, the generative process behind it is easy to justify. We can imagine it as the random placement of exactly E edges into the $N(N-1)/2$ entries of the matrix A, each with a probability q_{ij} of attracting an edge, with $\sum_{i<j} q_{ij} = 1$, yielding a multinomial distribution $P(A|q, E) = E! \prod_{i<j} q_{ij}^{A_{ij}}/A_{ij}!$, where, differently from Equation (11.8), the edge placements are not conditionally independent. But if we now sample the total number of edges E from a Poisson distribution $P(E|\bar{E})$ with average $\bar{E}$, by exploiting the relationship between the multinomial and Poisson distributions, we have $P(A|q) = \sum_E P(A|q, E)P(E|\bar{E}) = \prod_{i<j} e^{-\omega_{ij}} \omega_{ij}^{A_{ij}}/A_{ij}!$, where $\omega_{ij} = q_{ij}/\bar{E}$, which does amount to conditionally independent edge placements. Making $q_{ij} = \bar{E}\lambda_{b_i,b_j}$, and allowing self-loops, we arrive at Equation (11.8).

which is called the *evidence*, i.e. the total probability of the data under the model, which serves as a normalization constant in Equation (11.9). Equation (11.9) is known as Bayes' rule, and far from being only a simple mathematical step, it encodes how our prior beliefs about the model, i.e. before we observe any data – in the above represented by the *prior distributions* $P(\boldsymbol{b})$ and $P(\lambda|\boldsymbol{b})$ – are affected by the observation, yielding the so-called *posterior distribution* $P(\boldsymbol{b}|\boldsymbol{A})$. The overall approach outlined above has been proposed to the problem of network inference by several authors [5, 16, 37, 41, 43, 51, 64, 65, 71, 79, 84, 85, 89, 93, 95, 109], with different implementations that vary in some superficial details in the model specification, approximations used, and in particular in the choice of priors. Here we will not review or compare all approaches in detail, but rather focus on the most important aspects, while choosing a particular path that makes exact calculations possible.

The prior probabilities are a crucial element of the inference procedure, as they will affect the shape of the posterior distribution and, ultimately, our inference results. In more traditional scenarios, the choice of priors would be guided by previous observations of data that are believed to come from the same model. However, this is not an applicable scenario when considering networks, which are typically *singletons*, i.e. they are unique objects, instead of coming from a population (e.g. there is only one internet, one network of trade between countries, etc).[2] In the absence of such empirical prior information, we should try as much as possible to be guided by well-defined principles and reasonable assumptions about our data, rather than *ad hoc* choices. A central proposition we will be using is the *principle of maximum indifference* about the model before we observe any data. This will lead us to so-called *uninformative* priors,[3] which are maximum entropy distributions that ascribe the same probability to each possible parameter combination [48]. These priors have the property that they do not bias the posterior distribution in any particular way, and thus let the data "speak for itself." But as we will see in the following, the naive application of this principle will lead to adverse effects in many cases, and upon closer inspection we will often be able to identify aspects of the model that we should not be agnostic about. Instead, a more meaningful approach will be to describe higher-order aspects of the model with their own models. This can be done in a manner that preserves the unbiased nature of our results, while being able to provide a more faithful representation of the data.

We begin by choosing the prior for the partition, $\boldsymbol{b}$. The most direct uninformative prior is the "flat" distribution where all partitions into at most $B = N$ groups are equally likely, namely

$$P(\boldsymbol{b}) = \frac{1}{\sum_{b'} 1} = \frac{1}{a_N} \tag{11.12}$$

where a_N are the ordered Bell numbers [99], given by

$$a_N = \sum_{B=1}^{N} \left\{ {N \atop B} \right\} B! \tag{11.13}$$

---

[2] One could argue that most networks change in time, and hence belong to a time series, thus possibly allowing priors to be selected from earlier observations of the same network. This is a potentially useful way to proceed, but also opens a Pandora's box of dynamical network models, where simplistic notions of statistical stationarity are likely to be contradicted by data. Some recent progress has been made on the inference of dynamic networks [13, 27, 29, 59, 76, 87, 108, 113], but this field is still in relative infancy.

[3] The name "uninformative" is something of a misnomer, as it is not really possible for priors to truly carry "no information" to the posterior distribution. In our context, the term is used simply to refer to *maximum entropy priors*, conditioned on specific constraints.

where $\left\{ \begin{matrix} N \\ B \end{matrix} \right\}$ are the Stirling numbers of the second kind [98], which count the number of ways to partition a set of size N into B indistinguishable and nonempty groups (the $B!$ in the above equation recovers the distinguishability of the groups, which we require). However, upon closer inspection we often find that such flat distributions are not a good choice. In this particular case, since there are many more partitions into $B + 1$ groups than there are into B groups (if B is sufficiently smaller than N), Equation (11.12) will typically prefer partitions with a number of groups that is comparable to the number of nodes. Therefore, this uniform assumption seems to betray the principle of parsimony that we stated in the introduction, since it favors large models with many groups, before we even observe the data.[4] Instead, we may wish to be agnostic about the *number of groups itself*, by first sampling it from its own uninformative distribution $P(B) = 1/N$, and then sampling the partition conditioned on it

$$P(\mathbf{b}|B) = \frac{1}{\left\{ \begin{matrix} N \\ B \end{matrix} \right\} B!}, \tag{11.14}$$

since $\left\{ \begin{matrix} N \\ B \end{matrix} \right\} B!$ is the number of ways to partition N nodes into B labelled groups.[5] Since $\mathbf{b}$ is a parameter of our model, the number of groups B is a called a *hyperparameter*, and its distribution $P(B)$ is called a *hyperprior*. But once more, upon closer inspection we can identify further problems: If we sample from Equation (11.14), most partitions of the nodes will occupy all the groups approximately equally, i.e. all group sizes will be the approximately the same. Is this something we want to assume before observing any data? Instead, we may wish to be agnostic about this aspect as well, and choose to sample first the distribution of group sizes $\mathbf{n} = \{n_r\}$, where n_r is the number of nodes in group r, forbidding empty groups,

$$P(\mathbf{n}|B) = \binom{N-1}{B-1}^{-1}, \tag{11.15}$$

since $\binom{N-1}{B-1}$ is the number of ways to divide N nonzero counts into B nonempty bins. Given these randomly sampled sizes as a constraint, we sample the partition with a uniform probability

$$P(\mathbf{b}|\mathbf{n}) = \frac{\prod_r n_r!}{N!}. \tag{11.16}$$

This gives us finally

$$P(\mathbf{b}) = P(\mathbf{b}|\mathbf{n})P(\mathbf{n}|B)P(B) = \frac{\prod_r n_r!}{N!} \binom{N-1}{B-1}^{-1} N^{-1}. \tag{11.17}$$

At this point the reader may wonder if there is any particular reason to stop here. Certainly we can find some higher-order aspect of the group sizes $\mathbf{n}$ that we may wish to be agnostic about,

---

[4] Using constant priors such as Equation (11.12) makes the posterior distribution proportional to the likelihood. Maximizing such a posterior distribution is therefore entirely equivalent to a "non-Bayesian" maximum likelihood approach, and nullifies our attempt to prevent overfitting.

[5] We could have used simply $P(\mathbf{b}|B) = 1/B^N$, since B^N is the number of partitions of N nodes into B groups, which are allowed to be empty. However, this would force us to distinguish between the nominal and the actual number of groups (discounting empty ones) during inference [71], which becomes unnecessary if we simply forbid empty groups in our prior.

and introduce a *hyperhyperprior*, and so on, indefinitely. The reason why we should not keep recursively being more and more agnostic about higher-order aspects of our model is that it brings increasingly diminishing returns. In this particular case, if we assume that the individual group sizes are sufficiently large, we obtain asymptotically

$$\ln P(b) \approx -NH(n) + O(\ln N) \tag{11.18}$$

where $H(n) = -\sum_r (n_r/N)\ln(n_r/N)$ is the entropy of the group size distribution. The value $\ln P(b) \to -NH(n)$ is an information-theoretical limit that cannot be surpassed, regardless of how we choose $P(n|B)$. Therefore, the most we can optimize by being more refined is a marginal factor $O(\ln N)$ in the log-probability, which would amount to little practical difference in most cases.

In the above, we went from a purely flat uninformative prior distribution for b to a Bayesian hierarchy with three levels, where we sample first the number of groups, the groups sizes, and then finally the partition. In each of the levels we used maximum entropy distributions that are constrained by parameters that are themselves sampled from their own distributions at a higher level. In doing so, we removed some intrinsic assumptions about the model (in this case, number and sizes of groups), thereby postponing any decision on them until we observe the data. This will be a general strategy we will use for the remaining model parameters.

Having dealt with $P(b)$, this leaves us with the prior for the group to group connections, λ. A good starting point is an uninformative prior conditioned on a global average, $\bar{\lambda}$, which will determine the expected density of the network. For a continuous variable x, the maximum entropy distribution with a constrained average $\bar{x}$ is the exponential, $P(x) = e^{-x/\bar{x}}/\bar{x}$. Therefore, for λ we have

$$P(\lambda|b) = \prod_{r \leq s} e^{-n_r n_s \lambda_{rs}/(1+\delta_{rs})\bar{\lambda}} n_r n_s/(1 + \delta_{rs})\bar{\lambda}, \tag{11.19}$$

with $\bar{\lambda} = 2E/B(B+1)$ determining the expected total number of edges,[6] where we have assumed the local average $\langle \lambda_{rs} \rangle = \bar{\lambda}(1+\delta_{rs})/n_r n_s$, such that that the expected number of edges $e_{rs} = \lambda_{rs} n_r n_s/(1+\delta_{rs})$ will be equal to $\bar{\lambda}$, irrespective of the group sizes n_r and n_s [85]. Combining this with Equation (11.8), we can compute the integrated marginal likelihood of Equation (11.10) as

$$P(A|b) = \frac{\bar{\lambda}^E}{(\bar{\lambda}+1)^{E+B(B+1)/2}} \times \frac{\prod_{r<s} e_{rs}! \prod_r e_{rr}!!}{\prod_r n_r^{e_r} \prod_{i<j} A_{ij}! \prod_i A_{ii}!!}. \tag{11.20}$$

Just as with the node partition, the uninformative assumption of Equation (11.19) also leads to its own problems, but we postpone dealing with them to Section 11.6. For now, we have everything we need to write the posterior distribution, with the exception of the model evidence $P(A)$ given by Equation (11.11). Unfortunately, since it involves a sum over all possible partitions, it is not tractable to compute the evidence exactly. However, since it is just a normalization constant, we

---

[6] More strictly, we should treat $\bar{\lambda}$ just as another hyperparameter and integrate over its own distribution. But since this is just a global parameter, not affected by the dimension of the model, we can get away with setting its value directly from the data. It means we are pretending we know precisely the density of the network we are observing, which is not a very strong assumption. Nevertheless, readers who are uneasy with this procedure can rest assured that this can be completely amended once we move to microcanonical models in Section 11.6 (see footnote 15).

will not need to determine it when optimizing or sampling from the posterior, as we will see in Section 11.8. The numerator of Equation (11.9), which comprises of the terms that we can compute exactly, already contains all the information we need to proceed with the inference, and also has a special interpretation, as we will see in the next section.

The posterior of Equation (11.9) will put low probabilities on partitions that are not backed by sufficient statistical evidence in the network structure, and it will not lead us to spurious partitions such as those depicted in Figure 11.1. Inferring partitions from this posterior amounts to a so-called *nonparametric* approach; not because it lacks the estimation of parameters, but because the number of parameters itself, a.k.a. the *order* or *dimension* of the model, will be inferred as well. More specifically, the number of groups B itself will be an outcome of the inference procedure, which will be chosen in order to accommodate the structure in the data, without overfitting. The precise reasons why the latter is guaranteed might not be immediately obvious to those unfamiliar with Bayesian inference. In the following section we will provide an explanation by making a straightforward connection with information theory. The connection is based on a different interpretation of our model, which allows us to introduce some important improvements.

11.5 Microcanonical Models and the Minimum Description Length Principle

We can re-interpret the integrated marginal likelihood of Equation (11.20) as the joint likelihood of a *microcanonical* model given by[7]

$$P(A|b) = P(A|e,b)P(e|b),\tag{11.21}$$

where

$$P(A|e,b) = \frac{\prod_{r<s} e_{rs}! \prod_r e_{rr}!!}{\prod_r n_r^{e_r} \prod_{i<j} A_{ij}! \prod_i A_{ii}!!},\tag{11.22}$$

$$P(e|b) = \prod_{r<s} \frac{\bar\lambda^{e_{rs}}}{(\bar\lambda+1)^{e_{rs}+1}} \prod_r \frac{\bar\lambda^{e_{rs}/2}}{(\bar\lambda+1)^{e_{rs}/2+1}} = \frac{\bar\lambda^E}{(\bar\lambda+1)^{E+B(B+1)/2}},\tag{11.23}$$

and $e = \{e_{rs}\}$ is the matrix of edge counts between groups. The term "microcanonical" – borrowed from statistical physics – means that model parameters correspond to "hard" constraints that are *strictly* imposed on the ensemble, as opposed to "soft" constraints that are obeyed only on average. In the particular case above, $P(A|e,b)$ is the probability of generating a multigraph A where Equation (11.1) is always fulfilled, i.e. the total number of edges between groups r and s is always exactly e_{rs}, without any fluctuation allowed between samples (see [85] for a combinatorial derivation). This contrasts with the parameter λ_{rs} in Equation (11.8),

---

[7] Some readers may wonder why Equation (11.21) should not contain a sum, i.e. $P(A|b) = \sum_e P(A|e,b)P(e|b)$. Indeed, that is the proper way to write a marginal likelihood. However, for the microcanonical model there is only one element of the sum that fulfills the constraint of Equation (11.1), and thus yields a nonzero probability, making the marginal likelihood identical to the joint, as expressed in Equation (11.21). The same is true for the partition prior of Equation (11.17). We will use this fact in our notation throughout, and omit sums when they are unnecessary.

which determines only the *average* number of edges between groups, which fluctuates between samples. Conversely, the prior for the edge counts $P(e|b)$ is a mixture of geometric distributions with average $\bar{\lambda}$, which does allow the edge counts to fluctuate, guaranteeing the overall equivalence. The fact that Equation (11.21) holds is rather remarkable, since it means that – at least for the basic priors we used – these two kinds of model ("canonical" and microcanonical) cannot be distinguished from data, since their marginal likelihoods (and hence the posterior probability) are identical.[8]

With this microcanonical interpretation in mind, we may frame the posterior probability in an information-theoretical manner as follows. If a discrete variable x occurs with a probability mass $P(x)$, the asymptotic amount of information necessary to describe it is $-\log_2 P(x)$ (if we choose bits as the unit of measurement), by using an optimal lossless coding scheme such as Huffman's algorithm [57]. With this in mind, we may write the numerator of the posterior distribution in Equation (11.9) as

$$P(A|b)P(b) = P(A|e,b)P(e,b) = 2^{-\Sigma}, \qquad (11.24)$$

where the quantity

$$\Sigma = -\log_2 P(A,e,b) \qquad (11.25)$$

$$= -\log_2 P(A|e,b) - \log_2 P(e,b) \qquad (11.26)$$

is called the *description length* of the data [35, 91]. It corresponds to the asymptotic amount of information necessary to encode the data A *together* with the model parameters e and b. Therefore, if we find a network partition that maximizes the posterior distribution of Equation (11.20), we are also automatically finding one which minimizes the description length.[9] With this, we can see how the Bayesian approach outlined above prevents overfitting: As the size of the model increases (via a larger number of occupied groups), it will constrain itself better to the data, and the amount of information necessary to describe it when the model is known, $-\log_2 P(A|e,b)$, will decrease. At the same time, the amount of information necessary to describe the model itself, $-\log_2 P(e,b)$, will increase as it becomes more complex. Therefore, the latter will function as a *penalty*[10] that prevents the model from becoming overly complex, and the optimal choice will amount to a proper balance between both terms.[11] Among other things, this approach will allow

---

[8] This equivalence occurs for a variety of Bayesian models. For instance, if we flip a coin with a probability p of coming up heads, the integrated likelihood under a uniform prior after N trials in which m heads were observed is $P(x) = \int_0^1 p^m (1-p)^{N-m} \, dp = (N-m)!m!/(N+1)!$. This is the same as the "microcanonical" model $P(x) = P(x|m)P(m)$ with $P(x|m) = \binom{N}{m}^{-1}$ and $P(m) = 1/(N+1)$, i.e. the number of heads is sampled from a uniform distribution, and the coin flips are sampled randomly among those that have that exact number of heads.

[9] Sometimes the minimum description length principle (MDL) is considered as an alternative method to Bayesian inference. Although it is possible to apply the MDL in a manner that makes the connection with Bayesian inference difficult, as, for example, with the normalized maximum likelihood scheme [34, 97], in its more direct and tractable form it is fully equivalent to the Bayesian approach [35]. Note also that we do not in fact require the connection with microcanonical models made here, as the description length can be defined directly as $\Sigma = -\log_2 P(A,b)$, without referring explicitly to internal model parameters.

[10] Some readers may notice the similarity between Equation (11.26) and other penalty-based criteria, such as the Bayesian information criterion (BIC) [96] and the Akaike information criterion (AIC) [6]. Although all these criteria share the same overall interpretation, BIC and AIC rely on specific assumptions about the asymptotic shape of the model likelihood, which are known to be invalid for the SBM [110], unlike Equation (11.26) which is exact.

[11] An important result in information theory states that compressing random data is asymptotically impossible [14]. This lies at the heart of the effectiveness of the MDL approach in preventing overfitting, as incorporating randomness into the model description cannot be used to reduce the description length.

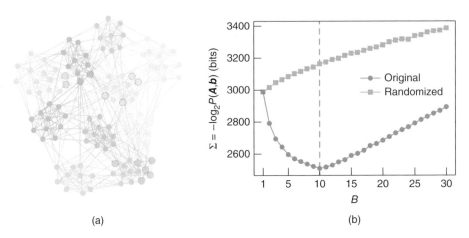

(a)   (b)

Figure 11.3 Bayesian inference of the SBM for a network of American college football teams [30]. (a) The partition that maximizes the posterior probability of Equation (11.9), or, equivalently, minimizes the description length of Equation (11.24). Nodes marked in red are not classified according to the known division into "conferences." (b) Description length as a function of the number of groups of the corresponding optimal partition, for both the original and randomized data.

us to properly estimate the dimension of the model – represented by the number of groups B – in a parsimonious way.

We now illustrate this approach with a real-world dataset of American college football teams [30], where a node is a team and an edge exists if two teams play against each other in a season. If we find the partition that maximizes the posterior distribution, we uncover $B = 10$ groups, as can be seen in Figure 11.3a. If we compare this partition with the known division of the teams into "conferences" [23, 24], we find that they match with a high degree of precision, with the exception of only a few nodes.[12] In Figure 11.3b we show the description length of the optimal partitions if we constrain them to have a pre-specified number of groups, which allows us to see how the approach penalizes both too simple and too complex models, with a global minimum at $B = 10$, corresponding to the most compressive partition. Importantly, if we now *randomize* the network, by placing all its edges in a completely random fashion, we obtain instead a trivial partition into $B = 1$ group, indicating that the best model for this data is indeed a fully random graph. Hence, we see that this approach completely avoids the pitfall discussed in Section 11.2 and does not identify groups in fully random networks, and that the division shown in Figure 11.3a points to a statistically significant structure in the data that cannot be explained simply by random fluctuations.

11.6   The "Resolution Limit" Underfitting Problem and the Nested SBM

Although the Bayesian approach outlined above is in general protected against overfitting, it is still susceptible to *underfitting*, i.e. when we mistake statistically significant structure for

---

[12] Care should be taken when comparing with "known" divisions in this manner, as there is no guarantee that the available metadata is in fact relevant for the network structure. See [47, 69, 77] for more detailed discussions.

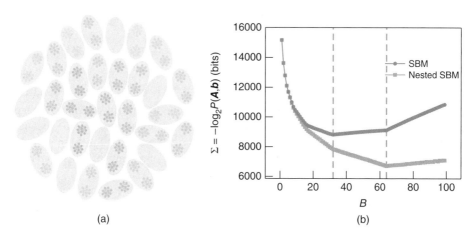

(a)   (b)

Figure 11.4 Inference of the SBM on a simple artificial network composed of 64 cliques of size 10, illustrating the underfitting problem. (a) The partition that maximizes the posterior probability of Equation (11.9) or, equivalently, minimizes the description length of Equation (11.24). The 64 cliques are grouped into 32 groups composed of two cliques each. (b) Minimum description length as a function of the number of groups of the corresponding partition, both for the SBM and its nested variant, which is less susceptible to underfitting, and puts all 64 cliques in their own groups.

randomness, resulting in the inference of an overly simplistic model. This happens whenever there is a large discrepancy between our prior assumptions and what is observed in the data. We illustrate this problem with a simple example. Consider a network formed of 64 isolated cliques of size 10, as shown in Figure 11.4a. If we employ the approach described in the previous section, and maximize the posterior of Equation (11.9), we obtain a partition into $B = 32$ groups, where each group is composed of two cliques. This is a fairly unsatisfying characterization of this network, and also somewhat perplexing, since the probability that the inferred SBM will generate the observed network, i.e. each of the 32 groups will simultaneously and spontaneously split in two disjoint cliques, is vanishingly small. Indeed, intuitively it seems we should do significantly better with this rather obvious example, and that the best fit would be to put each of the cliques in their own group. In order to see what went wrong, we need to revisit our prior assumptions, in particular our choice for $P(\lambda|b)$ in Equation (11.19) or, equivalently, our choice of $P(e|b)$ in Equation (11.23) for the microcanonical formulation. In both cases, they correspond to uninformative priors, which put approximately equal weight on all allowed types of large-scale structures. As argued before, this seems reasonable at first, since we should not bias our model before we observe the data. However, the implication of this choice is that we expect *a priori* the structure of the network at the aggregate group level, i.e. considering only the groups and the edges between them (not the individual nodes) to be fully random. This is indeed not the case in the simple example of Figure 11.4, and in fact it is unlikely to be the case for most networks that we encounter, which will probably be structured at a higher level as well. The unfavorable outcome of the uninformative assumption can also be seen by inspecting its effect on the description length of Equation (11.24). If we revisit our simple model with C cliques of size m, grouped uniformly into B groups of size C/B, and we assume that these values are sufficiently large so that Stirling's factorial approximation $\ln x! \approx x \ln x - x$ can be used, the description length becomes

$$\Sigma \approx -(E - N)\log_2 B + \frac{B(B + 1)}{2}\log_2 E, \qquad (11.27)$$

where $N = Cm$ is the total number of nodes and $E = C\binom{m}{2}$ is the total number of edges, and we have omitted terms that do not depend on B. From this, we see that if we increase the number of groups B, this incurs a quadratic penalty in the description length given by the second term of Equation (11.27), which originates precisely from our expression of $P(e|b)$: it corresponds to the amount of information necessary to describe all entries of a symmetric $B \times B$ matrix that takes independent values between 0 and E. Indeed, a slightly more careful analysis of the scaling of the description length [79, 85] reveals that this approach is unable to uncover a number of groups that is larger than $B_{max} \propto \sqrt{N}$, even if their existence is obvious, as in our example of Figure 11.4.[13]

Trying to avoid this limitation might seem like a conundrum, since replacing the uninformative prior for $P(e|b)$ amounts to making a more definite statement on the most likely large-scale structures that we expect to find, which we might hesitate to stipulate, as this is precisely what we want to discover from the data in the first place, and we want to remain unbiased. Luckily, there is in fact a general approach available to us to deal with this problem: we postpone our decision about the higher-order aspects of the model until we observe the data. In fact, we already saw this approach in action when we decided on the prior for the partitions; we do so by replacing the uninformative prior with a *parametric* distribution, whose parameters are in turn modelled by a another distribution, i.e. a *hyperprior*. The parameters of the prior then become *latent variables* that are learned from data, allowing us to uncover further structures, while remaining unbiased.

The microcanonical formulation allows us to proceed in this direction in a straightforward manner, as we can interpret the matrix of edge counts e as the adjacency matrix of a multigraph where each of the groups is represented as a single node. Within this interpretation, an elegant solution presents itself, where we describe the matrix e with *another* SBM, i.e. we partition each of the groups into meta-groups and the edges between groups are placed according to the edge counts between meta-groups. For this second SBM, we can proceed in the same manner and model it by a third SBM, and so on, forming a nested hierarchy, as illustrated in Figure 11.5 [82]. More precisely, if we denote by B_l, b_l and e_l the number of groups, the partition and the matrix of edge counts at level $l \in \{0, \dots, L\}$, we have

$$P(e_l|b_{l-1}, e_{l+1}, b_l) = \prod_{r<s}\left(\binom{n_r^l n_s^l}{e_{rs}^{l+1}}\right)^{-1}\prod_{r}\left(\binom{n_r^l(n_r^l + 1)/2}{e_{rs}^{l+1}/2}\right)^{-1}, \qquad (11.28)$$

with $\left(\binom{n}{m}\right) = \binom{n+m-1}{m}$ counting the number of m-combinations with repetitions from a set of size n. Equation (11.28) is the likelihood of a maximum-entropy multigraph SBM, i.e. every

---

[13] This same problem occurs for slight variations of the SBM and corresponding priors, provided they are uninformative, such as those in [16, 71, 95], and also with other penalty-based approaches that rely on a functional form similar to Equation (11.27) [106]. Furthermore, this limitation is conspicuously similar to the "resolution limit" present in the popular heuristic of modularity maximization [26], although it is not yet clear if a deeper connection exists between both phenomena.

multigraph occurs with the same probability, provided they fulfill the imposed constraints[14] [78]. The prior for the partitions is again given by Equation (11.17),

$$P(\boldsymbol{b}_l) = \frac{\prod_r n_r^l!}{B_{l-1}!} \left(\begin{array}{c} B_{l-1} - 1 \\ B_l - 1 \end{array} \right)^{-1} B_{l-1}^{-1},$$ (11.29)

with $B_{-1} = N$, so that the joint probability of the data, edge counts, and the hierarchical partition $\{\boldsymbol{b}_l\}$ becomes

$$P(A, \{\boldsymbol{e}_l\}, \{\boldsymbol{b}_l\}|L) = P(A|\boldsymbol{e}_1, \boldsymbol{b}_0) P(\boldsymbol{b}_0) \prod_{l=1}^{L} P(\boldsymbol{e}_l|\boldsymbol{b}_{l-1}, \boldsymbol{e}_{l+1}, \boldsymbol{b}_l) P(\boldsymbol{b}_l),$$ (11.30)

where we impose the boundary conditions $B_L = 1$ and $P(\boldsymbol{b}_L) = 1$. We can treat the hierarchy depth L as a latent variable as well, by placing a prior on it $P(L) = 1/L_{\max}$, where $L_{\max}$ is the maximum value allowed. But since this only contributes to an overall multiplicative constant, it has no effect on the posterior distribution, and thus can be omitted. If we impose $L = 1$, we recover the uninformative prior for $\boldsymbol{e} = \boldsymbol{e}_1$,

$$P(\boldsymbol{e}|\boldsymbol{b}_0) = \left(\left(\begin{array}{c} B(B+1)/2 \\ E \end{array} \right) \right)^{-1},$$ (11.31)

which is different from Equation (11.23) only in that the number of edges E is not allowed to fluctuate.[15] The inference of this model is done in the same manner as the uninformative one, by obtaining the posterior distribution of the hierarchical partition

$$P(\{\boldsymbol{b}_l\}|A) = \frac{P(A, \{\boldsymbol{b}_l\})}{P(A)} = \frac{P(A, \{\boldsymbol{e}_l\}, \{\boldsymbol{b}_l\})}{P(A)},$$ (11.32)

and the description length is given analogously by

$$\Sigma = -\log_2 P(A|\{\boldsymbol{e}_l\}, \{\boldsymbol{b}_l\}) - \log_2 P(\{\boldsymbol{e}_l\}, \{\boldsymbol{b}_l\}).$$ (11.33)

This approach has a series of advantages; in particular, we remain *a priori* agnostic with respect to what kind of large-scale structure is present in the network, having constrained ourselves simply in that it can be represented as a SBM at a higher level, and with the uninformative prior as a special case. Despite this, we are able to overcome the underfitting problem encountered with the uninformative approach: if we apply this model to the example of Figure 11.4, we can successfully distinguish all 64 cliques, and provide a lower overall description length for the data, as can be seen in Figure 11.4b. More generally, by investigating the properties of the model

---

[14] Note that we cannot use in the upper levels exactly the same model we use in the bottom level, given by Equation (11.22), as most terms in the subsequent levels will cancel out. This happens because the model in Equation (11.22) is based on the uniform generation of configurations, not multigraphs [85]. However, we are free to use Equation (11.28) in the bottom level as well.

[15] The prior of Equation (11.31) and the hierarchy in Equation (11.30) are conditioned on the total number of edges E, which is typically unknown before we observe the data. Similarly to the parameter $\bar{\lambda}$ in the canonical model formulation, the strictly correct approach would be to consider this quantity as an additional model parameter, with its prior distribution $P(E)$. However, in the microcanonical model there is no integration involved, and $P(E)$ – regardless of how we specify it – would contribute to an overall multiplicative constant that disappears from the posterior distribution after normalization. Therefore we can simply omit it.

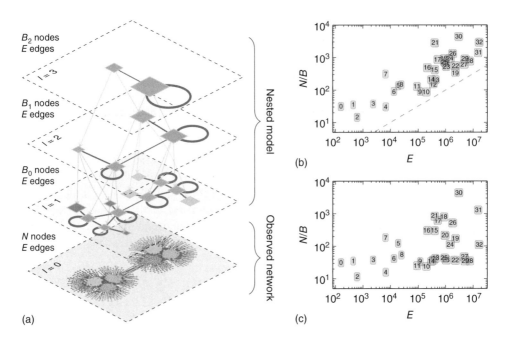

Figure 11.5 (a) Diagrammatic representation of the nested SBM described in the text, with $L = 3$ levels, adapted from [82]. (b) Average group sizes N/B obtained with the SBM using uninformative priors, for a variety of empirical networks, listed in [82]. The dashed line shows a slope $\sqrt{E}$, highlighting the systematic underfitting problem. (c) The same as in (b), but using the nested SBM, where the underfitting has virtually disappeared, with datasets randomly scattered in the allowed range.

likelihood, it is possible to show that the maximum number of groups that can be uncovered with this model scales as $B_{max} \propto N/\log N$, which is significantly larger than the limit with uninformative priors [82, 85]. The difference between both approaches manifests itself very often in practice, as shown in Figure 11.5b, where systematic underfitting is observed for a wide variety of network datasets, which disappears with the nested model, as seen in Figure 11.5c. Crucially, we achieve this decreased tendency to underfit without sacrificing our protection against overfitting: Despite the more elaborate model specification, the inference of the nested SBM is completely nonparametric, and the same Bayesian and information-theoretical principles still hold. Furthermore, as we have already mentioned, the uninformative case is a special case of the nested SBM, i.e. when $L = 1$, and hence it can only improve the inference (e.g. by reducing the description length), with no drawbacks. We stress that the number of hierarchy levels, as with any other dimension of the model, such as the number of groups in each level, is inferred from data and does not need to be determined *a priori*.

In addition to the above, the nested model also gives us the capacity of describing the data at multiple scales, which could potentially exhibit different mixing patterns. This is particularly useful for large networks, where the SBM might still give us a very complex description, which becomes easier to interpret if we concentrate first on the upper levels of the hierarchy. A good example is the result obtained for the internet topology at the autonomous systems level, shown in Figure 11.6. The lowest level of the hierarchy shows a division into a large number of groups,

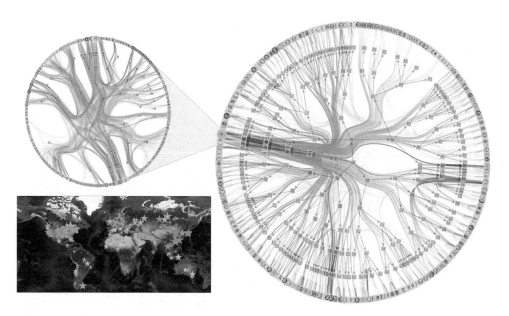

Figure 11.6 Fit of the (degree-corrected) nested SBM for the internet topology at the autonomous systems level, adapted from [82]. The hierarchical division reveals a core-periphery organization at the higher levels, where most routes go through a relatively small number of nodes (shown in the inset and in the map). The lower levels reveal a more detailed picture, where a large number of groups of nodes are identified according to their routing patterns (amounting largely to distinct geographical regions). The layout is obtained with an edge bundling algorithm by Holten [45], which uses the hierarchical partition to route the edges.

with a fairly complicated structure, whereas the higher levels show an increasingly simplified picture, culminating in a core-periphery organization as the dominating pattern.

11.7   Model Variations

Varying the number of groups and building hierarchies is not the only way we have of adapting the complexity of the model to the data. We may also change the internal structure of the model, and how the division into groups affects the placement of edges. In fact, the basic ansatz of the SBM is very versatile, and many variations have been proposed in the literature. In this section we review two important ones – SBMs with degree correction and group overlap – and review other model flavors in a summarized manner.

Before we go further into the model variations, we point out that the multiplicity of models is a strength of the inference approach. This is different from the broader field of network clustering, where a large number of available algorithms often yield conflicting results for the same data, leaving practitioners lost in how to select between them [32, 46]. Instead, within the inference framework we can in fact compare different models in a principled manner and select the best one according to the statistical evidence available. We proceed with a general outline of the model selection procedure before following with specific model variations.

11.7.1 Model Selection

Suppose we define two versions of the SBM, labeled $\mathscr{C}_1$ and $\mathscr{C}_2$, each with their own posterior distribution of partitions, $P(\boldsymbol{b}|A, \mathscr{C}_1)$ and $P(\boldsymbol{b}|A, \mathscr{C}_2)$. Suppose we find the most likely partitions $\boldsymbol{b}_1$ and $\boldsymbol{b}_2$, according to $\mathscr{C}_1$ and $\mathscr{C}_2$, respectively. How do we decide which partition is more representative of the data? The consistent approach is to obtain the so-called posterior odds ratio [48, 49]

$$\Lambda = \frac{P(\boldsymbol{b}_1, \mathscr{C}_1|A)}{P(\boldsymbol{b}_2, \mathscr{C}_2|A)} = \frac{P(A|\boldsymbol{b}_1, \mathscr{C}_1)P(\boldsymbol{b}_1)P(\mathscr{C}_1)}{P(A|\boldsymbol{b}_2, \mathscr{C}_2)P(\boldsymbol{b}_2)P(\mathscr{C}_2)}, \tag{11.34}$$

where $P(\mathscr{C})$ is our prior belief that variant $\mathscr{C}$ is valid. A value of $\Lambda > 1$ indicates that the choice $(\boldsymbol{b}_1, \mathscr{C}_1)$ is Λ times more plausible as an explanation for the data than the alternative, $(\boldsymbol{b}_2, \mathscr{C}_2)$. If we are *a priori* agnostic with respect to which model flavor is best, i.e. $P(\mathscr{C}_1) = P(\mathscr{C}_2)$, we have then

$$\Lambda = \frac{P(A|\boldsymbol{b}_1, \mathscr{C}_1)P(\boldsymbol{b}_1)}{P(A|\boldsymbol{b}_2, \mathscr{C}_2)P(\boldsymbol{b}_2)} = 2^{-\Delta\Sigma}, \tag{11.35}$$

where $\Delta\Sigma = \Sigma_1 - \Sigma_2$ is the description length difference between both choices. Hence, we should generally prefer the model choice that is most compressive, i.e. with the smallest description length. However, if the value of Λ is close to 1, we should refrain from forcefully rejecting the alternative, as the evidence in data would not be strongly decisive either way. In other words the actual value of Λ gives us the confidence with which we can choose the preferred model. The final decision, however, is subjective, since it depends on what we might consider plausible. A value of $\Lambda = 2$, for example, typically cannot be used to forcefully reject the alternative hypothesis, whereas a value of $\Lambda = 10^{100}$ might.

An alternative test we can make is to decide which model *class* is most representative of the data, when averaged over all possible partitions. In this case, we proceed in an analogous way by computing the posterior odds ratio

$$\Lambda' = \frac{P(\mathscr{C}_1|A)}{P(\mathscr{C}_2|A)} = \frac{P(A|\mathscr{C}_1)P(\mathscr{C}_1)}{P(A|\mathscr{C}_2)P(\mathscr{C}_2)}, \tag{11.36}$$

where

$$P(A|\mathscr{C}) = \sum_{\boldsymbol{b}} P(A|\boldsymbol{b}, \mathscr{C})P(\boldsymbol{b}) \tag{11.37}$$

is the model evidence. When $P(\mathscr{C}_1) = P(\mathscr{C}_2)$, Λ' is called the *Bayes factor*, with an interpretation analogous to Λ above, but where the statement is made with respect to all possible partitions, not only the most likely one. Unfortunately, as mentioned previously, the evidence $P(A|\mathscr{C})$ cannot be computed exactly for the models we are interested in, making this criterion more difficult to employ in practice (although approximations have been proposed, see e.g. [85]). We return to the issue of when it should we optimize or sample from the posterior distribution in Section 11.9, and hence which of the two criteria should be used.

11.7.2 Degree Correction

The underlying assumption of all variants of the SBM considered so far is that nodes that belong to the same group are statistically equivalent. As it turns out, this fundamental aspect results in a very unrealistic property. Namely, this generative process implies that all nodes that belong to the same group receive on average the same number of edges. However, a common property of many empirical networks is that they have very heterogeneous degrees, often broadly distributed over several orders of magnitudes [66]. Therefore, in order for this property to be reproduced by the SBM, it is necessary to group nodes according to their degree, which may lead to some seemingly odd results. An example of this was given in [50] and is shown in Figure 11.7a. It corresponds to a fit of the SBM to a network of political blogs recorded during the 2004 American presidential election campaign [2], where an edge exists between two blogs if one links to the other. If we guide ourselves by the layout of the figure, we identify two assortative groups, which happen to be those aligned with the Republican and Democratic parties. However, inside each group there is a significant variation in degree, with a few nodes with many connections and many with very few. Because of what just has been explained, if we perform a fit of the SBM using only $B = 2$ groups, it prefers to cluster the nodes into high-degree and low-degree groups, completely ignoring the party alliance.[16] Arguably, this is a bad fit of this network, since – similarly to the underfitting example of Figure 11.4 – the probability of the fitted SBM generating a network with such a party structure is vanishingly small. In order to solve this undesired behavior, Karrer and Newman [50] proposed a modified model, which they dubbed the degree-corrected SBM (DC-SBM). In this variation, each node i is attributed with a parameter θ_i that controls its expected degree, independently of its group membership. Given this extra set of parameters, a network is generated with probability

$$P(A|\lambda,\theta,b) = \prod_{i<j} \frac{e^{-\theta_i\theta_j\lambda_{b_i,b_j}}(\theta_i\theta_j\lambda_{b_i,b_j})^{A_{ij}}}{A_{ij}!} \times \prod_i \frac{e^{-\theta_i^2\lambda_{b_i,b_i}/2}(\theta_i^2\lambda_{b_i,b_i}/2)^{A_{ii}/2}}{(A_{ii}/2)!}, \qquad (11.38)$$

where λ_{rs} again controls the expected number of edges between groups r and s. Note that since the parameters λ_{rs} and θ_i always appear multiplying each other in the likelihood, their individual values may be arbitrarily scaled, provided their products remain the same. If we choose the parametrization $\sum_i \theta_i \delta_{b_i,r} = 1$ for every group r, then they acquire a simple interpretation: λ_{rs} is the expected number of edges between groups r ans s, $\lambda_{rs} = \langle e_{rs} \rangle$, and θ_i is proportional to the expected degree of node i, $\theta_i = \langle k_i \rangle / \sum_s \lambda_{b_i,s}$.

When inferring this model from the political blogs data – again forcing $B = 2$ – we obtain a much more satisfying result, where the two political factions are neatly identified, as seen in Figure 11.7b. As this model is capable of fully decoupling the community structure from the degrees, which are captured separately by the parameters λ and θ, respectively, the degree heterogeneity of the network does not interfere with the identification of the political factions.

Based on the above example, and on the knowledge that most networks possess heterogeneous degrees, we could expect the DC-SBM to provide a better fit for most of them. However, before we jump to this conclusion, we must first acknowledge that the seemingly increased quality of fit obtained with the SBM came at the expense of adding an extra set of parameters, θ [110]. However intuitive we might judge the improvement brought on by degree correction, simply adding more parameters to a model is an almost sure recipe for overfitting. Therefore, a more

---

[16] It is possible that unexpected results of this kind inhibited the initial adoption of SBM methods in the network science community, which focused instead on more heuristic community detection methods, save for a few exceptions (e.g. [11, 36, 37, 41, 43, 93]).

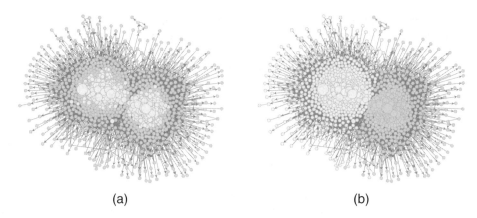

(a)   (b)

Figure 11.7   Inferred partition for a network of political blogs [2] using (a) the SBM and (b) the DC-SBM, in both cases forcing $B = 2$ groups. The node sizes are proportional to the node degrees. The SBM divides the network into low and high-degree groups, whereas the DC-SBM prefers the division into political factions.

prudent approach is once more to frame the inference problem in a Bayesian way, by focusing on the posterior distribution $P(b|A)$, and on the description length. For this, we must include a prior for the node propensities θ. The uninformative choice is the one which ascribes the same probability to all possible choices,

$$P(\theta|b) = \prod_r (n_r - 1)!\delta(\sum_i \theta_i \delta_{b_i,r} - 1). \tag{11.39}$$

Using again an uninformative prior for λ,

$$P(\lambda|b) = \prod_{r \leq s} e^{-\lambda_{rs}/(1+\delta_{rs})\overline{\lambda}}/(1 + \delta_{rs})\overline{\lambda} \tag{11.40}$$

with $\overline{\lambda} = 2E/B(B + 1)$, the marginal likelihood now becomes

$$P(A|b) = \int P(A|\lambda, \theta, b)P(\lambda|b)P(\theta|b) \, d\lambda d\theta$$

$$= \frac{\overline{\lambda}^E}{(\overline{\lambda} + 1)^{E+B(B+1)/2}} \times \frac{\prod_{r<s} e_{rs}! \prod_r e_{rr}!!}{\prod_{i<j} A_{ij}! \prod_i A_{ii}!} \times \prod_r \frac{(n_r - 1)!}{(e_r + n_r - 1)!} \times \prod_i k_i!, \tag{11.41}$$

where $k_i = \sum_j A_{ij}$ is the degree of node i, which can be used in the same way to obtain a posterior for b, via Equation (11.9). Once more, the model above is equivalent to a microcanonical formulation [85], given by

$$P(A|b) = P(A|k, e, b)P(k|e, b)P(e|b), \tag{11.42}$$

with

$$P(A|k, e, b) = \frac{\prod_{r<s} e_{rs}! \prod_r e_{rr}!! \prod_i k_i!}{\prod_{i<j} A_{ij}! \prod_i A_{ii}!! \prod_r e_r!!}, \tag{11.43}$$

$$P(k|e, b) = \prod_r \left(\binom{n_r}{e_r} \right)^{-1}, \tag{11.44}$$

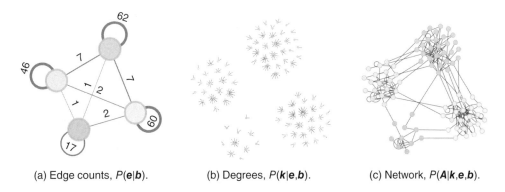

(a) Edge counts, $P(e|b)$.   (b) Degrees, $P(k|e,b)$.   (c) Network, $P(A|k,e,b)$.

Figure 11.8 Illustration of the generative process of the microcanonical DC-SBM. Given a partition of the nodes, the edge counts between groups are sampled (a), followed by the degrees of the nodes (b) and finally the network itself (c). Adapted from [85].

and $P(e|b)$ given by Equation (11.23). In the model above, $P(A|k,e,b)$ is the probability of generating a multigraph where the edge counts between groups *as well as* the degrees k are fixed to specific values (see Figure 11.8).[17] The prior $P(k|e,b)$ is the uniform probability of generating a degree sequence, where all possibilities that satisfy the constraints imposed by the edge counts e, namely $\sum_i k_i \delta_{b_i,r} = e_r$, occur with the same probability. The description length of this model is then given by

$$\Sigma = -\log_2 P(A,b) = -\log_2 P(A|k,e,b) - \log_2 P(k,e,b). \qquad (11.45)$$

Because uninformative priors were used to derive the above equations, we are once more subject to the same underfitting problem described previously. Luckily, from the microcanonical model we can again derive a nested DC-SBM, by replacing $P(e)$ by a nested sequence of SBMs, exactly in the same was as was done before [82, 85]. We also have the opportunity of replacing the uninformative prior for the degrees in Equation (11.44) with a more realistic option. As was argued in [85], degree sequences generated by Equation (11.44) result in exponential degree distributions, which are not quite as heterogeneous as what is often encountered in practice. A more refined approach, which is already familiar to us at this point, is to increase the Bayesian hierarchy and choose a prior that is conditioned on a higher-order aspect of the data, in this case the *frequency* of degrees, i.e.

$$P(k|e,b) = P(k|e,b,\eta)P(\eta|e,b), \qquad (11.46)$$

where $\eta = \{\eta_k^r\}$, with η_k^r being the number of nodes of degree k in group r. In the above, $P(\eta|e,b)$ is a uniform distribution of frequencies and $P(k|e,b,\eta)$ generates the degrees according to the sampled frequencies (we omit the respective expressions for brevity, and refer to [85] instead). Thus, this model is capable of using regularities in the degree distribution to inform the division into groups and is generally capable of better fits than the uniform model of Equation (11.44).

---

[17] The ensemble equivalence of Equation (11.42) is in some ways more remarkable than for the traditional SBM. This is because a direct equivalence between the ensembles of Equations (11.38) and (11.43) is not satisfied even in the asymptotic limit of large networks [28, 78], which does happen for Equations (11.8) and (11.22). Equivalence is observed only if the individual degrees k_i also become asymptotically large. However, when the parameters λ and θ are integrated out, the equivalence becomes exact for networks of any size.

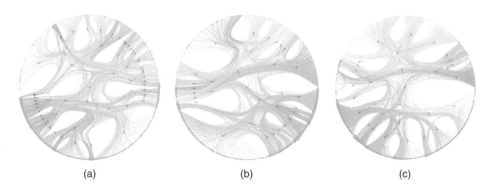

(a)   (b)   (c)

Figure 11.9 Most likely hierarchical partitions of a network of political blogs [2], according to the three model variants considered, as well as the inferred number of groups B_1 at the bottom of the hierarchy, and the description length Σ: (a) NDC-SBM, $B_1 = 42$, $\Sigma \approx 89{,}938$ bits, (b) DC-SBM, $B_1 = 23$, $\Sigma \approx 87{,}162$ bits, (c) DC-SBM with the degree prior of Equation (11.46), $B_1 = 20$, $\Sigma \approx 84{,}890$ bits. The nodes circled in blue were classified as "liberals" and the remaining ones as "conservatives" in [2] based on the blog contents. Adapted from [85].

If we apply this nonparametric approach to the same political blog network of Adamic and Glance [2], we find a much more detailed picture of its structure, revealing many more than two groups, as shown in Figure 11.9, for three model variants: the nested SBM, the nested DC-SBM, and the nested DC-SBM with the degree prior of Equation (11.46). All three model variants are in fact capable of identifying the same Republican/Democrat division at the topmost hierarchical level, showing that the non-degree-corrected SBM is not as inept in capturing this aspect of the data as the result obtained by forcing $B = 2$ might suggest. However, the internal divisions of both factions that they uncover are distinct from each other. If we inspect the obtained values of the description length with each model we see that the DC-SBM (in particular when using Equation (11.46)) results in a smaller value, indicating that it better captures the structure of the data, despite the increased number of parameters. Indeed, a systematic analysis carried out in [85] showed that the DC-SBM does in fact yield shorter description lengths for a majority of empirical datasets, thus ultimately confirming the original intuition behind the model formulation.

11.7.3   Group Overlaps

Another way we can change the internal structure of the model is to allow the groups to overlap, i.e. we allow a node to belong to more than one group at the same time. The connection patterns of the nodes are then assumed to be a mixture of the "pure" groups, which results in a richer type of model [5]. Following Ball *et al.* [7], we can adapt the Poisson formulation to overlapping SBMs in a straightforward manner,

$$P(A|\kappa, \lambda) = \prod_{i<j} \frac{e^{-\lambda_{ij}} \lambda_{ij}^{A_{ij}}}{A_{ij}!} \prod_{i} \frac{e^{-\lambda_{ii}/2}(\lambda_{ii}/2)^{A_{ii}/2}}{A_{ii}/2!}, \tag{11.47}$$

with

$$\lambda_{ij} = \sum_{rs} \kappa_{ir} \lambda_{rs} \kappa_{js},$$ (11.48)

where κ_{ir} is the probability with which node i is chosen from group r, so that $\sum_i \kappa_{ir} = 1$, and λ_{rs} is once more the expected number of edges between groups r and s. The parameters κ replace the disjoint partition b we have been using so far by a "soft" clustering into overlapping categories.[18] Note, however, that this model is a direct generalization of the non-overlapping DC-SBM of Equation (11.38), which is recovered simply by choosing $\kappa_{ir} = \theta_i \delta_{r,b_i}$. The Bayesian formulation can also be performed by using an uninformative prior for κ,

$$P(\kappa) = \prod_r (n-1)! \delta(\sum_i \kappa_{ir} - 1),$$ (11.49)

in addition to the same prior for λ in Equation (11.40). Unfortunately, computing the marginal likelihood using Equation (11.47) directly,

$$P(A|\kappa) = \int P(A|\kappa, \lambda) P(\lambda) \, d\lambda,$$ (11.50)

is not tractable, which prevents us from obtaining the posterior $P(\kappa|A)$. Instead, it is more useful to consider the auxiliary labelled matrix, or tensor, $G = \{G_{ij}^{rs}\}$, where G_{ij}^{rs} is a particular decomposition of A_{ij} where the two edge endpoints – or "half-edges" – of an edge (i,j) are labelled with groups (r,s), such that

$$A_{ij} = \sum_{rs} G_{ij}^{rs}.$$ (11.51)

Since a sum of Poisson variables is also distributed according to a Poisson, we can write Equation (11.47) as

$$P(A|\kappa, \lambda) = \sum_G P(G|\kappa, \lambda) \prod_{i \leq j} \delta_{\sum_{rs} G_{ij}^{rs}, A_{ij}},$$ (11.52)

with each half-edge labelling being generated by

$$P(G|\kappa, \lambda) = \prod_{i<j} \prod_{rs} \frac{e^{-\kappa_{ir}\lambda_{rs}\kappa_{js}} (\kappa_{ir}\lambda_{rs}\kappa_{js})^{G_{ij}^{rs}}}{G_{ij}^{rs}!} \times \prod_i \prod_{rs} \frac{e^{-\kappa_{ir}\lambda_{rs}\kappa_{is}/2} (\kappa_{is}\lambda_{rs}\kappa_{is}/2)^{G_{ii}^{rs}/2}}{(G_{ii}^{rs}/2)!}.$$ (11.53)

We can now compute the marginal likelihood as

$$P(G) = \int P(G|\kappa, \lambda) P(\kappa) P(\lambda|\bar{\lambda}) \, d\kappa d\lambda,$$

$$= \frac{\bar{\lambda}^{-E}}{(\bar{\lambda}+1)^{E+B(B+1)/2}} \frac{\prod_{r<s} e_{rs}! \prod_r e_{rr}!!}{\prod_{rs} \prod_{i<j} G_{ij}^{rs}! \prod_i G_{ii}^{rs}!!} \times \prod_r \frac{(N-1)!}{(e_r + N - 1)!} \times \prod_{ir} k_i^r!,$$ (11.54)

---

[18] Note that, differently from the non-overlapping case, here it is possible for a node not to belong to any group, in which case it will never receive an incident edge.

which is very similar to Equation (11.41) for the DC-SBM. With the above, and knowing from Equation (11.51) that there is only one choice of A that is compatible with any given G, i.e.

$$P(A|G) = \prod_{i \leq j} \delta_{\sum_{rs} G_{ij}^{rs}, A_{ij}}, \tag{11.55}$$

we can sample from (or maximize) the posterior distribution of the half-edge labels G, just like we did for the node partition b in the nonoverlapping models,

$$P(G|A) = \frac{P(A|G)P(G)}{P(A)} \propto P(G) \times \prod_{i \leq j} \delta_{\sum_{rs} G_{ij}^{rs}, A_{ij}}, \tag{11.56}$$

where the product in the last term only accounts for choices of G which are compatible with A, i.e. fulfill Equation (11.51). Once more, the model of Equation (11.54) is equivalent to its microcanonical analogue [84],

$$P(G) = P(G|k, e)P(k|e)P(e), \tag{11.57}$$

where

$$P(G|k, e) = \frac{\prod_{r<s} e_{rs}! \prod_r e_{rr}!! \prod_{ir} k_i^r!}{\prod_{rs} \prod_{i<j} G_{ij}^{rs}! \prod_i G_{ii}^{rs}!! \prod_r e_r!}, \tag{11.58}$$

$$P(k|e) = \prod_r \left(\binom{e_r}{N} \right)^{-1} \tag{11.59}$$

and $P(e)$ given by Equation (11.23). The variables $k = \{k_i^r\}$ are the labelled degrees of the labelled network G, where k_i^r is the number of incident edges of type r a node i has. The description length becomes likewise

$$\Sigma = -\log_2 P(G) = -\log_2 P(G|k, e) - \log_2 P(k|e) - \log_2 P(e). \tag{11.60}$$

The nested variant can be once more obtained by replacing $P(e)$ in the same manner as before, and $P(k|e)$ in a manner that is conditioned on the labelled degree frequencies and degree of overlap, as described in detail in [84].

Equipped with this more general model, we may ask ourselves again if it provides a better fit of most networks, like we did for the DC-SBM in the previous section. Indeed, since the model is more general, we might conclude that this is a inevitability. However, this could be a fallacy, since more general models also include more parameters and hence are more likely to overfit. Indeed, previous claims about the existence of "pervasive overlap" in networks, based on nonstatistical methods [3], seemed to be based to some extent on this problematic logic. Claims about community overlaps are very different from, for example, the statement that networks possess heterogeneous degrees, since community overlap is not something that can be observed directly; instead it is something that must be *inferred*, which is precisely what our Bayesian approach is designed to do, in a methodologically correct manner. An example of such a comparison is shown in Figure 11.10, for a small network of political books. This network, when analyzed using the nonoverlapping SBM, seems to be composed of three groups, easily interpreted as "left wing," "right wing" and "center," as the available metadata corroborates. If

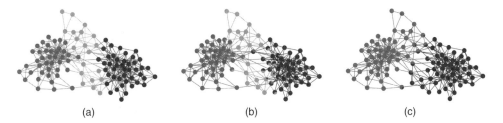

(a)   (b)   (c)

Figure 11.10 Network of co-purchases of books about US politics [54], with groups inferred using (a) the non-overlapping DC-SBM, with description length $\Sigma \approx 1938$ bits, (b) the overlapping SBM with description length $\Sigma \approx 1931$ bits, and (c) the overlapping SBM forcing only $B = 2$ groups, with description length $\Sigma \approx 1946$ bits.

we fit the overlapping SBM, we observe a mixed division into the same kinds of group. If we force the inference of only two groups, we see that some of the "center" nodes are split between "right wing" or "left wing." The latter might seem like a more pleasing interpretation, but looking at the description length reveals that it does not improve the description of the data. The best model in this case does seem to be the overlapping SBM with $B = 3$ groups. However, the difference in the description length between all model variants is not very large, making it difficult to fully reject any of the three variants. A more systematic analysis done in [84] revealed that for most empirical networks, in particular larger ones, the overlapping models do not provide the best fits in the majority of cases, and yield larger description lengths than the nonoverlapping variants. Hence it seems that the idea of overlapping groups is less pervasive than that of degree heterogeneity, at least according to our modeling ansatz.

It should be emphasized that we can always represent a network generated by an overlapping SBM by one generated with the nonoverlapping SBM with a larger number of groups representing the individual types of mixtures. Although model selection gives us the most parsimonious choice between the two, it does not remove the equivalence. In Figure 11.11 we show how networks generated by the overlapping SBM can be better represented by the nonoverlapping SBM (i.e. with a smaller description length) as long as the overlapping regions are sufficiently large.

11.7.4 Further Model Extensions

The simple and versatile nature of the SBM has spawned a large family of extensions and generalizations incorporating various types of more realistic features. This includes, for example, versions of the SBM that are designed for networks with continuous edge covariates (a.k.a. edge weights) [4, 86], multilayer networks that are composed of different types of edges [18, 74, 83, 101, 103], networks that evolve in time [13, 27, 29, 59, 76, 87, 108, 113], networks that possess node attributes [75] or are annotated with metadata [47, 69], networks with uncertain structure [58], as well as networks that do not possess a discrete modular structure at all, and are instead embedded in generalized continuous spaces [70]. These model variations are too numerous to be described here in any detail, but it suffices to say that the general Bayesian approach outlined here, including model selection, also applicable to these variations, without any conceptual difficulty.

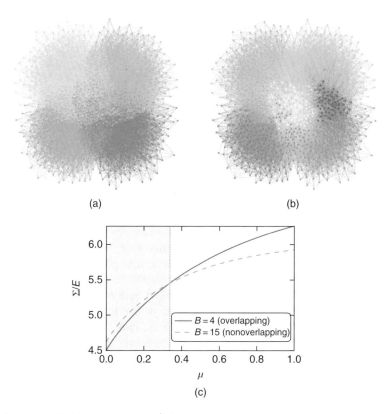

Figure 11.11 (a) Artificial network sampled from an assortative overlapping SBM with $B = 4$ groups and expected mixture sizes given by $n_{\vec{b}} \propto \mu^{|\vec{b}|}$, with $\mu \in [0, 1]$ controlling the degree of overlap (see [83] for details). (b) the same network as in (a), but generated according to an equivalent nonoverlapping SBM with $B = 15$ groups. (c) Description length per edge Σ/E for the same models in (a) and (b), as a function of the degree of overlap μ, showing a cross-over where the nonoverlapping model is preferred. Adapted from [83].

11.8   Efficient Inference Using Markov Chain Monte Carlo

Although we can write exact expressions for the posterior probability of Equation (11.9) (up to a normalization constant) for a variety of model variants, the resulting distributions are not simple enough to allow us to sample from them – much less find their maximum – in a direct manner. In fact, fully characterizing the posterior distribution or finding its maximum is, for most models like the SBM, typically a NP-hard problem. What we can do, however, is to employ Markov chain Monte Carlo (MCMC) [68], which can be done efficiently, and in an asymptotically exact manner, as we now show. The central idea is to sample from $P(\boldsymbol{b}|\boldsymbol{A})$ by first starting from some initial configuration $\boldsymbol{b}_0$ (in principle arbitrary) and making move proposals $\boldsymbol{b} \rightarrow \boldsymbol{b}'$ with a probability $P(\boldsymbol{b}'|\boldsymbol{b})$, such that, after a sufficiently long time, the equilibrium distribution is given exactly by $P(\boldsymbol{b}|\boldsymbol{A})$. In particular, given any arbitrary move proposals $P(\boldsymbol{b}'|\boldsymbol{b})$ – with the only condition that they fulfill ergodicity, i.e. that they allow every state to be visited eventually – we can guarantee that the desired posterior distribution is eventually reached by employing the Metropolis–Hastings criterion [42, 60], which dictates we should accept a given move proposal $b \rightarrow b'$ with a probability a given by

$$a = \min \left(1, \frac{P(b'|A)}{P(b|A)} \frac{P(b|b')}{P(b'|b)} \right),$$ (11.61)

otherwise the proposal is rejected. The ratio $P(b|b')/P(b'|b)$ in Equation (11.61) enforces a property known as *detailed balance* or *reversibility*, i.e.

$$T(b'|b)P(b|A) = T(b|b')P(b'|A),$$ (11.62)

where $T(b'|b)$ are the final transition probabilities after incorporating the acceptance criterion of Equation (11.61). The detailed balance condition of Equation (11.62) together with the ergodicity property guarantee that the Markov chain will converge to the desired equilibrium distribution $P(b|A)$. Importantly, we note that when computing the ratio $P(b'|A)/P(b|A)$ in Equation (11.61), we do not need to determine the intractable normalization constant of Equation (11.9), since it cancels out, and thus it can be performed exactly.

The above gives a generic protocol that we can use to sample from the posterior whenever we can compute the numerator of Equation (11.9). If instead we are interested in maximizing the posterior, we can introduce an "inverse temperature" parameter β, by changing $P(b|A) \rightarrow P(b|A)^{\beta}$ in the above equations, and making $\beta \rightarrow \infty$ in slow increments; what is known as *simulated annealing* [53]. The simplest implementation of this protocol for the inference of SBMs is to start from a random partition b_0, and use move proposals where a node i is randomly selected, and then its new group membership b'_i is chosen randomly between all $B + 1$ choices (where the remaining choice means we populate a new group),

$$P(b'_i|b) = \frac{1}{B + 1}.$$ (11.63)

By inspecting Equations (11.20), (11.41), (11.54) and (11.17) for all SBM variants considered, we notice that the ratio $P(b'|A)/P(b|A)$ can be computed in time $O(k_i)$, where k_i is the degree of node i, independently of other properties of the model such as the number of groups B. Note that this is not true for all alternative formulations of the SBM, e.g. for the models in [16, 33, 71, 90, 95] computing such an update requires $O(k_i + B)$ time [the heat-bath move proposals of [71] increases this even further to $O(B(k_i + B))$], thus making them very inefficient for large networks, where the number of groups can reach the order of thousands or more. Hence, when using these move proposals, a full sweep of all N nodes in the network can be done in time $O(E)$, independent of B.

Although fairly simple, the above algorithm suffers from some shortcomings that can seriously degrade its performance in practice. In fact, it is typical for naive implementations of the Metropolis–Hastings algorithm to perform very badly, despite its theoretical guarantees. This is because the asymptotic properties of the Markov chain may take a very long time to be realized, and the equilibrium distribution is never observed in practical time. Generally, we should expect good convergence times only when (i) the initial state b_0 is close enough to the most likely states of the posterior and (ii) the move proposals $P(b'|b)$ resemble the shape of the posterior. Indeed, it is a trivial (and not very useful) fact that if the starting state b_o is sampled directly from the posterior, and the move proposals match the posterior exactly, $P(b'|b) = P(b'|A)$, the Markov chain would be instantaneously equilibrated. Hence if we can approach this ideal scenario, we should be able to improve the inference speeds. Here we describe two simple strategies in achieving such an improvement which have been shown to yield a significance performance impact [80]. The first one is to replace the fully random move proposals of Equation (11.63) by a more informative choice. Namely, we use the current information about the model being inferred to guide our next move. We do so by selecting the membership of a node i being moved according to

$$P(b_i = r|\boldsymbol{b}) = \sum_s P(s|i) \frac{e_{sr} + \varepsilon}{e_s + \varepsilon(B+1)}, \tag{11.64}$$

where $P(s|i) = \sum_j A_{ij} \delta_{b_j,s}/k_i$ is the fraction of neighbors of node i that belong to group s, and $\varepsilon > 0$ is an arbitrary parameter that enforces ergodicity, but with no other significant impact in the algorithm, provided it is sufficiently small (however, if $\varepsilon \to \infty$ we recover the fully random moves of Equation (11.63)). What this move proposal means is that we inspect the local neighborhood of the node i and see which groups s are connected to this node, and we use the typical neighborhood r of the groups s to guide our placement of node i (see Figure 11.12a). The purpose of these move proposals is not to waste time with attempted moves that will almost surely be rejected, as will typically happen with the fully random version. We emphasize that the move proposals of Equation (11.64) do not bias the partitions toward any specific kind of mixing pattern; in particular they do not prefer assortative versus non-assortative partitions. Furthermore, these proposals can be generated efficiently, simply by following three steps: (i) sample a random neighbor j of node i and inspect its group membership $s = b_j$, (ii) with probability $\varepsilon(B+1)/(e_s + \varepsilon(B+1))$ sample a fully random group r (which can be a new group, and (iii) sample a group label r with a probability proportional to the number of edges leading to it from group s, e_{sr}. These steps can be performed in time $O(k_i)$, again independently of B, as long as a continuous book-keeping is made of the edges which are incident to each group, and therefore it does not affect the overall $O(E)$ time complexity.

The second strategy is to choose a starting state that lies close to the mode of the posterior. We do so by performing a Fibonacci search [52] on the number of groups B, where for each value we obtain the best partition from a larger partition with $B' > B$ using an agglomerative heuristic, composed of the following steps taken alternatively: (i) we attempt the moves of Equation (11.64) until no improvement to the posterior is observed and (ii) we merge groups

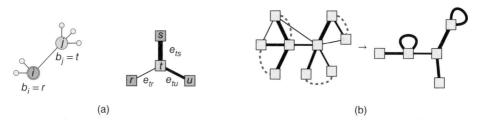

(a)   (b)

Figure 11.12 Efficient MCMC strategies. (a) Move proposals are made by inspecting the neighborhood of node i and selecting a random neighbor j. Based on its group membership $t = b_j$, the edge counts between groups are inspected (right) and the move proposal $b_i = s$ is made with probability proportional to e_{ts}. (b) The initial state of the MCMC is obtained with an agglomerative heuristic, where groups are merged together using the same proposals described in (a).

together, achieving a smaller number of groups $B'' \in [B, B']$, stopping when $B'' = B$. We do the last step by treating each group as a single node and using Equation (11.64) as a merge proposal, and selecting the ones that least decrease the posterior (see Figure 11.12b). As shown in [80], the overall complexity of this initialization algorithm is $O(E\log^2 N)$, and thus can be employed for very large networks.

The approach above can be adapted to the overlapping model of Section 11.7.3, where instead of the partition b, the move proposals are made with respect to the individual half-edge labels [84]. For the nested model, we have instead a hierarchical partition $\{b_l\}$, and we proceed in each step of the Markov chain by randomly choosing a level l and performing the proposals of Equation (11.64) on that level, as described in [85].

The combination of the two strategies described above makes the inference procedure quite scalable, and has been successfully employed on networks on the order of 10^7 to 10^8 edges, and up to $B = N$ groups. The MCMC algorithm described in this section, for all model variants described, is implemented in the `graph-tool` library [81], freely available under the GPL license at http://graph-tool.skewed.de.

11.9   To Sample or To Optimize?

In the examples so far, we have focused on obtaining the most likely partition from the posterior distribution, which is the one that minimizes the description length of the data. But is this in fact the best approach? In order to answer this, we need first to quantify how well our inference is doing by comparing our estimate $\hat{b}$ of the partition to the true partition that generated the data b^*, by defining a so-called *loss function*. For example, if we choose to be very strict, we may reject any partition that is strictly different from b^* on equal measure, using the indicator function

$$\Delta(\hat{b}, b^*) = \prod_i \delta_{\hat{b}_i, b_i^*}, \tag{11.65}$$

so that $\Delta(\hat{b}, b^*) = 1$ only if $\hat{b} = b^*$, otherwise $\Delta(\hat{b}, b^*) = 0$. If the observed data A and parameters b are truly sampled from the model and priors, respectively, the best assessment we can make for b^* is given by the posterior distribution $P(b|A)$. Therefore, the average of the indicator over the posterior is given by

$$\overline{\Delta}(\hat{b}) = \sum_b \Delta(\hat{b}, b) P(b|A). \tag{11.66}$$

If we maximize $\overline{\Delta}(\hat{b})$ with respect to $\hat{b}$, we obtain the so-called maximum *a posteriori* (MAP) estimator

$$\hat{b} = \underset{b}{\operatorname{argmax}} \; P(b|A), \tag{11.67}$$

which is precisely what we have been using so far and it is equivalent to employing the MDL principle. However, using this estimator is arguably overly optimistic, as we are unlikely to find the true partition with perfect accuracy in any but the most ideal cases. Instead, we may relax our expectations and consider instead the overlap function

$$d(\hat{b}, b^*) = \frac{1}{N} \sum_i \delta_{\hat{b}_i, b_i^*}, \tag{11.68}$$

which measures the *fraction* of nodes that are correctly classified. If we maximize now the average of the overlap over the posterior distribution

$$\overline{d}(\hat{\boldsymbol{b}}) = \sum_{\boldsymbol{b}} d(\hat{\boldsymbol{b}}, \boldsymbol{b}) P(\boldsymbol{b}|\boldsymbol{A}),$$ (11.69)

we obtain the *marginal estimator*

$$\hat{b}_i = \operatorname*{argmax}_{r} \pi_i(r),$$ (11.70)

where

$$\pi_i(r) = \sum_{\boldsymbol{b}\setminus b_i} P(b_i = r, \boldsymbol{b}\setminus b_i|\boldsymbol{A})$$ (11.71)

is the marginal distribution of the group membership of node i, summed over all remaining nodes.[19] The marginal estimator is notably different from the MAP estimator in that it leverages information from the entire posterior distribution to inform the classification of any single node. If the posterior is tightly concentrated around its maximum, both estimators will yield compatible answers. In this situation the structure in the data is very clear, and both estimators agree. Otherwise, the estimators will yield different aspects of the data, in particular if the posterior possesses many local maxima. For example, if the data has indeed been sampled from the model we are using, the multiplicity of local maxima can be just a reflection of the randomness in the data, and the marginal estimator will be able to average over them and provide better accuracy [63, 112].

In view of the above, one could argue that the marginal estimator should be generally preferred over MAP. However, the situation is more complicated for data which are not sampled from model being used for inference (i.e. the model is *misspecified*). In this situation, multiple peaks of the distribution can point to very different partitions that are all statistically significant. These different peaks function as alternative explanations for the data that must be accepted on equal footing, according to their posterior probability. The marginal estimator will in general mix the properties of all peaks into a consensus classification that is not representative of any single hypothesis, whereas the MAP estimator will concentrate only on the most likely one (or an arbitrary choice if they are all equally likely). An illustration of this is given by the well-known Zachary's karate club network [111], which captures the social interactions between members of a karate club amidst a conflict between the club's administrator and an instructor, which lead to a split of the club in two disjoint groups. The measurement of the network was done before the final split actually happened, and it is very often used as an example of a network exhibiting community structure. If we analyze this network

---

[19] The careful reader will notice that we must have in fact a trivial constant marginal $\pi_i(r) = 1/B$ for every node i, since there is a symmetry of the posterior distribution with respect to re-labelling of the groups, in principle rendering this estimator useless. In practice, however, our samples from the posterior distribution (e.g. using MCMC) will not span the whole space of label permutations in any reasonable amount of time, and instead will concentrate on a mode around one of the possible permutations. Since the modes around the label permutations are entirely symmetric, the node marginals obtained in this manner can be meaningfully used. However, for networks where some of the groups are not very large, *local* permutations of individual group labels are statistically possible during MCMC inference, leading to degeneracies in the marginal $\pi_i(r)$ of the affected nodes, resulting in artefacts when using the marginal estimator. This problem is exacerbated when the number of groups changes during MCMC sampling.

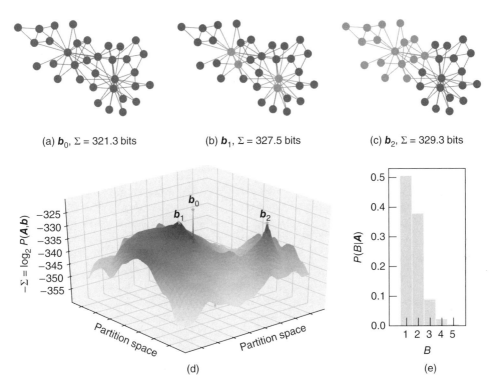

(a) $\boldsymbol{b}_0$, $\Sigma = 321.3$ bits   (b) $\boldsymbol{b}_1$, $\Sigma = 327.5$ bits   (c) $\boldsymbol{b}_2$, $\Sigma = 329.3$ bits

Figure 11.13 Posterior distribution of partitions of Zachary's karate club network using the DC-SBM. (a)–(c) show three modes of the distribution and their respective description lengths; (d) 2D projection of the posterior obtained using multidimensional scaling [15]; (e) marginal posterior distribution of the number of groups B.

with the DC-SBM, we obtain three partitions that occur with very high probability from the posterior distribution: a trivial $B = 1$ partition, corresponding to the configuration model without communities (Figure 11.13a), a "leader-follower" division into $B = 2$ groups, separating the administrator and instructor, together with two close allies, from the rest of the network (Figure 11.13b), and finally a $B = 2$ division into the aforementioned factions that anticipated the split (Figure 11.13c). If we would guide ourselves strictly by the MDL principle (i.e. using the MAP estimator), the preferred partition would be the trivial $B = 1$ one, indicating that the most likely explanation of this network is a fully random graph with a pre-specified degree sequence, and that the observed community structure emerged spontaneously. However, if we inspect the posterior distribution more closely, we see that other divisions into $B > 1$ groups amount to around 50% of the posterior probability (see Figure 11.13e). Therefore, if we consider all $B > 1$ partitions *collectively*, they give us little reason to completely discard the possibility that the network does in fact posses some group structure. Inspecting the posterior distribution even more closely, as shown in Figure 11.13d, reveals a multimodal structure clustered around the three aforementioned partitions, giving us three very different explanations for the data, none of which can be decisively discarded in favor of the others, at least not according to the evidence available in the network structure alone.

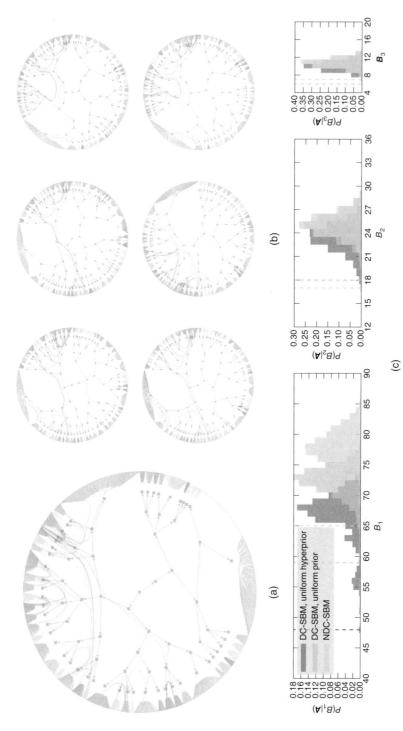

Figure 11.14 Hierarchical partitions of a network of collaboration between scientists [67]. (a) Most likely hierarchical partition according to the DC-SBM with a uniform hyperprior. (b) Uncorrelated samples from the posterior distribution. (c) Marginal posterior distribution of the number of groups at the first three hierarchical levels, according to the model variants described in the legend. The vertical lines mark the value obtained for the most likely partition. Adapted from [85].

The situation encountered for the karate club network is a good example of the so-called *bias-variance trade-off* that we are often forced to face: If we choose to single-out a singe partition as a unique representation of the data, we must invariably *bias* our result toward any of the three most likely scenarios, discarding the remaining ones at some loss of useful information. Otherwise, if we choose to eliminate the bias by incorporating the entire posterior distribution in our representation, by the same token it will incorporate a larger variance, i.e. it will simultaneously encompass diverging explanations of the data, leaving us without an unambiguous and clear interpretation. The only situation where this trade-off is not required is when the model is a perfect fit to the data, such that the posterior is tightly peaked around a single partition. Therefore, the variance of the posterior serves as a good indication of the quality of fit of the model, providing another reason to include it in the analysis.

It should also be remarked that when using a nonparametric approach, where the dimension of the model is also inferred from the posterior distribution, the potential bias incurred when obtaining only the most likely partition usually amounts to an *underfit* of the data, since the uncertainty in the posterior typically translates into the existence of a more conservative partition with fewer groups.[20] Instead, if we sample from the posterior distribution, we will average over many alternative fits, including those that model the data more closely with a larger number of groups. However, each individual sample of the posterior will tend to incorporate more randomness from the data, which will disappear only if we average over all samples. This means that single samples will tend to overfit the data, and hence we must resist looking at them individually. It is only in the aforementioned limit of a perfect fit that we are guaranteed not to be misled one way or another. An additional example of this is shown in Figure 11.14 for a network of collaborations among scientists. If we infer the best nested SBM, we find a specific hierarchical division of the network. However, if we sample hierarchical divisions from the posterior distribution, we typically encounter larger models, with a larger number of groups and deeper hierarchy. Each individual sample from the posterior is likely to be an overfit, but collectively they give a more accurate picture of the network in comparison with the most likely partition, which probably over-simplifies it. As already mentioned, this discrepancy, observed for all three SBM versions, tells us that neither of them is an ideal fit for this network.

The final decision on which approach to take depends on the actual objective and resources available. In general, sampling from the posterior will be more suitable when the objective is to generalize from observation and make predictions (see next section and [104]), and when computational resources are ample. Conversely, if the objective is to make a precise statement about the data, e.g. in order to summarize and interpret it, and the computational resources are scarce, maximizing the posterior tends to be more adequate.

11.10 Generalization and Prediction

When we fit a model like the SBM to a network, we are doing more than simply dividing the nodes into statistically equivalent groups; we are also making a statement about a possible mechanism that generated the network. This means that, to the extent that the model is a good representation of the data, we can use it generalize and make predictions about what has *not* been

---

[20] This is different from *parametric* posteriors, where the dimension of the model is externally imposed in the prior and the MAP estimator tends to overfit [63, 112].

observed. This has been most explored for the prediction of missing and spurious links [11, 37]. This represents the situation where we know or stipulate that the observed data is noisy, and may contain edges that in fact do not exist, or does not contain edges that do exist. With a generative model like the SBM, we are able to ascribe probabilities to existing and non-existing edges of being spurious or missing, respectively, as we now describe.

Following [104], the scenario we will consider is the situation where there exists a complete network G which is decomposed in two parts,

$$G = A^O + \delta A \tag{11.72}$$

where A^O is the network that we observe, and the δA is the set of missing and spurious edges that we want to predict, where an entry $\delta A_{ij} > 0$ represents a missing edge and $\delta A_{ij} < 0$ a spurious one. Hence, our task is to obtain the posterior distribution

$$P(\delta A | A^O). \tag{11.73}$$

The central assumption we will make is that the complete network G has been generated using some arbitrary version of the SBM, with a marginal distribution

$$P_G(G|b). \tag{11.74}$$

Given a generated network G, we then select δA from some arbitrary distribution that models our source of errors

$$P_{\delta A}(\delta A | G). \tag{11.75}$$

With the above model for the generation of the complete network and its missing and spurious edges, we can proceed to compute the posterior of Equation (11.73). We start from the joint distribution

$$P(A^O, \delta A | G) = P(A^O | \delta A, G) P_{\delta A}(\delta A | G) \tag{11.76}$$

$$= \delta(G - (A^O + \delta A)) P_{\delta A}(\delta A | G), \tag{11.77}$$

where we have used the fact $P(A^O | \delta A, G) = \delta(G - (A^O + \delta A))$ originating from Equation (11.72). For the joint distribution conditioned on the partition, we sum the above over all possible graphs G, sampled from our original model,

$$P(A^O, \delta A | b) = \sum_G P(A^O, \delta A | G) P_G(G|b) \tag{11.78}$$

$$= P_{\delta A}(\delta A | A^O + \delta A) P_G(A^O + \delta A | b). \tag{11.79}$$

The final posterior distribution of Equation (11.73) is therefore

$$P(\delta A | A^O) = \frac{\sum_b P(A^O, \delta A | b) P(b)}{P(A^O)} \tag{11.80}$$

$$= \frac{P_{\delta A}(\delta A | A^O + \delta A) \sum_b P_G(A^O + \delta A | b) P(b)}{P(A^O)}, \tag{11.81}$$

with $P(A^O)$ being a normalization constant, independent of δA. This expression gives a general recipe to compute the posterior, where one averages the marginal likelihood $P_G(A^O + \delta A|b)$ obtained by sampling partitions from the prior $P(b)$. However, this procedure will typically take an astronomical time to converge to the correct asymptotic value, since the largest values of $P_G(A^O + \delta A|b)$ will be far away from most values of b sampled from $P(b)$. A much better approach is to perform importance sampling, by rewriting the posterior as

$$P(\delta A|A^O) \propto P_{\delta A}(\delta A|A^O + \delta A) \sum_b P_G(A^O + \delta A|b) \frac{P_G(A^O|b)}{P_G(A^O|b)} P(b) \tag{11.82}$$

$$\propto P_{\delta A}(\delta A|A^O + \delta A) \sum_b \frac{P_G(A^O + \delta A|b)}{P_G(A^O|b)} P_G(b|A^O), \tag{11.83}$$

where $P_G(b|A^O)$ is the posterior of partitions obtained by pretending that the observed network came directly from the SBM. We can sample from this posterior using MCMC as described in Section 11.8. As the number of entries in δA is typically much smaller than the number of observed edges, this importance sampling approach will tend to converge much faster. This allows us to compute $P(\delta A|A^O)$ in practical manner, up to a normalization constant. However, if we want to compare the relative probability between specific sets of missing/spurious edges, $\{\delta A_i\}$, via the ratio

$$\lambda_i = \frac{P(\delta A_i|A^O)}{\sum_j P(\delta A_j|A^O)}, \tag{11.84}$$

this normalization constant plays no role. The above still depends on our chosen model for the production of missing and spurious edges, given by Equation (11.75). In the absence of domain-specific information about the source of noise, we must consider all alternative choices $\{\delta A_i\}$ to be equally likely *a priori*, so that the we can simply replace $P_{\delta A}(\delta A|A^O + \delta A) \propto 1$ in Equation (11.83), although more realistic choices can also be included.

In Figure 11.15 we show the relative probabilities of two hypothetical missing edges for the American college football network, obtained with the approach above. We see that a particular missing edge between teams of the same conference is almost a hundred times more likely than one between teams of different conference.

The use of the SBM to predict missing and spurious edges has been employed in a variety of applications, such as the prediction of novel interactions between drugs [38], conflicts in social networks [94], as well to provide user recommendations [31, 40], and in many cases has outperformed a variety of competing methods.

11.11 Fundamental Limits of Inference: The Detectability–Indetectability Phase Transition

Besides defining useful models and investigating their behavior in data, there is another line of questioning which deals with how far it is possible to go when we try to infer the structure of networks. Naturally, the quality of the inference depends on the statistical evidence available in the data, and we may therefore ask if it is possible at all to uncover *planted* structures, i.e. structures that we impose ourselves, with our inference methods, and if so, what is the best performance we can expect. Research in this area has exploded in recent years [63, 112] after

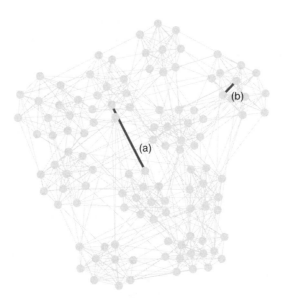

Figure 11.15 Two hypothetical missing edges in the network of American college football teams. The edge (a) connects teams of different conferences, whereas (b) connects teams of the same conference. According to the nested DC-SBM, their posterior probability ratios are $\lambda_a \approx 0.013(1)$ and $\lambda_b \approx 0.987(1)$.

it was shown by Decelle *et al.* [19, 20] that not only may it be impossible to uncover planted structures with the SBM, but the inference undergoes a "phase transition" where it becomes possible only if the structure is strong enough to cross a non-trivial threshold. This result was obtained using methods from statistical physics, which we now describe.

The situation we will consider is a "best case scenario," where all parameters of the model are known, with the exception of the partition b, this in contrast to our overall approach so far, where we considered all parameters to be unknown random variables. In particular, we will consider only the prior

$$P(b|\gamma) = \prod_i \gamma_{b_i}, \tag{11.85}$$

where γ_r is the probability of a node belonging in group r. Given this, we wish to obtain the posterior distribution of the node partition, using the SBM of Equation (11.8),

$$P(b|A, \lambda, \gamma) = \frac{P(A|b, \lambda)P(b|\gamma)}{P(A|\lambda, \gamma).} = \frac{e^{-\mathcal{H}(b)}}{Z} \tag{11.86}$$

which was written above in terms of the "Hamiltonian"

$$\mathcal{H}(b) = -\sum_{i<j}(A_{ij} \ln \lambda_{b_i,b_j} - \lambda_{b_i,b_j}) - \sum_i \ln \gamma_{b_i}, \tag{11.87}$$

drawing an analogy with Potts-like models in statistical physics [107]. The normalization constant, called the "partition function," is given by

$$Z = \sum_b e^{-\mathcal{H}(b)}. \tag{11.88}$$

Far from being an unimportant detail, the partition function can be used to determine all statistical properties of our inference procedure. For example, if we wish to obtain the marginal posterior distribution of node i, we can do so by introducing the perturbation $\mathscr{H}'(b) = \mathscr{H}(b) - \mu\delta_{b_i,r}$ and computing the derivative

$$P(b_i = r|A, \lambda, \gamma) = \frac{\partial \ln Z}{\partial \mu}\bigg|_{\mu=0} = \sum_b \delta_{b_i,r}\frac{e^{-\mathscr{H}(b)}}{Z}. \tag{11.89}$$

Unfortunately, it does not seem possible to compute the partition function Z in closed form for an arbitrary graph A. However, there is a special case for which we *can* compute the partition function, namely when A is a *tree*. This is useful for us, because graphs sampled from the SBM will be "locally tree-like" if they are sparse (i.e. the degrees are small compared to the size of the network $k_i \ll N$), and the group sizes scale with the size of the system, i.e. $n_r = O(N)$ (which implies $B \ll N$). Locally tree-like means that typical loops will have length $O(N)$, and hence at the immediate neighborhood of any given node the graph will look like a tree. Although being locally tree-like is not quite the same as being a tree, the graph will become increasing *closer* to being a tree in the "thermodynamic limit" $N \to \infty$. Because of this, many properties of locally tree-like graphs will become asymptotically identical to trees in this limit. If we assume that this limit holds, we can compute the partition function by pretending that the graph is close enough to being a tree, in which case we can write the so-called Bethe free energy (we refer to [19, 62] for a detailed derivation)

$$\mathscr{F} = -\ln Z = -\sum_i \ln Z^i + \sum_{i<j} A_{ij}\ln Z^{ij} - E \tag{11.90}$$

with the auxiliary quantities given by

$$Z^{ij} = N\sum_{r<s}\lambda_{rs}(\psi_r^{i \to j}\psi_s^{j \to i} + \psi_s^{i \to j}\psi_r^{j \to i}) + N\sum_r \lambda_{rr}\psi_r^{i \to j}\psi_r^{j \to i} \tag{11.91}$$

$$Z^i = \sum_r n_r e^{-h_r}\prod_{j\in\partial i}\sum_r N\lambda_{rb_i}\psi_r^{j \to i}, \tag{11.92}$$

where ∂i means the neighbors of node i. In the above equations, the values $\psi_r^{i \to j}$ are called "messages" and they must fulfill the self-consistency equations

$$\psi_r^{i \to j} = \frac{1}{Z^{i \to j}}\gamma_r e^{-h_r}\prod_{k\in\partial i\backslash j}\left(\sum_s N\lambda_{rs}\psi_s^{k \to i}\right) \tag{11.93}$$

where $k \in \partial i\backslash j$ means all neighbors k of i excluding j, the value $Z^{i \to j}$ is a normalization constant enforcing $\sum_r \psi_r^{i \to j} = 1$, and $h_r = \sum_i\sum_r\lambda_{rb_i}\psi_r^i$ is a local auxiliary field. Equations (11.93) are called the *belief-propagation* (BP) equations [62], and the entire approach is also known under the name "cavity method" [61]. The values of the messages are typically obtained by iteration, where we start from some initial configuration (e.g. a random one) and compute new values from the right-hand side of Equation (11.93), until they converge asymptotically. Note that the messages are only defined on edges of the network, and an update involves inspecting the values at the neighborhood of the nodes, where the messages can be interpreted as carrying information about the marginal distribution of a given node, if the same is removed from the network (hence the names "belief propagation" and "cavity method"). Each iteration of the BP equations can be done in time $O(EB^2)$, and the convergence is often obtained only after a few iterations, rendering the whole computation fairly efficient, provided B is reasonably small. After the messages have been obtained, they can be used to compute the node marginals,

$$P(b_i = r|A, \lambda, \gamma) = \psi_r^i = \frac{1}{Z^i}\gamma_r \prod_{j \in \partial i}\left[\sum_s (N\lambda_{rs})^{A_{ij}}e^{-\lambda_{rs}}\psi_s^{j \to i}\right], \qquad (11.94)$$

where Z^i is a normalization constant.

This whole procedure gives a way of computing the marginal distribution $P(b_i = r|A, \lambda, \gamma)$ in a manner that is asymptotically exact, if A is sufficiently large and locally tree-like. Since networks that are sampled from the SBM fulfill this property,[21] we may proceed with our original question and test if we can recover the true value of b we used to generate a network. For the test, we use a simple parametrization named the planted partition model (PP) [12, 21], where $\gamma_r = 1/B$ and

$$\lambda_{rs} = \lambda_{in}\delta_{rs} + \lambda_{out}(1 - \delta_{rs}), \qquad (11.95)$$

with λ_{in} and λ_{out} specifying the expected number of edges between nodes of the same groups and of different groups, respectively. If we generate networks from this ensemble, use the BP equations to compute the posterior marginal distribution of Equation (11.94) and compare its maximum values with the planted partition, we observe, as shown in Figure 11.16, that it is recoverable only up to a certain value of $\varepsilon = N(\lambda_{in} - \lambda_{out})$, above which the posterior

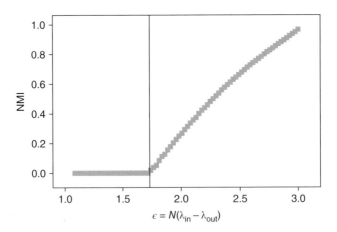

Figure 11.16 Normalized mutual information (NMI) between the planted and inferred partitions of a PP model with $N = 10^5$, $B = 3$ and $\langle k \rangle = 3$, and $\varepsilon = N(\lambda_{in} - \lambda_{out})$. The vertical line marks the detectability threshold $\varepsilon^* = B\sqrt{\langle k \rangle}$.

---

[21] Real networks, however, should not be expected to be locally tree-like. This does not invalidate the results of this section, which pertain strictly to data sampled from the SBM. However, despite not being exact, the BP algorithm yields surprisingly accurate results for real networks, even when the tree-like property is violated [19].

distribution is fully uniform. By inspecting the stability of the fully uniform solution of the BP equations, the exact threshold can be determined [19],

$$\varepsilon^* = B\sqrt{\langle k \rangle}, \tag{11.96}$$

where $\langle k \rangle = N \sum_{rs} \lambda_{rs}/B^2$ is the average degree of the network. The existence of this threshold is remarkable because the ensemble is only equivalent to a completely random one if $\varepsilon = 0$, yet there is a non-negligible range of values $\varepsilon \in [0, \varepsilon^*]$ for which *the planted structure cannot be recovered even though the model is not random*. This might seem counter-intuitive, if we argue that making N sufficiently large should at some point give us enough data to infer the model with arbitrary precision. The hole in logic lies in the fact that the number of parameters – the node partition b – also grows with N, and that we would need the effective sample size, i.e. the number of edges E, to grow *faster* than N to guarantee that the data is sufficient. Since for sparse graphs we have $E = O(N)$, we are never able to reach the limit of sufficient data. Thus, we should be able to achieve asymptotically perfect inference only for dense graphs (e.g. with $E = O(N^2)$) or by inferring simultaneously from many graphs independently sampled from the same model. Neither situation, however, is representative of what we typically encounter when we study networks.

The above result carries important implications into the overall field of network clustering. The existence of the "detectable" phase for $\varepsilon > \varepsilon^*$ means that, in this regime, it is possible for algorithms to discover the planted partition in polynomial time, with the BP algorithm doing so optimally. Furthermore, for $B > 4$ (or $B > 3$ for the dissortative case with $\lambda_{in} < \lambda_{out}$) there is another regime in a range $\varepsilon^* < \varepsilon < \varepsilon^\dagger$, where BP converges to the planted partition only if the messages are initialized close enough to the corresponding fixed point. In this regime, the posterior landscape exhibits a "glassy" structure, with exponentially many maxima that are almost as likely as the planted partition, but are completely uncorrelated with it. The problem of finding the planted partition in this case is possible, but conjectured to be NP-hard.

Many systematic comparisons of different community detection algorithms were done in a manner that was oblivious to these fundamental facts regarding detectability and hardness [55, 56], even though their existence had been conjectured before [88, 92] and hence should be re-framed with it in mind. Furthermore, we point out that although the analysis based on the BP equations is mature and widely accepted in statistical physics, they are not completely rigorous from a mathematical point of view. Because of this, the result of Decelle *et al.* [19] leading to the threshold of Equation (11.96) has initiated intense activity from mathematicians in search of rigorous proofs, which have subsequently been found for a variety of relaxations of the original statement (see [1] for a review) and remains an active area of research.

11.12 Conclusion

In this chapter we gave a description of the basic variants of the stochastic blockmodel (SBM), and a consistent Bayesian formulation that allows us to infer them from data. The focus has been on developing a framework to extract the large-scale structure of networks while avoiding both overfitting (mistaking randomness for structure) and underfitting (mistaking structure for randomness), and doing so in a manner that is analytically tractable and computationally efficient.

The Bayesian inference approach provides a methodologically correct answer to the very central question in network analysis of whether patterns of large-scale structure can in fact be supported by statistical evidence. Besides this practical aspect, it also opens a window into the fundamental limits of network analysis itself, giving us a theoretical underpinning we can use to understand more about the nature of network systems.

Although the methods described here go a long way to allowing us to understand the structure of networks, some important open problems remain. From a modeling perspective, we know that for most systems the SBM is quite simplistic and falls very short of giving us a mechanistic explanation for them. We can interpret the SBM as being to network data what a histogram is to spatial data [73], and thus while it fulfills the formal requirements of being a generative model, it will never deplete the modeling requirements of any particular real system. Although it is naive to expect to achieve such a level of success with a general model like the SBM, it is yet still unclear how far we can go. For example, it remains to be seen how tractable it is to incorporate local structures – like densities of subgraphs – together with the large-scale structure that the SBM prescribes.

From a methodological perspective, although we can select between the various SBM flavors given the statistical evidence available, we still lack good methods to assess the quality of fit of the SBM at an absolute level. In particular, we do not yet have a systematic understanding of how well the SBM is able to reproduce properties of empirical systems, and what would be the most important sources of deficiencies, and how these could be overcome.

In addition to these outstanding challenges, there are areas of development that are more likely to undergo continuous progress. Generalizations and extensions of the SBM to cover specific cases are essentially open ended, such as the case of dynamic networks, and we can perhaps expect more realistic models to appear. Furthermore, since the inference of the SBM is in general a NP-hard problem, and thus most probably lacks a general solution, the search for more efficient algorithmic strategies that work in particular cases is also a long term goal that is likely to attract further attention.

References

1. E. Abbe. Community detection and stochastic block models: recent developments. *arXiv:1703.10146 [cs, math, stat]*, Mar. 2017. arXiv: 1703.10146.

2. L. A. Adamic and N. Glance. The political blogosphere and the 2004 U.S. election: divided they blog. In *Proceedings of the 3rd International Workshop on Link Discovery*, LinkKDD '05, pages 36–43, New York, NY, USA, 2005. ACM. ISBN

3. Y.-Y. Ahn, J. P. Bagrow, and S. Lehmann. Link communities reveal multiscale complexity in networks. *Nature*, 466(7307):761–764, 2010.

4. C. Aicher, A. Z. Jacobs, and A. Clauset. Learning latent block structure in weighted networks. *Journal of Complex Networks*, 3(2):221–248, 2015.

5. E. M. Airoldi, D. M. Blei, S. E. Fienberg, and E. P. Xing. Mixed Membership Stochastic Blockmodels. *J. Mach. Learn. Res.*, 9:1981–2014, 2008.

6. H. Akaike. A new look at the statistical model identification. *IEEE Transactions on Automatic Control*, 19(6):716–723, 1974.

7. B. Ball, B. Karrer, and M. E. J. Newman. Efficient and principled method for detecting communities in networks. *Physical Review E*, 84(3):036103, 2011.

8. G. Bianconi, P. Pin, and M. Marsili. Assessing the relevance of node features for network structure. *Proceedings of the National Academy of Sciences*, 106(28):11433–11438, 2009.

9. M. Boguñá and R. Pastor-Satorras. Class of correlated random networks with hidden variables. *Physical Review E*, 68(3):036112, 2003.

10. B. Bollobás, S. Janson, and O. Riordan. The phase transition in inhomogeneous random graphs. *Random Structures & Algorithms*, 31(1): 3–122, 2007.

11. A. Clauset, C. Moore, and M. E. J. Newman. Hierarchical structure and the prediction of missing links in networks. *Nature*, 453(7191):98–101, 2008.

12. A. Condon and R. M. Karp. Algorithms for graph partitioning on the planted partition model. *Random Structures & Algorithms*, 18(2): 116–140, 2001.

13. M. Corneli, P. Latouche, and F. Rossi. Exact ICL maximization in a non-stationary temporal extension of the stochastic block model for dynamic networks. *Neurocomputing*, 192:81–91, 2016.

14. T. M. Cover and J. A. Thomas. *Elements of Information Theory*. Wiley-Interscience, 99th edition, 1991.

15. T. F. Cox and M. A. A. Cox. *Multidimensional Scaling*, 2nd edition. Chapman and Hall/CRC, Boca Raton, 2000.

16. E. Côme and P. Latouche. Model selection and clustering in stochastic block models based on the exact integrated complete data likelihood. *Statistical Modelling*, 15(6):564–589, 2015.

17. J.-J. Daudin, F. Picard, and S. Robin. A mixture model for random graphs. *Statistics and Computing*, 18(2):173–183, 2008.

18. C. De Bacco, E. A. Power, D. B. Larremore, and C. Moore. Community detection, link prediction, and layer interdependence in multilayer networks. *Physical Review E*, 95(4):042317, 2017.

19. A. Decelle, F. Krzakala, C. Moore, and L. Zdeborová. Asymptotic analysis of the stochastic block model for modular networks and its algorithmic applications. *Physical Review E*, 84(6):066106, 2011.

20. A. Decelle, F. Krzakala, C. Moore, and L. Zdeborová. Inference and Phase Transitions in the Detection of Modules in Sparse Networks. *Physical Review Letters*, 107(6):065701, 2011.

21. M. E. Dyer and A. M. Frieze. The solution of some random NP-hard problems in polynomial expected time. *Journal of Algorithms*, 10(4): 451–489, 1989.

22. P. Erdős and A. Rényi. On random graphs, I. *Publicationes Mathematicae (Debrecen)*, 6:290–297, 1959.

23. T. S. Evans. Clique graphs and overlapping communities. *Journal of Statistical Mechanics: Theory and Experiment*, 2010(12):P12037, 2010.

24. T. S. Evans. American College Football Network Files. *FigShare*, July 2012.

25. S. Fortunato. Community detection in graphs. *Physics Reports*, 486(3-5): 75–174, 2010.

26. S. Fortunato and M. Barthélemy. Resolution limit in community detection. *Proceedings of the National Academy of Sciences*, 104(1):36–41, 2007.

27. W. Fu, L. Song, and E. P. Xing. Dynamic Mixed Membership Blockmodel for Evolving Networks. In *Proceedings of the 26th Annual International Conference on Machine Learning*, ICML '09, pages 329–336, New York, NY, USA, 2009. ACM.

28. D. Garlaschelli, F. d. Hollander, and A. Roccaverde. Ensemble nonequivalence in random graphs with modular structure. *Journal of Physics A: Mathematical and Theoretical*, 50(1):015001, 2017.

29. A. Ghasemian, P. Zhang, A. Clauset, C. Moore, and L. Peel. Detectability thresholds and optimal algorithms for community structure in dynamic networks. *Physical Review X*, 6(3):031005, 2016.

30. M. Girvan and M. E. J. Newman. Community structure in social and biological networks. *Proceedings of the National Academy of Sciences*, 99 (12):7821–7826, June 2002.

31. A. Godoy-Lorite, R. Guimerà, C. Moore, and M. Sales-Pardo. Accurate and scalable social recommendation using mixed-membership stochastic block models. *Proceedings of the National Academy of Sciences*, 113(50): 14207–14212, 2016.

32. B. H. Good, Y.-A. de Montjoye, and A. Clauset. Performance of modularity maximization in practical contexts. *Physical Review E*, 81(4):046106, 2010.

33. P. K. Gopalan and D. M. Blei. Efficient discovery of overlapping communities in massive networks. *Proceedings of the National Academy of Sciences*, 110(36):14534–14539, 2013.

34. P. Grünwald. A tutorial introduction to the minimum description length principle. *arXiv:math/0406077*, June 2004.

35. P. D. Grünwald. *The Minimum Description Length Principle*. The MIT Press, 2007.

36. R. Guimerà and L. A. Nunes Amaral. Functional cartography of complex metabolic networks. *Nature*, 433(7028):895–900, 2005.

37. R. Guimerà and M. Sales-Pardo. Missing and spurious interactions and the reconstruction of complex networks. *Proceedings of the National Academy of Sciences*, 106(52):22073–22078, 2009.

38. R. Guimerà and M. Sales-Pardo. A network inference method for large-scale unsupervised identification of novel drug-drug interactions. *PLoS Comput Biol*, 9(12):e1003374, 2013.

39. R. Guimerà, M. Sales-Pardo, and L. A. N. Amaral. Modularity from fluctuations in random graphs and complex networks. *Physical Review E*, 70(2):025101, 2004.

40. R. Guimerà, A. Llorente, E. Moro, and M. Sales-Pardo. Predicting human preferences using the block structure of complex social networks. *PLoS One*, 7(9):e44620, 2012.

41. M. B. Hastings. Community detection as an inference problem. *Physical Review E*, 74(3):035102, 2006.

42. W. K. Hastings. Monte Carlo sampling methods using Markov chains and their applications. *Biometrika*, 57(1):97–109, 1970.

43. J. M. Hofman and C. H. Wiggins. Bayesian approach to network modularity. *Physical Review Letters*, 100(25):258701, 2008.

44. P. W. Holland, K. B. Laskey, and S. Leinhardt. Stochastic blockmodels: First steps. *Social Networks*, 5(2):109–137, 1983.

45. D. Holten. Hierarchical edge bundles: visualization of adjacency relations in hierarchical data. *IEEE Transactions on Visualization and Computer Graphics*, 12(5):741–748, 2006.

46. D. Hric, R. K. Darst, and S. Fortunato. Community detection in networks: Structural communities versus ground truth. *Physical Review E*, 90(6): 062805, 2014.

47. D. Hric, T. P. Peixoto, and S. Fortunato. Network structure, metadata, and the prediction of missing nodes and annotations. *Physical Review X*, 6(3): 031038, Sept. 2016.

48. E. T. Jaynes. *Probability Theory: The Logic of Science*. Cambridge University Press, Cambridge, New York, 2003.

49. H. Jeffreys. *Theory of Probability*. Oxford University Press, Oxford, New York, 3rd edition, 2000.

50. B. Karrer and M. E. J. Newman. Stochastic blockmodels and community structure in networks. *Physical Review E*, 83(1):016107, 2011.

51. C. Kemp and J. B. Tenenbaum. Learning systems of concepts with an infinite relational model. *In Proceedings of the 21st National Conference on Artificial Intelligence*, 2006.

52. J. Kiefer. Sequential Minimax Search for a Maximum. *Proceedings of the American Mathematical Society*, 4(3):502, 1953.

53. S. Kirkpatrick, C. D. Gelatt Jr, and M. P. Vecchi. Optimization by simulated annealing. *Science*, 220(4598):671, 1983.

54. V. Krebs. Analyzing one network to reveal another. *Bulletin of Sociological Methodology/Bulletin de Méthodologie Sociologique*, 79(1):61–70, 2003.

55. A. Lancichinetti and S. Fortunato. Benchmarks for testing community detection algorithms on directed and weighted graphs with overlapping communities. *Physical Review E*, 80(1):016118, 2009.

56. A. Lancichinetti, S. Fortunato, and F. Radicchi. Benchmark graphs for testing community detection algorithms. *Physical Review E*, 78(4):046110, 2008.

57. D. J. C. MacKay. *Information Theory, Inference and Learning Algorithms*. Cambridge University Press, 2003.

58. T. Martin, B. Ball, and M. E. J. Newman. Structural inference for uncertain networks. *Physical Review E*, 93(1):012306, 2016.

59. Matias Catherine and Miele Vincent. Statistical clustering of temporal networks through a dynamic stochastic block model. *Journal of the Royal Statistical Society: Series B (Statistical Methodology)*, 79(4):1119–1141, 2016.

60. N. Metropolis, A. W. Rosenbluth, M. N. Rosenbluth, A. H. Teller, and E. Teller. Equation of state calculations by fast computing machines. *Journal of Chemical Physics*, 21(6):1087, 1953.

61. M. Mezard. *Spin Glass Theory And Beyond: An Introduction To The Replica Method And Its Applications*. WSPC, Singapore, New Jersey, 1986.

62. M. Mezard and A. Montanari. *Information, Physics, and Computation*. Oxford University Press, 2009.

63. C. Moore. The Computer Science and Physics of Community Detection: Landscapes, Phase Transitions, and Hardness. Feb. 2017. arXiv: 1702.00467.

64. M. Mørup and M. N. Schmidt. Bayesian community detection. *Neural Computation*, 24(9):2434–2456, 2012.

65. M. Mørup, M. N. Schmidt, and L. K. Hansen. Infinite multiple membership relational modeling for complex networks. In *2011 IEEE International Workshop on Machine Learning for Signal Processing*, pages 1–6, 2011.

66. M. Newman. *Networks: An Introduction*. Oxford University Press, 2010.

67. M. E. J. Newman. Finding community structure in networks using the eigenvectors of matrices. *Physical Review E*, 74(3):036104, 2006.

68. M. E. J. Newman and G. T. Barkema. *Monte Carlo Methods in Statistical Physics*. Oxford University Press, oxford, New York, 1999.

69. M. E. J. Newman and A. Clauset. Structure and inference in annotated networks. *Nature Communications*, 7:11863, June 2016.

70. M. E. J. Newman and T. P. Peixoto. Generalized communities in networks. *Physical Review Letters*, 115(8):088701, 2015.

71. M. E. J. Newman and G. Reinert. Estimating the number of communities in a network. *Physical Review Letters*, 117(7):078301, 2016.

72. K. Nowicki and T. A. B. Snijders. Estimation and prediction for stochastic blockstructures. *Journal of the American Statistical Association*, 96(455): 1077–1087, 2001.

73. S. C. Olhede and P. J. Wolfe. Network histograms and universality of blockmodel approximation. *Proceedings of the National Academy of Sciences*, 111(41):14722–14727, 2014.

74. S. Paul and Y. Chen. Consistent community detection in multi-relational data through restricted multi-layer stochastic blockmodel. *Electronic Journal of Statistics*, 10(2):3807–3870, 2016.

75. L. Peel. Active discovery of network roles for predicting the classes of network nodes. *Journal of Complex Networks*, 3(3):431–449, 2015.

76. L. Peel and A. Clauset. Detecting change points in the large-scale structure of evolving networks. In *Twenty-Ninth AAAI Conference on Artificial Intelligence*, Feb. 2015.

77. L. Peel, D. B. Larremore, and A. Clauset. The ground truth about metadata and community detection in networks. *Science Advances*, 3(5): e1602548, 2017.

78. T. P. Peixoto. Entropy of stochastic blockmodel ensembles. *Physical Review E*, 85(5):056122, 2012.

79. T. P. Peixoto. Parsimonious module inference in large networks. *Physical Review Letters*, 110(14):148701, 2013.

80. T. P. Peixoto. Efficient Monte Carlo and greedy heuristic for the inference of stochastic block models. *Physical Review E*, 89(1):012804, 2014.

81. T. P. Peixoto. The `graph-tool` python library. *figshare*, 2014. Available at https://figshare.com/articles/graph_tool/1164194.

82. T. P. Peixoto. Hierarchical block structures and high-resolution model selection in large networks. *Physical Review X*, 4(1):011047, 2014.

83. T. P. Peixoto. Inferring the mesoscale structure of layered, edge-valued, and time-varying networks. *Physical Review E*, 92(4):042807, 2015.

84. T. P. Peixoto. Model selection and hypothesis testing for large-scale network models with overlapping groups. *Physical Review X*, 5(1):011033, 2015.

85. T. P. Peixoto. Nonparametric Bayesian inference of the microcanonical stochastic block model. *Physical Review E*, 95(1):012317, 2017.

86. T. P. Peixoto. Nonparametric weighted stochastic block models. *Physical Review E*, 97(1):012306, 2018.

87. T. P. Peixoto and M. Rosvall. Modelling sequences and temporal networks with dynamic community structures. *Nature Communications*, 8(1):582, 2017.

88. J. Reichardt and M. Leone. (Un)detectable cluster structure in sparse networks. *Physical Review Letters*, 101(7):078701, 2008.

89. J. Reichardt, R. Alamino, and D. Saad. The interplay between microscopic and mesoscopic structures in complex networks. *PLoS One*, 6(8):e21282, 2011.

90. M. A. Riolo, G. T. Cantwell, G. Reinert, and M. E. J. Newman. Efficient method for estimating the number of communities in a network. *Physical Review E*, 96(3):032310, 2017.

91. J. Rissanen. Modeling by shortest data description. *Automatica*, 14(5): 465–471, 1978.

92. P. Ronhovde and Z. Nussinov. Multiresolution community detection for megascale networks by information-based replica correlations. *Physical Review E*, 80(1):016109, 2009.

93. M. Rosvall and C. T. Bergstrom. An information-theoretic framework for resolving community structure in complex networks. *Proceedings of the National Academy of Sciences*, 104(18):7327–7331, 2007.

94. N. Rovira-Asenjo, T. Gumí, M. Sales-Pardo, and R. Guimerà. Predicting future conflict between team-members with parameter-free models of social networks. *Scientific Reports*, 3, 2013.

95. M. Schmidt and M. Morup. Nonparametric Bayesian modeling of complex networks: An introduction. *IEEE Signal Processing Magazine*, 30(3): 110–128, 2013.

96. G. Schwarz. Estimating the dimension of a model. *Annals of Statistics*, 6 (2):461–464, 1978.

97. Y. M. Shtar'kov. Universal sequential coding of single messages. *Problemy Peredachi Informatsii*, 23(3):3–17, 1987.

98. N. J. A. Sloane. *The on-line encyclopedia of integer sequences: A008277.* 2003.

99. N. J. A. Sloane. *The on-line encyclopedia of integer sequences: A000670.* 2003.

100. T. A. B. Snijders and K. Nowicki. Estimation and prediction for stochastic blockmodels for graphs with latent block structure. *Journal of Classification*, 14(1):75–100, 1997.

101. N. Stanley, S. Shai, D. Taylor, and P. J. Mucha. Clustering network layers with the strata multilayer stochastic block model. *IEEE Transactions on Network Science and Engineering*, 3(2):95–105, 2016.

102. B. Söderberg. General formalism for inhomogeneous random graphs. *Physical Review E*, 66(6):066121, 2002.

103. T. Vallès-Català, F. A. Massucci, R. Guimerà, and M. Sales-Pardo. stochastic block models reveal the multilayer structure of complex networks. *Physical Review X*, 6(1):011036, 2016.

104. T. Vallès-Català, T. P. Peixoto, R. Guimerà, and M. Sales-Pardo. On the consistency between model selection and link prediction in networks. May 2017. arXiv: 1705.07967.

105. Y. J. Wang and G. Y. Wong. Stochastic blockmodels for directed graphs. *Journal of the American Statistical Association*, 82(397):8–19, 1987.

106. Y. X. R. Wang and P. J. Bickel. Likelihood-based model selection for stochastic block models. *Annals of Statistics*, 45(2):500–528, 2017.

107. F. Y. Wu. The Potts model. *Reviews of Modern Physics*, 54(1):235–268, 1982.

108. K. Xu and A. Hero. Dynamic stochastic blockmodels for time-evolving social networks. *IEEE Journal of Selected Topics in Signal Processing*, 8 (4):552–562, 2014.

109. X. Yan. Bayesian model selection of stochastic block models. 2016. arXiv: 1605.07057.

110. X. Yan, C. Shalizi, J. E. Jensen, F. Krzakala, C. Moore, L. Zdeborová, P. Zhang, and Y. Zhu. Model selection for degree-corrected block models. *Journal of Statistical Mechanics: Theory and Experiment*, 2014(5): P05007, 2014.

111. W. W. Zachary. An information flow model for conflict and fission in small groups. *Journal of Anthropological Research*, 33(4):452–473, 1977.

112. L. Zdeborová and F. Krzakala. Statistical physics of inference: thresholds and algorithms. *Advances in Physics*, 65(5):453–552, 2016.

113. X. Zhang, C. Moore, and M. E. J. Newman. Random graph models for dynamic networks. *European Physical Journal B*, 90(10):200, 2017.

12

Structured Networks and Coarse-Grained Descriptions: A Dynamical Perspective

Michael T. Schaub[1,2], Jean-Charles Delvenne[3], Renaud Lambiotte[4,5], and Mauricio Barahona[6,7]

[1]Institute for Data, Systems and Society, Massachusetts Institute of Technology
[2]Department of Engineering Sciences, University of Oxford
[3]ICTEAM and CORE, Université Catholique de Louvain
[4]Mathematical Institute, University of Oxford
[5]Entités de recherche: Institut Namurois des systémes complexes (naXys) University of Namur
[6]Department of Mathematics, Imperial College London
[7]EPSRC Centre for Mathematics of Precision Healthcare, Imperial College London

12.1 Introduction

The language of networks and graphs has become a ubiquitous tool to formalize and analyze systems and relational data across scientific disciplines, from biology to physics, from computer science to sociology [50]. Accordingly, scholars from a variety of areas have investigated such networks from different angles, developing diverse computational and mathematical toolboxes in order to analyze and ascribe meaning to the different patterns found in specific networks of interest. Modular structures are one of the most commonly studied features of networks in this context [24, 25, 58, 67, 74]. Yet, as highlighted by the lack of a common terminology (modules, partitions, blocks, communities, and clusters are but a few terms commonly found to denote various notions of modular structure in the literature), *why* scholars are interested in modular structures and *how* these structures are construed can be broadly different. Hence the perspective adopted when studying the modular structure in networks must depend on the

Advances in Network Clustering and Blockmodeling, First Edition.
Edited by Patrick Doreian, Vladimir Batagelj, and Anuška Ferligoj.
© 2020 John Wiley & Sons Ltd. Published 2020 by John Wiley & Sons Ltd.

context and specific application in mind [72, 74]. In the following, we focus on one particular motivation: namely, the rich interplay between network structure and a dynamics acting on top of the network as a means of identifying modules in the network or describing the effect that modules can have on the dynamical behavior of a system.

Why a Dynamical Perspective?

One of the main motivations for identifying modular structures in networks is that they provide a simplified, coarse-grained description of the system structure. Think of a social network, in which we might be able to decompose the system into (overlapping) groups of people such as circles of friends. We may then represent the system in terms of the interactions between these different groups, thereby reducing the complexity of the description. The hope is not only to arrive at a more compact structural description but also that the obtained modules can be interpreted as building blocks with a functional meaning.

For instance, consider the well-known Karate Club network studied by Zachary [85], representing the social interactions between members of a karate club that eventually split into two factions after a dispute. An interesting feature of this network is the fact that the split of the club is commensurate with the graph structure: if we apply graph partitioning methods to this network, the partition into two groups found is commonly well aligned with the split that occurred in reality. While the example of the Karate Club is by no means to be taken as a general indication of the relationship between structure and function, or between network structure and any other type of external data [55], it highlights the ultimate rationale for the detection of modules is often to gain insight into the system behavior. For instance, we might be interested in how rumors spread in a social network, or opinions are formed. To understand such processes, we need to take into account the system structure but we also need an understanding of the *dynamics* that acts on top of this structure, since the system behavior is the result of the interplay between the *structure* of a network and the *dynamics* acting on top of it. We thus aim to gain a reduced description of the system that takes into account both its structure and dynamics.

Dynamics on Networks or Dynamics of Networks?

We should make a distinction here between the dynamics of the network structure itself, which we call *structural dynamics* in the sequel, and dynamical processes that happen on top of a *fixed* network structure.

On the one hand, a social network can be subject to a structural dynamics over time as people become acquainted or start to dislike each other so that links and nodes appear, disappear or change weight (e.g., if we see who follows whom on twitter, who declares to be friends on Facebook, etc). The study of how these structures vary over time can be of central importance, e.g. for the spread of pathogens that can spread faster or slower depending on contact patterns. See [32] for an overview and further references on these topics.

On the other hand, data may often be naturally interpreted as a dynamics evolving and supported on a latent, unobserved fixed network. For instance, communication patterns between different people (e.g. on an online social network, an email or a mobile phone call network) may be thought as a type of point process that activates latent links at particular times [87]. The sequence of activation patterns may not be completely random at each step, but have a certain type of path dependence or memory (e.g. travellers traversing a network of flight connections from one to another city [64]). Hence, while the information recorded is temporal, the underlying network itself may be interpreted as a quasi-static object on which a path-dependent dynamics occurs.

There are of course other systems in which the dynamics on the network and the structural dynamics of the network influence each other leading to an evolution of the network structure that reflects the prevalent dynamic patterns on it. For instance, neuronal networks are known to have high plasticity and adjust their weight structure (links) based on the activity of their nodes (neurons), a feature that is commonly associated with learning.

Whether one should focus on structural dynamics, dynamics on top of a network, or both is therefore dependent on what the network representation aims to capture. In reality, all of these viewpoints are ultimately abstractions and thus attempts to capture different aspects of real world systems which hopefully provide additional insight into their behavior.

Network Dynamics: the Scope of this Chapter

Our focus here will be on dynamical processes acting on top of networks. We thereby assume that the underlying (latent) network structure is known and approximately constant over the time scales of the observed dynamics. Hence, we largely omit the issue of structural dynamics, even though this may not be justified in certain applications. In the sequel, we will show that this approach is fruitful in many contexts, yielding insights that go beyond purely structural network analysis. While clearly important, the joint treatment of structural dynamics in conjunction with dynamics on networks has received less attention in the literature and requires a more elaborate mathematical machinery that goes beyond the scope of this chapter. Furthermore, we do not consider here the question of how and why the networks have arisen in the first place. (A reader interested in these questions may refer to some of the other chapters in this book.) We will therefore assume that the observed network is well defined, i.e. we treat it as an empirical reality with low uncertainty. The dynamical perspective adopted here is especially useful in such cases: the network is specified, but the emergent behavior (our object of interest) might be hard to grasp due to the complexity of the system.

More explicitly, think again of the Karate Club example. From a statistical perspective, one might want to answer the question of why the structure of the network is as observed. We may adopt a generative model (e.g. a stochastic blockmodel) and assume that the observed network is a random realization from this model. We could then attempt to find a classification of the nodes such that the observed link probabilities between blocks of nodes reflect the observed block structure, hence explaining parsimoniously the main features in the data [76]. Using this perspective, we assume that if we could repeat the experiment that created the network multiple times, the realization of the network would be different each time, and we want our model to correspond to the simplest generative process consistent with those observations. In many circumstances this is a hypothetical question, however, as we only have access to a single observed network, and thus need to assume that our class of models (e.g. stochastic blockmodels) provides a suitable approximate depiction of all important features of the network.

Here we ask a complementary question: given the particular network we observe and an endowed dynamics taken place on it, are there partitions aligned with this process? For the Karate Club this could give an indication of whether the split of the club was facilitated by how its particular network structure influenced the opinion formation process in this social network. Irrespective of the network's genesis, these types of questions are of interest in many areas and underpin our perspective in this chapter.

Outline of this Chapter

We divide this chapter into three parts. In the first part, we introduce the general mathematical setup for the types of dynamics we consider throughout the chapter. We provide two guiding examples, namely consensus dynamics and difussion processes (random walks), motivating their connection to social network analysis, and provide a brief discussion on the general dynamical framework and its possible extensions.

In the second part, we focus on the influence of graph structure on the dynamics taking place on the network, focusing on three concepts that allow us to gain insight into this notion. First, we describe how time scale separation can appear in the dynamics on a network as a consequence of graph structure. Second, we discuss how the presence of particular symmetries in the network give rise to invariant dynamical subspaces that can be precisely described by graph partitions. Third, we show how this dynamical viewpoint can be extended to study dynamics on networks with signed edges, which allow us to discuss connections to concepts in social network analysis, such as structural balance.

In the third part, we discuss how to use dynamical processes unfolding on the network to detect meaningful network substructures. We then show how different such measures can be related to seemingly different methods for community detection and coarse-graining proposed in the literature. We conclude with a brief summary and highlight interesting open future directions.

Our account is geared towards conveying intuition rather than covering technical details. We provide pointers to additional literature with detailed results throughout the text.

Notation

For simplicity, in the following we consider mainly undirected, connected graphs with n nodes (vertices) and m links (edges). Our ideas extend to directed graphs, however, and we provide appropriate references to the literature for the interested reader as we go along. The topology of a graph is encoded in the weighted adjacency matrix $\mathbf{A} \in \mathbb{R}^{n \times n}$, where A_{ij} is the weight of the link between node i and node j. Clearly, for an undirected graph $\mathbf{A} = \mathbf{A}^\top$. Typically, most graphs are unsigned (i.e. $A_{ij} \geq 0$, $\forall i, j$). The weighted out-degrees (or strengths) of the nodes are given by the vector outdeg $= \mathbf{d} = \mathbf{A}\mathbf{1}$, where $\mathbf{1}$ is the $n \times 1$ vector of ones. For a given vector $\mathbf{v}$, we will sometimes define the associated diagonal matrix diag($\mathbf{v}$) with elements v_i on the diagonal and zero elsewhere. For instance, we define the diagonal matrix of degrees $\mathbf{D} = \text{diag}(\mathbf{d})$ and denote the total weight of the edges by $w = \mathbf{1}^\top \mathbf{D}\mathbf{1}/2 = \mathbf{1}^\top \mathbf{d}/2$.

The combinatorial graph Laplacian is defined as $\mathbf{L} = \mathbf{D} - \mathbf{A}$, while the normalized graph Laplacian is defined as $\mathbf{L}_N = \mathbf{D}^{-1/2}\mathbf{L}\mathbf{D}^{-1/2} = \mathbf{I} - \mathbf{D}^{-1/2}\mathbf{A}\mathbf{D}^{-1/2}$. Both these Laplacians are symmetric positive semi-definite, with a simple zero eigenvalue when the graph is connected [16, 28]. When describing diffusion processes on graphs, it is also useful to define the (asymmetric) random walk Laplacian $\mathbf{L}_{RW} = \mathbf{D}^{-1}\mathbf{L}$, which is isospectral with the normalized Laplacian for undirected graphs.

We will also consider *signed graphs*, where the weights A_{ij} can be positive or negative. In the case of signed graphs, we define the vector of absolute degrees $\mathbf{d}_S = |\mathbf{A}|\mathbf{1}$, where the absolute value is taken element-wise, with the corresponding absolute degree matrix $\mathbf{D}_S = \text{diag}(\mathbf{d}_S)$. For signed networks, we will define the signed Laplacian $\mathbf{L}_S = \mathbf{D}_S - \mathbf{A}$, which is also positive semi-definite. The signed Laplacian reduces to the combinatorial Laplacian in the case of an unsigned graph.

A (hard) partition $\mathscr{P}$ of a graph of n nodes into k cells $\{\mathscr{C}_i\}_{i=1}^{k}$ can be encoded by an indicator matrix $\mathbf{C} \in \{0, 1\}^{n \times k}$, with entries $C_{ij} = 1$ if node i is part of cell $\mathscr{C}_j$ and $C_{ij} = 0$ otherwise. Hence the columns of $\mathbf{C}$ are the indicator vectors $\mathbf{c}^{(i)}$ of the cells:

$$\mathbf{C} := [\mathbf{c}^{(1)}, \dots, \mathbf{c}^{(k)}]. \tag{12.1}$$

12.2   Part I: Dynamics on and of Networks

12.2.1   General Setup

In its most general form, we are interested in dynamical systems of the form:

$$\dot{\mathbf{x}}(t) = \mathscr{A}(\mathbf{x}(t), t)\, \mathbf{x}(t) + \mathscr{B}(t)\, \mathbf{u}(t) \qquad \mathscr{A} \in \mathbb{R}^{\ell \times \ell}, \quad \mathscr{B} \in \mathbb{R}^{\ell \times p} \tag{12.2a}$$

$$\mathbf{y}(t) = \mathscr{D}(t)\, \mathbf{x}(t) \qquad\qquad\qquad\qquad \mathscr{D} \in \mathbb{R}^{c \times \ell}, \tag{12.2b}$$

where $\mathbf{x} \in \mathbb{R}^{\ell}, \mathbf{y} \in \mathbb{R}^{c}, \mathbf{u} \in \mathbb{R}^{p}$ are the state, the observed state, and the input vectors of the system, respectively. Discrete time versions are also of interest [19, 20], but we will stick to the continuous time version henceforth.

In the context of networked systems, the system of ordinary differential equations Equation (12.2) arises by endowing each node with one or more state variables, whose union corresponds to the state vector $\mathbf{x}$. In general, the matrix $\mathscr{A}$ is linked to the network: a time-varying, state-dependent coupling between the state variables of the agents (nodes). A set of exogenous inputs, described by the vector $\mathbf{u}$, acts on the state variables through the input matrix $\mathscr{B}$. In such a system, we may not be able to observe and measure all the system states. This is captured by the fact that the output $\mathbf{y}$ is a linear transformation of $\mathbf{x}$. This framework can naturally account for weighted, signed or other types of interactions. Furthermore, the fact that each node can be endowed with several state variables allows for the modeling of higher order dynamics (e.g. higher order Markov processes) [21, 30, 64, 65]. Note that this form also allows for the inclusion of exogenous inputs, a factor usually neglected in standard network analyses, although it has recently gained prominence for the problem of controlling networks [44].

The system Equation (12.2) formally describes the full coupled dynamics *of* and *on* a network, since the network (encapsulated in the matrix $\mathscr{A}(\mathbf{x}(t), t)$) is both state and time-dependent. However, such systems are difficult to analyze in general. When the coupling $\mathscr{A}(t)$ is only time-dependent, the system describes the dynamics on a time-varying network. Such linear time-varying models have a long history in systems and control theory, and there is a rich literature pertaining to their analysis [12, 35] at the expense of more advanced mathematical machinery. Although a growing literature in *dynamical social network analysis* melding such concepts from control and dynamical systems with social network analysis (e.g. for opinion formation [59]) has recently emerged, such models have been comparably less studied within the scope of network theory.

Dynamics on Fixed Networks

To simplify our exposition, we will here assume that the (latent) coupling remains constant over time, i.e. $\mathscr{A}, \mathscr{B}, \mathscr{D}$ have no explicit time dependence. This is what we have termed *dynamics on a (fixed) network.* It is important to remark that this assumption does not imply that each link is constantly activated over time, but that it is available for a potential interaction [21, 30, 64, 65]. We will also assume that $\ell = p = c = n$, which implies that there is only one state variable per node:

$$\dot{\mathbf{x}} = \mathscr{A}\, \mathbf{x} + \mathscr{B}\, \mathbf{u}(t) \qquad \mathscr{A} \in \mathbb{R}^{n \times n}, \quad \mathscr{B} \in \mathbb{R}^{n \times n} \tag{12.3a}$$

$$\mathbf{y} = \mathscr{D}\, \mathbf{x} \qquad\qquad\qquad \mathscr{D} \in \mathbb{R}^{n \times n}. \tag{12.3b}$$

In the following, we consider examples of this simpler form, which provide rich insights into problems of interest in practical applications. Specifically, we first consider consensus dynamics and its variants (motivated by opinion formation), followed by diffusion processes and random walk dynamics (motivated by information propagation).

12.2.2   Consensus Dynamics

Consensus is one of the most popular and well studied dynamics on networks [33, 52, 53, 61, 62, 84], and can be thought of as a linear version of synchronization [9, 34]. The attractiveness of consensus lies in its analytic tractability and simplicity, which nevertheless provides a good first description of some fundamental behaviors. For instance, in the socio-economic domain, consensus provides a model for opinion formation in a society of individuals, whereas in engineering systems, consensus constitutes a basic building block for efficient distributed computation of global functions in networks of sensors, robots, or other agents [33]. For a recent survey of consensus processes with a particular focus on opinion formation, we refer the reader to Proskurnikov *et al.* [59].

To define the standard consensus dynamics, consider a given connected network of n nodes with adjacency matrix $\mathbf{A}$. Let us endow each node with a scalar state variable $x_i \in \mathbb{R}$. The consensus dynamics on such a network is defined as:

$$\dot{\mathbf{x}} = -\mathbf{L}\mathbf{x}, \qquad \text{(consensus dynamics)} \tag{12.4}$$

where $\mathbf{L}$ is the graph Laplacian. Clearly, the consensus dynamics amounts to

$$\dot{x}_i = -D_{ii}x_i + \sum_j A_{ij}x_j = -\sum_j A_{ij}(x_i - x_j), \ \forall i,$$

i.e. each node adjusts its state such that the difference to its neighbors is reduced. The name of the dynamics derives from the fact that for any given initial state $\mathbf{x_0} = \mathbf{x}(0)$, the differential equation Equation (12.4) drives the state to a global consensus state, where the state variables of all nodes are equal to the arithmetic average of the initial node states: $x_i = x_*$ $\forall i$, where $x_* = \mathbf{1}^\top \mathbf{x_0}/n$ as $t \to \infty$. Relative to our framework Equation (12.3), the standard consensus dynamics Equation (12.4) corresponds to $\mathscr{A} = -\mathbf{L}$, $\mathscr{D} = \mathbf{I}$, $\mathbf{u} = \mathbf{0}$.

Intuitively, this dynamics may be interpreted as an opinion formation process on a network of agents who, in the absence of further inputs, will eventually agree on the same value of their state (opinion), namely, the average opinion of their initial states. Figure 12.1b shows an example of the consensus dynamics on the Karate Club network starting from a random initial condition for the agents and converging asymptotically towards the common, final opinion. Yet the network structure plays a role in the form in which this opinion is approached: the opinions of each of

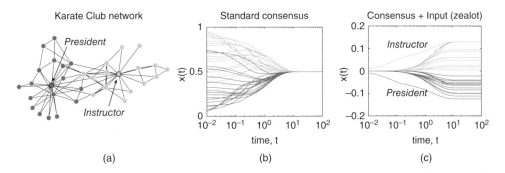

Figure 12.1   Consensus dynamics on the Karate Club network. (a) The Karate Club network originally analyzed by Zachary [85] with nodes colored according to the split that occurred in the real case.
(b) Consensus dynamics on the Karate Club network starting from a random initial condition. As time
progresses, the states of the individual nodes become more aligned and eventually reach the consensus
value equal to the arithmetic average of the initial condition. Note that above in the time scale given by the
eigenvalue $1/\lambda_2(\mathbf{L}) \approx 1/0.47$, the agents converge into two groups that reflect the observed split before
converging to global consensus (see Section 12.3.1). (c) If an external input is applied to the system (see
text), the opinion dynamics will in general not converge to a single value but lead to a dispersed set of final
opinions, which still reflect the split observed in reality.

the two factions (as recorded by their eventual split in real life) converge earlier towards a group
opinion with higher cohesion.

While in the absence of external inputs the standard consensus dynamics converges to a fixed
point, the framework Equation (12.3) allows us to explore the influence of inputs over time
$\mathbf{u}(t) \neq \mathbf{0}$, e.g. by external agents, media, etc. In that case, the asymptotic convergence of the
dynamics to an eventual consensus is not guaranteed. For instance, some agents may behave
like zealots, who do not update their opinion as described above, but give more weight to their
own opinion [1, 47, 60]. Let us consider the Karate Club network with a constant external input:

$$\dot{\mathbf{x}} = -\mathbf{L}\mathbf{x} + \mathbf{u}, \qquad (12.5)$$

with $u_{\text{president}} = -1$ for the president, $u_{\text{instructor}} = +1$ for the instructor, and all other nodes have
no input $u_i = 0$. This can be thought of as a simplified model of a zealot-like behavior of these
two agents. In this case, there is no final consensus reached within the system: the final opinion
of each of the agents is dispersed between the extreme positions taken by the instructor and
the president (Figure 12.1c). Importantly, the final opinions of the agents are well aligned with
the split that eventually occurred in the Karate Club, in which half of the members joined the
instructor to form a new club and the other half stayed with the president.

These results highlight how the graph properties (encapsulated by the graph Laplacian L) can
shape and constrain the dynamics on the network, and thus influence the observed behavior of
the system.

Discussion: More Detailed Consensus Models

The consensus dynamics studied here is chosen for its simplicity. Of course, real-world
systems the process of opinion formation is much more complex. For instance, opinions can be

interlinked and part of a belief system [26, 59], update and gossiping processes may be nonlinear or asynchronous [34, 53, 84], and noisy external inputs may influence the process [83]. All these factors lead to a much more complex dynamics. In particular, opinions may not converge to a single value, or might stabilize to different values in different parts of the network. See [59] and references therein for a discussion on the so-called social cleavage problem.

12.2.3 Diffusion Processes and Random Walks

Random walks are another important dynamical process which can naturally evolve on graphs. Random walks are often taken as a (simple) proxy for diffusive processes and, like consensus processes, these type of models have found applications in various domains, including information diffusion in social networks. Other applications of such processes include searching on networks, web browsing, dimensionality reduction via diffusion maps, respondent driven sampling, and, indeed, community detection [2, 46]. Perhaps the most popular example is the celebrated PageRank algorithm, which is used for the ranking of webpages, and can be seen as an application of random walks on directed networks.

The dynamics of a continuous-time unbiased random walk on a network with combinatorial Laplacian $\mathbf{L}$ is governed by the Kolmogorov forward equation:

$$\dot{\mathbf{p}}^{\mathsf{T}} = -\mathbf{p}^{\mathsf{T}}\mathbf{L}_{\mathrm{RW}} \qquad \text{(random walk)} \qquad (12.6)$$

where $\mathbf{L}_{\mathrm{RW}} = \mathbf{D}^{-1}\mathbf{L}$ is the random walk Laplacian. This equation describes the diffusion of a random particle on the network; specifically, the time evolution of the probability mass function $\mathbf{p}(t)$ of an n-dimensional random vector $\mathbf{x}(t)$ with components $x_i(t) = 1$ if the particle is at node i at time t and zero otherwise. For a connected, undirected network, this dynamics will converge to the stationary distribution $\boldsymbol{\pi} = \mathbf{d}/2w$. For the general case of a directed graph, see [20, 40, 41].

An illustration of the time evolution of such a random walk on the Karate Club network is displayed in Figure 12.2. As time progresses, the process converges towards the distribution $\boldsymbol{\pi}$, which is proportional to the degree vector. It is known that the degree is also a simple centrality measure for the nodes of a graph [41], and this observed behavior highlights how the centrality of the instructor and the president (the highest degree nodes) is also of dynamical importance. Figure 12.2b also illustrates the notion of information propagation: two random walks started one from the instructor or alternatively from the president spread initially mostly within their natural groups before eventually spreading across the whole network.

Discussion: Random Walks and Consensus as Dual Processes

It is worth remarking that the random walk dynamics Equation (12.6) may be seen as the dual of a non-standard consensus dynamics of the form $\dot{\mathbf{x}} = -\mathbf{D}^{-1}\mathbf{L}\mathbf{x}$, which (unlike the standard consensus Equation (12.4)) converges to a different final value: a *degree-weighted* average of the initial node-states. Conversely, a combinatorial Laplacian dynamics of the form $\dot{\mathbf{p}}^{\mathsf{T}} = -\mathbf{p}^{\mathsf{T}}\mathbf{L}$ can also be seen as describing a different random walk, which is the dual of the standard consensus Equation (12.4).

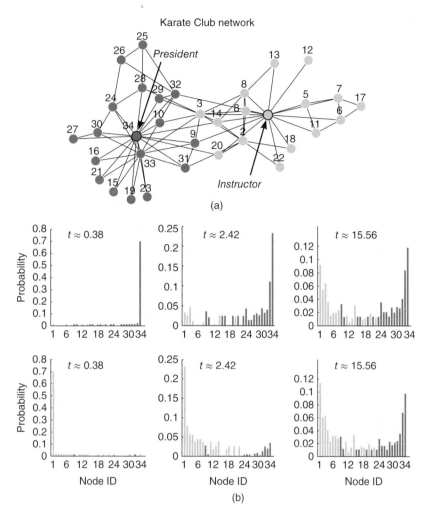

Figure 12.2   Illustration of the evolution of a random walk dynamics on the Karate Club network. (a) The Karate Club network with labelled nodes. The factions of the ultimate split observed in reality are indicated by color (grayscale). (b) The evolution of the probability distribution of the random walk over time exemplified at three time snapshots from two different initial conditions: the random walker starts at time $t = 0$ at the president node 34 (upper three panels) or at the instructor node 1 (lower three panels). As time evolves, the probability of the walker being found on the other nodes becomes more spread out on the graph and eventually reaches the stationary distribution $\pi = \mathbf{d}/2w$. Note that for short times, the probability is spread mostly within the corresponding factions (i.e. president for the top panels; instructor for the bottom panels). Beyond the slowest time scale in this dynamics given by $1/\lambda(\mathbf{L}_{\mathrm{RW}}) \approx 1/0.13$, the random walk becomes well mixed and the information spreads across factions (see Section 12.3.1 for more details).

12.3 Part II: The Influence of Graph Structure on Network Dynamics

12.3.1 Time Scale Separation in Partitioned Networks

Standard Time Scale Separation

The classic concept of time scale separation in a dynamical system follows from the simple system:

$$\frac{dx}{dt} = f(x, y), \tag{12.7a}$$

$$\varepsilon^{-1}\frac{dy}{dt} = g(x, y), \tag{12.7b}$$

where f, g are bounded functions and ε is a small constant relative to those bounds. By a simple rescaling we may assume that f, g are of order 1 and $\varepsilon \ll 1$. In this system, $x(t)$ changes much more rapidly than $y(t)$ since $dy/dt = \varepsilon g(x, y)$, which is small by construction. An alternative, equivalent statement follows from defining a slow time variable $\tau := \varepsilon t$, hence Equation (12.6) can be rewritten as $dy/d\tau = g(x, y)$. From this rewriting, it follows that there is a *separation of time scales* in the dynamics, where y evolves according to the slow time scale τ and x evolves according to the faster t.

When a time scale separation is present in a system, it can be exploited to simplify its analysis in various ways. If we are interested in the short-term behavior of the system, we may effectively treat y as a fixed parameter and ignore its time evolution, leading to an effective one-dimensional system description based on the *fast* time scale, for instance describing the (fast) convergence of x to the fixed point $x_*(y)$. On the other hand, if we are interested in the long-term behavior of the system, then it is y we are most interested in. Since the dynamics of x is faster than that of y, we may assume that $x \simeq x_*(y)$ at all times beyond an initial transient, leading to a one-dimensional system for the evolution of $y(t)$. Using these simplifications will, of course, lead to errors when comparing the approximation to the actual time-evolution. However, the error can be bounded through time scale separation theory.

In summary, when there are two well-separated time scales in the system (t and $\tau = \varepsilon t$) the dynamics almost decouples into two different regimes: for the fast behavior, we may simply concentrate on x, whereas for the slow, long-term behavior we may focus on y and forget about the detailed dynamics of x. When several *distinct* time scales are present, we can similarly approximate the dynamics over particular time scales by reduced dynamics that can be obtained by finding quasi-invariant subspaces in the original system. These concepts emerge naturally in the study of networked dynamics, as we discuss below.

Time Scales in Consensus Dynamics on Networks

Time scale separation can appear naturally in the context of dynamics on networks. For simplicity, we will describe the results here in the context of consensus dynamics, though translating these ideas to diffusion processes is straightforward.

Given an initial condition $\mathbf{x}_0$, linear systems theory tells us that the solution to the consensus dynamics Equation (12.4) is given by

$$\mathbf{x}(t) = \exp(-\mathbf{L}t)\,\mathbf{x}_0,$$

where $\exp(\cdot)$ denotes the matrix exponential. Writing the full solution in this manner obscures the time scales present in the evolution of $\mathbf{x}(t)$, and how they get mixed via the network interactions represented by the Laplacian $\mathbf{L}$. To reveal the characteristic time scales, we use the spectral decomposition of the Laplacian (if the graph is undirected), or a slightly more careful, yet related, treatment for directed graphs [41].

For simplicity of exposition, let us assume that the graph is undirected (i.e. $\mathbf{L} = \mathbf{L}^\top$). Let us denote the eigenvectors of $\mathbf{L}$ by $\mathbf{v}_i$ with associated eigenvalues λ_i in increasing order $0 = \lambda_1 \leq \lambda_2 \leq \dots \leq \lambda_n$. The spectral decomposition of the Laplacian is then

$$\mathbf{L} = \sum_{i=1}^{n} \lambda_i \mathbf{v}_i \mathbf{v}_i^\top .$$

Accordingly, the solution of the consensus dynamics can be written in the spectral basis as

$$\mathbf{x}(t) = \sum_{i=1}^{n} e^{-\lambda_i t} \, \mathbf{v}_i \mathbf{v}_i^\top \mathbf{x}_0 = \sum_{i=1}^{n} (\mathbf{v}_i^\top \mathbf{x}_0) e^{-\lambda_i t} \, \mathbf{v}_i .$$

In this rewriting, the time scales of the process become apparent: they are dictated by the eigenvalues of the Laplacian matrix, with each eigenvector (eigenmode) decaying with a characteristic time scale $\tau_i = 1/\lambda_i$. Hence, if there are large differences (gaps) between eigenvalues, the system will have time scale separation. For instance, if the k smallest eigenvalues $\{\lambda_1 = 0, \dots, \lambda_k\}$ are well separated from the remaining eigenvalues such that $\lambda_k \ll \lambda_{k+1}$, the eigenmodes associated with $\{\lambda_{k+1}, \dots, \lambda_n\}$ become negligible for $t > 1/\lambda_{k+1}$ and it follows that the system can be effectively described by the k smallest eigenmodes for $t > 1/\lambda_{k+1}$. More technically, we say that the first k eigenvectors form a dominant invariant subspace of the dynamics and there exists an associated lower dimensional ($k < n$) description of the dynamics on the network which is valid after the time scale $1/\lambda_{k+1}$. A natural question is: how is the time scale separation that emerges from the spectral properties of the Laplacian related to the network structure? In the case of networks, time scale separation is typically associated with a lower dimensional description of the dynamics which is aligned with *localized substructures* in the graph, as we illustrate through the following example.

Example: A Modular Partitioned Network Structure Induces Time Scale Separation

To illustrate the discussion above, let us consider a network composed of k modules with strong in-block coupling and weaker inter-block coupling, as given by the adjacency matrix

$$\mathbf{A} = \begin{pmatrix} \mathbf{A}_1 & & & \\ & \mathbf{A}_2 & & \\ & & \ddots & \\ & & & \mathbf{A}_k \end{pmatrix} + \mathbf{A}_{\text{random}} := \mathbf{A}_{\text{structure}} + \mathbf{A}_{\text{random}}. \tag{12.8}$$

Here $\mathbf{A}_{\text{random}}$ can be interpreted as a realisation of an Erdös–Rényi (ER) random graph with unstructured, sparse connectivity, whereas the $\mathbf{A}_i$ are the adjacency sub-matrices of the individual modules that have higher random connectivity inside. Here, we model these modules also as ER graphs that possess a higher connectivity probability than $\mathbf{A}_{\text{random}}$ (Figure 12.3). How does the structure present in the graph affect the spectrum and eigenvectors of the corresponding Laplacian, $\mathbf{L} = \mathbf{L}_{\text{structure}} + \mathbf{L}_{\text{random}}$?

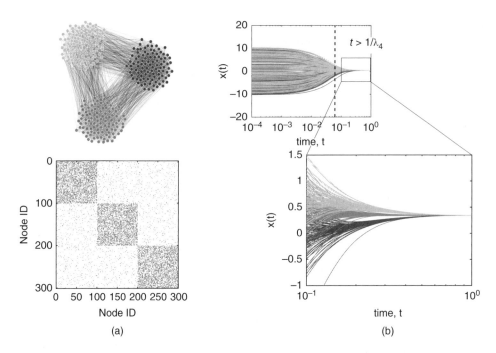

Figure 12.3 Consensus dynamics on a structured network. (a) Visualization of the network and adjacency matrix of an unweighted structured network with three groups of the form Equation (12.8). (b) The consensus dynamics on this network displays time scale separation: after $t \approx 1/\lambda_4 = 0.06$, approximate consensus is reached within each group (groups indicated by color/grayscale) followed by global consensus across the whole network. A similar effect can be observed in the consensus dynamics and random walk on the Karate Club network in Figures 12.1 and 12.2.

Let us first consider the case where $\mathbf{A}_{\text{random}} = 0$, i.e. the graph consists of k disconnected components. In that case, it is easy to see that we have a repeated eigenvalue $\lambda = 0$ with multiplicity k and the associated k-dimensional eigenspace can be spanned by the k indicator vectors $\mathbf{c}^{(j)} \in \mathbb{R}^n$ defined in Equation (12.1), which are localized on the blocks in the graph.

To gain insight into the case where $\mathbf{A}_{\text{random}} \neq 0$, we invoke matrix perturbation theory and random matrix theory [10, 78, 81, 82]. For a network of the form Equation (12.8), a form of the Davis–Kahan theorem [8, 42, 63, 81] provides bounds on the angular distance between the subspace $\mathscr{S}$ spanned by the block-vectors $\{\mathbf{c}^{(j)}\}_{i=1}^{k}$ and the subspace $\mathscr{S}'$ spanned by the eigenvectors of $\mathbf{L}$ associated with the k smallest eigenvalues.

Intuitively, Davis–Kahan states that if the random component is small, then $\mathscr{S} \approx \mathscr{S}'$. The implication is that the dominant invariant subspaces will be commensurate with the structural decomposition of the network in terms of the block-vectors. Hence the long-term dynamics will directly reflect the structural decomposition of the network. In other words, the time scale separation in such a networked system takes an intuitive meaning: quasi-consensus is reached more quickly within each block, while global consensus is only reached on a longer time scale.

These points are illustrated numerically in Figure 12.3, where we show how the consensus dynamics evolves on a network of the form Equation (12.8) consisting of three groups with 100 nodes each. As shown in Figure 12.3b, the dynamics becomes effectively low dimensional after around $t \simeq 1/\lambda_4$, beyond which it is well approximated by the three dominant eigenmodes aligned with the intrinsic blocks in the network.

Discussion: Time Scale Separation Beyond Homogeneous Block Structures

It is important to emphasise that time scale separation may also be induced by network structures that are not block-homogeneous. Many networks contain natural non-clique-like substructures (e.g. ring-shaped), which may act effectively as a dynamical substructure over a particular time scale [69, 70]. The presence of such substructures will too affect the observed consensus dynamics on the network.

Furthermore, our discussion is not limited to the case where the structure of the network is block-diagonal, but can be extended seamlessly to networks consisting of a low-rank structure plus a random noise component [10, 49, 56, 86]. Such networks encompass stochastic block-models [42, 63], although the spectral properties of a realization of a stochastic blockmodel may not be concentrated around their expectation when the network is very sparse [18, 37, 48]. For more details on stochastic blockmodels, see [6, 31, 76] and some of the other chapters in this book.

12.3.2 Strictly Invariant Subspaces of the Network Dynamics and External Equitable Partitions

Let us now consider another type of network structure that induces a specific form of exact coarse-graining of the dynamical process acting on the network [51, 71]: the so-called *external equitable partition* (EEP).

In order to introduce EEPs, we first recall the well known graph-theoretic notion of equitable partition [28]. An equitable partition splits the graph into a set $\{\mathscr{C}_i\}$ of non-overlapping groups of nodes called cells that fulfil the following condition: *for each node $v \in \mathscr{C}_i$, the number of connections to nodes in cell $\mathscr{C}_j$ is only dependent on i, j.* Stated differently, the nodes inside each cell of an equitable partition have the same out-degree pattern with respect to every cell. The EEP is a relaxed version of the equitable partition: the requirement on equal out-degree need only hold for the number of connections between *different* cells $\mathscr{C}_i, \mathscr{C}_j$, where $i \neq j$. EPs and EEPs are closely related to so-called orbit partitions and to symmetry properties of a graph, and may be detected using tools from computational group theory [54, 66, 77]. An example of a graph with an EEP is shown in Figure 12.4a.

The presence of an EEP in a network has important consequences for a dynamics taking place on it. To see this, we consider the algebraic characterization [13, 15, 22] of an EEP of a graph of n nodes into k cells encoded by the $n \times k$ indicator matrix $\mathbf{C}$ Equation (12.1). Associated with the EEP there is a quotient graph, a coarse-grained version of the original graph, such that each cell becomes a node and the weights between these new nodes are the total out-degrees between the cells in the original graph (Figure 12.4a). It can then be verified that

$$\mathbf{LC} = \mathbf{CL}^\pi, \tag{12.9}$$

where $\mathbf{L}^\pi$ is the $k \times k$ Laplacian of the quotient graph induced by $\mathbf{C}$:

$$\mathbf{L}^\pi = (\mathbf{C}^\top \mathbf{C})^{-1} \mathbf{C}^\top \mathbf{LC} = \mathbf{C}^+ \mathbf{LC}. \tag{12.10}$$

Here the $k \times n$ matrix $\mathbf{C}^+$ is the (left) Moore–Penrose pseudoinverse of $\mathbf{C}$. Observe that multiplying a vector $\mathbf{x} \in \mathbb{R}^n$ by $\mathbf{C}^\top$ from the left sums up the components within each cell, and that $\mathbf{C}^\top\mathbf{C}$ is a diagonal matrix with the number of nodes per cell on the diagonal. Hence, $\mathbf{C}^+ = (\mathbf{C}^\top\mathbf{C})^{-1}\mathbf{C}^\top$ can be simply interpreted as a *cell averaging operator* [51]. After straightforward algebraic manipulations it is easy to show that

$$\mathbf{C}^+\mathbf{L} = \mathbf{L}^\pi\mathbf{C}^+, \tag{12.11}$$

which summarizes the relationship between the cell averaging operator $\mathbf{C}^+$ and the Laplacians of the original and quotient graphs. Note that although the Laplacian of the original (undirected) graph $\mathbf{L}$ is symmetric, the Laplacian of the quotient graph $\mathbf{L}^\pi$ will be asymmetric in general.

Dynamical Implications of EEPs

The definition of the EEP Equation (12.9) implies an invariance of the partition encoded by $\mathbf{C}$ with respect to the Laplacian $\mathbf{L}$. Specifically, if we apply the Laplacian to the indicator matrix $\mathbf{C}$ we obtain a linearly rescaled (by $\mathbf{L}^\pi$) version of $\mathbf{C}$. A similar invariance of the cell averaging operator $\mathbf{C}^+$ with respect to $\mathbf{L}$ underpins Equation (12.11).

Let us expand on some of the consequences of this invariance. Equation (12.9) implies that the columns of $\mathbf{C}$ span an invariant subspace of $\mathbf{L}$. As the invariant subspaces of $\mathbf{L}$ are expressible in terms of its eigenvectors, it follows that there exists a set of eigenvectors of $\mathbf{L}$ whose components are constant on each cell of the partition. Furthermore, it can be shown that the eigenvalues associated with the eigenvectors spanning the invariant subspace are shared with $\mathbf{L}^\pi$ [51]. If $\mathbf{L}$ has degenerate eigenvalues, an eigenbasis can be chosen such that it is localized on the cells of the partition [51].

These algebraic properties of an EEP have implications for the dynamics dictated by $\mathbf{L}$, as we illustrate now for consensus dynamics [51]. First, an EEP is consistent with *partial consensus* such that the agreement within a cell (if present) is preserved. Specifically, let the initial state vector be given by $\mathbf{x} = \mathbf{C}\mathbf{y}$ for some arbitrary $\mathbf{y}$, so that every node within cell $\mathscr{C}_i$ has the same initial value y_i of their opinion variable. Then the nodes inside each cell remain identical *for all times* and the dynamics of the cell variables is governed by the quotient graph

$$\dot{\mathbf{x}} = \mathbf{C}\dot{\mathbf{y}} \quad \text{with} \quad \dot{\mathbf{y}} = -\mathbf{L}^\pi\mathbf{y}, \tag{12.12}$$

which follows directly from Equation (12.9). In words, if consensus has been reached within each EEP cell, then the lower dimensional Laplacian matrix $\mathbf{L}^\pi$ (with dimensionality equal to the number of cells k) describes the full dynamics of the system (Figure 12.4b). This dynamical invariance thus provides a simpler model of the system.

A second consequence of the presence of an EEP is that the dynamics of the cell-averaged states $\langle\mathbf{x}\rangle_{\mathscr{C}_i}$ is exactly described by the quotient graph:

$$\frac{d\langle\mathbf{x}\rangle_{\mathscr{C}_i}}{dt} = -\mathbf{L}^\pi\langle\mathbf{x}\rangle_{\mathscr{C}_i} \quad \text{where} \quad \langle\mathbf{x}\rangle_{\mathscr{C}_i} := \mathbf{C}^+\mathbf{x}. \tag{12.13}$$

This dynamical coarse-graining follows directly from Equation (12.11). Hence the cell-averaged dynamics is also governed by the lower dimensional quotient Laplacian (Figure 12.4c), and if we are only interested in such cell averages we can reduce our model significantly.

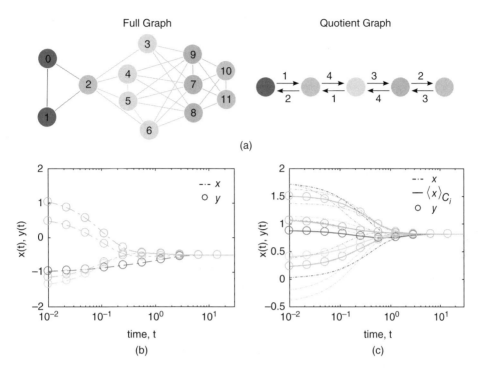

Figure 12.4 The external equitable partition and its dynamical implications. (a) A graph with $n = 12$ nodes (left) with an external equitable partition into five cells (indicated with colors) and its associated quotient graph according to the EEP (right). (b) Invariance of the EEP: the consensus dynamics on the full graph (Equation (12.4)) from an initial condition $\mathbf{x} = \mathbf{Cy}$ is shown with dash-dotted lines, whereas the associated quotient dynamics (Equation (12.12)) governing $\mathbf{y}$ is shown with circles. If all states within each cell are equal (i.e. cluster-synchronized), the dynamics will always remain cluster-synchronized and are described by the dynamics of the quotient graph for all times. (c) Cell-averaging dynamics of the EEP: for consensus dynamics, the quotient graph dynamics (circles) also describes the cell-averaged dynamics (solid lines) of the unsynchronized full graph dynamics (dash-dotted lines), as given by Equation (12.13).

Finally, a third implication of the EEP structure relates to the system with inputs. It can be shown [51] that all the results for the autonomous consensus dynamics with no inputs can be equivalently rephrased for the system with inputs

$$\dot{\mathbf{x}} = -\mathbf{Lx} + \mathbf{u}(t), \tag{12.14}$$

when the input $\mathbf{u}(t) = \mathbf{Cv}(t)$, $\mathbf{v}(t) \in \mathbb{R}^c$ is consistent with the cells of an EEP. In that case, the nodes inside each cell remain identical for all times, as in Equation (12.12).

Remark: While we have focused here on the implications of an EEP for linear consensus dynamics, invariant partitions like the EEP play a similar role for other linear and *nonlinear* dynamics (e.g. Kuramoto synchronization). See [71] for an extended discussion including synchronization dynamics, as well as dynamics on signed networks.

Discussion: Differences and Relationships between EEPs and Time Scale Separation

Let us discuss briefly the difference between the presence of an EEP in a network, and time scale separation in the system. In our context, both concepts can be related to the notion of (strictly or almost) invariant subspaces in the dynamics. However, the link between structure and dynamics that each of them represents can be very different.

The presence of an EEP is related to symmetries in the graph, which translate into the fact that a set of Laplacian eigenvectors have components that are constant on each cell in the graph. However, these eigenvectors can be associated to any eigenvalue of the graph, whether fast or slow. In broad terms, for the EEP the shape of the eigenvectors with respect to the cells is important, but the eigenvalues themselves are not relevant.

This notion is therefore different to the time scale separation discussed in Section 12.3.1, where the defining criterion focuses on the eigenvalues, more precisely, on the existence of gaps between eigenvalues that separate them into groups associated with different time scales. In our particular example of a planted partition model in Figure 12.3, the associated eigenvectors were indeed approximately constant on each cell (i.e. on each block of nodes) and would tend to align with the cells as the random part decreases. Hence in this case both the (approximate) EEP structure and the time scale separation are well aligned. However, this may not always the case. We may indeed have an EEP in which the set of eigenvectors corresponding to the cells are precisely the slowest eigenmodes, but this is not necessary. Conversely, the eigenvectors corresponding to the slowest time scales do not have to be exactly constant on every cell, or may not be block-structured in general [69, 70]. Therefore the notions of EEP and time scale separation are distinct but not mutually exclusive.

12.3.3   Structural Balance: Consensus on Signed Networks and Polarized Opinion Dynamics

Signed Networks and Structural Balance

In social networks, relationships can be friendly or hostile, or reflect either trust or distrust between individuals. The positive or negative character of links between agents is a central concept associated with the emergence of conflict and tension in social psychology, and has been studied within the classic literature in social network theory [14, 29]. More recently, the study of networks with *signed interactions* has gained popularity in the context of online social networks [43] and online cooperation [79]. More broadly, networks with signed edges are also essential to model biological systems and their dynamics. Examples include genes that promote or repress the expression of other genes in genetic regulatory networks [17], or neurons that can excite or inhibit the firing of other neurons in neuronal networks, thereby shaping the global dynamics of the system [27, 70].

Following Cartwright [14], a network is *structurally balanced* if the product of the signs along any closed path in the network is positive. This implies that only consistent social relationships are allowed in triangles of three nodes: either all interactions are positive or there are exactly two negative links, which may be interpreted in the sense that "the enemy of my enemy is my friend" [29]. Equivalently, a network is structurally balanced if it can be split into two factions, where each faction contains only positive interactions internally, while the connections between the two factions are purely antagonistic [4], as illustrated in Figure 12.5. It has been shown that

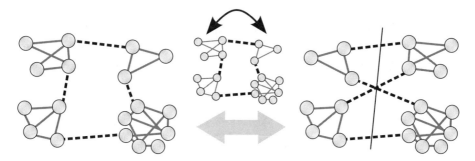

Figure 12.5 Example of a structurally balanced graph. A structurally balanced graph (left) with positive edges (solid, red), and negative edges (dashed, blue) can be redrawn in a different way (right) so that it can be split into two groups such that the negative edges connect nodes of different groups and links between nodes in the same group are positive.

many social networks are empirically close to structural balance [23], suggesting that there might be a structural dynamics *driving* social networks towards structural balance [45, 80].

Consensus Dynamics on Signed Networks and Polarization

There are several mathematical formulations that incorporate positive and negative network interactions. Here we consider systems where the interactions are mediated through a signed Laplacian matrix defined as [4, 39]:

$$\mathbf{L}_S = \mathbf{D}_S - \mathbf{A}, \tag{12.15}$$

where $\mathbf{D}_S = \text{diag}(|\mathbf{A}|\mathbf{1})$ is the diagonal absolute degree matrix, and the adjacency matrix $\mathbf{A}$ can contain both positive and negative weights. Like the standard Laplacian, the signed Laplacian $\mathbf{L}_S$ is positive semidefinite and its spectrum contains a zero eigenvalue when the graph is connected and structurally balanced [4, 39]. To see this, note that the signed Laplacian can be decomposed as:

$$\mathbf{L}_S = \mathbf{B}\mathbf{W}_{\text{abs}}\mathbf{B}^\top, \tag{12.16}$$

where $\mathbf{W}_{\text{abs}} = \text{diag}(|\mathbf{w}_e|)$ is the diagonal matrix containing the absolute edge weights and $\mathbf{B} \in \mathbb{R}^{n \times m}$ is the node-to-edge incidence matrix:

$$B_{ie} = \begin{cases} 1 & \text{if } i \text{ is the tail of edge } e, \\ -\text{sign}(e) & \text{if } i \text{ is the head of edge } e. \end{cases}$$

When the graph is connected but not structurally balanced, $\mathbf{L}_S$ is positive definite.

The following interesting alternative characterization of a structurally balanced graph was highlighted by Altafini [4, 5]: a network is structurally balanced if there exists a ± 1 diagonal matrix $\mathbf{\Sigma}$ such that the matrix

$$\mathbf{L}_{\mathbf{\Sigma}} = \mathbf{\Sigma}\mathbf{L}_S\mathbf{\Sigma} \tag{12.17}$$

contains only negative elements on the off-diagonal. In other words, $\mathbf{L}_{\mathbf{\Sigma}}$ is a standard Laplacian matrix for another (associated) graph with only positive weights. The matrix $\mathbf{\Sigma}$ is called switching equivalence, signature similarly, or gauge transformation in the literature [5]. Choosing $\boldsymbol{\Sigma} = \mathbf{I}$ makes it clear that a network with only positive weights is itself balanced, which emphasizes that $\mathbf{L}_S$ is a proper generalization of the standard graph Laplacian. Using this characterization, one can efficiently determine whether a network is structurally balanced [23] and obtain the corresponding switching equivalence matrix $\boldsymbol{\Sigma}$.

We are now in a position to study the opinion dynamics on a signed network given by:

$$\dot{\mathbf{x}} = -\mathbf{L}_S\mathbf{x}. \tag{12.18}$$

The above equation is often called signed consensus dynamics and we will adopt this name here, but it is important to note that this dynamics does not have to lead to a single consensus for all agents [3, 4, 60]. Intuitively, this should be clear: since the agents can be repelled by the opinions of their neighbors, the eventual dynamics need not converge to a unique consensus, even in the absence of external inputs. This is in contrast to the standard consensus case shown in Figure 12.1.

A remarkable feature of structurally balanced networks is that the signed consensus dynamics Equation (12.18) converges asymptotically to a *polarized state*, i.e. a state where the final value for all nodes is the same in magnitude, but the nodes are divided into two sets with opposite sign (Figure 12.5). More precisely, there is an eigenvector of $\mathbf{L}_S$ associated with eigenvalue 0 of the form $\boldsymbol{\sigma} = [\sigma_1, \ldots, \sigma_N]^T$ where all $\sigma_i \in \{-1, +1\}$ such that the dynamics Equation (12.18) converges to the final state

$$\lim_{t \to \infty} \mathbf{x}(t) = \frac{\boldsymbol{\sigma}^T\mathbf{x}_0}{n}\boldsymbol{\sigma}. \tag{12.19}$$

The sign pattern of $\boldsymbol{\sigma}$ corresponds precisely to the switching equivalence transformation $\boldsymbol{\Sigma} = \text{diag}(\boldsymbol{\sigma})$ and σ_i is denoted the polarization of node i. Note that the overall sign of $\boldsymbol{\sigma}$ is arbitrary, so that the polarization of a node is only meaningful relative to that of other nodes. An example of this behavior is shown in Figure 12.6.

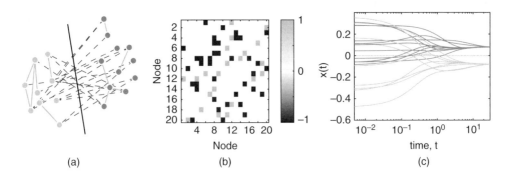

(a)   (b)   (c)

Figure 12.6   Polarized opinion dynamics in structurally balanced networks. (a) Visualization of a structurally balanced network with positive (solid) and negative (dashed) links. As indicated by the drawing, the network can be partitioned into two subsets such that all negative links are between nodes in different sets. (b) The adjacency matrix of the same structurally balanced network. Fom this representation it is no immediately obvious whether or not the network is structurally balanced. (c) Starting from a random initial condition, the signed consensus dynamics (Equation (12.18)) on this structurally balanced network converges asymptotically to a state of two polarized opinions (Equation (12.19)), where the final opinions are exactly the same in magnitude, but opposite in sign for the two groups shown in (a).

Discussion: Dynamics of Networks Towards Structural Balance

Our discussions up to now have focused mostly on the effects of graph structure on the dynamics taking place on a fixed network. Let us now briefly discuss a form of structural dynamics (where the network structure itself changes) in the context of signed networks.

Albeit commonly explained from a static perspective (i.e. parity of the cyclic paths in the network), structural balance is essentially a dynamical theory: it posits that networks tend to *evolve* towards a state of structural balance. An important question is therefore how such a structural dynamics of network evolution could look like. Antal *et al.* [7] provided one of the first explorations of this issue in the discrete time setting, followed by the continuous time version by Kułakowski *et al.* [38], which was analyzed in detail by Marvel and collaborators [45]. Here we focus on the following variation of this model proposed by Traag *et al.* [80]:

$$\dot{\mathbf{X}} = \mathbf{X}\mathbf{X}^\top \quad \text{with} \quad \mathbf{X}(0) = \mathbf{X}_0, \tag{12.20}$$

where $\mathbf{X} \in \mathbb{R}^{n \times n}$ denotes the matrix of opinions that agents have of each other.

It can be shown [80] that this model converges to a structurally balanced network for almost all initial conditions, i.e. we reach either a split into two factions of opposing opinions or a state in which all nodes have positive opinion about each other. More precisely, the normalized solution of Equation (12.20) converges for some time t^* to

$$\lim_{t \to t^*} \frac{\mathbf{X}(t)}{\| \mathbf{X} \|_F} = \mathbf{v}\mathbf{v}^T,$$

where $\| \cdot \|_F$ denotes the Frobenius norm and $\mathbf{v}$ is a real-valued vector whose sign pattern indicates the two factions. However, an important shortcoming of this model is the fact that while the *sign pattern* converges to a balanced network, the *magnitude* of the opinions diverge after a finite time t^*, unless certain technical conditions are fulfilled [80]. A second shortcoming is the assumption of homogeneity in the network an in the agents: each agent has an opinion about every other agent in the network (all-to-all connectivity) and all agents follow the same update rule. These assumptions are unrealistic in many real-world scenarios. However, extended models that relax these simplifications are far more difficult to analyze rigorously, thus providing an interesting object for further study.

12.4 Part III: Using Dynamical Processes to Reveal Network Structure

As we have seen in the previous sections, graph structure can have a notable impact on a dynamics acting on a network. What about the converse? Can we choose a dynamics, let it evolve on a graph, and learn a modular or coarse-grained description of the network based on some of its properties (time scale separation, quasi-invariance, etc)?

In the following, we provide a high-level, algorithmic point of view of how this may be achieved. Interestingly, many community detection methods and graph partitioning heuristics in the literature can be seen as particular cases of this general abstract viewpoint [19, 20]. Although not every method for community detection is best interpreted in terms of dynamical quantities [72], the dynamics-based perspective presented here can serve as a general framework to establish differences and similarities between the plethora of existing methods. Our perspective reveals some surprising relationships between measures and methods that have been proposed in seemingly different ares in the literature.

12.4.1   A Generic Algorithmic Framework for Dynamics-Based Network Partitioning and Coarse Graining

Given a complex network, one important task is to find relevant communities (i.e. groups of nodes that can act as a coarse-grained description of the particular network) in an unsupervised manner. There is a very large array of such methods. However, as we will make more precise through an example below, one can interpret many of the community detection algorithms through the following generic algorithmic recipe, which includes four ingredients:

1. A *dynamical process* $\mathbf{x} : t \mapsto \mathbf{x}(t)$ acting on a given network.
2. An *inference operator* $\mathscr{F}$, which is used to characterize some statistical information about the dynamical process.
3. A *selection and summary operator* $\mathscr{S}$, which assigns nodes into groups and clusters (or coarse-grains) the dynamical process.
4. A *quality function* $\mathscr{Q}$, which scores the clustering

While more general scenarios can be used, in the cases we discuss below these ingredients take the following form. Given a graph, we use a dynamical process $\mathbf{x}(t)$ to assign to each node a particular time trajectory or a time-evolving probability measure. We then apply the inference operator $\mathscr{F}$ on this time-course data to produce a node-to-node time-dependent similarity measure. The similarity measure is then aggregated according to a given grouping of nodes by the selection operator $\mathscr{S}$. The proposed clustering is then evaluated by the quality function $\mathscr{Q}$, which can then be optimized (Figure 12.7).

Example of the Framework: Markov Stability for Community Detection

Let us exemplify these ideas through the Markov Stability framework, as discussed in [19, 20, 69]. We use this example to demonstrate the applicability of the proposed algorithmic framework, since it can be shown that the Markov Stability framework encompasses a variety

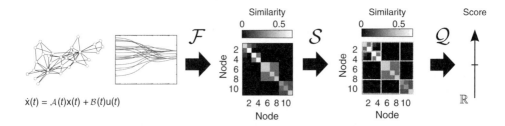

Figure 12.7   Schematic of the algorithmic framework for partitioning of networks based on dynamics. A dynamics $\mathbf{x}(t)$ operating on a network with inputs $\mathbf{u}(t)$ is filtered through the inference operator $\mathscr{F}$ to provide a similarity matrix between the nodes, from which a partitioning (or coarse-graining) of the nodes is obtained by a selection and summary operator $\mathscr{S}$. The quality of this partition is summarized by a numerical metric through the quality operator $\mathscr{Q}$. This abstraction of dynamics-based community detection covers many existing methods in the literature and opens up a wide array of possibilities suited to different purposes.

of community detection methods as particular cases [19, 20, 69], including modularity, spectral clustering, and various Potts models. For a discussion that includes processes beyond diffusion and further relations to control-theoretic notions we refer to [73].

Markov Stability is a quality function that scores a given non-overlapping partition $\mathscr{P}$ of a graph into node groups $\mathscr{C}_i$. The Markov Stability [19, 40] score is parametrically dependent on a time parameter t and is given by

$$\text{Markov Stability}(t, \mathscr{P}) := r(t, \mathscr{P}) = \sum_{\mathscr{C}_i \in \mathscr{P}} \Pr(\mathscr{C}_i, 0, t) - \Pr(\mathscr{C}_i, 0, \infty) \qquad (12.21)$$

where $\Pr(\mathscr{C}_i, 0, t)$ is the probability for a random walker to be in the same cell $\mathscr{C}_i$ at time zero and at time t. For an extended discussion of this framework, see [19, 20, 40, 41, 69].

We now see how Markov Stability fits within the above four ingredient recipe:

1. To an undirected, connected graph with Laplacian $\mathbf{L}$ we associate a diffusion dynamics (random walk) given by

$$\dot{\mathbf{p}}^{\mathsf{T}} = -\mathbf{p}^{\mathsf{T}} \mathbf{L}_{\text{RW}}, \qquad (12.22)$$

which describes the evolution of the probability mass function $\mathbf{p}(t)$ of an n-dimensional random indicator vector $\mathbf{x}(t)$ with $x_k(t) = 1$ if the particle is at node i at time t and zero otherwise. Recall that $\mathbf{L}_{\text{RW}} = \mathbf{D}^{-1}\mathbf{L}$ is the random walk Laplacian.

2. The inference operator $\mathscr{F}$ is set to be the *autocovariance function* of this process, which computes the correlation between the random state vector $\mathbf{x}(\tau)$ at stationarity with itself after a time t, $\mathbf{x}(\tau + t)$):

$$\mathscr{F}(\mathbf{x}(t)) = \text{cov}[\mathbf{x}(\tau), \mathbf{x}(\tau + t)] = \mathbf{\Pi P}(t) - \boldsymbol{\pi}\boldsymbol{\pi}^{\mathsf{T}}. \qquad (12.23)$$

$\boldsymbol{\pi} = \mathbf{d}/2w$ denotes the stationary distribution of the diffusion.

3. For a given partition of the graph with indicator matrix $\mathbf{C}$, we choose a selection and summary operator $\mathscr{S}$ that aggregates the entries of an $n \times n$ matrix defined on the nodes of the graph and aggregates the entries inside the k cells in the partition to return a $k \times k$ matrix. This is achieved simply by the pre- and post-multiplication $\mathscr{S}(\cdot) = \mathbf{C}^{\mathsf{T}}(\cdot)\,\mathbf{C}$. Applying $\mathscr{S}$ to the autocovariance matrix Equation (12.23) yields

$$\mathscr{S}(\mathscr{F}(\mathbf{x}(t))) = \mathbf{C}^{\mathsf{T}}(\mathbf{\Pi P}(t) - \boldsymbol{\pi}\boldsymbol{\pi}^{\mathsf{T}})\mathbf{C} := \mathbf{R}(t, \mathbf{C}), \qquad (12.24)$$

where $\mathbf{R}(t, \mathbf{C})$ is the *clustered autocovariance* of the dynamics on the network. Due to the linearity of the chosen dynamics and the selection operator, the clustered autocovariance matrix is also the autocovariance of the coarse-grained indicator vectors $\mathbf{y}(t) = \mathbf{C}^{\mathsf{T}}\mathbf{x}(t)$ [20]. Whence, with a slight abuse of notation, we conclude that in this case

$$\mathscr{S}(\mathscr{F}(\mathbf{x}(t))) = \mathscr{F}(\mathscr{S}(\mathbf{x}(t))) = \mathbf{R}(t, \mathbf{C}).$$

4. The quality function in Markov Stability is chosen as $\mathscr{Q}(\cdot) = \text{trace}(\cdot)$, i.e. the trace of the matrix. When applied to the clustered autocovariance matrix $\mathbf{R}(t, \mathbf{C})$, it sums over the k autocovariances of each group for a given time t. Hence, the quality of the partition is the sum of the contributions from the individual groups.

In summary, Markov Stability can be written suggestively as the following function composition:

$$r(t, \mathbf{C}) = \mathcal{Q}(\mathcal{S}(\mathcal{F}(\mathbf{x}(t)))) = \text{trace } \mathbf{C}^\top (\mathbf{\Pi P}(t) - \boldsymbol{\pi}\boldsymbol{\pi}^\top)\mathbf{C}. \tag{12.25}$$

It is easy to verify that this expression is equivalent to Equation (12.21). To find good partitions, our objective will be to maximize $r(t, \mathbf{C})$ over the space of partitions $\{\mathcal{P}\}$ for each value of t. This is usually a hard combinatorial optimization, which is approached computationally through different heuristics.

12.4.2   Extending the Framework by using other Measures

Thinking of community detection as the composition of operators immediately opens up a range of extensions by varying the individual ingredients. We discuss now some of these possibilities, focusing mainly on $\mathcal{F}$ and $\mathcal{Q}$, the inference and quality operators (the choice of dynamical process $\mathbf{x}(t)$ has been discussed at length in Refs. [19, 40, 41]). Considering these extensions allows us to establish further connections with other community detection methods and heuristics proposed in the literature.

12.4.2.1   Varying the Dynamical Process x(t)

From our discussion above, it is clear that different dynamical processes will lead to different community detection and coarse-graining algorithms. We do not dwell on this topic here, as extended studies in [20, 40, 41] have discussed a variety of measures based on different dynamics (discrete and continuous time), and their connection with a variety of methods and heuristics proposed in the literature. Notably, our approach leads naturally to dynamical interpretations of different variants of modularity based on distinct statistical null models which can be understood both as stationary points of the dynamics and as centrality measures of the network. Similarly, generalized modularity measures, such as Potts models, can be understood as linearisations/linear interpolations of Markov Stability [20, 41]. Many further extensions are possible and in particular translating these ideas to non-Markovian dynamics appears to be a fruitful avenue for future research.

12.4.2.2   Varying the Inference Operator $\mathcal{F}$

A well-known issue with the covariance is the fact that its absolute value is not easy to interpret, as it depends on the magnitude of the random variables involved (e.g. a simple rescaling of the variables can increase or decrease the covariance). A standard way to discount the magnitude of the random variables is to normalize them, leading to *correlation measures*. A classic example is the Pearson correlation, in which the random variables are rescaled to have unit variance.

Accordingly, we can change the inference operator to $\mathcal{F} = \text{corr}(\cdot)$, where the cross-correlation operator between to random vectors $\mathbf{x_1}, \mathbf{x_2}$ is defined as:

$$\text{corr}(\mathbf{x_1}, \mathbf{x_2}) = \mathbf{S}_{x_1}^{-1/2}(\mathbb{E}(\mathbf{x_1}\mathbf{x_2}^\top) - \mathbb{E}(\mathbf{x_1})\mathbb{E}(\mathbf{x_2}^\top))\mathbf{S}_{x_2}^{-1/2}. \tag{12.26}$$

Here $\mathbf{S}_{x_1}$ and $\mathbf{S}_{x_2}$ are diagonal matrices containing the variances of the components of the vectors $\mathbf{x}_1$ and $\mathbf{x}_2$, respectively, and $\mathbb{E}$ denotes the expectation operator. Substituting the inference operator in this way leads to a correlation-based equivalent of Markov Stability:

$$r_{corr}(t, \mathbf{C}) = \text{trace } \mathbf{C}^\top \text{ corr}(\mathbf{x}_1(\tau), \mathbf{x}_1(\tau + t))\mathbf{C}$$

$$= \text{trace } \mathbf{C}^\top \mathbf{S}^{-1/2}(\mathbf{\Pi P}(t) - \boldsymbol{\pi}\boldsymbol{\pi}^\top)\mathbf{S}^{-1/2}\mathbf{C}, \tag{12.27}$$

where (like in Markov Stability) we assume that the random diffusion process is at stationarity, and thus we need to normalize by $\mathbf{S} = \mathbf{\Pi}(\mathbf{I} - \mathbf{\Pi})$ the diagonal matrix with variances of Bernoulli random variables of mean π_i on the diagonal.

Interestingly, using correlation instead of covariance was considered by Shen *et al.* [75] in the context of modularity community detection for directed graphs. Their simplified formula for undirected networks is

$$\text{corr. modularity} = \sum_{\mathscr{C}} \sum_{i,j \in \mathscr{C}} \frac{A_{ij} - \frac{d_i d_j}{2w}}{\sqrt{d_i\left(1 - \frac{d_i}{2w}\right)}\sqrt{d_j\left(1 - \frac{d_j}{2w}\right)}}. \tag{12.28}$$

It is easy to see that this measure (proposed in [75] from a combinatorial viewpoint) corresponds to the Markov Stability with dynamics governed by a *discrete-time* random walk with transition matrix $\mathbf{P}(t) = (\mathbf{D}^{-1}\mathbf{A})^t$ evaluated at time $t = 1$, and with the Pearson correlation as the inference operator $\mathscr{F}$:

$$r_{corr}(1, \mathbf{C}) = \text{trace } \mathbf{C}^\top \mathbf{S}^{-1/2}[\mathbf{\Pi}\mathbf{D}^{-1}\mathbf{A} - \boldsymbol{\pi}\boldsymbol{\pi}^\top]\mathbf{S}^{-1/2}\mathbf{C} = \text{corr. modularity}.$$

Indeed, in the original formulation of Markov Stability [19], it was already noted that a version of Markov Stability based on correlations is related to normalised cut for graph partitioning.

Other Extensions:

Although we have focused here on the Pearson correlation, many other metrics are possible. A particular area of interest is the investigation of inference operators $\mathscr{F}$ based on mutual information. It is also worth remarking that a number of variants arise from changing the order of application of the selection and inference operators, which will not commute in general $\mathscr{S}(\mathscr{F}(\cdot)) \neq \mathscr{F}(\mathscr{S}(\cdot))$, leading to different notions of clusterings. These directions will be the focus of future research.

12.4.2.3 Varying the Selection and Summary Operator $\mathscr{S}$

The choice of operator $\mathscr{S}$ opens up a broad range of directions, which are beyond the scope of this chapter. We mention only one direction. Our focus here has been on *hard* graph partitions, i.e. forcing a hard split of the nodes into non-overlapping groups. We could, however, allow for overlapping (possibly probabilistic) membership. In terms of $\mathscr{S}$, this would amount to relaxing the assumptions of the indicator matrix $\mathbf{C}$: instead of requiring that $C_{ij} = 1$ if node i belongs to community j and zero otherwise, we may merely require that $\sum_j C_{ij} = 1$. In that case, the node i may belong to multiple groups with a certain probability. This relaxation is equivalent to requiring that the matrix $\mathbf{C}$ is row-stochastic, instead of binary, thus opening the possibility of using different optimization techniques.

12.4.2.4   Varying the Quality Function $\mathscr{Q}$

We now consider in some detail the choice of the quality function $\mathscr{Q}$, for which an array of possibilities is also available. Our algorithmic framework with the standard choices for Markov Stability for the operators are $\mathscr{F} = \mathrm{cov}(\cdot)$, $\mathscr{S} = \mathbf{C}^{\top}(\cdot)\mathbf{C}$ on a diffusion dynamics $\mathbf{x}(t)$ on the network leading to the clustered autocovariance matrix $\mathbf{R}(t, \mathbf{C})$ given in Equation (12.24). To this matrix, we then apply the quality function $\mathscr{Q}(\cdot) = \mathrm{trace}(\cdot)$, which sums over all the community autocovariances, a measure akin to taking an average of the individual autocovariances.

There are of course other options for the quality function. For instance, one could take into account the off-diagonal terms of the matrix $\mathbf{R}(t, \mathbf{C})$ and define a metric that searches for maximally diagonal matrices. Such an alternative operator quality function would be especially relevant in conjunction with a different, nonlinear inference operator $\mathscr{F}$ (e.g. mutual information). Another option is to define the quality of the partition in terms not of the average autocovariance over the groups, but of the weakest group autocovariance of the partition, i.e. we would favor partitions where no bad groups exist. This could be achieved by using the quality function

$$\mathscr{Q} = \min\{\mathrm{diag}(\cdot)\}.$$

We consider this quality function in more detail as it provides an interesting connection with another measure for community detection proposed by Piccardi [57], as we now show.

Let us consider an unbiased random walk discrete-time dynamics $\mathbf{x}(t)$ with transition matrix $\mathbf{P}(t) = (\mathbf{D}^{-1}\mathbf{A})^{t}$ and stationary distribution $\boldsymbol{\pi} = \mathbf{d}/2w$ and define the inference operator to be $\mathscr{F} = \mathrm{cov}(\cdot)$. To make the connection clearer and to avoid a bias towards partitions that contain few communities of large size, we need to set our selection operator to perform a normalized block-averaging

$$\mathscr{S}(\cdot) = (\mathbf{C}^{\top}\mathbf{\Pi}\mathbf{C})^{-1}\mathbf{C}^{\top}(\cdot)\mathbf{C},$$

where $(\mathbf{C}^{\top}\mathbf{\Pi}\mathbf{C}) = \mathrm{diag}([\pi_{\mathscr{C}_1}, \ldots, \pi_{\mathscr{C}_k}])$ and $\pi_{\mathscr{C}_i} = \sum_{\mathscr{C}_i} d_i/2w$ in this case. Thus, the $\mathscr{S}$ operator normalizes each group according to the starting probability of the diffusion process within each community. Combining these operators gives us a clustered normalized autocovariance matrix, to which we then apply the quality function $\mathscr{Q} = \min\{\mathrm{diag}(\cdot)\}$ to obtain the following clustering measure:

$$r_{\min}(t, \mathbf{C}) = \min_{i} ((\mathbf{C}^{\top}\mathbf{\Pi}\mathbf{C})^{-1}\mathbf{C}^{\top}(\mathbf{\Pi}\mathbf{P}(t) - \boldsymbol{\pi}\boldsymbol{\pi}^{\top})\mathbf{C})_{ii}. \qquad (12.29)$$

Piccardi [57] proposed a technique for community detection based on the idea that the outflow from a community under a random walk should be small. To that end, he defined a lumped Markov chain with transition matrix:

$$\mathbf{U} = (\mathrm{diag}(\boldsymbol{\pi}\mathbf{C}))^{-1}\mathbf{C}^{\top}\mathbf{\Pi}\mathbf{D}^{-1}\mathbf{A}\mathbf{C} \qquad (12.30)$$

and stationary distribution $\boldsymbol{\pi}_{\ell} = \boldsymbol{\pi}\mathbf{C}$, i.e. the appropriately summed version of the stationary distribution of the original Markov process. The best split into communities (denoted an α-partition) is found through the optimization

$$\max_{\mathbf{C}} k \quad \text{subject to} \quad U_{ii} \geq \alpha, \quad i = 1, 2, \ldots, k \qquad (12.31)$$

where $\alpha \in [0, 1]$ is a parameter set *a priori* and defining the minimal amount of flow retained in each community.

It can be shown that a partition $\mathbf{C}$ with $r_{\min}(1, \mathbf{C}) \geq \beta_1$ is also an α-partition with a guaranteed value of α. This follows directly from the definitions in Equations (12.29) and (12.30):

$$r_{\min}(1, \mathbf{C}) \geq \beta_1 \;\Rightarrow\; U_{ii} \geq \beta_1 + [(\mathbf{C}^{\mathsf{T}}\mathbf{\Pi}\mathbf{C})^{-1}\mathbf{C}^{\mathsf{T}}\boldsymbol{\pi}^{\mathsf{T}}\boldsymbol{\pi}\mathbf{C}]_{ii} = \beta_1 + \pi_{\ell,i} := \gamma_i. \qquad (12.32)$$

Therefore a partition with $r_{\min}(1, \mathbf{C}) \geq \beta_1$ is also an α-partition with $\alpha = \max_i \gamma_i$, and optimizing $r_{\min}(t, \mathbf{C})$ in the space of all partitions for any t encompasses α-partitions as a special case (for $t = 1$). For a more detailed discussion of these connections, see [68].

12.5 Discussion

This chapter has focused on the relationship between structure and dynamics in complex networks, concentrating on how the notions of time scale separation, external equitable partitions, and structural balance are related to coarse-graining and community detection in networks. We have exemplified these concepts through consensus dynamics and random walks, both of which have been studied extensively linked to social network analysis.

Our discussion is underpinned by the fact that both time scale separation and EEPs are intimately related to the algebraic notion of invariant subspaces. Such invariance provides us with a better understanding of the emerging dynamics on a network, and highlights structural features in the network that are of dynamical importance. For instance, in a diffusion process time scale separation implies regions in the network where the diffusion is trapped far longer than one would expect. Hence, over a particular time scale, a diffusive signal emanating from a node inside such a group will likely reach nodes inside the same group, so that these nodes are *dynamically* almost decoupled from the rest of the network. In another example, the presence of certain graph symmetries in the network implies the presence of EEPs, graph partitions that are invariant under the dynamics. As a consequence, the averages of the consensus dynamics over the cells of the EEP can be described by the reduced dynamics of the lower-dimensional quotient graph.

Alternatively, instead of looking only at the impact of the structure on the dynamics, we can also ask the opposite question: can we use information gathered from dynamics taking place on the network to reveal important structures in the graph? We have proposed an algorithmic perspective to exploit this dynamics-driven route to the analysis of network structure and have shown that, interestingly, by adopting such a viewpoint we recover many methods of community detection proposed in the literature from seemingly different perspectives. This dynamical interpretation for such diverse methods enables us to place them with a unifying framework, providing additional insights that might not have been apparent from their initial definition.

Open Directions

There is a large literature on complex networks and dynamical processes acting on top of those, and hence there are many topics of interest related to the interplay between structure and dynamics that we have not discussed here.

A fruitful area to extend these ideas is to go beyond (first-order) Markovian and diffusion-based dynamics. Indeed, many dynamics relevant to real-world phenomena, such as epidemic spreading, are not of a diffusive type, yet being able to obtain dynamically important modules in such systems is crucial, e.g. to design effective vaccination strategies.

Considering higher-order or non-Markovian models, allowing for the presence of memory, would, for instance, allow us to study bursty dynamics or other path-dependent dynamical processes [21].

Furthermore, it will be important to connect the concept of dynamical modules considered here in connection with model order reduction tools formally studied in control theory. Such a link will allow us to quantify how far the modules describe the original dynamics in a precise manner.

Ultimately, a key aim is to describe not only the dynamics on the network in a modular fashion, but also to take into account changes of the network itself and interactions between different types of dynamics acting on the network. Indeed, a key area for future work (barely considered here) is that of systems where the topology of the network changes, as in social contact networks, and the inclusion of multiple kinds of interactions within one system, formalized mathematically through multiplex networks [11, 36].

Acknowledgements

MTS gratefully acknowledges funding from the European Union's Horizon 2020 Research and Innovation Programme under the Marie Sklodowska-Curie grant agreement No 702410. Most of this work was completed while MTS was at Université Catholique de Louvain and University of Namur. JCD and RL acknowledge support from FRS-FNRS; the Flagship European Research Area Network (FLAG-ERA) Joint Transnational Call "FuturICT 2.0" and the ARC (Action de Recherche Concertée) on Mining and Optimisation of Big Data Models funded by the Wallonia- Brussels Federation. MB acknowledges funding from the EPSRC, through the award EP/N014529/1 funding the EPSRC Centre for Mathematics of Precision Healthcare. The funders had no role in the design of this study. The results presented here reflect solely the authors' views.

References

1. D. Acemoğlu, G. Como, F. Fagnani, and A. Ozdaglar. Opinion fluctuations and disagreement in social networks. *Mathematics of Operations Research*, 38(1):1–27, 2013.
2. D. Aldous and J. Fill. *Reversible Markov chains and random walks on graphs*. Available at https://www.stat.berkeley.edu/users/aldous/RWG/book.pdf.
3. C. Altafini. Dynamics of opinion forming in structurally balanced social networks. *PloS One*, 7(6):e38135, 2012.
4. C. Altafini. Consensus problems on networks with antagonistic interactions. *IEEE Transactions on Automatic Control*, 58(4):935–946, 2013.
5. C. Altafini. Stability analysis of diagonally equipotent matrices. *Automatica*, 49(9):2780–2785, 2013.
6. C. J. Anderson, S. Wasserman, and K. Faust. Building stochastic blockmodels. *Social Networks*, 14(1):137–161, 1992.
7. T. Antal, P. L. Krapivsky, and S. Redner. Dynamics of social balance on networks. *Physical Review E*, 72(3):036121, 2005.
8. A. S. Bandeira. Random Laplacian matrices and convex relaxations. *Foundations of Computational Mathematics*, 1–35, 2015.
9. M. Barahona and L. M. Pecora. Synchronization in small-world systems. *Physical Review Letters*, 89:054101, 2002.
10. F. Benaych-Georges and R. R. Nadakuditi. The eigenvalues and eigenvectors of finite, low rank perturbations of large random matrices. *Advances in Mathematics*, 227(1):494–521, 2011.

11. S. Boccaletti, G. Bianconi, R. Criado, C. del Genio, J. Gómez-Gardeñes, M. Romance, I. Sendiña-Nadal, Z. Wang, and M. Zanin. The structure and dynamics of multilayer networks. *Physics Reports*, 544(1):1–122, 2014.
12. R. W. Brockett. *Finite dimensional linear systems*. SIAM, 2015.
13. D. M. Cardoso, C. Delorme, and P. Rama. Laplacian eigenvectors and eigenvalues and almost equitable partitions. *European Journal of Combinatorics*, 28(3):665–673, 2007.
14. D. Cartwright and F. Harary. Structural balance: a generalization of Heider's theory. *Psychological Review*, 63(5):277, 1956.
15. A. Chan and C. D. Godsil. Symmetry and eigenvectors. In *Graph Symmetry*, pages 75–106. Springer, 1997.
16. F. R. Chung. *Spectral graph theory*, volume 92. American Mathematical Society 1997.
17. E. Davidson and M. Levin. Gene regulatory networks. *Proceedings of the National Academy of Sciences*, 102(14):4935–4935, 2005.
18. A. Decelle, F. Krzakala, C. Moore, and L. Zdeborová. Inference and phase transitions in the detection of modules in sparse networks. *Physical Review Letters*, 107(6):065701, 2011.
19. J.-C. Delvenne, S. N. Yaliraki, and M. Barahona. Stability of graph communities across time scales. *Proceedings of the National Academy of Sciences*, 107(29):12755–12760, 2010.
20. J.-C. Delvenne, M. T. Schaub, S. N. Yaliraki, and M. Barahona. The stability of a graph partition: A dynamics-based framework for community detection. In A. Mukherjee, M. Choudhury, F. Peruani, N. Ganguly, and B. Mitra, editors, *Dynamics On and Of Complex Networks, Volume 2*, Modeling and Simulation in Science, Engineering and Technology, pages 221–242. Springer, New York, 2013.
21. J.-C. Delvenne, R. Lambiotte, and L. E. C. Rocha. Diffusion on networked systems is a question of time or structure. *Nature Communications*, 6: 7366, 2015.
22. M. Egerstedt, S. Martini, M. Cao, K. Camlibel, and A. Bicchi. Interacting with networks: How does structure relate to controllability in single-leader, consensus networks? *IEEE Control Systems*, 32(4):66–73, 2012.
23. G. Facchetti, G. Iacono, and C. Altafini. Computing global structural balance in large-scale signed social networks. *Proceedings of the National Academy of Sciences*, 108(52):20953–20958, 2011.
24. S. Fortunato. Community detection in graphs. *Physics Reports*, 486(3-5): 75–174, 2010.
25. S. Fortunato and D. Hric. Community detection in networks: A user guide. *Physics Reports*, 659:1–44, 2016.
26. N. E. Friedkin, A. V. Proskurnikov, R. Tempo, and S. E. Parsegov. Network science on belief system dynamics under logic constraints. *Science*, 354 (6310):321–326, 2016.
27. W. Gerstner, W. M. Kistler, R. Naud, and L. Paninski. *Neuronal dynamics: From single neurons to networks and models of cognition*. Cambridge University Press,2014.
28. C. Godsil and G. F. Royle. *Algebraic graph theory*, volume 207. Springer Science & Business Media, 2013.
29. F. Heider. Attitudes and cognitive organization. *Journal of Psychology*, 21(1):107–112, 1946.
30. T. Hoffmann, M. A. Porter, and R. Lambiotte. Generalized master equations for non-poisson dynamics on networks. *Physics Review E*, 86: 046102, 2012.
31. P. W. Holland, K. B. Laskey, and S. Leinhardt. Stochastic blockmodels: First steps. *Social Networks*, 5(2):109–137, 1983.
32. P. Holme and J. Saramäki. Temporal networks. *Physics Reports*, 519(3): 97–125, 2012.
33. A. Jadbabaie, J. Lin, and A. Morse. Coordination of groups of mobile autonomous agents using nearest neighbor rules. *IEEE Transactions on Automatic Control*, 48(6):988–1001, 2003.
34. A. Jadbabaie, N. Motee, and M. Barahona. On the stability of the Kuramoto model of coupled nonlinear oscillators. In *Proceedings of the American Control Conference, 2004.*, volume 5, pages 4296–4301, 2004. URL http://ieeexplore.ieee.org/xpls/abs_all.jsp?arnumber=1383983&tag=1.
35. T. Kailath. *Linear systems*, volume 156. Prentice-Hall, Englewood Cliffs, NJ, 1980.
36. M. Kivela, A. Arenas, M. Barthelemy, J. P. Gleeson, Y. Moreno, and M. A. Porter. Multilayer networks. *Journal of Complex Networks*, 2(3):203–271, 2014.
37. F. Krzakala, C. Moore, E. Mossel, J. Neeman, A. Sly, L. Zdeborová, and P. Zhang. Spectral redemption in clustering sparse networks. *Proceedings of the National Academy of Sciences*, 110(52):20935–20940, 2013.
38. K. Kułakowski, P. Gawroński, and P. Gronek. The Heider balance: A continuous approach. *International Journal of Modern Physics C*, 16(05): 707–716, 2005.
39. J. Kunegis, S. Schmidt, A. Lommatzsch, J. Lerner, E. W. De Luca, and S. Albayrak. Spectral analysis of signed graphs for clustering, prediction and visualization. In *SIAM Data Mining*, volume 10, pages 559–559. SIAM, 2010.
40. R. Lambiotte, J.-C. Delvenne, and M. Barahona. Laplacian dynamics and multiscale modular structure in networks. arxiv:0812.1770, pages 1–29, 2009.

41. R. Lambiotte, J. Delvenne, and M. Barahona. Random walks, markov processes and the multiscale modular organization of complex networks. *IEEE Transactions on Network Science and Engineering*, 1(2):76–90, 2014.
42. J. Lei and A. Rinaldo. Consistency of spectral clustering in stochastic block models. *Annals of Statistics*, 43(1):215–237, 2015.
43. J. Leskovec, D. Huttenlocher, and J. Kleinberg. Signed networks in social media. In *Proceedings of the SIGCHI Conference on Human Factors in Computing Systems*, pages 1361–1370. ACM, 2010.
44. Y.-Y. Liu, J.-J. Slotine, and A.-L. Barabási. Controllability of complex networks. *Nature*, 473(7346):167, 2011.
45. S. A. Marvel, J. Kleinberg, R. D. Kleinberg, and S. H. Strogatz. Continuous-time model of structural balance. *Proceedings of the National Academy of Sciences*, 108(5):1771–1776, 2011.
46. N. Masuda, M. A. Porter, and R. Lambiotte. Random walks and diffusion on networks. *Physics Reports*, 716–717:1–58, 2017.
47. M. Mobilia, A. Petersen, and S. Redner. On the role of zealotry in the voter model. *Journal of Statistical Mechanics: Theory and Experiment*, 2007 (08):P08029, 2007.
48. E. Mossel, J. Neeman, and A. Sly. A proof of the block model threshold conjecture. *Combinatorica*, 38(3):665–708, 2018.
49. R. R. Nadakuditi and M. E. J. Newman. Graph spectra and the detectability of community structure in networks. *Physical Review Letters*, 108(18),2012.
50. M. E. J. Newman. *Networks: An Introduction*. Oxford University Press, Oxford, New York, 2010.
51. N. O'Clery, Y. Yuan, G.-B. Stan, and M. Barahona. Observability and coarse graining of consensus dynamics through the external equitable partition. *Physical Review E*, 88(4),2013.
52. R. Olfati-Saber and R. Murray. Consensus problems in networks of agents with switching topology and time-delays. *IEEE Transactions on Automatic Control*, 49(9):1520–1533, 2004.
53. R. Olfati-Saber, J. Fax, and R. Murray. Consensus and cooperation in networked multi-agent systems. *Proceedings of the IEEE*, 95(1):215–233, 2007.
54. L. M. Pecora, F. Sorrentino, A. M. Hagerstrom, T. E. Murphy, and R. Roy. Cluster synchronization and isolated desynchronization in complex networks with symmetries. *Nature Communications*, 5:4079, 2014.
55. L. Peel, D. B. Larremore, and A. Clauset. The ground truth about metadata and community detection in networks. *Science Advances*, 3(5): e1602548, 2017.
56. T. P. Peixoto. Eigenvalue spectra of modular networks. *Physical Review Letters*, 111(9):098701, 2013.
57. C. Piccardi. Finding and testing network communities by lumped Markov chains. *PLoS One*, 6(11):e27028, 11 2011.
58. M. Porter, J. Onnela, and P. Mucha. Communities in networks. *Notices of the AMS*, 56(9):1082–1097, 1164–1166, 2009.
59. A. V. Proskurnikov and R. Tempo. A tutorial on modeling and analysis of dynamic social networks. Part I. *Annual Reviews in Control*, 43:65–79, 2017.
60. A. V. Proskurnikov, A. S. Matveev, and M. Cao. Opinion dynamics in social networks with hostile camps: Consensus vs. polarization. *IEEE Transactions on Automatic Control*, 61(6):1524–1536, 2016.
61. W. Ren, R. W. Beard, and E. M. Atkins. A survey of consensus problems in multi-agent coordination. In *Proceedings of the 2005 American Control Conference*, pages 1859–1864. IEEE, 2005.
62. W. Ren, R. W. Beard, and E. M. Atkins. Information consensus in multivehicle cooperative control. *IEEE Control Systems*, 27(2):71–82, 2007.
63. K. Rohe, S. Chatterjee, and B. Yu. Spectral clustering and the high-dimensional stochastic blockmodel. *Annals of Statistics*, 1878–1915, 2011.
64. M. Rosvall, A. V. Esquivel, A. Lancichinetti, J. D. West, and R. Lambiotte. Memory in network flows and its effects on spreading dynamics and community detection. *Nature Communications*, 5:4630, 2014.
65. V. Salnikov, M. T. Schaub, and R. Lambiotte. Using higher-order Markov models to reveal flow-based communities in networks. *Scientific Reports*, 6:23194, 2016.
66. R. J. Sanchez-Garcia. Exploiting symmetry in network analysis. arXiv preprint arXiv:1803.06915, 2018.
67. S. E. Schaeffer. Graph clustering. *Computer Science Review*, 1(1):27–64, 2007.
68. M. T. Schaub. *Unraveling complex networks under the prism of dynamical processes: relations between structure and dynamics*. PhD thesis, Imperial College London, 2014.
69. M. T. Schaub, J.-C. Delvenne, S. N. Yaliraki, and M. Barahona. Markov dynamics as a zooming lens for multiscale community detection: non clique-like communities and the field-of-view limit. *PLoS One*, 7(2):e32210, 2012.
70. M. T. Schaub, Y. N. Billeh, C. A. Anastassiou, C. Koch, and M. Barahona. Emergence of slow-switching assemblies in structured neuronal networks. *PLoS Comput Biol*, 11(7):e1004196, 2015.

71. M. T. Schaub, N. O'Clery, Y. N. Billeh, J.-C. Delvenne, R. Lambiotte, and M. Barahona. Graph partitions and cluster synchronization in networks of oscillators. *Chaos: An Interdisciplinary Journal of Nonlinear Science*, 26 (9):094821, 2016.

72. M. T. Schaub, J.-C. Delvenne, M. Rosvall, and R. Lambiotte. The many facets of community detection in complex networks. *Applied Network Science*, 2(1):4, 2017.

73. M. T. Schaub, J.-C. Delvenne, R. Lambiotte, and M. Barahona. Multiscale dynamical embeddings of complex networks. arXiv:1804.03733, 2018. URL https://arxiv.org/abs/1804.03733.

74. S. Shai, N. Stanley, C. Granell, D. Taylor, and P. J. Mucha. Case studies in network community detection. arXiv:1705.02305, 2017.

75. H.-W. Shen, X.-Q. Cheng, and B.-X. Fang. Covariance, correlation matrix, and the multiscale community structure of networks. *Physical Review E*, 82:016114, 2010.

76. T. A. Snijders. Statistical models for social networks. *Annual Review of Sociology*, 37:131–153, 2011.

77. F. Sorrentino, L. M. Pecora, A. M. Hagerstrom, T. E. Murphy, and R. Roy. Complete characterization of the stability of cluster synchronization in complex dynamical networks. *Science Advances*, 2(4):e1501737, 2016.

78. G. W. Stewart. *Matrix algorithms volume 2: eigensystems*, volume 2. SIAM, 2001.

79. M. Szell, R. Lambiotte, and S. Thurner. Multirelational organization of large-scale social networks in an online world. *Proceedings of the National Academy of Sciences*, 107(31):13636–13641, 2010.

80. V. A. Traag, P. Van Dooren, and P. De Leenheer. Dynamical models explaining social balance and evolution of cooperation. *PloS One*, 8(4): e60063, 2013.

81. R. van Handel. Structured random matrices. In *Convexity and Concentration*, pages 107–156. Springer, 2017.

82. U. Von Luxburg. A tutorial on spectral clustering. *Statistics and Computing*, 17(4):395–416, 2007.

83. G. F. Young, L. Scardovi, and N. E. Leonard. Robustness of noisy consensus dynamics with directed communication. In *American Control Conference (ACC), 2010*, pages 6312–6317. IEEE, 2010.

84. W. Yu, G. Chen, and M. Cao. Consensus in directed networks of agents with nonlinear dynamics. *IEEE Transactions on Automatic Control*, 56 (6):1436–1441, 2011.

85. W. W. Zachary. An information flow model for conflict and fission in small groups. *Journal of Anthropological Research*, 33:452–473, 1977.

86. X. Zhang, R. R. Nadakuditi, and M. E. Newman. Spectra of random graphs with community structure and arbitrary degrees. *Physical Review E*, 89 (4):042816, 2014.

87. Q. Zhao, M. A. Erdogdu, H. Y. He, A. Rajaraman, and J. Leskovec. Seismic: A self-exciting point process model for predicting tweet popularity. In *Proceedings of the 21th ACM SIGKDD International Conference on Knowledge Discovery and Data Mining*, pages 1513–1522. ACM, 2015.

13

Scientific Co-Authorship Networks

Marjan Cugmas[1], Anuška Ferligoj[1,2], and Luka Kronegger[1]
[1]FDV, University of Ljubljana
[2]NRU HSE, Moscow

13.1 Introduction

Network studies of science offer researchers great insights into the dynamics of how knowledge is created and the social structure of scientific society. The flow of ideas and the scientific community's overall cognitive structure is observed through citations among scientific contributions, usually manifested as patents or papers published in scientific journals. The social structure of this society consists of relationships among scientists and their collaborations. De Haan [9] suggests six operationalized indicators of collaborative relations between scientists: co-authorship, shared editorship of publications, shared supervision of PhD projects, writing a research proposal together, participation in formal research programs, and shared organization of scientific conferences.

Due to the accessibility and ease of acquiring data through bibliographic databases, most scientific collaboration analyses are performed with co-authorship data, which play a particularly important role in research into the collaborative social structure of science. Co-authorship networks are personal networks in which the vertices represent authors, and two authors are connected by a tie if they co-authored one or more publications. These ties are necessarily symmetric. The study of community structures through scientific co-authorship is especially valuable because scientific (sub)disciplines can often display local properties that vary greatly from the properties of the scientific network as a whole.

Yet co-authorship data have some flaws. The wide pallet of relationships among scientists does not result in common publications [20, 24, 28]. Laudel [24] reports that about half of all scientific collaborations are invisible in formal communication channels because they lead to neither co-authored publications or formal acknowledgments in scientific texts. On the other hand, we also know that co-authorship sometimes indicates false positive relations arising

Advances in Network Clustering and Blockmodeling, First Edition.
Edited by Patrick Doreian, Vladimir Batagelj, and Anuška Ferligoj.
© 2020 John Wiley & Sons Ltd. Published 2020 by John Wiley & Sons Ltd.

from resource-related issues [30]. Despite this, as an indicator of scientific collaboration, co-authorship data form a reasonable compromise between quality and cost.

There are relatively few applications of blockmodeling to co-authorship networks. This may be due to the method's limitations regarding the size of the analyzed networks. One of the earliest applications can be found in [13], where the results of blockmodeling (clustering of relational data) of a co-authorship network of Slovenian sociologists and the results of clustering with a relational constraint (clustering of attribute and relational data) on the same network were compared according to the researchers' publication performance. As expected, the methods produced different results. This indicates that the decision about which method is used depends on the research problem under study. The unexpected result of the analysis in [13] was a core-periphery structure, with seven cores and a periphery, obtained when blockmodeling the co-authorship network.

Further investigation of the multicore-periphery structure was presented in [21], where the authors analyzed a network structure's development over time. In their analysis of the co-authorship networks of four scientific disciplines (physics, mathematics, biotechnology, and sociology) measured in four consecutive 5-year time spans, they observed that a multicore-periphery structure was present from early on in the development of each scientific discipline. They also found that, although the number of cores increases with the growth of a discipline, the cores' sizes did not change. The structure's description as constituting multiple cores and a periphery was extended with multiple elements: a weakly connected semi-periphery, a completely empty periphery and bridging cores, describing clusters of authors connecting two or more cores from the central part of the network. The authors described four levels of network complexity in the network structure's evolution through time:

1. simple core-periphery form: simple cores, semi-periphery, periphery
2. weakly consolidated core-periphery form: simple cores, bridging individuals, semi-periphery, periphery
3. consolidated core-periphery form: simple cores, bridging cores, semi-periphery, periphery
4. strongly consolidated core-periphery form: simple cores, bridging cores, bridging individuals, semi-periphery, periphery.

The multi-core–semi-periphery–periphery structure was also confirmed in a relatively small co-authorship network constructed from the curricula vitae (CVs) and bibliographies of teaching staff at the Faculty of Humanities and Education Science's Department of Library Science (DHUBI) at the National University of La Plata in Argentina [6].

Besides describing the overall structure, [21] attempted the first (visual) presentation to follow individual units in the transitions between blockmodels over time to reveal differences in the network dynamics across the disciplines analyzed here.

13.2 Methods

Considerable attention has been paid to studying the relationship between collaboration on one side and the quality of research and speed of diffusion of scientific knowledge on the other [1, 15, 17, 25]. While much research has considered the structure of co-authorship blockmodels [2, 14, 29], not so much has examined the stability of long-term collaborations.

Here, we show that blockmodeling can be used to reveal the global structure of co-authorship networks and how the stability of the blockmodels so obtained can be operationalized and measured. This is especially important when seeking to explain the stability of research teams.

13.2.1   Blockmodeling

The goal of blockmodeling is to reduce a large, complex network to a smaller, comprehensible, and interpretable structure [12]. It may not only be used to find groups of highly linked units within a network, but also the relationships between the groups. While it can reveal much information about the global co-authorship structure, obtaining the solution (especially in the case of direct blockmodeling) can be very computationally expensive when networks with a higher number of units are involved.

Blockmodeling can be either direct or indirect. Indirect blockmodeling is based on a dissimilarity matrix among units. The calculated dissimilarity measure must be consistent with a chosen equivalence between units. The studies by Kronegger *et al.* [21] and Cugmas *et al.* [8] used the corrected Euclidean distance, which is consistent with the structural equivalence [5]. The process of hierarchically clustering units can be visualized in a dendrogram in which the units (or clusters) and the dissimilarity between the units (or clusters) are represented. Kronegger *et al.* [21] and Cugmas *et al.* [8] defined the number of positions based on such visualizations.

In contrast, unlike indirect blockmodeling, direct blockmodeling can be achieved through a local optimization procedure [5], e.g. using an iterative method where for each displacement of a unit from one group into another, the value of the criterion function is calculated, defined as the difference between the ideal and empirical clustering where the ideal clustering has to express a blockmodel's assumed structure. It turns out that this procedure can be very time-consuming if a higher number of units in the network is analyzed. Cugmas *et al.* [8] also report that the algorithm implemented in `Pajek` has some difficulties detecting very small, structurally equivalent cores, particularly in the case of scientific disciplines with a very large number of researchers. To mitigate these characteristics, they removed the periphery and the structurally equivalent cliques from the network before applying the procedure. They later merged them to obtain the final solution.

13.2.2   Measuring the Obtained Blockmodels' Stability

The main result of blockmodeling is a partition which assigns a researcher to a certain core, semi-periphery, or periphery. In the case of temporal co-authorship networks (where time is seen as a discrete variable), blockmodeling can be applied for each time period separately such that one partition for each time period is obtained.[1] A very important characteristic of temporal co-authorship networks is that some researchers (called newcomers) join the network in a later time period while others (called outgoers) leave the network in the later time period. Besides the presence of newcomers and outgoers, the splitting of cores and merging of cores can also be seen as separate features that indicate the lower stability of the obtained blockmodels or cores.

---

[1] Along with the methods for generalized blockmodeling of multilevel networks [36], which can also be used to block-model temporal networks, different versions of stochastic blockmodeling exist for temporal networks [3, 26, 34, 35].

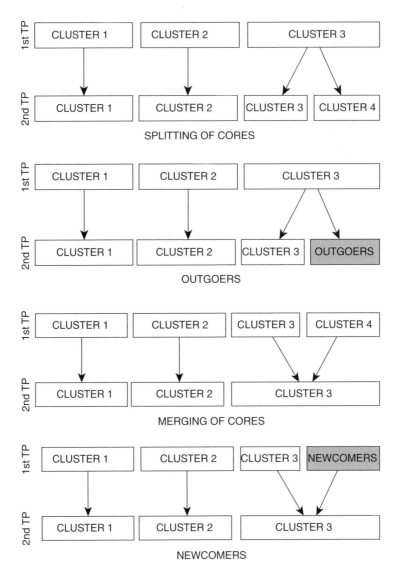

Figure 13.1 The features that can be used as indicators of less stable clusterings.

Nevertheless, a combination of different features usually appears simultaneously; a visualization of each feature is presented in Figure 13.1. Each visualization is divided into two parts: the white rectangles at the top visualize the clusters (which are cores obtained by blockmodeling in the case of co-authorship blockmodels) from the partition $U = \{u_1, ..., u_r\}$ obtained on the set of units from the first time period while the rectangles on the bottom visualize the clusters from the partition $V = \{v_1, ..., v_c\}$ obtained on the set of units from the second time period. Gray rectangles are added to the clusters and visualize the outgoers and newcomers. The links between the rectangles visualize the clusters' stability.

Adjusted Rand Index Based on the two assumptions that the merging and splitting of clusters indicate a lower level of cluster stability in time and that no newcomers or no outgoers are present (or, at least, they are ignored), one can use one of the most popular indices for comparing partitions, the Adjusted Rand Index [19, 32]. Here, the adjective "adjusted" refers to the necessary correction for chance since the expected value is usually not 0 in the case of two random and independent partitions. This correction allows the values of the index obtained from different partitions to be compared. We focus on the Rand Index [31], defined as

$$RI = \frac{a+d}{a+b+c+d}$$

where a stands for the number of pairs of researchers classified in the same cluster in both time periods, b stands for the number of pairs of researchers classified in the same clusters in the first period but in different clusters in the second period, c stands for the number of researchers classified in different clusters in the first, but in the same cluster in the second period and, finally, d stands for the number of pairs of researchers classified in different clusters in both the first and second time periods. Following this definition, the Rand Index can be interpreted in the co-authorship network context as the proportion of all possible pairs of researchers classified in the same or in different clusters in both time periods out of all possible pairs of researchers.

Wallace indices There are situations when the merging and splitting of clusters has to be considered differently. Therefore, one of two Wallace indices can be used: in the case of the Wallace Index 1 (W1), only the splitting of clusters is considered a feature that indicates lower cluster stability, while with the Wallace Index 2 (WI2) only the merging of cores is considered a feature indicating the lower stability of clusters. Formally, W1 is defined as

$$W1 = \frac{a}{a+b}$$

where a and b are defined the same as in the case of the Rand Index. W1 can be interpreted as the proportion of all researcher pairs placed in the same core in both time periods out of the number of all possible researcher pairs placed in the same core in the first period. Similarly, W2 is defined as

$$W2 = \frac{a}{a+c}$$

and interpreted as the ratio between the number of all possible researcher pairs classified in the same cluster in both periods and the number of all possible researcher pairs classified in the same cluster in the second period.

Modified Rand Index and Wallace indices As mentioned, it is common in temporal co-authorship networks for some researchers to join the network and some to leave the network in later time periods. When this happens, one can either simply ignore those researchers when calculating the Rand or Wallace indices, or treat the newcomers and outgoers as features indicating a lower level of stability of the cores. When the latter is assumed, one has to form new partitions $U' = \{u_1, u_2, ..., u_{r+1}\}$ and $V' = \{v_1, v_2, ..., v_{c+1}\}$ with the new clusters of newcomers u_{r+1} and outgoers v_{c+1} added to the partitions U and V. Then, the Modified Rand Index (MRI), the Modified W1, and the Modified W2 are calculated in the same way as RI, W1, and W2 where the values in the numerator consider the partitions U' and V'. The modified Rand Index and the modified Wallace indices can be further modified so that only newcomers or only outgoers are considered as features indicating lower core stability (for more details, see [7]) (see Figure 13.2).

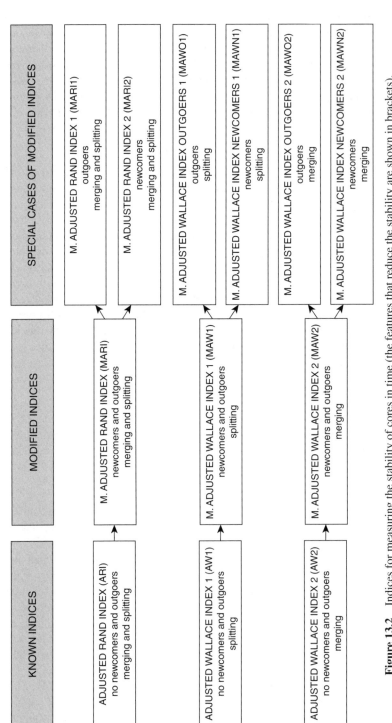

Figure 13.2 Indices for measuring the stability of cores in time (the features that reduce the stability are shown in brackets).

Along with the modified Rand Index and the modified Wallace indices, Cugmas and Ferligoj [7] proposed a correction for chance (based on Monte Carlo simulations) that allows one to compare the values of indices obtained in different scientific disciplines. With non-adjusted indices, the number of clusters (cores, newcomers, and outgoers) and the number of researchers also influence the expected value of an index in the case of two random and independent partitions. The expected value of two random and independent partitions in the case of adjusted indices equals 0, and the maximum value of an index is 1. It should be highlighted that higher values of the presented indices indicate greater cluster stability, while lower values indicate less stability.

13.3   The Data

The data for this research were obtained from the Co-operative Online Bibliographic System and Services (COBISS) and the Slovenian Current Research Information System (SICRIS) maintained by the Institute of Information Science (IZUM) and the Slovenian Research Agency (SRA).

SICRIS provides data about all researchers who have an ID assigned by the SRA, including their educational background and field of research according to the SRA's classification scheme. There are seven scientific fields and 72 scientific disciplines defined in this classification scheme. The seventh scientific field is Interdisciplinary Studies and is not included in the analysis since it has never gained full recognition as a separate research field in Slovenia [14].

The analyzed data are based on complete personal bibliographies of each researcher (constructed based on SICRIS and COBISS). The network boundaries are therefore defined only by those researchers registered as a researcher at the SRA. Among such researchers, those who published at least one scientific bibliographic unit between 1990 and 2010 are included. The bibliographic units considered as a scientific publication by the SRA are listed in Table 13.1.

Compared to the analysis conducted by Kronegger *et al.* [21], who studied four selected scientific disciplines in four time periods, the current analysis is performed on data for two consecutive 10-year periods between 1991 and 2010. The difference in the length of the periods

Table 13.1   Number of published scientific bibliographic units by type for two time periods

Type of scientific bibliographic unit	1991–2000 (independence)	2001–2010 (joining the EU)
Original scientific article	26531	47905
Review article	4895	5738
Short scientific article	969	2530
Published scientific conference contribution (invited lecture)	3427	5279
Published scientific conference contribution	28670	41138
Independent scientific component part in monograph	6417	14759
Scientific monograph	1725	2912
Scientific or documentary films, sound or video recording	44	133
Complete scientific database or corpus	73	182
Patent	381	710
Total	73132	121286

mainly affects the size and density of the generated co-authorship networks and, in terms of the stability of the research teams, results in a lower level of stability. However, the two selected periods reflect a time of major changes in scientific research and development policies in Slovenia. The first period is marked by the independence of Slovenia, meaning that the country had started adopting and implementing its own science policies, while the second period is marked by it joining the European Union and adopting European Union standards. By the end of this period, Slovenia had already partly integrated its national science system into the European one.

Between 1991 and 2000, 73,132 scientific bibliographic units were published and a further 121,286 scientific units were published between 2001 and 2010. The most common units are published scientific conference contributions and original scientific articles. The distribution of different types of bibliographic units varies among scientific disciplines. For example, published scientific conference contributions are very common in scientific disciplines from the technical sciences while original scientific articles are frequent among scientific disciplines within the social sciences and humanities. There are differences at the level of scientific disciplines according to the distribution of types of scientific bibliographic units which can be published by one or several researchers. Kronegger *et al.* [22] studied the differences between scientific disciplines according to collaboration patterns in time and confirmed the scientific discipline geography is more similar to scientific disciplines in the scientific fields of natural sciences and mathematics than the scientific field of the humanities, where it is found according to the SRA's classification scheme. Even within several scientific disciplines one can expect certain differences in types of co-authorships. In the case of sociology, [29] concluded that quantitative work is more likely to be co-authored than non-quantitative work.

Although many co-authorship networks are analyzed in this study, we present sociology co-authorship networks as an example. The units represent researchers and a link between two researchers exists if they published at least one scientific bibliographic unit in co-authorship. Therefore, only symmetric links are possible in the case of co-authorship networks. There are also some researchers without any link who are later classified in the so-called periphery, as detailed in the next section. However, it should be pointed out that the absence of links is not necessarily the consequence of only single-authored scientific bibliographic units by a certain researcher, but can also be the outcome of co-authoring only with researchers who do not have a researcher ID, for example with researchers from abroad. Isolated researchers are present in both time periods. The next important network characteristic common to almost all scientific disciplines is that the co-authorship networks grow in time.

13.4   The Structure of Obtained Blockmodels

Based on four scientific disciplines, Kronegger *et al.* [21] showed that the structure of co-authorship networks consists of the multi-core, semi-periphery, and periphery. To confirm that this structure is also present in other scientific disciplines, Cugmas *et al.* [8] used indirect blockmodeling to detect the approximate number of cores and direct blockmodeling to obtain the final solution, as described in Section 13.2.1. The assumed blockmodel structure was confirmed in all scientific disciplines included in the analysis. Most disciplines that were excluded (indicated in Figure 13.3 by asterisks) were removed due to the small number of researchers in the first or second time period or absence of co-authorship in the current period. One such discipline is theology, which did not have a single co-authored scientific bibliographic item

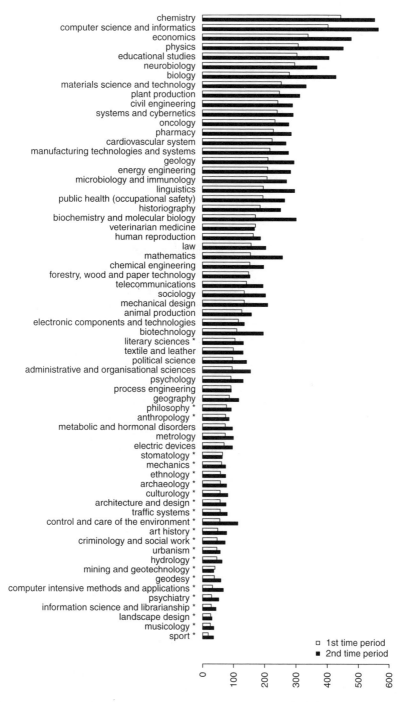

Figure 13.3 A list of scientific disciplines with the number of researchers in the first and second periods (an asterisk indicates scientific disciplines excluded from the analysis due to the small number of researchers in the first or second time period or absence of co-authorship in the current period).

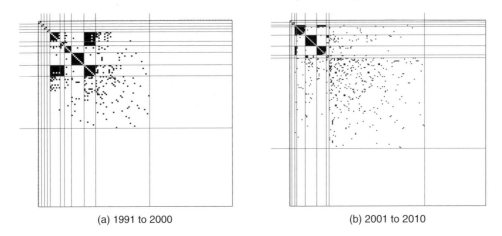

(a) 1991 to 2000   (b) 2001 to 2010

Figure 13.4 Structure of the sociology co-authorship blockmodel for the first and second time periods.

published in the first period. It can also be seen in Figure 13.3 that the number of researchers who published at least one scientific bibliographic item is increasing over time in almost all scientific disciplines. The average growth in the number of researchers publishing at least one scientific bibliographic item in the second period is 34%. Only in the disciplines veterinary medicine, stomatology (i.e. oral medicine), and mining and geotechnology is there a drop in the number of researchers from the first to the second period observed.

Figure 13.4 visualizes two empirical blockmodels for the scientific discipline sociology. The first blockmodel corresponds to the first period while the second blockmodel is for the second period. The rows and columns of each blockmodel contain the IDs of the researchers, where the black dots in the cells denote co-authorships between two given researchers. A clear multi-core–semi-periphery–periphery structure can be seen in the case of sociology in both time periods. Along with the already described multi-core, semi-periphery, and periphery, in the blockmodel in the first period a so-called bridging core is seen (as a full off-diagonal block). The bridging core is a group of researchers who collaborate with each other very systematically and also with researchers from at least two other cores. They are called "bridging" since they connect two or more cores. They are relatively common in Slovenian scientific disciplines. There was a minimum of one bridging core in at least one time period in 20 scientific disciplines of all that were analyzed.

The visualization in Figure 13.5 emphasizes the researchers' transitions between the cores (including the semi-periphery and periphery) obtained for the two periods: the upper part visualizes the partition of researchers for the first period while the bottom part visualizes the partition of researchers for the second period. It is shown in Figure 13.4 that the share of researchers classified in the periphery is decreasing in sociology, which cannot be seen in the visualization of researchers' transitions in time in Figure 13.5b. This is caused by the newcomers and outgoers. Figure 13.5a reveals a high share of researchers who were not classified in the cores in both time periods (e.g. many researchers were classified in the periphery in the first and second periods). Furthermore, many newcomers were classified in the semi-periphery or periphery in the second period. A similar pattern of many new researchers who were not connected to any previously existing authors was also found in other studies [2].

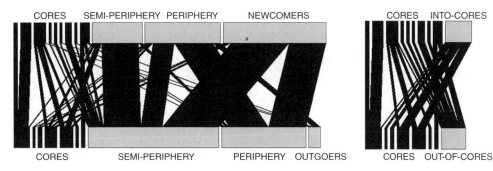

CORES   SEMI-PERIPHERY PERIPHERY   NEWCOMERS   CORES   INTO-CORES

CORES   SEMI-PERIPHERY   PERIPHERY OUTGOERS   CORES   OUT-OF-CORES

(a)Transitions between the cores, semi-periphery   (b) Transitions between the
and periphery   cores, into-cores and
                                                                                      out-of-cores

Figure 13.5 Visualization of researchers' transitions in two time periods for sociology.

Since the main interest of study is the stability of the cores of the obtained blockmodels, researchers not classified in the cores in at least one period can be removed from the visualization. Therefore, a new visualization is presented in Figure 13.5b consisting of two parts (one for each period) without the semi-periphery, periphery, newcomers, and outgoers. Instead, researchers classified in the cores in the first but not the second period are now called "out-of-cores" researchers and, similarly, researchers not classified in the cores in the first period but classified in the core in the second period are now called "into-cores" researchers. Focusing on the core part of the sociology example, it can be observed that cores 1 and 2 merged in the second period, while core 3 splits into three cores in the second period. There are also many cores which disappear in the second period (out-of-cores researchers) and a lot of researchers not classified in the cores in the first but classified in the cores in the second period. These into-cores researchers usually join the existing cores in the second period.

Visualizations of researchers' transitions between the cores, into-cores, and out-of-cores in the two periods are made for all analyzed scientific disciplines (Figure 13.6). A relatively high share of into-cores and out-of-cores researchers in all analyzed scientific disciplines and some merging and splitting of cores in the core part of the visualized transitions can be seen. Here, the into-cores and out-of-cores researchers are shown to be the primary source of instability of the core part of scientific disciplines. Although the share of into-cores researchers exceeds the share of out-of-cores researchers in almost all analyzed scientific disciplines, some scientific disciplines reveal that the share of out-of-cores prevails over the share of into-cores researchers.

The number and size of the cores, the size of the semi-periphery, and the size of the periphery vary across scientific disciplines (see Figure 13.7 and Tables 13.2 and 13.3). For example, the discipline administrative and organizational sciences has six cores in the first period and 16 cores in the second. As shown below, most of the existing cores in the second time period emerged from the non-core part of the network. The newly emerged cores are smaller in size (on average 2.64 researchers) than the cores in the first time period. These observations indicate that scientific collaboration might have been seen as more beneficial by the researchers from the field administrative and organizational sciences. Keep in mind that this is a scientific discipline with relatively few researchers (96 researchers in the first and 155 researchers – with at least one published scientific bibliographic unit – in the second time period).

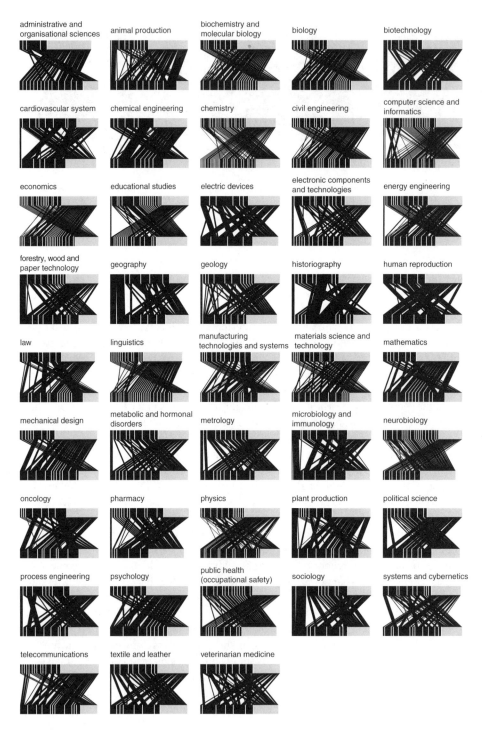

Figure 13.6 Visualization of researchers' transitions between the cores in the two periods for all analyzed scientific disciplines (the black rectangles on the top and bottom correspond to the cores, the gray rectangles on the top correspond to the group of into-cores while the gray rectangles on the bottom correspond to the group of out-of-cores).

Table 13.2 The considered scientific disciplines and their characteristics in the two periods – part 1

Scientific discipline and scientific field	1991–2000					2001–2010					% of researchers that left the cores
	N	Number of cores	Semi-per. (%)	Per. (%)	Average core size	N	Number of cores	Semi-per. (%)	Per. (%)	Average core size	
1. Natural sciences and mathematics											
Biochemistry and molecular biology	171	11	45.61	30.41	4.56	301	17	45.85	33.55	4.13	68.29
Biology	279	12	46.59	37.63	4.40	427	18	55.04	27.40	4.69	70.45
Chemistry	443	22	64.79	12.87	4.95	553	29	67.09	11.93	4.30	68.69
Geology	211	14	44.55	34.12	3.75	294	10	43.54	42.86	5.00	64.44
Mathematics	155	9	34.19	51.61	3.14	257	13	51.36	32.30	3.82	59.09
Pharmacy	228	15	51.75	25.00	4.08	285	8	71.58	14.04	6.83	77.36
Physics	307	15	61.89	15.96	5.23	451	15	65.63	12.42	7.62	52.94
2. Engineering sciences and technologies											
Chemical engineering	153	8	60.13	24.18	4.00	197	10	66.50	18.78	3.62	62.50
Civil engineering	242	15	65.29	15.70	3.54	289	18	61.94	13.49	4.44	69.57
Computer science and informatics	401	25	51.87	27.43	3.61	565	21	63.36	21.24	4.58	62.65
Electric devices	70	8	31.43	25.71	5.00	97	6	38.14	26.80	8.50	56.67
Electronic components and technologies	116	14	36.21	23.28	3.92	135	15	40.74	25.93	3.46	61.70
Energy engineering	210	15	57.14	16.67	4.23	283	14	60.07	15.90	5.67	70.91
Manufacturing technologies and systems	217	14	45.62	28.57	4.67	276	14	52.90	23.55	5.42	50.00
Materials science and technology	253	13	60.08	17.39	5.18	332	15	62.05	13.86	6.15	61.40
Mechanical design	135	11	55.56	21.48	3.44	210	13	56.19	17.62	5.00	64.52
Metrology	74	12	39.19	20.27	3.00	100	11	35.00	30.00	3.89	56.67
Process engineering	92	11	52.17	11.96	3.67	93	11	48.39	17.20	3.56	66.67
Systems and cybernetics	240	12	54.17	21.67	5.80	291	14	58.42	17.53	5.83	48.28
Telecommunications	142	15	39.44	32.39	3.08	195	13	45.13	21.03	6.00	50.00
Textile and leather	100	11	55.00	12.00	3.67	131	9	61.07	11.45	5.14	63.64
3. Medical sciences											
Cardiovascular system	224	11	54.91	16.52	7.11	268	8	67.54	12.69	8.83	57.81
Human reproduction	164	7	53.66	21.95	8.00	187	7	55.61	11.76	12.20	42.50
Metabolic and hormonal disorders	74	12	13.51	45.95	3.00	97	11	38.14	28.87	3.56	66.67
Microbiology and immunology	208	10	54.81	14.90	7.88	270	8	67.78	8.89	10.50	47.62
Neurobiology	296	22	51.01	25.68	3.45	366	12	69.67	16.67	5.00	81.16
Oncology	233	10	54.08	17.60	8.25	277	11	56.32	13.36	9.33	45.45
Public health (occupational safety)	195	17	34.87	41.54	3.07	264	12	56.06	29.55	3.80	80.43

Table 13.3 The considered scientific disciplines and their characteristics in the two periods – part 2

Scientific discipline and scientific field	N	1991–2000				N	2001–2010				
		Number of cores	Semi-per. (%)	Per. (%)	Average core size		Number of cores	Semi-per. (%)	Per. (%)	Average core size	% of researchers that left the cores
4. Biotechnical sciences											
Animal production	127	11	64.57	6.30	4.11	158	12	48.73	7.59	6.90	35.14
Biotechnology	111	9	49.55	18.92	5.00	196	9	58.16	13.78	7.86	60.00
Forestry, wood and paper technology	149	12	62.42	11.41	3.90	153	10	53.59	11.76	6.62	43.59
Plant production	247	10	68.83	12.96	5.62	312	11	66.35	8.33	8.78	35.56
Veterinarian medicine	171	10	60.23	8.19	6.75	168	8	61.31	5.36	9.33	51.85
5. Social sciences											
Administrative and organisational sciences	96	6	17.71	64.58	4.25	155	16	34.19	41.94	2.64	58.82
Economics	338	20	49.11	32.84	3.39	477	22	61.01	21.17	4.25	73.77
Educational studies	303	19	38.28	39.60	3.94	404	17	51.24	33.91	4.00	76.12
Law	157	10	24.20	47.13	5.62	204	15	31.37	35.78	5.15	40.00
Political science	98	12	23.47	48.98	2.70	142	9	45.77	31.69	4.57	74.07
Psychology	92	9	29.35	48.91	2.86	131	10	38.17	36.64	4.12	65.00
Sociology	135	11	28.15	42.22	4.44	203	9	48.77	31.03	5.86	45.00
6. Humanities											
Geography	87	8	33.33	35.63	4.50	117	8	53.85	24.79	4.17	40.74
Historiography	186	11	9.14	74.73	3.33	251	11	23.11	60.56	4.56	43.33
Linguistics	196	15	10.71	71.94	2.62	296	25	22.64	55.07	2.87	61.76

There is usually a higher number of cores in disciplines that have more researchers, which is expected given the personal limits of each researcher to cooperate with a limited number of coauthors and produce a limited number of publications [11, 21]. One such example is physics with 307 researchers (with at least one published scientific bibliographic unit) in the first and 451 researchers in the second period. There are 15 cores revealed in both the first and second period. As this is the case for almost all scientific disciplines, the average core size is bigger in the second period than in the first. The decrease in the average core size is usually the consequence of many cores of size two having emerged (e.g. biochemistry and molecular biology, chemistry, law). These can consist of any kind of researchers, for example a core of size two can consist of a student and his/her mentor. However, it is assumed that any kind of scientific collaboration that leads to the creation of a scientific bibliographic unit requires very intensive collaboration – the exchange of knowledge and ideas. Pairs of scientists who collaborate as researchers are also very common in the field of social network analysis. One example is Borgatti and Everett.

However, some studies found that the type (e.g. natural vs. social or office vs. lab or theoretical vs. empirical) of a scientific discipline affects the size of research teams [18, 23, 27]. Here, the biggest average core size in the first period is observed in oncology (8.3 researchers) and human reproduction (8.0 researchers), while the lowest average core size in the first period is observed in linguistics (2.6 researchers) and psychology (2.9 researchers). In general, the overall average number of cores is similar in both periods (around 11 cores), while the overall average core size is increasing in time (from 4.4 to 5.6 researchers, $p < 0.01$), as confirmed by Amat and Perruchas [4].

Following the distinction between the natural and technical sciences on one side, and the social sciences and humanities on the other, it may be concluded that the average core size is growing, especially in the natural and technical sciences (from 4.6 to 6.1 researchers, $p < 0.01$), while in the social sciences and humanities the growth of the average core size (from 3.8 to 4.2 researchers, $p = 0.30$) is not statistically significant. In general, the average core size is lower in the social sciences and humanities in the two periods (for 0.95 researchers in the first and 1.85 researchers in the second time period; $p < 0.05$ and $p < 0.01$ subsequently) (Figure 13.7).

Solo authors or authors who only published in co-authorship with authors from outside the discipline are classified in the periphery. The average share of these authors among the analyzed scientific disciplines decreases over time (from 29% to 23% average share of the periphery,

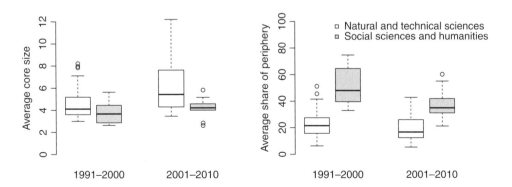

Figure 13.7   The average core size and the average size of the periphery by field and time period.

$p = 0.06$). The biggest reduction in the percentage of the periphery in the second period is observed in criminology and social work (a 65% decrease). In some scientific disciplines, the percentage of the periphery increased in the second period. These are mainly disciplines from the natural and technical fields. However, the periphery size is greater in the social sciences and humanities (the average size is 44%) than in scientific disciplines classified in the natural and technical sciences (the average size is 21%) ($p < 0.01$). In addition, the average share of the periphery decreases from the first to the second period, especially in the natural and technical sciences (from 51% to 37%, $p < 0.05$), while the difference in the average share of the periphery is not statistically significant ($p = 0.11$) in the social sciences and humanities (the periphery share decreased from 23% to 19%) (Figure 13.7).

13.5   Stability of the Obtained Blockmodel Structures

In this section, the stability of cores is studied according to different operationalizations of core stability. Although the presented visualizations of researchers' transitions between two time periods (Figure 13.6) provide a very efficient tool for studying the stability of the cores obtained but whose interpretation is complex, the values of the indices proposed in Section 13.2 are calculated. These indices are more objective operationalizations of core stability and allow us to compare the values calculated for different scientific disciplines. The scientific disciplines are then clustered according to the calculated indices. The groups of scientific disciplines thus obtained are further analyzed.

In the second part, the operationalization of the stability of cores is restricted to one of the described indices for measuring core stability, namely, as applied in Cugmas *et al.* [8], only the splitting of cores and the out-of-cores researchers are seen as features, indicating the cores are less stable. The differences in the mean stability of cores among different scientific fields are studied using linear regression. Some further controlling explanatory variables are also included in the model.

First, the values of each presented index for each analyzed scientific discipline are shown in Table 13.4 and provide the basis for all further analyses. In this table, one sees that the values of the Adjusted Rand Index and the adjusted Wallace indices are relatively large, while the others are relatively small. This is due to the high share of into-cores and out-of-cores researchers who are not considered when calculating the values of the Adjusted Rand Index and the adjusted Wallace indices for each scientific discipline. The high values of the first three indices and the low values of the others confirm that the into-cores researchers and out-of-cores researchers are the biggest source of the obtained cores' instability.

13.5.1   Clustering of Scientific Disciplines According to Different Operationalizations of Core Stability

Based on the calculated standardized indices (see Table 13.4 for non-standardized values of the indices), the analyzed scientific disciplines are clustered using Ward's agglomerative clustering method and squared Euclidean distance. Three clusters were chosen on using the GAP Statistics [33] and the obtained dendrogram. By observing the means of the calculated indices for each cluster (see Table 13.4), the obtained clusters can be ordered from the least stable (cluster 1) to

Table 13.4 The values of different indices for measuring the stability of cores for all analyzed scientific disciplines by obtained clusters

Discipline	ARI	AW1	AW2	MARIO	MWO1	MWO2	MARII	MWII	MWOI
Cluster 1 (unstable)									
Biochemistry and molecular biology	−0.16	−0.11	−0.27	−0.02	0.00	0.00	−0.03	0.00	0.00
Geology	0.09	0.11	0.07	0.00	0.02	0.00	0.00	0.01	0.03
Psychology	0.04	0.07	0.03	0.00	0.02	0.00	−0.09	0.00	0.01
Pharmacy	0.50	0.69	0.39	0.01	0.03	0.01	0.09	0.01	0.03
Physics	0.40	0.67	0.28	0.03	0.15	0.04	0.03	0.02	0.08
Neurobiology	0.43	0.68	0.32	0.00	0.08	0.01	0.03	0.01	0.09
Materials science and technology	0.33	0.26	0.45	0.01	0.06	0.02	0.02	0.01	0.04
Public health (occupational safety)	0.37	0.32	0.42	0.00	0.03	0.00	0.00	0.00	0.02
Biology	0.38	0.48	0.31	0.01	0.09	0.02	−0.02	0.00	0.04
Educational studies	0.32	0.34	0.31	0.00	0.04	0.01	−0.02	0.01	0.07
Linguistics	0.43	0.29	0.82	0.00	0.16	0.03	0.03	0.00	0.08
Electric devices	0.50	0.46	0.56	0.03	0.11	0.06	0.04	0.03	0.07
Metrology	0.42	0.44	0.40	0.01	0.16	0.04	0.00	0.02	0.09
Cluster 1 means	0.31	0.36	0.31	0.01	0.07	0.07	0.01	0.01	0.05
Cluster 2 (average)									
Mathematics	1.00	1.00	1.00	0.05	0.27	0.08	0.01	0.01	0.09
Civil engineering	0.86	0.76	1.00	0.01	0.14	0.02	0.01	0.01	0.06
Energy engineering	0.81	0.75	0.88	0.01	0.15	0.02	0.05	0.01	0.09
Systems and cybernetics	0.70	0.57	0.92	0.04	0.19	0.09	0.12	0.04	0.15
Computer science and informatics	0.63	0.57	0.71	0.01	0.19	0.03	0.04	0.02	0.14
Telecommunications	0.69	0.89	0.56	0.04	0.35	0.08	0.06	0.02	0.09
Electronic components and technologies	0.62	0.47	0.91	0.01	0.13	0.03	0.11	0.03	0.19
Mechanical design	1.00	1.00	1.00	0.04	0.23	0.06	0.04	0.01	0.09
Process engineering	0.84	0.72	1.00	0.02	0.14	0.04	0.1	0.04	0.17
Textile and leather	0.80	0.80	0.80	0.03	0.18	0.04	0.00	0.03	0.10
Human reproduction	0.40	0.93	0.26	0.08	0.28	0.14	0.12	0.06	0.14
Metabolic and hormonal disorders	0.73	1.00	0.57	0.02	0.07	0.01	−0.02	0.01	0.06
Chemistry	0.60	0.46	0.89	0.01	0.17	0.02	0.04	0.01	0.17
Forestry, wood and paper technology	0.64	0.69	0.60	0.06	0.34	0.14	0.10	0.05	0.15
Animal production	0.49	0.51	0.47	0.06	0.19	0.13	0.06	0.01	0.06
Veterinarian medicine	0.52	0.68	0.43	0.04	0.15	0.05	0.13	0.05	0.09
Biotechnology	0.73	1.00	0.57	0.04	0.14	0.05	−0.01	0.01	0.04
Economics	0.71	0.64	0.80	0.01	0.14	0.01	0.01	0.01	0.07
Administrative and organisational sciences	0.80	0.67	1.00	0.06	0.11	0.09	0.05	0.01	0.19
Law	0.58	0.80	0.45	0.09	0.29	0.17	−0.14	0.06	0.22
Political science	0.86	1.00	0.75	0.01	0.13	0.02	−0.03	0.01	0.05
Historiography	0.57	0.68	0.49	0.08	0.39	0.20	0.05	0.07	0.09
Cluster 2 means	0.71	0.75	0.73	0.04	0.20	0.17	0.04	0.03	0.12

(*continued overleaf*)

Table 13.4 (*continued*)

Discipline	ARI	AW1	AW2	MARIO	MWOI	MWO2	MARII	MWII	MWOI
Cluster 3 (stable)									
Plant production	0.90	0.84	0.97	0.15	0.45	0.34	0.11	0.05	0.19
Oncology	0.89	0.85	0.93	0.11	0.42	0.23	0.12	0.11	0.35
Chemical engineering	1.00	1.00	1.00	0.08	0.33	0.14	0.09	0.08	0.30
Manufacturing technologies and systems	0.95	0.90	1.00	0.06	0.43	0.12	0.11	0.07	0.25
Microbiology and immunology	0.88	0.86	0.91	0.09	0.32	0.16	0.25	0.16	0.26
Cardiovascular system	1.00	1.00	1.00	0.06	0.30	0.10	0.32	0.18	0.30
Sociology	0.52	0.55	0.50	0.06	0.36	0.16	0.25	0.14	0.23
Geography	0.36	0.29	0.49	0.03	0.15	0.12	0.22	0.12	0.21
Cluster 3 means	0.81	0.79	0.85	0.08	0.35	0.02	0.18	0.11	0.26

the most stable (cluster 3). Cluster 2 is named average since the values of all indices are closest to the global means compared to the other groups. Table 13.5 summarizes some descriptive statistics of other blockmodels' characteristics:

- *The percentage of the into-cores (% into-cores) and out-of-cores (% out-of-cores) researchers.* The percentage of into-cores researchers is defined as the ratio between the number of researchers not in the cores in the first period and the number of researchers classified in the cores in the first period. In contrast, the percentage of out-of-cores researchers is defined as the ratio between the number of researchers who joined the cores in the second period and the number of researchers classified in cores in the second period. Since Slovenian scientific disciplines are generally growing, the average share of into-cores researchers is less than the share of out-of-cores researchers. However, a higher percentage of into-cores than out-of-cores researchers is typical for the unstable cluster of scientific disciplines.
- *The overall average core size (core size) and the overall number of researchers across clusters of scientific disciplines (# of res.).* The average core size is relatively small, with the smallest in the case of the unstable cluster (3.9 researchers) and the highest in the case of the most stable cluster (5.8 researchers). While a higher average core size is typical for more stable scientific disciplines, a higher number of researchers per discipline is related to less stable scientific disciplines.
- *The number of scientific disciplines.* The average cluster, according to the values of the stability measures, has the highest number of scientific disciplines, followed by the unstable and the stable cluster.

In the Slovenian Research Agency's classification scheme, scientific fields are further divided into several scientific disciplines and then into scientific sub-disciplines. Based on this, most scientific disciplines from the fields of engineering sciences and technologies (nine out of 14), biotechnological sciences (four out of five), and the social sciences (four out of seven) were classified in the unstable cluster. Most (five out of seven) scientific disciplines from the natural sciences and mathematics were classified in the average cluster and three out of seven scientific

Table 13.5 Basic descriptive statistics of the obtained clusters (averages on the level of clusters are reported)

Cluster	% into cores	% out of cores	core size	# of res.
Cluster 1 ($N = 13$) (unstable)	72	67	3.9	322
Cluster 2 ($N = 22$) (average)	69	58	4.2	274
Cluster 3 ($N = 8$) (stable)	53	48	5.8	272

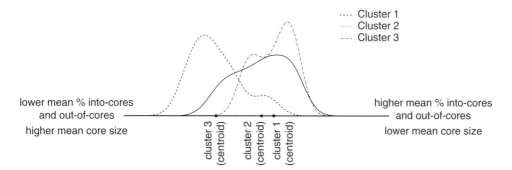

Figure 13.8 Distribution of standardized values of the first canonical discriminant FUNCTION by clusters.

disciplines from the field of medical sciences were classified in the most stable cluster. We can say the most stable scientific disciplines come from the medical sciences and the most unstable ones from the technical field and social sciences. Similarly, Melin [27] concluded that researchers from the medical sciences field almost always work in teams and from time to time collaborate with other teams. Kyvik [23] reports that the greatest number of multi-authored papers in Norway is in medicine.

Since a scientific discipline's affiliation with a certain cluster is a categorical variable, one can check if the basic characteristics presented in Table 13.5 can be used to predict the cluster to which a given scientific discipline belongs. To do this, discriminant analysis can be used. Since there are three clusters of scientific disciplines, two discriminant functions can be calculated based on the four explanatory variables presented in Table 13.5. Only the first discriminant function is statistically significant ($p < 0.01$), meaning that based on the four explanatory variables one can separate well between the stable cluster (cluster 3) on one side and unstable clusters (clusters 1 and 2) on the other. The discriminant functions are defined as linear combinations of explanatory variables. In Figure 13.8, the first discriminant function is visualized. Here, the highest values of the first discriminant function are characterized by a higher mean percentage of into-cores (0.74) and percentage out-of-cores researchers (0.20) and a lower average number of researchers in the cores (−0.31). The value of the standardized canonical coefficient of the explanatory variable "number of researchers" is relatively low (−0.09) and therefore not mentioned in Figure 13.8. The centroids for each cluster are also marked, along with the distribution of the standardized discriminant function for the disciplines by clusters.

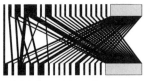

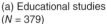

(a) Educational studies   (b) Textile and leather   (c) Microbiology and immunology
(*N* = 379)   (*N* = 123)   (*N* = 226)

Figure 13.9 Visualizations of researchers' transitions between the cores, into-cores, and out-of-cores for the two periods for the scientific discipline closest to the centroid in each cluster (the black rectangles on the top and on the bottom correspond to the cores, the gray rectangles on the top correspond to the group of into cores while the gray rectangles on the bottom correspond to the group of out of cores).

A scientific discipline from each cluster of the scientific disciplines was chosen to represent the cluster (the closest one to the centroid). The representative of the unstable cluster is the scientific discipline educational studies (see Figure 13.9). Here, many into-cores and out-of-cores researchers can be seen. Most pairs of researchers classified in the same core at the first time point were not classified in the same core in the second period. The representative of the average cluster is the scientific discipline of textile and leather. Here, the share of out-of-cores and into-cores researchers is lower. Some relatively large cores which remain relatively stable in the second period can also be observed. This is more typical for the representative of the stable cluster, namely microbiology and immunology.

13.5.2   Explaining the Stability of Cores

To analyze the differences in the stability of cores among scientific fields, Cugmas *et al.* [8] classified the fields into two categories: the fields natural sciences and mathematics, engineering sciences and technologies, medical sciences, biotechnical sciences in the category "natural and technical sciences", and social sciences and humanities in the category "social sciences and humanities". The selected features that reduced the stability of the cores were the splitting of clusters and out-of-cores researchers and, therefore, the stability of clusters was measured by the MAW1. They show there is no statistically significant difference in the average core stability.

Given the high level of variability in the characteristics of the co-authorship networks and the blockmodel structures across scientific disciplines, the stability of the cores must be controlled by some additional network and blockmodel characteristics. Therefore, to explain the differences in core stability across scientific disciplines, as controlling explanatory variables Cugmas *et al.* [8] also included in the linear model [2] the characteristics of the networks (number of researchers, growth from the first period to the second period in the number of researchers and the growth of the density) and the obtained blockmodels (average core size, percentage of cores, presence of a bridging core at the first time point, percentage of outgoers).

---

[2] The Least Squares Method was used to estimate the parameters' values. The correlations among independent variables were observed before the parameters were estimated. After that, the Variance Inflation Factor was checked to further investigate the potential problems of multicollinearity. The distribution of the residuals was also examined to identify any problems of heteroscedasticity or other unsatisfied assumptions.

The main results are presented in Table 13.6. Here, the humanities is used as the reference field since many studies suggest that the social sciences are becoming more similar to the natural and technical sciences regarding publishing behavior [22, 23]. In Table 13.6 (Model 1), one can see there are no statistically significant differences between the humanities and other scientific fields when the percentage of out-of-cores is not included in the model. However, when the percentage of out-of-cores is included in the model, the differences in the mean stability of cores between the humanities and the engineering sciences and technologies and the humanities and the medical sciences become statistically significant ($p < 0.10$). Here, the scientific disciplines of both fields are seen as more stable than the humanities. Since the percentage of out-of-cores researchers forms part of the core stability index, the statistically significant differences between the mentioned scientific fields are mainly the consequence of the splitting of cores.

The effects of some controlling explanatory variables are statistically significant at ($p < 0.10$) as well. When the variable percentage of out-of-cores researchers is not included in the model (see Model 1 in Table 13.6), the growth of the density and the average core size in the first

Table 13.6 The impact of the characteristics of the network, blockmodel, and disciplines on the stability of the cores

	Model 1			Model 2		
	b	SE(b)	p	b	SE(b)	p
Intercept	0.0906	0.2027	0.66	0.8349	0.1840	0.00
Number of researchers (first time period)	−0.0002	0.0003	0.58	0.0001	0.0002	0.77
Growth of number of researchers (first and second time periods)	0.0010	0.0015	0.53	0.0004	0.0010	0.72
Growth of density (first and second time periods)	0.0015	0.0010	0.04	0.0091	0.0005	0.07
Average core size (first time period)	0.0625	0.0177	0.00	0.0053	0.0152	0.73
Percentage of cores (first and second time periods)	−0.0054	0.0049	0.28	−0.0069	0.0033	0.05
Presence of the bridge (first time period)	0.0404	0.0450	0.38	−0.0005	0.0313	0.99
Percentage of out-of-cores		Not included		−1.0160	0.1667	0.00
Humanities (reference category)						
Natural science and math.	−0.1511	0.0892	0.10	0.0378	0.0680	0.58
Engineering sciences and tech.	−0.0120	0.0834	0.89	0.1339	0.0615	0.04
Medical sciences	−0.0850	0.0954	0.38	0.1421	0.0748	0.07
Biotechnical sciences	−0.0353	0.1008	0.72	0.0338	0.0694	0.63
Social sciences	−0.0707	0.0844	0.41	0.0847	0.0626	0.19
Number of obs. (disciplines):			*43*			*43*
AdjustedR² :			*0.23*			*0.65*
F Statistics:		*2.151 (11; 31) (p < 0.05)*			*7.375 (12; 30) (p < 0.01)*	
Method of estimation:		*Least Squares Method*			*Least Squares Method*	

time period are statistically significant. The density is defined as the share of all realized ties from all possible ties. The value is typically greater in the case of smaller networks with a low percentage of researchers in the periphery and many cores with a lot of researchers included. Therefore, together with the variable average core size, it can be argued that in the case of greater density there are more researchers who co-authored only occasionally (semi-periphery) and more complete cores with a higher number of researchers. The probability of creating ties with new researchers is therefore lower and the stability of the cores is higher. Similarly, De Haan et al. [10] mentioned that the size of a research group affects the persistence of collaboration.

When the percentage of out-of-cores researchers is included in the model (see Model 2), the growth of density and the percentage of cores are statistically significant ($p < 0.10$) along with the controlling explanatory variable percentage of out-of-cores researchers, which is highly statistically significant ($p < 0.01$). Since the latter is part of the definition of the response variable, the percentage of explained variance of the stability of cores is much higher in the model that includes percentage of out-of-cores researchers (Adjusted $R^2 = 0.65$) compared to the model where this variable is not included (Adjusted $R^2 = 0.23$).

13.6   Conclusions

It is crucial to understand how modern science works to ensure that appropriate research and development policies are adopted that lead to improved scientific output. Modern information databases containing information about scientific bibliographic units can help in understanding the formation and maintenance of co-authorships among researchers. Although the borderline of scientific collaboration is unclear and there is no accurate way to measure it [20], co-authorships can be seen as a rough operationalization of scientific collaboration, which is one of the primary results of scientific collaboration and represents one of the most formal manifestations of scientific communication [16].

The co-authorship patterns were studied through co-authorship networks. These are networks where the vertices present authors (or researchers) and a link between them exists if they co-authored at least one scientific bibliographic unit. Kronegger et al. [21] analyzed the co-authorship networks of four Slovenian scientific disciplines (physics, mathematics, biotechnology, and sociology) in four periods (from 1990 to 2010). By observing the links among researchers from different scientific disciplines they confirmed that different co-authorship cultures exist between "lab" and "office" scientific disciplines. Publishing in co-authorship is more common in "lab" sciences while solo-authored scientific units are more common in "office" scientific disciplines where teamwork is not so crucial for the research. Hu et al. [18] classified four scientific disciplines into two groups: theoretical disciplines and experimental disciplines. They observed a stronger correlation between collaboration and productivity in experimental disciplines compared to theoretical ones.

However, one of the chief interests of the study by Kronegger et al. [21] was the global network structure. To analyze this, they used generalized blockmodeling on network slices in four 5-year consecutive periods. They confirmed the network structure of multi-cores, semi-periphery, and periphery was present in all scientific disciplines. It can happen that the mentioned structure is not so outstanding at the earliest time points in some scientific disciplines. They defined the core as a group of researchers who very systematically co-author with each other, but who usually do not collaborate with researchers from the other cores. The semi-periphery consists of authors who collaborate with others within the network, but in a less systematic way. It is not possible to cluster researchers from the semi-periphery into several well-separated clusters. The last part, the periphery, is the biggest part of the analyzed networks. Here are authors who published at least one bibliographic unit but as a single author or with researchers from abroad (with researchers not registered at the Slovenian Research Agency). Besides the main three types of mentioned positions, they observed so-called bridging cores. These are groups of researchers who collaborate with at least two other cores.

Cugmas *et al.* [8] extended the analysis at the level of all Slovenian scientific disciplines. Like Kronegger *et al.* [21], they analyzed data for the period between 1991 and 2010, but only analyzed the data in two 10-year periods. The wider time span has an effect on the network density. Despite this, there are some scientific disciplines without any links in the first or second period, e.g. theology. These kinds of scientific disciplines were removed from the analysis, leaving 43 out of 72 scientific disciplines for further analysis. The assumed multi-core–semi-periphery–periphery structure was confirmed as being present in all analyzed scientific disciplines. In many of them, bridging cores were also found. On average, the number of researchers is increasing in time, also reflected in the higher average core size, which is higher in the second period in both scientific disciplines from the fields of the natural and technical sciences and scientific disciplines from the social sciences and humanities. Here, the average core size is smaller in the social sciences and humanities in both time periods. The differences may be affected by the fact that authors from abroad are not included in the analysis since the rate of co-authored publications with researchers from abroad is higher in the natural and technical sciences than in the social sciences and humanities. As reported by Kronegger *et al.* [21], the main part of co-authorship networks is represented by authors from the periphery, which is generally decreasing over time.

Another important property of co-authorship networks is that the cores can emerge in time, disappear, split, or merge. To measure the stability of cores, operationalized with these four rules in different ways, several indices were proposed. The value of each was calculated for each scientific discipline and, based on this, the scientific disciplines were clustered in three clusters. The observation of these clusters reveals that, according to the values of the proposed indices, they are mainly characterized by different levels of cluster stability and can therefore be ordered from least to most stable. The majority of scientific disciplines were classified in the stable–unstable cluster (22 scientific disciplines) while only a few were classified in the most stable cluster (eight scientific disciplines). It turns out that the average percentage of researchers classified in the cores in both periods is increasing along with the stability of the clusters. On the other hand, the percentage of researchers leaving the cores in the first time period and the percentage of researchers joining the cores in the second period is decreasing with the average stability of cores with the obtained clusters. The average core size is higher in the most stable cluster of scientific disciplines, indicating the existence of well-established scientific research teams in these scientific disciplines.

A higher average number of researchers is associated with less stable cores. There are several explanations for this phenomenon, including the fact there are many opportunities to collaborate with different researchers in bigger scientific disciplines. The others are chiefly related to national research and development policies (e.g. the Young Researchers Program) and the nature of the work in such scientific disciplines (e.g. lab vs. office scientific disciplines or natural and technical sciences vs. social sciences and humanities).

To explain the differences between the natural and technical sciences and the social sciences and humanities, Cugmas *et al.* [8] performed a linear regression in which several network- and blockmodel-related variables (number of researchers in the scientific discipline, growth in number of researchers, growth in density, average core size, average percentage of cores, presence of a bridge) were included in the model as explanatory variables, while the stability of cores (response variable) was operationalized by the MWI1, where the splitting of cores and out-of-cores researchers reduces the value of an index and thus indicates lower core stability. There were no statistically significant differences in the mean stability of cores between the natural and technical sciences on one hand and the social sciences and humanities on the other. This could be caused by many differences in the publication culture within these two groups of scientific disciplines (which is also a consequence of the characteristics of the Slovenian Research Agency's classification scheme of scientific fields, disciplines, and sub-disciplines). In fact, even within some scientific disciplines the publication cultures vary widely. Moody [29] found that quantitative work is more likely to be co-authored than non-quantitative work in sociology.

Yet, when the analysis is performed on the level of scientific disciplines, the scientific discipline natural sciences and mathematics is statistically significantly ($p < 0.10$) less stable than the field of the humanities. The growth of density and the average core size are also statistically significant ($p < 0.05$) and positively correlated with the stability of the cores. When the additional variable percentage of out-of-cores researchers is included in the model, the difference in the average stability of cores between the humanities and medical sciences becomes statistically significant ($p < 0.10$). Here, it must be highlighted that when the variable percentage out-of-cores researchers is included in the model, only the splitting of cores is seen as a feature indicating a less stable core.

Acknowledgements

This work was supported in part by the Slovenian Research Agency (research program P5-0168, research project J5-7551 and 'Young Researchers' program) and by the Russian Academic Excellence Project '5–100'.

References

1. A. Abbasi, J. Altmann, and L. Hossain. Identifying the effects of co-authorship networks on the performance of scholars: A correlation and regression analysis of performance measures and social network analysis measures. *Journal of Informetrics*, 5(4):594–607, 2011.
2. A. Abbasi, L. Hossain, and L. Leydesdorff. Betweenness centrality as a driver of preferential attachment in the evolution of research collaboration networks. *Journal of Informetrics*, 6(3):403–412, 2012.
3. E. M. Airoldi, D. M. Blei, S. E. Fienberg, and E. P. Xing. Combining stochastic block models and mixed membership for statistical network analysis. In *Statistical Network Analysis: Models, Issues, and New Directions*, pages 57–74. Springer, 2007.
4. C. B. Amat and F. Perruchas. Evolving cohesion metrics of a research network on rare diseases: a longitudinal study over 14 years. *Scientometrics*, 108(1):1–16, 2015.
5. V. Batagelj, A. Ferligoj, and P. Doreian. Direct and indirect methods for structural equivalence. *Social Networks*, 14(1):63–90, 1992.
6. Z. Chinchilla-Rodríguez, A. Ferligoj, S. Miguel, L. Kronegger, and F. de Moya-Anegón. Blockmodeling of co-authorship networks in library and information science in Argentina: a case study. *Scientometrics*, 93(3):699–717, 2012.

7. M. Cugmas and A. Ferligoj. Comparing two partitions of non-equal sets of units. *Metodološki Zvezki – Advances in Methodology and Statistics*, 15(1):1–21, 2018.

8. M. Cugmas, A. Ferligoj, and L. Kronegger. The stability of co-authorship structures. *Scientometrics*, 106(1):163–186, 2016.

9. J. De Haan. Authorship patterns in dutch sociology. *Scientometrics*, 39(2):197–208, 1997.

10. J. De Haan, F. Leeuw, and C. Remery. Accumulation of advantage and disadvantage in research groups. *Scientometrics*, 29(2):239–251, 1994.

11. D. de Solla Price. *Little science, big science ... and beyond*. Columbia University Press, New York, 1986.

12. P. Doreian, V. Batagelj, and A. Ferligoj. *Generalized blockmodeling*, volume 25. Cambridge University Press, 2005.

13. A. Ferligoj and L. Kronegger. Clustering of Attribute and/or Relational Data. *Metodološki Zvezki – Advances in Methodology and Statistics*, 6(2): 135–153, 2009.

14. A. Ferligoj, L. Kronegger, F. Mali, T. A. Snijders, and P. Doreian. Scientific collaboration dynamics in a national scientific system. *Scientometrics*, 104(3):985–1012, 2015.

15. K. Frenken, W. Hölzl, and F. de Vor. The citation impact of research collaborations: the case of european biotechnology and applied microbiology (1988–2002). *Journal of Engineering and Technology Management*, 22(1):9–30, 2005.

16. B. Groboljšek, A. Ferligoj, F. Mali, L. Kronegger, and H. Iglič. The role and significance of scientific collaboration for the new emerging sciences: The case of Slovenia. *Teorija in Praksa*, 51(5):866, 2014.

17. A. Hollis. Co-authorship and the output of academic economists. *Labour Economics*, 8(4):503–530, 2001.

18. Z. Hu, C. Chen, and Z. Liu. How are collaboration and productivity correlated at various career stages of scientists? *Scientometrics*, 101(2): 1553–1564, 2014.

19. L. Hubert and P. Arabie. Comparing partitions. *Journal of Classification*, 2(1):193–218, 1985.

20. J. S. Katz and B. R. Martin. What is research collaboration? *Research Policy*, 26(1):1–18, 1997.

21. L. Kronegger, A. Ferligoj, and P. Doreian. On the dynamics of national scientific systems. *Quality & Quantity*, 45(5):989–1015, 2011.

22. L. Kronegger, F. Mali, A. Ferligoj, and P. Doreian. Classifying scientific disciplines in slovenia: A study of the evolution of collaboration structures. *Journal of the Association for Information Science and Technology*, 66(2):321–339, 2015.

23. S. Kyvik. Changing trends in publishing behaviour among university faculty, 1980-2000. *Scientometrics*, 58(1):35–48, 2003.

24. G. Laudel. What do we measure by co-authorship? *Research Evaluation*, 11(1):3–15, 2002.

25. S. Lee and B. Bozeman. The impact of research collaboration on scientific productivity. *Social Studies of Science*, 35(5):673–702, 2005.

26. C. Matias and V. Miele. Statistical clustering of temporal networks through a dynamic stochastic block model. arXiv preprint arXiv:1506.07464, 2015.

27. G. Melin. Pragmatism and self-organization: Research collaboration on the individual level. *Research Policy*, 29(1):31–40, 2000.

28. G. Melin and O. Persson. Studying research collaboration using co-authorship. *Scientometrics*, 36(3):363–377, 1996.

29. J. Moody. The structure of a social science collaboration network: Disciplinary cohesion from 1963 to 1999. *American Sociological Review*, 69(2):213–238, 2004.

30. B. Ponomariov and C. Boardman. What is co-authorship? *Scientometrics*, 109(3):1–25, 2016.

31. W. M. Rand. Objective criteria for the evaluation of clustering methods. *Journal of the American Statistical Association*, 66(336):846–850, 1971.

32. G. Saporta and G. Youness. Comparing two partitions: Some proposals and experiments. In *Compstat*, pages 243–248. Springer, 2002.

33. R. Tibshirani, G. Walther, and T. Hastie. Estimating the number of clusters in a data set via the gap statistic. *Journal of the Royal Statistical Society: Series B (Statistical Methodology)*, 63(2):411–423, 2001.

34. E. P. Xing, W. Fu, L. Song, et al. A state-space mixed membership blockmodel for dynamic network tomography. *Annals of Applied Statistics*, 4(2):535–566, 2010.

35. K. S. Xu and A. O. Hero III. Dynamic stochastic blockmodels: Statistical models for time-evolving networks. In *International Conference on Social Computing, Behavioral-Cultural Modeling, and Prediction*, pages 201–210. Springer, 2013.

36. A. Žiberna. Blockmodeling of multilevel networks. *Social Networks*, 39: 46–61, 2014.

14

Conclusions and Directions for Future Work

Patrick Doreian[3,4], Anuška Ferligoj[3,5], and Vladimir Batagelj[1,2,5]

[1]IMFM Ljubljana
[2]IAM, University of Primorska, Koper
[3]FDV, University of Ljubljana
[4]University of Pittsburgh
[5]NRU HSE, Moscow

As noted in the opening chapter, our aim in designing this book was to have a sustained examination of the general topic of *network clustering*. A clustering is a general term that means a set of clusters. Clustering also refers to a process for establishing a clustering. There are several types of clusterings, including a partition (a set of clusters that do not overlap and cover the whole set of units), a hierarchy (usually represented by a dendrogram), a pyramid, a fuzzy clustering, a clustering with overlapping clusters, and a clustering with disjoint clusters not covering the whole set of units. We use the more general term, clustering, throughout our discussion of the contributions contained in the book's chapters. However, when contributing authors use the terms partition and partitions, we use these terms.

Another goal for us was to make sure we included multiple perspectives and approaches to the problem of network clustering. As the foregoing chapters show, this topic is a highly diverse realm, both technically and substantively. We have much to learn from each other on both the technical and substantive fronts, at least when freed from the restrictions imposed by academic departments, fields, sub-fields, and different specific approaches. In an ideal world, generating knowledge transcends such constraints.

Even though many of the contributing authors ended their contributions with open problems, we use this concluding chapter to make additional suggestions regarding potential future work. Our suggestions and speculations take two forms. One is to focus on issues raised in the

Advances in Network Clustering and Blockmodeling, First Edition.
Edited by Patrick Doreian, Vladimir Batagelj, and Anuška Ferligoj.
© 2020 John Wiley & Sons Ltd. Published 2020 by John Wiley & Sons Ltd.

individual chapters. The other deals with linking ideas considered in separate chapters of this volume.

14.1   Issues Raised within Chapters

We start with Chapter 2. Given that the network clustering literature has expanding at a rapid rate and will continue to do so, it makes sense to obtain the citation network for this literature for 2020. No doubt the identified network will be much larger. More importantly, it will include additional fields, new contributors, new perspectives, and new issues. All will merit further attention. We think the coupling of ideas from multiple fields will be particularly important.

The authors of this chapter identified a set of nine link islands as shown in Figure 2.7. They are an example of a partition not including all the units in the network. While they focused on four of them, the others could be examined further. They all have distinct structures, which raises the question of whether the details of network structure in citation networks have import for the generation of scientific knowledge and its transmission over time. More generally, this concern could be folded into the general issue of the impacts of network structure on network processes and the design of networks to achieve specific objectives. This is a general topic of considerable importance for the study of clustering *all* networks. While three of these link islands had little to do directly with the core clustering topics of this book, the general issue remains: How does the structure of a citation network inform our understanding of how knowledge is generated and transmitted?

The authors, both in this chapter and in their earlier work [3], suggest that the institutional structure of science has a very large impact on the generation of scientific knowledge and the generation of scientific citation networks. In this context, journal citation networks seem particularly important. Some are controlled by publishers such as Elsevier, Springer, and Wiley. But many are sponsored by professional associations defined for disciplines and specific scientific interest groups. Many of these associations are focused on promoting their scientific interests and those of their members. This includes the nature of publishing strategies and the promotion of their journals. Journal citation networks seem worthy of greater attention. There are two distinct but overlapping aspects to this. One features journal-to-journal networks in specific areas of scientific inquiry, as described in Chapter 2, with the second being the study of journal networks across all disciplines. See, for example, [12] which is part of a long term effort studying journal-to-journal networks. Bibliographic coupling, a relatively old idea introduced in 1963 [11], also merits further attention, especially with the other link islands identified by these authors. Also, it will be desirable to do more with fractional bibliographic coupling, especially with the Jaccard islands they identify.

In Chapter 3, the optimization approach to the clustering problem was advocated. Using an appropriate criterion function, we can express our clustering goals, including the reduction of complexity, understanding network structures, and modeling networks. Together, optimization and the sought goals help define the nature of a "good" clustering. Additional knowledge about the specific clustering problem can be expressed using the feasibility predicate which defines the set of feasible clusterings.

Also mentioned in this chapter was that most of the existing social network clustering approaches are essentially based on structural equivalence. In [8] a generalized blockmodeling based on different types of equivalences is described. Current procedures for generalized blockmodeling can be applied to networks with up to some hundreds of nodes. An important task is to develop efficient methods for generalized blockmodeling of large sparse networks.

Regarding Chapter 4, we were intrigued by the content of Figure 4.2 because it illustrated two community partitions highlighting different aspects of networks as reflected in these partitions. This seems particularly important. We have long thought that multiple partitions of the same network have value and that the notion of having, or even wanting, a single partition of a network as the "best" one makes little sense. This idea also was expressed in multiple chapters in this book, albeit in different substantive contexts. If anything, this suggests that having multiple partitions of the same network has considerable merit and examining them closely is important for understanding the interplay of network structures and the network processes generating them. Of course, this observation extends to multiple clusterings of the same network.

Given that the authors of this chapter noted the value of examining differences among the four perspectives regarding community detection, as they outlined them, examining these differences further is another important task. One of their four approaches has a consideration of a dynamical perspective which merits further attention. This is important, and we comment further on this in the next section because the issue of dynamics was raised in multiple chapters. Such ideas need to be considered in conjunction.

Chapter 5 presented a different approach to the partitioning networks that were considered therein, one that is very fast. To examine how the algorithm works, the author used planted community structures to explore the operation of label propagation algorithms. This is an important idea not only in its own right but in a far more general context. While it is abundantly clear that networks have diverse structures, most often obscured in the construction of simple networks indices such as modularity and centrality measures, examining the global partition structure of networks is of great importance. This suggests a need to examine a more extended set of planted structures. In turn, this raises the issue of generating a catalog of network structures with different global forms from which planted structures could be selected. Many discussions in this book examined related ideas regardless of whether these structures are planted, used as demonstration examples, or were designed to examine certain structures in the context of networks clustering.

The inclusion of node preferences in Chapter 5 is important also and forms a step towards the inclusion of node properties in the network clustering algorithms explicitly considered therein. This is a line of inquiry that blockmodeling folk need to consider given their preoccupation with clustering networks without being attentive to nodal attributes. Depending on the substantive concerns of analysts, the set of constraints provided in this chapter could be expanded along with increasing the number of node preference regimes that could be included.

Two further items in Chapter 5 have the potential for opening new avenues of inquiry. One is to use label propagation methods on a much wider set of empirical networks, a topic to which we return later. The other is the consideration of having partitions with overlapping clusters and groups. This issue, having importance for blockmodeling as well as community detection, was raised in other chapters in this volume.

But as noted in Chapter 1, and discussed extensively in Chapter 6, clustering, or even mere partitioning – and studying – valued networks is far from being a straightforward task. Yet it must be done. The authors of Chapter 6 claim that when pre-specified blockmodels are needed, generalized blockmodeling approaches are preferable. While we agree, we would argue that more work is needed to construct a wider range of pre-specified blockmodels. This includes going far beyond using only structural equivalence to include many other equivalence types. The generalized blockmodeling approach is designed to facilitate the creation of many different types of equivalences. As a concern for network clustering is present in so many fields, it is highly likely that the specifics of network structures in these fields will generate the construction of new equivalence types. In general, such constructions must be driven by substantive concerns which will vary by fields.

The authors of Chapter 6 consider a very small number of valued networks, albeit to a very useful effect as the resulting partitions that were established are interesting both in terms of the partitions produced and the substantive interpretations that followed. However, in a much broader context, the number of considered valued networks must be expanded greatly both by those using the procedures outlined in this chapter and in the employment of other methods. As the authors note, their extension of partitioning networks is a natural - and necessary - extension. Partitioning and, more generally, clustering many more valued networks can only expand our understanding of dealing with valued networks.

At face value, the content of Chapter 7 may strike some readers as having limited importance by being confined to relatively small networks. But consistent with the idea that *research design matters* is the notion that *data quality matters*. It matters greatly how data are obtained, especially for recording data accurately and not discarding useful information. The authors focused on actor non-response and provided methods for dealing with this problem. This is very useful for recovering a full network accurately, especially for delineating the global structure of such networks. This line of inquiry has been applied also to the study of network centrality indices [13].

Yet attention to both errors in the recording of ties and item non-response may have even greater importance. Detecting actor non-response is straightforward. Discerning the presence of the other two forms of measurement error is much more difficult. Doing this is a task of great importance.

It will be straightforward to conduct studies of these types of measurement errors with techniques similar to those used in Chapter 7. The value of such efforts would depend on our ability to detect such measurement errors. One of the data sets considered in Chapter 7 came from a study asking questions about seeking advice from others and providing advice from others in an organization. If the data are accurate, the transpose of one relation would correspond to the other. Discrepancies between the two networks provide clues regarding item-specific measurement error. It would be useful to have a collection of data sets with such "reversed" relations to detect differences as measurement error and get an estimate of the amount of inaccuracy in the reporting of such ties.

Having high quality data is of critical importance when studying the structure of networks and the processes generating these structures. This applies to all network data sets regardless of their sizes. While it may be tempting to think that measurement error is irrelevant for large networks, especially very large networks, we think this view would be mistaken. As was shown in Chapter 2, it is critical to "clean" the data, even for large networks. If the techniques described in Chapter 7 cannot be extended easily to much larger networks, then other ways of assessing the presence of measurement error and treating such errors must be developed. Error-prone data are not a good source for understanding network structure nor for understanding the processes generating them.

While on the topic of data quality, a more general statement can be made. An important base for developing data analytic methods is the availability of a collection of data sets specific for each selected problem. Information about data sets for network analytic problems can be obtained, for example, at the Colorado Index of Complex Networks (ICON) [1]. Unfortunately for some combinations of network "dimensions" (mode, weighted, node attributes, linked, temporal, spatial) the corresponding data sets are very scarce or non-existent. For example, for the linked networks discussed in Chapter 10, only a few interesting network collections are available. This needs to be expanded. Also, to study and develop methods for temporal network clustering [2] it is necessary to have some temporal networks with node attributes.

Chapter 8 is devoted to partitioning signed networks. By taking a formal approach, the authors laid the foundations for further work in this area, foundations we hope others can build upon. In this context, further developments regarding weak structural balance will be useful. Within the blockmodeling approach, the authors note that the criterion function used for partitioning signed networks contains a parameter, α, allowing for differential weighting of two types of inconsistencies. One is the presence of negative ties in positive blocks and the other concerns having positive ties in negative blocks. The formulation was a natural extension with $\alpha = 0.5$ being used most often. But it created what can be called the alpha problem: How can values for α be selected? While using $\alpha = 0.5$ made considerable sense, it can be view as an arbitrary choice, especially if the numbers of negative and positive ties differ greatly. Using different values for this parameter, most often, leads to different partitions of signed networks regardless of the number of clusters. The problem is simple to state: Is there a principled way of selecting values for α? Some attention has been given to this but without any clear resolution emerging thus far.

The authors couple the partitioning of signed networks to community detection approaches. In doing so, they expand the concept of modularity, defined initially for unsigned networks, to deal with the presence of signed ties. This is particularly useful, and we look forward to future efforts extending this line of analysis. Similarly, using spectral methods and considering the constant Potts model was useful. A book review article of two books, produced by physicists studying networks, introduced the idea of "The Invasion of the Physicists" [4]. Bonacich was very clear that there were some good ideas in this literature, despite the very colonial claim that the physicists had invented a totally new field, called network science, to which members of the social network field needed to be attentive. There were two reactions to this invasion. One was outrage, a very narrow parochial response. The other was to think that the "old" social network analytic field needed to pay attention. Chapter 8 expresses this attentiveness.

In the design of this volume, only one chapter was devoted to signed networks. But this type of network was mentioned in other chapters. We consider this further in the next section when we look at links between many ideas expressed in different chapters of this volume.

The authors of Chapter 9 are most explicit in affirming the idea of making connections between different parts of the overall network clustering literature. Their idea of extending the modularity concept, defined in the community detection literature, to two-mode networks has great appeal. We hope that their approach to two-mode data and using both projections from such data will end, once and for all, the debates about the loss of information when projections are made. The authors make a connection to signed two-mode networks with suggestions for future work which we hope will be heeded. They also consider the use of spectral methods and advocate the use of two-mode stochastic blockmodels. Clearly these ideas make explicit links to other chapters when the use of spectral methods was raised.

Chapter 9 presents multiple partitions of a classic two-mode data set. This is fully consistent with our idea of having multiple partitions of the same network that have legitimacy if they can be interpreted in substantive terms. Our hope is that the ideas expressed in this chapter can be extended to larger – even much larger – two-mode networks. We concur with their view that "the complexities of this type of data in terms of collecting, analyzing, and interpreting remain challenging and deeply fascinating." They have provided practitioners with some very useful advice.

Chapter 10 presented ideas associated with blockmodeling linked networks conceived as collections of different one-mode networks coupled through two-mode networks. The strong message of this chapter, after multiple comparisons, is that the true-linked approach for analyzing linked networks has great promise. As only two empirical examples were considered, it would be useful to have other such networks analyzed in the same fashion. One of the open problems outlined in Chapter 10 concerns combining multiple criteria, expressed as criterion functions. The author links this idea to other work in the literature on multi-criteria partitioning, another observation fully consistent with the overall inclusive and integrative approach stressed in this volume. On the topic of multi-criteria partitioning of networks, some work has been done that we think will proved useful (see [9] and [5]). We note that having one-mode networks for different time points opens this approach to temporal dynamics, which we consider later in this chapter. Another link is made to stochastic blockmodeling, the topic of Chapter 11.

In our view, it does not matter if readers take a frequentist approach or Bayesian approach to analyzing data. Adherents of both perspectives will learn a great deal from considering the contents of Chapter 11. While the prose in this chapter comes perilously close to insisting that the Bayesian approach is the only viable approach, we do not think this is the author's intent. If so, the frequentists need to pay close attention to the contents of this chapter.

One of the key, and in our view, fundamental ideas expressed in this chapter is the idea of coupling generative mechanisms, as parts of general processes, to the coarse-grained modular structure, regardless of how fine-grained or coarse-grained are such depictions. We would extend this to a general statement about the coupling of network structures and the processes generating them. This idea is particularly relevant for many blockmodeling aficionados focused on depicting the macro structure of networks without considering the underlying network processes generating the identified network structures. In general terms, this is particularly important.

It is abundantly clear that the notion of a "model" is critical, for models can vary greatly, and that considering variations in models is important when thinking about network clustering. Without doubt, as the author of this chapter notes, having a multiplicity of models is a strength - but not only in a probabilistic sense. We cleave to a view that having multiple clustering models fitted to a network is important and that multiple such clusterings have the potential to suggest important ideas about network structure. Discerning the "best" such model may be a quixotic quest. But we agree with the author that *any delineated clustering must have solid evidence that the fitted model is appropriate and justified.*

Chapter 12, as noted in the opening chapter, picks up the idea of delineating coarse-grained structures of networks but within an explicit dynamic perspective. Their focus is on the "rich interplay between network structure and dynamics acting on top of the network" as a way of gauging the dynamical behavior of a network system. While we know that empirical networks have actors joining the network and other actors departing, it seems very useful to consider changes in the properties of actors located in a specific network. Not all problems can be solved at the same time. Examining change over a fixed network has considerable merit, especially if this can be generalized, as the authors note.

As a general framework for studying change, using a differential equation model for continuous data, or a difference equation model for discrete data, is especially appropriate. The authors of Chapter 12 focus on consensus dynamics in a specific empirical network along with a discussion of random walks in networks. Coupling them as dual processes is especially useful.

It is well known that there are slow processes with a relaxed time scale for their operation and fast processes where the dynamics operate far more quickly. At face value, this applies across the entire network.

The idea of having a modular partitioned network for which the time scales for different parts of a network differ is especially intriguing. The authors consider a specific modular structure in the form of a diagonal blockmodel. This could be extended to other blockmodel structures. The ideas presented in Chapter 12 on the dynamics for signed networks are considered in more detail in the next section. The idea of incorporating dynamical processes to reveal network structure, as outlined in this chapter, has additional appeal. While the substantive content of Chapter 13 is focused on scientific coauthorship networks, the issue of dynamics is present also. The technical context is blockmodeling with a focus on identifying cores and discerning their stability over time as well as the instability of cores. The presence of cores is a critical feature of the structure of coauthorship relations within scientific communities. The presentation of useful indices for measuring the stability of cores is particularly useful, as are the visualizations of the movement of researchers between well-identified positions in a co-authorship network. It seems reasonable to think that these tools could be applied fruitfully to many other types of networks when there are temporal changes.

14.2   Linking Ideas Found in Different Chapters

A wide variety of approaches and methods related to network clustering have been presented in different chapters. While they could be viewed as rivals, it seems more fruitful to think of them as sources for ideas that could be coupled in a fruitful fashion. One general idea is to think of identifying, or creating, data sets where the different methods and algorithms could be used to cluster networks. Of course, there will be no one network for which all the methods could be applied. Differences regarding valued ties, signed ties, the number of relations studied, and the sizes of networks makes this impossible. But for various network types, a subset of the methods presented in this book could be mobilized. This notion can be coupled to the idea of having a catalog of different structures for which different methods, as shown in this volume, could be applied to useful effect.

The goal would not be to find a so-called winner but to examine the insights generated by each of the methods. As we know, every approach has its assumptions and some constraints implied by these assumptions. It follows that using as many different methods as possible allows us to assess the value of the clusterings that are obtained. They can be compared in a systematic fashion. A core feature of this effort must be grounded in the substantive concerns specific to the fields within which network data are collected. Establishing clusterings is not the final product, for they must be interpreted to help researchers understand the structures of networks and, if possible, the processes generating them, or outcomes generated over networks.

Perhaps a less ambitious path would be to combine ideas from different methods to construct different ways of thinking about tools and establishing additional methods. As a modest step in this direction, we consider linking ideas expressed in the specific chapters of this book. In our view, there is some value in drawing clear distinctions between different approaches to better understand both their strengths and limitations.

There may be researchers who think that any approach not labeled as community detection has little value. And there are others thinking that blockmodeling covers all possible network clustering approaches. Both views are mistaken. We have doubts about universalistic claims for single approaches. There are claims that community detection covers everything [10]. Some authors of this volume note that community detection is a special case of blockmodeling, a view on which we have commented already. Others claim that the whole field of network clustering is subsumed within stochastic blockmodeling [14].

Chapter 2 and 4 consider some of these issues. Clearly, blockmodeling and community detection have some common features. They also have different literatures, as documented in Chapter 2, suggesting there are some real differences between them. We doubt this is merely a matter of perception. It would be useful to have clear exploration of their similarities and differences to establish a succinct statement about them. The goal would not be to see which is "better", a foolish quest, but to explore ways of combining ideas from both approaches to strengthen each of them.

One technique used in Chapter 2 was the identification of coherent parts of networks without establishing a partition of the entire network. Both the community detection and blockmodeling approaches seek partitions of entire networks. We think it would be useful to compare the results of identifying link islands and complete partitions to couple the interpretations that each approach yields. There may be other ways of identifying smaller parts of networks that could be included in such comparisons.

The authors of Chapters 4 and 9 examine closely the role of modularity as it pertains to network clustering. In doing so, they expand the formulation of this concept and provide additional operational equations. This idea is considered in other chapters also. It would be useful to develop these ideas further in a sustained effort. It seems important to examine whether modularity could be used in the formulation of some criterion functions used in blockmodeling. Considering whether some of the blockmodeling criterion functions could be used to reformulate definitions of modularity would be another avenue for exploration. Again, the objective is one of having ideas flow between different approaches.

Many chapters consider criterion functions that are optimized when delineating clusterings, with a wide variety of them being employed. It would be useful to examine the ways that variations across them have an impact on the clusters that are established. Also, the relationships among different criterion functions need to be studied in detail.

Chapter 5 examined label propagation with a suggestion that this approach could be used for signed networks. Chapter 8 focused on clustering signed networks. Could label propagation be useful for solving some of the clustering problems for signed networks? We certainly hope so. Chapter 8 settled for partitions consistent with strong balance, leaving open the issue of using weak balance for such networks in future work. One example in the literature [7] showed there were severe problems with the standard signed blockmodeling approach due to its handling of positive ties. The signed community detection approach had major problems with its handling of negative ties. Label propagation for signed networks has the potential to solve both problems.

There is a clear divide between deterministic approaches to network clustering and probabilistic methods. Chapter 11 is resolutely within the latter approach using Bayesian ideas. The authors of Chapter 9 and 12 make explicit links to stochastic blockmodeling, an idea meriting detailed attention. More generally, it would be useful to have a systematic assessment of the results stemming from the uses of deterministic and probabilistic approaches. Again, the goal would not be to establish which is the better approach but to see if there are complementary results and to explore any differences to see why the clustering results differ.

We turn to consider network dynamics. Chapter 11 and 12 are focused on dynamic change, albeit in very different ways. In thinking about change in networks, the mechanisms involved in network processes are critical. Being able to specify them and understand their operation has the utmost importance. This is the clear message in both chapters despite taking starkly different approaches. In the spirit of wanting to couple different ideas, we wonder if this can be done with these two approaches. At face value, this could be useful even though the technical issues involved will be fearsome. A step in this direction is provided in [6].

The authors of Chapter 12 tackle also signed networks within the rubric of structural balance. This is an ambitious approach with the potential of being very fruitful. However, it requires abandoning the notion that the network is fixed. The study of signed networks involves the examination of changes in the signs and strengths of ties. If actors drop out of such networks, or others join, the task becomes far more complex. We note that Chapter 4 also has a dynamic component that could be incorporated into this discussion.

14.3 A Brief Summary and Conclusion

Many ideas and issues have been raised throughout the book. As noted above, many avenues have been opened for future work on network clustering. We finish by stressing two very general ideas. One is the importance of the exchange of ideas between different approaches with the goal of strengthening them. The second is the coupling of network processes and network structures to help us understand both. Our hope is that this book will help promote these general issues as well as all the ideas contained in its chapters. If it helps in doing this, then this volume will have the impact we hoped it would have.

References

1. ICON – The Colorado Index of Complex Networks. August 2018. URL https://icon.colorado.edu/.
2. V. Batagelj and S. Praprotnik. An algebraic approach to temporal network analysis based on temporal quantities. *Social Network Analysis and Mining*, 6(1):28, 2016.
3. V. Batagelj, P. Doreian, A. Ferligoj, and N. Kejžar. *Understanding Large Temporal Networks and Spatial Networks: Exploration, Pattern Searching, Visualization and Network Evolution*. Wiley Series in Computational and Quantitative Social Science Series. Wiley, 2014.
4. P. Bonacich. The invasion of the physicists. *Social Networks*, 26:285–288, 2004.
5. M. Brusco, P. Doreian, D. Steinley, and C. Satorino. Multiobjective blockmodeling for social network analysis. *Psychometrika*, 78:498–525, 2013.
6. M. Cugmas, A. Ferligoj, and A. Žiberna. Generating global network structures by triad types. PLOS One, https://doi.org/10.1371/journal.pone.0197514, 2018.
7. P. Doreian and A. Mrvar. Structural balance and signed international relations. *Journal of Social Structure*, 16:1–49, 2016.
8. P. Doreian, V. Batagelj, and A. Ferligoj. *Generalized Blockmodeling*. Structural Analysis in the Social Sciences. Cambridge University Press, Cambridge, 2005.
9. A. Ferligoj and V. Batagelj. Direct multicriteria clustering algorithms. *Journal of Classification*, 9:43–61, 1992.
10. S. Fortunato. Community detection in graphs. *Physics Reports*, 486: 75–174, 2010.
11. M. M. Kessler. Bibliographic coupling between scientific papers. *American Documentation*, 14:10–25, 1963.
12. L. Leydesdorff, C. S. Wagner, and L. Bornmann. Betweenness and diversity in journal citation networks as measures of interdisciplinarity – a tribute to Eugene Garfield. *Scientometrics*, 114:567–592, 2018.
13. A. Žnidaršič, A. Ferligoj, and P. Doreian. Stability of centrality measures in valued networks regarding different actor non-response treatments and macro-network structures. *Network Science*, 6:1–33, 2017.
14. J.-G. Young, G. St-Onge, P. Desrosiers, and L. J. Dubé. Universality of the stochastic block model. *Physical Review E*, 98(3):032309, 2018.

TOPIC INDEX

actor non-response, 4, 193, 392
affiliation matrix, 253
algorithm
    bipartite recursive induced modules
       (BRIM), 253
    CONCOR, 155, 158, 161
    COPRA, 140
    DEMON, 140
    genetic, 8
    greedy, 109
    Kernighan–Lin, 107
    REGE, 155, 162
    relocation, 164
    SLPA, 140
    SpeakEasy, 140
    spectral, 108
alpha problem, 393
approach
    conversion, 268
    cut-based, 114, 116
    deterministic, 396
    deviational, 159
    direct, 4, 90
    fractional, 42, 48, 98
    frequentist, 394
    heuristic, 3
    indirect, 4, 5, 89, 160, 178
    islands, 90
    linked, 276, 277, 282, 394
    multilevel, 269
    spectral, 6, 107, 111, 393
    stochastic, 396
    stored data, 77
    stored dissimilarity matrix, 77
    weighted sum, 270

balance
    extended, 257
    relaxed, 257
    theory, 257
Bayesian inference, 7, 111, 294
    evidence, 294
    hyperprior, 296
    model selection, 306
       AIC, 299
       Bayes factor, 306
       BIC, 299
       posterior odds ratio, 306
    posterior distribution, 7, 294
    prior probabilities, 7, 295
belief propagation, 325
bias-variance trade-off, 321
bibliographic coupling, 2, 50, 145, 390
bibliographic unit, 370, 384
bipartite spectral co-partitioning, 258
bipartition, 109
block, 1, 11, 90, 94, 106, 157, 163
    negative, 393
    positive, 393
    regular, 158
blockmodel, 2, 12, 40, 110, 238, 255
    stability, 365
    stochastic, 7, 106
blockmodeling, 1, 2, 4, 5, 8, 11, 15, 30, 35, 75, 189,
      201, 395, 396
    binary, 165, 272
       generalized, 170
    deterministic, 7
    deviational
       generalized, 166
       indirect, 164

Advances in Network Clustering and Blockmodeling, First Edition.
Edited by Patrick Doreian, Vladimir Batagelj, and Anuška Ferligoj.
© 2020 John Wiley & Sons Ltd. Published 2020 by John Wiley & Sons Ltd.

PERSON INDEX

Advances in Network Clustering and Blockmodeling, First Edition.
Edited by Patrick Doreian, Vladimir Batagelj, and Anuška Ferligoj.
© 2020 John Wiley & Sons Ltd. Published 2020 by John Wiley & Sons Ltd.